Polycarbonates
Polyacetals
Polyesters
Cellulose Esters

Engineering Thermoplastics

Polycarbonates Polyacetals Polyesters Cellulose Esters

Edited by Ludwig Bottenbruch

With contributions from

S. Anders, R. Binsack, L. Bottenbruch, F. Brinkschröder, U. Grigo, A. Horbach, A. Kaminski, R. Kappenstein, K. Kircher, Ch. Leuschke, F. Müller, H.-H. Müller, P.R. Müller, H.-D. Sabel, H. Schlaf, P. Unger, U. Ziegler

With 360 Figures and 96 Tables

Hanser Publishers, Munich Vienna New York

Hanser/Gardner Publications, Inc., Cincinnati

The Editor: *Dr. Ludwig Bottenbruch*, Wöhlerstr. 5, D-47800 Krefeld

Distributed in the USA and in Canada by
Hanser/Gardner Publications, Inc.
6600 Clough Pike, Cincinnati, Ohio 45244-4090, USA
Fax: (513) 527-8950
Phone: (513) 527-8977 or 1-800-950-8977

Distributed in all other countries by
Carl Hanser Verlag
Postfach 86 04 20, D-81631 München, Germany
Fax: +49 (89) 98 12 64

Library of Congress Cataloging-in-Publication Data

Polycarbonate, polyacetale, polyester, celluloseester. English.
Engineering thermoplastics: Polycarbonates, polyacetals, polyesters, cellulose esters / edited by Ludwig Bottenbruch ; with contributions from S. Anders ... [et al.].
p. cm.
Translation of: Polycarbonate, polyacetale, polyester, celluloseester.
Includes bibliographical references and index.
ISBN 1-56990-183-X
1. Thermoplastics. 2. Cellulose esters. 3. Polyesters. 4. Polycarbonates. 5. Acetal resins. I. Anders, S. (Siegfried) II. Title
TA455.P5P66 1996
658.4'23--dc20 96-31005
CIP

Die Deutsche Bibliothek - CIP-Einheitsaufnahme

Polycarbonates, polyacetals, polyesters, cellulose esters: engineering thermoplastics ; with 96 tables / ed. by Ludwig Bottenbruch. With contributions from S. Anders ... - Munich ; Vienna ; New York : Hanser ; Cincinnati : Hanser/Gardner, 1996
Einheitssacht.: Technische Thermoplaste <engl.>
Teilausg.
ISBN 3-446-17473-7
NE: Bottenbruch, Ludwig [Hrsg.]; Anders, Siegfried; EST

Typeset in Ireland by Datapage International Ltd., Dublin
Printed in Germany by C. H. Beck'sche Druckerei, Nördlingen

The Authors and their Contributions

Dr.-Ing. *Siegfried Anders,* Bayer AG, Geschäftsbereich Kunststoffe, D-51368 Leverkusen

3.8.1 Injection Molding

Dr. rer. nat. *Rudolf Binsack*, Bayer AG, Geschäftsbereich Kunststoffe, D-47812 Krefeld

2 Thermoplastic Polyesters

Dr. rer. nat. *Ludwig Bottenbruch*, Wöhlerstr. 5, D-47800 Krefeld

1 Introduction

Dipl.-Ing. *Franz Josef Brinkschröder* †, Bayer AG, Geschäftsbereich Kunststoffe, D-51368 Leverkusen

3.8.2 Extrusion Procedure for Polycarbonate
3.8.3 Extrusion Blow Molding of Hollow Profiles

Dr. rer. nat. *Ulrich Grigo*, Bayer AG, Geschäftsbereich Kunststoffe, D-51368 Leverkusen

3.1 Polycarbonates, Introduction
3.2 Chemical Structure
3.3 Preparation of Aromatic Polycarbonates
3.4 Modification of Polycarbonates through Additives
3.5 Polycarbonate Blends
3.7 Properties of Bisphenol A Polycarbonate

Dr. rer. nat. *Alfred Horbach*, Heckerhof 48, D-47800 Krefeld

3.6 Polycarbonate Analysis

Dipl.-Ing. *Axel Kaminski*, Bayer AG, Geschäftsbereich Kunststoffe, D-51368 Leverkusen

3.8.1 Injection Molding

Dipl.-Ing. *Reinhold Kappenstein*, Bayer AG, Geschäftsbereich Kunststoffe, D-51368 Leverkusen

3.8.1 Injection Molding

Dr. rer. nat. *Klaus Kircher*, Bayer AG, Geschäftsbereich Kunststoffe, D-51368 Leverkusen

3.7 Properties of Bisphenol A Polycarbonate
3.8 Processing of Polycarbonate
3.9 Processing
3.10 Application Areas for Polycarbonate
3.11 Economic Significance
3.12 Commercial Products

Ing. (grad.) *Christian Leuschke*, Bayer AG, Geschäftsbereich Kunststoffe,
D-41538 Dormagen

5.5.8 UV/IR-Absorbers
5.10.1 Health Specifications
5.11.3 Injection Blow Molding

Dr. rer.nat. *Friedemann Müller*, Bayer AG, Geschäftsbereich Kunststoffe,
D-41538 Dormagen

5 Organic Cellulose Esters

Dr. rer. nat. *Heinz-Herbert Müller*, Bayer AG, G3 Polyurethane, D-47812 Krefeld

3.6 Polycarbonate Analysis

Dr. rer. nat. *Peter Rudolf Müller*, Bayer Corp., Pittsburgh/USA

3.7 Properties of Bisphenol A Polycarbonate

Dr. rer. nat. *Hans-Dieter Sabel*, Hoechst AG, GB-B, Technische Kunststoffe,
D-65929 Frankfurt a. M.

4.4 Finishing of Acetal Resins
4.5 Standard Products
4.7 Properties of Polyacetals (Sections 4.7.4 to 4.7.10)

Dr. rer. nat. *Helmut Schlaf*, Hoechst AG, GB-B, Technische Kunststoffe,
D-65929 Frankfurt a. M.

4.1 Polyacetals, Introduction
4.2 Chemical Structure of Polyacetals
4.3 Preparation of Polyacetals
4.7 Properties of Polyacetals (Section 4.7.11)
4.11 Commercial Products
4.12 Market Development

Dr.-Ing. *Peter Unger*, Hoechst AG, GB-B, Technische Kunststoffe, D-65929 Frankfurt a. M.

4.8 Processing of Polyacetals
4.9 Working and Secondary Operations
4.10 Applications

Dr. rer. nat. *Ursula Ziegler*, Hoechst AG, GB-B, Technische Kunststoffe,
D-65929 Frankfurt a. M.

4.6 Modified Products
4.7 Properties of Polyacetals (Sections 4.7.1 to 4.7.3)

Preface

The economical and technical development of thermoplastics have, by far, exceeded the development of all other polymeric materials. The so-called engineering thermoplastics are of special importance among the various product classes.

Engineering thermoplastics have given special impulses to the development of plastics, the new products increasing the application range of plastics considerably. They have gained extraordinary importance over the last 20 years. This is demonstrated by their economical development as well as by their wide range of application. Many technical developments would not have been possible without engineering thermoplastics, although the specific amount of plastic per application is often rather small. This shows that engineering thermoplastics and new technologies should be developed together.

A comprehensive survey of the production, properties, processing and applications of polyesters, polycarbonates, polyacetals and cellulose esters – a major segment of the engineering thermoplastics – is given in this Volume. The authors tried to present newer product and application developments. The ecological importance of recycling was also taken into account.

The publication of a handbook like this requires solid preparation, continuous contact with the participating authors and an encompassing coordination. In this context, the editor owes special gratitude to *Dr. Rudolf Binsack* who helped in all these assignments as well as in the editorial process. Especially he likes to thank the authors who were successful in providing the product surveys in a very short time. As always, many discussions stimulate a book like this and he likes to thank many employees of the participating companies for their critical remarks.

He likes to thank the translaters, especially *Dr. Harald Pielartzik* and his wife *Barbara Irmisch-Pielartzik*, who translated Chapters 1, 2, 3 and 5, and *Dr. John B. Starr, Jr.* for the translation of Chapter 4. He also likes to thank the Bayer Corporation, Pittsburgh, USA, and Hoechst Celanese, Summit, USA, for supporting the translation.

Mrs. *Helene Weiß* from Hanser Verlag not only made a readable but also a beautiful book of all manuscripts. I owe her a special gratitude.

Summer 1996 *Ludwig Bottenbruch*

Contents

1 Introduction

L. Bottenbruch

Engineering thermoplastics are polymers that are grouped together not because they are similar in chemical structure or morphology but rather because of their specific properties or superior property profiles. This combination of properties allows engineering polymers to be used in rigorous applications, particularly in the electrical, electronics, and automotive fields, but also in the areas of medicine, optics, athletics, and construction.

These applications demand good short- and long-term heat distortion resistance, creep resistance, notched and unnotched impact strength, and dimensional stability, as well as good electrical, optical, and acoustical characteristics. Also needed is stability in hostile environments, including compatibility with living tissue. Additionally good processibility is essential. Recently the recycling characteristics of engineering polymers have increased in importance. Manufacturers and processors of plastics are working on intelligent and lasting solutions.

Most applications require a combination of specific properties. This often leads to compromises in which the price/performance ratio plays a key role.
Engineering polymers form a pyramid (see Figure 1.1): specialty engineering polymers with superior properties occupy the apex, and engineering polymers with average properties form the base. The pyramid has three other characteristics. From bottom to top, the costs and prices increase and the volumes decrease.

At the base of the pyramid one would place polybutylene terephthalate, polyethylene terephthalate, polycarbonate, polyacetal, and cellulose ester, which are addressed in this Volume. Synthesis and development of these polymers go back to the early applications of thermoplastics. Table 1.1 lists a few of the engineering thermoplastics covered in this book, along with the year of their discovery and the year of entry into the market.

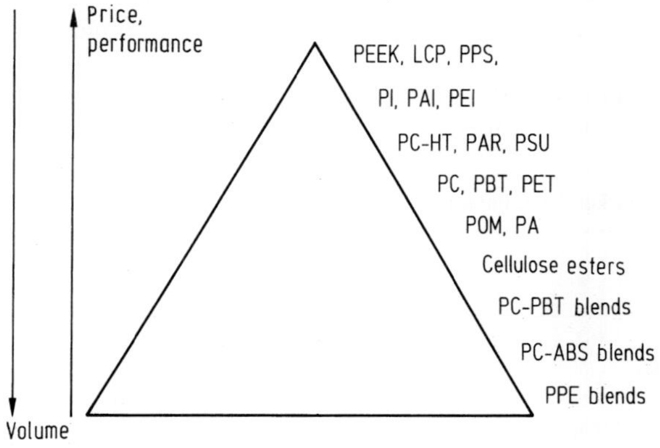

Figure 1.1. Engineering thermoplastics.

The year of discovery of these polymers does not always correspond to the start of the development of a new material. The laboratory development of a technically feasible route can require significant time. The practical synthesis of polyphenylenesulfide was

Table 1.1 Historical Data on Engineering Thermoplastics

Name of Thermoplastic	Discovery	Market Entry
Polyamides	1935/38	1948
Polycarbonates	1953	1958
Polyacetales	1925	1958
Polyphenylene ether-other blends	1963	1966
Polyethyleneterephthalate	1941	1966
Polybutylenterephthalate	1941	1970
Polycarbonate blends	1964	1973
Polysulfones	1958	1965
Polyetherimides	1972	1982
Polyarylenesulfides	1948	1972
Polycarbonate with very high heat distortion temperature	1988	1991

possible only through the use of *N*-methylpyrrolidine by Phillips. Another example is the introduction of a thermally stable end group into polyacetal by Celanese and DuPont. The production of polymers worldwide had increased to over 100 million tons by 1989. That includes approximately 82 million tons of thermoplastics and approximately 18 million tons of thermosets, polyurethane, and coatings [1]. The engineering thermoplastics described in this book make up approximately 2.5 million tons, and more than 1.5 million tons of polyethylene terephthalate is produced for bottles and films, as well. The engineering polymers market is segmented as shown in Table 1.2.

The stormy development of engineering polymers is shown in Figure 1.2. Between 1972 and 1980 the annual growth rate in the industrialized nations—United States, Western Europe (WE), Japan—was 7.9%; between 1980 and 1990 it was 8.3%. A few thermoplastics, such as polycarbonate and blends of polycarbonate, achieved significant higher annual growth rates of 15.7 and 10.9%, respectively, during this time period.

Although the 2.5 million tons of engineering polymers produced in 1994 represented only 3% of the total thermoplastic production, this small percentage does not reflect the significance of these materials. The weight or volume fraction of an engineering polymer in a car, a radio, or a television set is often quite low. Nevertheless, the polymeric part of the apparatus frequently adds performance or makes the economical production of the device possible.

For example, polycarbonate based on bisphenol A is found in a series of applications that take advantage of its optical properties: twin wall sheet, compact disks (CDs), and

Table 1.2 Segmentation of the Market for Engineering Polymers

×	Amount (× 1000 tons)	Percentage
Polyamides	1,300	38.6
Polycarbonate (including blends)	870	25.9
Polyacetales	450	13.4
Polybutylene and polyethylenterephthalates*	320	9.5
Polyphenylene ether blends	240	7.1
Cellulose esters	60	1.8
Various engineering thermoplastics (high temperature thermoplastics)	125	3.7
Total engineering thermoplastics	3,365	100

* Excluding PET for film or bottles.

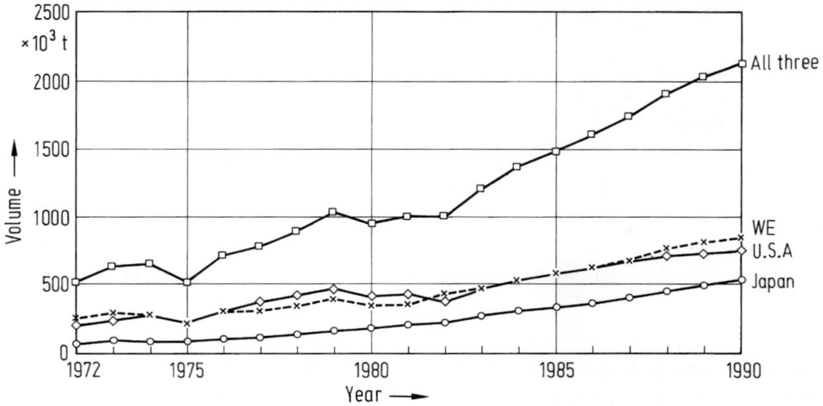

Figure 1.2. Consumption of engineering thermoplastics in Western Europe, the United States, and Japan.

lenses for glasses where the high refractive index and thinness are desirable. Light transmission, despite the high damping of the high refractive index, is also a decisive advantage.

Especially in CDs it has become clear that polycarbonate is irreplaceable. Minimal birefringence, high heat distortion temperature, and excellent dimensional stability have made possible the phenomenal growth of CDs worldwide:

Year	CD Production (millions)
1988	450
1989	650
1990	950
1991	1300

Another example of the irreplaceability of engineering thermoplastics consists of electrical tools, whose unprecedented growth is based on lower price and improved lifetime, achievable only through integrated manufacturing and because of the physical properties of glass-reinforced engineering thermoplastics.

The last category of key applications of engineering thermoplastics comprises outer shells and caps of dialysis devices made from a special polycarbonate. No other transparent materials have the same combination of superior properties and good compatibility with human blood.

The chemical routes employed in the synthesis of engineering polymers are based on various reactions. In general, condensation reactions are employed. These require high purity and reproducible equivalence of monomers to ensure the achievement of the desired molecular weight.

The picture of engineering polymers is not static. On the contrary, it is ever changing. There are two reasons for this: new applications can demand new property combinations, and new polymeric materials stimulate the development of new applications.

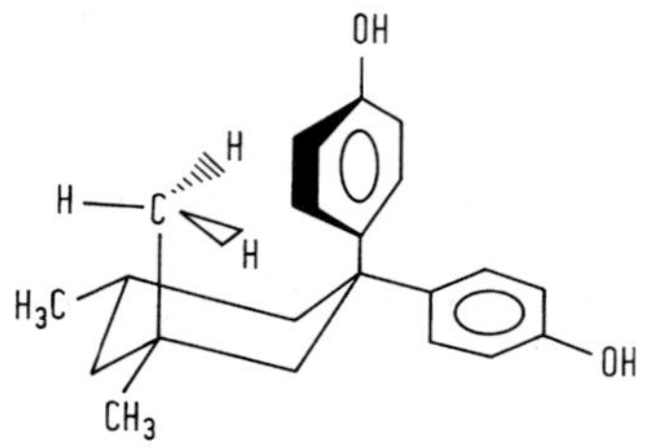

Figure 1.3. Three-dimensional structure of bisphenol trimethyl carbonate (TMC).

In general, new polymeric materials are developed in one of two ways:

- By modification of known engineering polymers through block copolymers, through additives, through fillers and reinforcers, and through polymer blends.
- By synthesis of new polymers from new monomers (in which case the polymer must show a new property profile compared to existing materials).

The histories of polycarbonate are outlined below:

1953 Polycarbonate based on bisphenol A
1956 Various molecular weight grades
1964 Self-extinguishing polycarbonates through addition of catalytic amounts of sulfonic acid salts (for electric and electronic applications)
1976 Polyfunctional monomers to give branched polymer for extrusion
1979 Block copolymers with polydimethylsiloxane blocks with higher toughness at low temperatures (improved flammability behavior)
1980 Polycarbonate with new end groups and better molecular weight control for improved melt flow without loss of toughness (CDs)
1982 Blends of polycarbonate and acrylonitrile–butadiene–styrene for improved processing and better price/performance ratio for certain applications
1985 Halogen-free, self-extinguishing polycarbonate/ABS blends (numerous electric and electronic applications)
1988 New polycarbonates from a sterically hindered bisphenol based on dihydroisophorone (bisphenol TMC; see Figure 1.3), which has a heat distortion temperature of 239 °C and can be easily processed; copolymers with bisphenol A have heat distortion temperatures between those of bisphenol A polycarbonate and the new polycarbonate [2]

The importance of engineering polymers will increase in the next ten years, not only in the western industrialized nations but also in developing nations. In developing nations, particularly in South Korea and Taiwan, the consumption of engineering polymers will grow at a yearly rate of 8–9% to 600,000 tons.

The potential applications for engineering polymers are impressive. In addition to the possibilities just listed, two points should be considered:

- Designers still think in terms of classical materials such as metals and ceramics without considering the advantages of function and economics of engineering polymers.
- Many applications are established in standard polymers, although engineering polymers may offer a better solution in terms of function and economy.

Parallel to the growing consumption of engineering polymers is the develoment of chemical routes and, when there are no other possibilities, thermal recycling routes. Here, as is the case with other materials, standardization of engineering thermoplastics for certain applications is increasingly important. Through better and more streamlined production and processing routes, the economics are improved and the ecological advantages are increased.

The further development of engineering thermoplastics, the development of modifiers, and the development of new engineering thermoplastics hinges on the close and early cooperation of the polymer manufacturer, the polymer processor, and the polymer user which, in retrospect, has been necessary for the development of all materials.

References for Chapter 1

1. *Freitag, D., Binsack, R.:* Wege zu neuen Polymerwerkstoffen. 2. Symposium Materialforschung, Dresden, August 1991.
2. *Freitag, D., Westeppe, U.:* Angew. Makromol. Chem., Rapid Communications 12 (1991) 95–99.

2 Thermoplastic Polyesters

R. Binsack

2.1 Introduction

2.1.1 Definition and Nomenclature

The thermoplastic polyesters considered in this chapter, which are products of aromatic dicarboxylic acids and aliphatic diols, find use in injection molding and extrusion applications. Fibers and monofilaments of thermoplastic polyesters are not considered. The production of polyester film is discussed only briefly.

This section describes in detail the thermoplastic polyester polybutylene terephthalate (PBT) and considers its significance as a construction material. Polyethylene terephthalate (PET), which to date has not achieved the commercial significance of PBT, is treated in less detail.

Copolyesters such as amorphous polyesters, block copolymers, and low melting copolyesters, which differ significantly in properties from PBT and PET, are briefly discussed. Aromatic copolyesters from aromatic dicarboxylic acids and aromatic diols such as liquid crystalline polyesters (LCPs) are handled in a separate chapter.

PBT and PET are polyesters from terephthalic acid and 1,4-butanediol or ethylene glycol and have the following general structure:

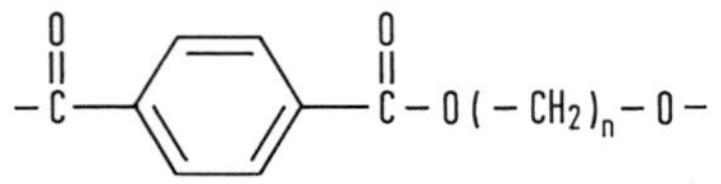

PBT: $n = 4$
PET: $n = 2$

2.1.2 Historical Development

The first fiber-forming polyesters were prepared by Carothers and Hill in 1932 using the melt condensation of dicarboxylic acids and aliphatic diols [1]. Because of their low melting points and ready dissolution in organic solvents, these polymers were not considered to be a significant advance. The development of polyester fibers having high melting points and more chemical resistance was possible through the use of terephthalic acid. This was discovered simultaneously in the early 1940s by Schlack (PBT) at Agfa Wolfen [2] and by Whinfield and Dickson (PET) at Calico Printers [3]. The technical development of PET fibers began after DuPont and ICI purchased the rights to the technology of Dacron and Terylene, respectively. A series of licensee arrangements by ICI led to the production of fiber in most industrialized countries. In Germany PET fiber was produced by Hoechst (Trevira) and Vereinigte Glanzstoff-Fabriken (Diolen).

In addition to fiber spinning, the mid-1950s saw the increased significance of PET film (Mylar for DuPont and Hostaphan from Kalle).

As a design material, polyethylene terephthalate was first commercialized in 1966 through the introduction of Arnite by Akzo. An injection-moldable PET was now commercial which, because of nucleating agents, had a crystallization rate that made economic injection molding possible [4–6]. Afterward a number of firms brought injection-moldable PET to the market: Hostadur from Hoechst, Crastin from Ciba

Geigy, and FR-PET from Teijin. A breakthrough occurred in 1970 with the introduction of glass-reinforced PBT by Celanese (X-917).

In the following years myriad grades of PBT were introduced by firms in Europe and the United States: Hostadur from Hoechst, Ultradur from BASF, Pocan from Bayer, Vestodur from Hüls, Crastin from Ciba Geigy, Arnite from Akzo, Celanex from Celanese, Tenite from Eastman Kodak, Valox from GE, and Vituf from Goodyear. In Japan, which lacked a domestic source of 1,4-butanediol, market growth was fueled by material produced in the United States. PBT grew faster than PET despite the higher raw material cost of the former because PBT offered improved processibility.

From the beginning of the 1980s PET became a more significant force in the injection molding market as DuPont (Rynite) and Mobay (Petlon) introduced more readily processible products.

After the ban of acrylonitirle-containing barrier materials by the U.S. Food and Drug Administration (FDA), biaxially oriented beverage bottles were introduced in the United States (1976) and in Europe (1980).

The application spectrum of polyesters was significantly broadened in 1972 when DuPont commercialized the first polyester thermoplastic elastomer (Hytrel). Shortly thereafter, Toyobo and Akzo followed with similar products: Pelprene and Arnitel, respectively.

With the introduction of impact-modified polybutylene terephthalate (a blend with graft elastomers: Pocan S 1506 by Bayer), it became possible to paint automobile bumpers "in-line."

The application window of polyesters was also broadened with the commercialization by GE of poly(cyclohexyldimethylene)terephthalate. This material (Valox 9730M) exhibits a heat distortion temperature of 254 °C [7].

References for Section 2.1

1. *Carothers, W. H.; Hill, J. W.:* J. Am. Chem. Soc. 54 (1932) p. 1559
2. DBP 922 255 (1942) Bobingen AG für textil-Faser *(Schlack, P.)*
3. GB-PS 578 079 and US-PS 2 465 319, Du Pont (both patents with priority of 1941, *Whinfield, J. R.; Dickson, J. T.*)
4. *Krevelen, D. W. van,* et al.: Int. Congr. 1966. Processing Polymers to Products. Amsterdam 17.–19. October 1966, N. V. 't Readthuys c/o Royal Netherlands Industries Fair, Utrecht, p. 167
5. *Pflüger, R.:* Kunststoffe 57 (1967) p. 31
6. *Asmus, K. D.,* et al.: Kunststoffe 59 (1969) p. 266
7. *Schlack, M.:* Plast. World (1987) Nr. 6, p. 88

2.2 Chemical Structure of Polyesters

The semicrystalline homopolyesters polybutylene terephthalate and polyethylene terephthalate gained the greatest technical importance among the thermoplastic polyesters. The copolyesters, except the elastomer block copolyetheresters, are not important despite their broad variety.

2.2.1 Homopolyesters

Polybutylene terephthalate and polyethylene terephthalate are linear polyesters made from terephthalic acid or terephthalic acid dimethylester and 1,4-butanediol or ethylene glycol, respectively. The basic molecular structure is as follows:

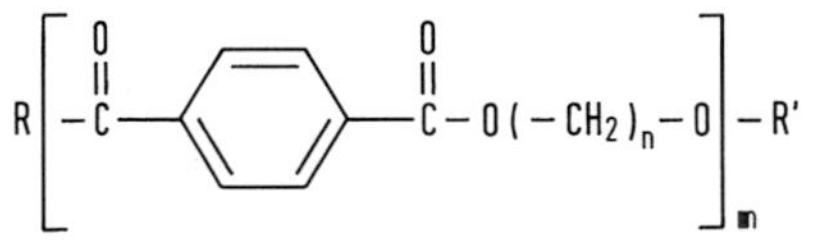

PBT: n = 4
PET: n = 2

The polymerization degree P_m of technically interesting polyesters is in general higher than 90 and, depending on the technical use, lower than 200. The preparation of polyesters is a two-step process for PBT as well as for PET.

Step 1: Transesterification of terephthalic acid dimethylester with 1,4-butanediol or ethylene glycol, respectively, with simultaneous oligomerization. The terephthalic acid reacts with the glycol directly in this first step.

Step 2: Melt polycondensation of the (trans)esterified precondensates or polycondensation in vacuum to form a polyester. To obtain high molecular weights, polycondensation in solid phase follows for certain applications.

The end groups R and R′ of the polyester are hydroxy, carboxy, or methoxy groups; their composition depends on the method of chemical preparation and on the obtained degree of polymerization.

2.2.2 Copolyesters

When other dicarboxylic acids or other glycols are added, together with terephthalic acid dimethylester and butane-1,4-diol, respectively, using ethylene glycol in the production of PBT and PET lowers the melting temperature, the maximum degree of crystallization, and the crystallization speed. The diagram for copolyesters made of terephthalic acid, adipic acid, and ethylene glycol (Figure 2.1) shows the strong decrease in the melting temperatures of these copolyesters.

In some cases melting temperature, crystallinity, and crystallization speed of the copolyesters can be increased in comparison to the homopolyesters PBT or PET by incorporating small amounts of special codiols [2–6].

2.2.2.1 Amorphous Copolyesters

In contrast to the semicrystalline homopolyesters PBT and PET, amorphous copolyesters are transparent, even as thick parts. Other advantages of amorphous copolyesters are high toughness, low shrinkage, and high stability in size. To increase glass transition temperature and thermostability, chain flexibility must be decreased. This can be accomplished, for example, by increasing the aromatic part of the dicarboxylic acid, by stiffening cycloaliphatic diols, by increasing intermolecular forces with polar groups, or by the use of special heterocyclic diols [7, 8]. Known technical amorphous copolyesters are based on, for example, terephthalic acid, isophthalic acid, 1,4-dimethylol-cyclohexane (Kodar A 150 by Eastman Kodak) or on terephthalic acid, ethylene glycol, 1,4-dimethylol-cyclohexane (e.g. Kodar PETG 6763).

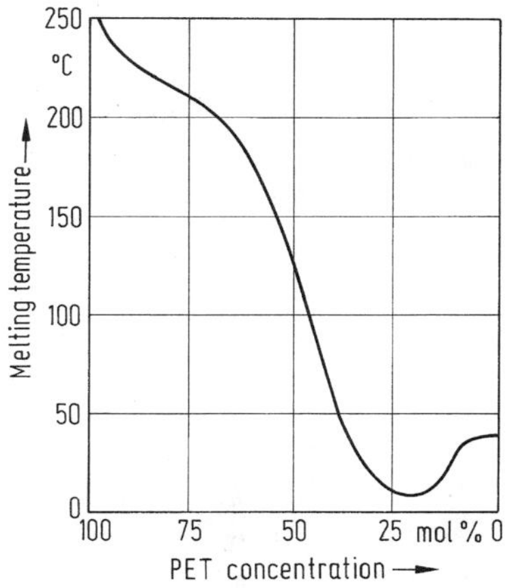

Figure 2.1. Melting temperatures of copolyesters of terephthalic acid, adipic acid, and ethylene glycol [1].

2.2.2.2 Block Copolyether Esters

The polycondensation of aromatic dicarboxylic acids, short-chain alkylene glycols, and low molecular weight poly(alkylene ether) glycols can be varied, and the preparation of many interesting polymers is possible [9–12]. Of all these polymers, block copolyether esters made of terephthalic acid (dimethylester), butane-1,4-diol, and poly(tetramethylene ether) glycol gained the greatest technical importance.

These block copolyether esters have repeating hard crystalline tetramethylene terephthalate blocks and soft, amorphous poly(tetramethylene ether)terephthalate blocks in the same macromolecule in statistical distribution. The block copolyether esters have the following structure:

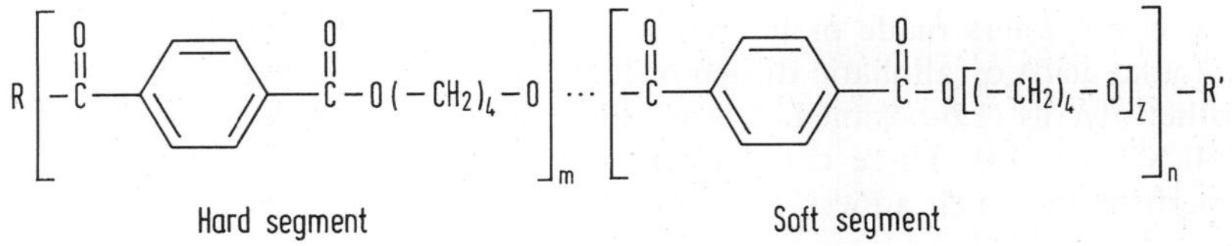

In general, the block length of the soft segments is given by the molecular weight of the poly(tetramethylene ether) glycol used, which is between 600 and 2000 for technical products, In contrast, the average block length of the hard segments depends on the concentration of tetramethylene terephthalate in the polymer, at a constant molecular weight of poly(tetramethylene ether) glycol. Figure 2.2 shows this relationship for a block copolyether ester from terephthalic acid dimethylester, 1,4-butanediol, and poly(tetramethylene ether) glycol with the molecular weight 1000.

Partial substitution in the hard segments of terephthalic acid by isophthalic or sebacic acid causes an increase in tensile strength, tear resistance strength, and resistance to solvents, especially for elastomer mixtures. It also prevents brittleness, especially for elastomer mixtures and the hard block copolymer esters, in certain oils [9, 12].

As for PBT and PET, the preparation of block copolyether ester is a two-step process. First terephthalic acid dimethylester reacts with poly(tetramethylene ether) glycol and 1,4-butanediol. The ratio of soft to hard segments is determined by the amount of poly(tetramethylene ether) glycol. Excess 1,4-butanediol is used for better

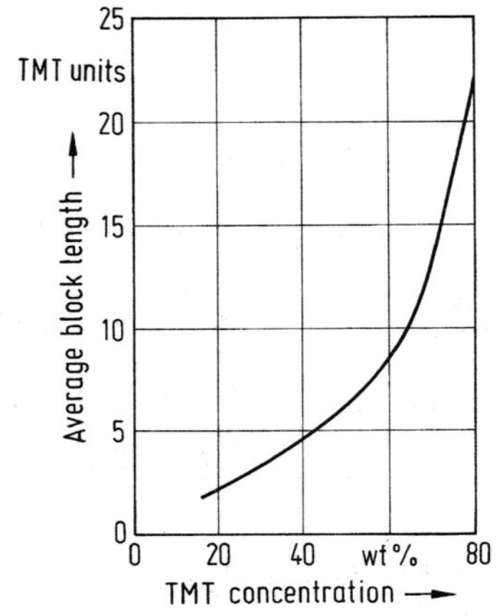

Figure 2.2. Dependence of the average block length of the hard segments on the concentration of tetramethylene terephthalate (TMT) [10].

transesterification. In the second step, which follows transesterification, the prepared oligomer undergoes polycondensation under reduced pressure. The excess 1,4-butanediol is distilled off. Technically usable block copolyether esters have an average molecular weight of approximately 25,000 [10].

The block copolyether esters are morphologically biphasic. They are an amorphous matrix with rubber–elastic properties above the glass transition temperature and crystalline ranges that serve as crosslinking loci below their melting point. The amorphous phase is built of soft segments. The crystalline phase is based on hard segments. The presence of these two phases causes the typical elastomeric properties of block copolyether esters.

2.2.2.3 Low Melting Copolyesters

Certain copolyesters made of terephthalic acid and other aromatic dicarboxylic acids (isophthalic acid) or aliphatic dicarboxylic acids (adipic acid) as well as 1,4-butanediol and other glycols (1,6-hexanediol) are still crystalline but have low melting points (i.e., 40–130 °C) [13, 14]. These copolyesters are used as physically removed, solvent-free, high performance melt adhesives to coat laminated leather, textiles, and resins, respectively.

2.2.2.4 Other Copolyesters

Modification of PBT and PET to gain special properties is frequently described in the literature: Examples include condensation with certain dye-retaining groups for better dying of polyester fibers and increasing toughness by grafting of monomers like styrene [15], methyl methacryalate [16], or acrylic acid [17] on polyethylene terephthalate fibers or crosslinking of PBT [18]. Nevertheless, these measurements have failed to gain technical significance when PBT and PET are used, as thermoplastics as described for applications in this chapter.

References for Section 2.2

1. *Ludewig, H.:* Polyesterfasern, Chemie und Technologie. Akademie-Verl., Berlin 1975, p. 11
2. DE-PS 24 31 072 (1974) Bayer AG

3. DE-OS 25 07 674 (1975) Bayer AG
4. DE-OS 25 07 776 (1975) Bayer AG
5. DE-OS 25 45 720 (1975) Bayer AG
6. *Bier, P.*, et al.: Angew. Makromol. Chem. 65 (1977) p. 1
7. *Häberlein, H.; Korbanka, H.:* Angew. Makromol. Chem. 33 (1973) p. 111
8. *Habermeier, J.*, et al.: Chem. Z. 98 (1974) p. 184
9. *Hoeschele, G. K.; Witsiepe, W. K.:* Angew. Makromol. Chem. 29/30 (1973) p. 267
10. *Cella, R. J.:* Polyesters, Elastomeric. In: *Mark, H. F.; Bikales, N. M.* (Eds.): Encyclopedia of Polymer Science and Technology. Vol. 2. Wiley, New York 1977, p. 485
11. *Elias, H. G.; Vohwinkel, F.:* Neue polymere Werkstoffe für die industrielle Anwendung. 2. Ed. Hanser, Munich 1983
12. *Domininghaus, H.:* Die Kunststoffe und ihre Eigenschaften. 2. Ed. VDI-Verl., Düsseldorf 1986
13. *Temin, S. C.:* Adhesive Compositions, Polyesters. In: *Mark, H. F.; Bikales, N. M.; Overberger, C. G.; Menges, G.* (Eds.): Encyclopedia of Polymer Science and Engineering. Vol. 1. Wiley, New York 1985, p. 572
14. DE-OS 24 35 863 (1974) Dynamit Nobel AG
15. *Vlagiu, I.; Stanett, V.:* J. Macromol. Sci., Chem. A 7 (1973) 8, p. 1677
16. *Hebeish, A.*, et al. J. Appl. Polym. Sci. 26 (1981) p. 3253
17. *Hebeish, A.*, et al. J. Appl. Polym. Sci. 26 (1981) p. 3245
18. DE-OS 19 00 270 (1968) Societé Rhodiaceta

2.3 Preparation of Polyesters

2.3.1 Raw Materials

This section describes the technical preparation of the most important raw materials and their properties, as well as the raw material specifications.

2.3.1.1 Terephthalic Acid and Terephthalic Acid Dimethylester

The technical preparation of PBT is entirely based on terephthalic acid dimethylester, while PET is increasingly produced from very pure terephthalic acid (fiber grade quality). Compared to terephthalic acid dimethylester, terephthalic acid has some advantages for the production of PET:

- lower price
- needs no transesterification catalyst
- higher polycondensation speed
- small amount of carboxylic end groups
- no recovery of ethanol necessary

Because of these benefits, the world production of terephthalic acid has been increasing [1].

Capacities for 1994, in millions of tons per year, were as follows:

	Terephthalic Acid Dimethylester	Terephthalic Acid
World	5.3	10.8
West Europe	1.7	1.2

The most important process for the preparation of terephthalic acid dimethylester is the Witten-Hercules process. In two steps liquid phase *p*-xylol is oxidized with air in the presence of cobalt-manganese salts at 150–170 °C and 4–8 bar. In the first step oxidation leads to *p*-toluic acid. This is esterified with methanol at 140 °C and a pressure of 30–40 bar to yield *p*-toluic acid methylester. In the second step, oxidation to terephthalic acid monomethylester takes place. This ester is esterified with methanol to form terephthalic acid dimethylester. Technically two steps—oxidation and esterification—are combined in two reactors [2]:

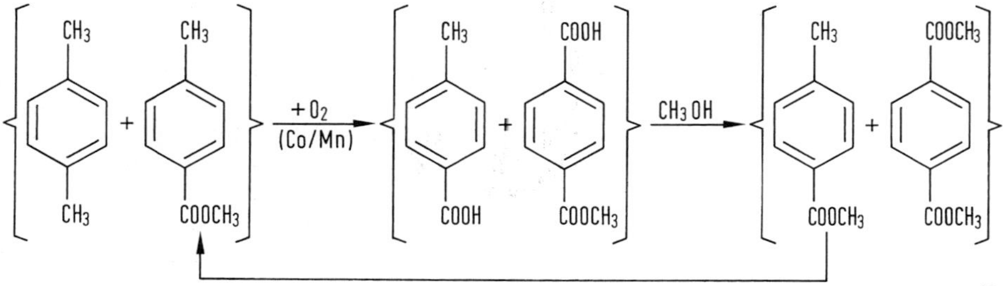

(2.1)

Terephthalic acid dimethylester with a purity of 99.99% is obtained by recrystallization and distillation.

Terephthalic acid dimethylester can also be produced from raw terephthalic acid by esterification with methanol: DuPont, Eastman Kodak, Mitsui, and Mitsubishi have developed processes for its manufacture.

Terephthalic acid is produced mainly by the Amoco process. Inhibition of the oxidation of the second methyl group of *p*-xylol is neutralized by additional activators based on bromine compounds. The *p*-xylol is oxidized continuously with air in acetic acid in the presence of cobalt and manganese salts as oxidation catalysts and in the presence of bromine compounds as activators at 215–235 °C and 25–35 bar. The raw terephthalic acid, containing 4-carboxybenzaldehyde as the main contaminant, is purified by hydrogenation and crystallization. 4-Carboxybenzaldehyde is reduced to *p*-methylbenzoic acid and eliminated by recrystallization. The final terephthalic acid has a purity of 99.99% [3].

Other technical processes differ in several process parameters.

To produce light-colored and high molecular weight polyester, the following demands, incorporating values established by the American Public Health Association (APHA), are made on terephthalic acid dimethylester and terephthalic acid:

Terephthalic Acid Dimethylester	
Freezing point	≥ 140.6 °C
Acid number	≤ 0.03 mg KOH/g
Acid number after 4 h at 175 °C	≤ 0.06 mg KOH/g
Color number after melting	≤ 25 APHA
Color number in H_2SO_4	≤ 10 APHA
Iron content	≤ 1 ppm
Water content	≤ 0.02%

Terephthalic Acid	
Acid number	675 ± 2 mg KOH/g
Ash content	≤ 15 ppm
Metals	≤ 10 ppm
Fe, Ca, Al, Na, K	≤ 2 ppm
Co, Mo, Cr, Ni, Ti, Mg	≤ 1 ppm

4-Carboxybenzaldehyde	≤ 25 ppm
Water content	≤ 0.5%
Color number in dimethylformamide	≤ 10 APHA*

* Amount, in milligrams of platinum [dissolved as potassium hexachloroplatinate(IV) with cobalt(II) chloride hexahydrate in a ratio of 1246:1 in 1000 mL aqueous HCl] that shows approximately the same color in the same layer thickness as the sample.

2.3.1.2 1,4-Butanediol

The biggest volume of industrial 1,4-butanediol is made from acetylene in a two-step process via 2-butyne-1,4-diol. First copper acetylide catalyzes the addition of formaldehyde in an aqueous solution of acetylene at 80–100 °C and 2–6 bar. The intermediate propargyl alcohol is not isolated. The aqueous butynediol solution is hydrogenated with excess hydrogen at 80–106 °C and 300 bar in the presence of copper and manganese-modified nickel catalysts on silicone dioxide. Pure 1,4-butanediol is obtained by fractional distillation of the reaction mixture containing methanol, propanol, and butanol as the most important side products [4]:

$$HC \equiv CH \xrightarrow[\text{(Cu-acetylide)}]{+2\ CH_2O} HOCH_2-C \equiv C-CH_2OH \xrightarrow[\text{(Cu/Mn/Ni)}]{+2\ H_2} HO(CH_2)_4OH \qquad (2.2)$$

Other technical processes to produce 1,4-butanediol start from the following raw materials:

- 1,4-dichloro-2-butene and 3,4-dichloro-1-butane (Toyo Soda)
- allyl alcohol and synthetic gas (Daicel)
- butadiene and acetic acid (Mitsubishi)

To produce light-colored, high molecular weight PBT, the following demands are made on 1,4-butanediol:

Purity	≥ 99%
Water content	0.01–0.02%
Freezing range	19.5–19.8 °C
Boiling range (27 mbar, 5–95 mL/DIN 53 406)	137–138 °C
Refractive index n_D^{25}	1.4435–1.4445
Density at 20 °C (DIN 51 757)	1.01–1.02 g/cm^3
Color number (DIN 53 409)	5–10 APHA
Inflammation point (DIN 51 758)	134 °C

Capacity in 1994:	World	613,000 tons
	Western Europe	190,000 tons

2.3.1.3 Ethylene Glycol

Ethylene glycol is produced industrially by the reaction of ethylene oxide with excess water. The production of diglycol and oligomer glycols is kept low by a high water excess (10–30 times the molar amount) as well as by avoiding back-mixing in the reactor. The typical product distribution between ethylene glycol, diethylene glycol, and triethylene glycol is a mass relation of 30:4:1. The glycols are separated by purification under vacuum in distillation columns connected in series.

To produce light-colored, high molecular weight PET, the following demands are made on ethylene glycol:

Purity	≥ 99.9%
Water content	≤ 0.05%
Diethylene glycol content	≤ 0.03–0.04%
Ash content	≤ 0.005 g/100 mL
Iron content	≤ 0.05 ppm
Acid content as acetic acid	≤ 0.005%
Acetaldehyde content	≤ 7 ppm
Density at 20 °C	1.1151–1.1156 g/cm^3
Color number	≤ 5 APHA
After 4 h boiling	≤ 10 APHA
UV permeability	
at 220 nm	≥ 70%
at 275 nm	≥ 90%
at 350 nm	≥ 98%
Capacity in 1995: World	10,500,000 tons
Western Europe	1,200,000 tons

2.3.1.4 Poly(tetramethylene ether) Glycol

The raw material for the soft segments in polyester elastomers is poly(tetramethylene ether) glycol, which can be produced continuously or discontinuously by cationic polymerization of tetrahydrofuran. Although some oxonium salts, Lewis acids, and strong proton acids catalyze the polymerization, the industrial process works best with fuming sulfuric acid. Fluorosulfuric acid has the advantage of being able to produce poly(tetramethylene ether) glycols of every desired molecular weight, even over 1,200.

Poly(tetramethylene ether) glycols are offered with following specifications (DuPont, Quaker Oats):

Acid number	≤ 0.05 mg KOH/g
Water content	≤ 0.03%
Color number	≤ 90 APHA
Capacity in 1993: World	127,000 tons
Western Europe	41,000 tons

2.3.2 Chemical Fundamentals of Polyester Production

PBT and PET are produced in two steps. First dihydroxybutylterephthalate or dihydroxyethylterephthalate is synthesized by transesterification of dimethylterephthalate with 1,4-butanediol or ethylene glycol. Although the preparation of PBT via esterification has been described, the transesterification process is the only one that is used on an industrial scale. However, the esterification of terephthalic acid with ethylene glycol to produce PET is gaining more and more importance.

In the second step the obtained dihydroxyalkylterephthalate undergoes polycondensation at higher temperature to form the polyester, while glycol is removed. An advanced polycondensation in the solid state just below the melting temperature of the polyester is necessary to obtain very high molecular weights.

2.3.2.1 Transesterification Processes

The transesterification of dimethylterephthalate with 1,4-butanediol or ethylene glycol takes place above 150 °C and is carried out operationally up to 220 °C. The temperature should increase slowly to avoid sublimination of nontransesterified dimethylterephthalate. The methanol by-product is distilled off during the reaction.

The transesterification is accelerated catalytically. Although many usable catalysts are described in patents, in general, soluble titanium compounds, especially titanium alcoholates, are used for 1,4-butanediol and acetates of calcium, manganese, cobalt, cadmium, lead, and zinc, as well as lithium glycolate in the amount of 0.005–0.1 wt % (related to dimethylterephthalate) are used for ethylene glycol.

Hydroxyalkylterephthalic acid methylester, dihydroxyalkylterephthalate, and oligomers are intermediate steps during transesterification:

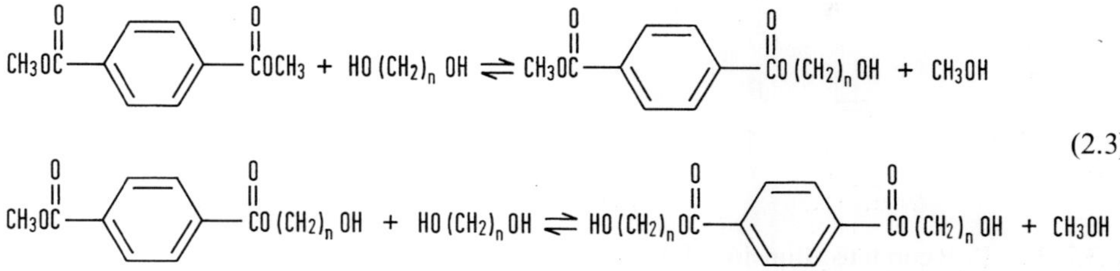

(2.3)

While the concentration of dimethylterphthalate decreases quickly, the concentration of hydorxybutylterephthalic acid methylester reaches a maximum after a short reaction time. The maximum of dihydroxyalkylterephthalate is attained much later. Like monohydroxyesters and dihydroxyesters, oligomers are already formed after a short transesterification time (Figure 2.3).

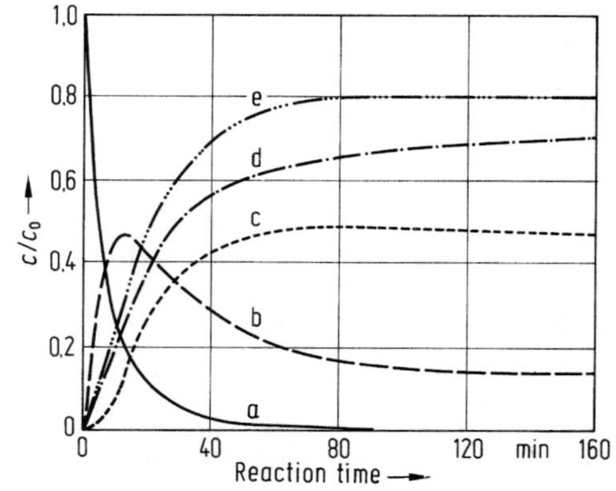

Figure 2.3. Transesterification of dimethylterephthalate (1 mol) with butane-1,4-diol (2.2 mol) at 150 °C; titanium tetrabutylate as catalyst [5]: a, dimethylterephthalate (DMT); b, hydroxybutylmethyl terephthalate; c, dihydroxybutylterephthalate; d, DMT—part bound to oligoester; e, transesterification grade.

2.3.2.2 Esterification Processes

The first patent to produce PBT from terephthalic acid was published in 1949 [6] and several other patents and publications about esterification of these two monomers followed soon. This process has had no industrial importance for the production of PBT until now [7]. The relatively high loss of 1,4-butanediol in the form of tetrahydrofuran offsets the economical advantages of terephthalic acid and of the unnecessary methanol recovery during the transesterification step. The increased formation of tetrahydrofuran—compared to the transesterificatiion process—can be explained by the terephthalic acid catalyzed dehydration of 1,4-butanediol.

In contrast to PBT, the esterification process for PET became more important with the availability of ultrapure terephthalic acid. As already described in Section 2.3.1.1, the process has economical and process engineering advantages.

Because of the poor solubility of terephthalic acid, esterification temperatures of 220–260 °C are used. The water by-product is distilled off. In contrast to the transesterification, 10% molar excess ethylene glycol is enough to obtain the dimer esterification product [1]:

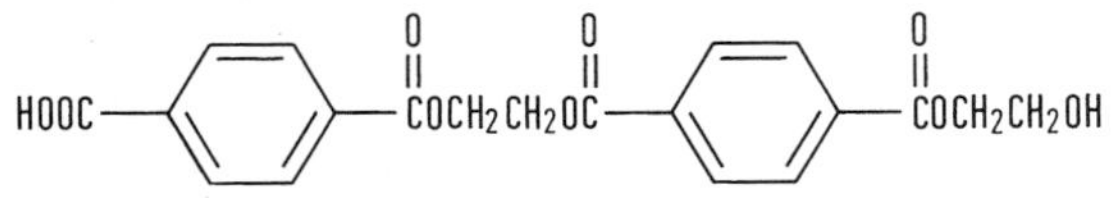

2.3.2.3 Polycondensation in Melts

In a vacuum and at an elevated temperature, excess glycol is distilled from the esterification or transesterification product that already contains the oligomers. Then the polyester is formed in a catalytically accelerated polytransesterification reaction with separation of 1,4-butanediol or ethylene glycol. Because the polytransesterification is an equilibrium reaction, the glycol by-product must be removed as quickly as possible to accelerate polycondensation. The glycol is removed from the melt in vacuum by using intensive mixing to create a surface area as large as possible. The polycondensation temperatures are 250–270 °C for PBT and 270–285 °C for PET.

Transesterification catalysts used to accelerate polycondensation can be reused. For PBT, soluble titanium compounds can also be used for polycondensation; special polycondensation catalysts based on antimony, germanium, titanium, and lead compounds are used for PET, however, because the transesterification catalysts favor the thermal degradation of the formed polyester and the forming of diglycols. Therefore, the glycols must be inactivated by phosphorous acid or its esters before polycondensation takes place. To prevent the yellowing of PET while using titanium tetrabutylate, metaphosphoric acid and its alkali metal salts are added [8].

2.3.2.4 Polycondensation in the Solid Phase

The melt polycondensation of PBT, carried out under common industrial conditions, leads to products with a molecular weight of 25,000–35,000, while the application of the same process to PET leads to products with a molecular weight of 15,000–25,000. Certain injection molding and extrusion applications require high quality, higher molecular weight polyesters with improved mechanical and thermal properties. Polyesters with a molecular weight of more than 40,000 (PBT) or 30,000 (PET) can rarely be prepared under conditions of melt polycondensation because of the long reaction times and because at high polycondensation temperatures, thermal degradation occurs more rapidly than formation. Higher molecular weight polyesters can be obtained more easily by polycondensation in the solid phase below the polyester's melt temperature because at the lower temperatures, the speed of formation is higher than the speed of degradation.

The speed of polycondensation in the solid phase depends mainly on the diffusion of the glycol by-product to the surface of the polyester particle and on the polycondensation temperature. To accelerate the diffusion, polycondensation is carried out in vacuum as well as in an inert gas stream (nitrogen). Polyesters of small grain size favor the diffusion of glycol.

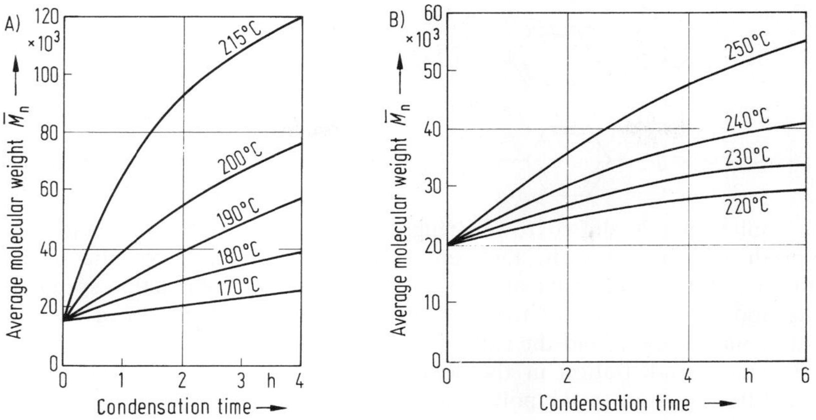

Figure 2.4. The influence of polycondensation time and temperature on polycondensation in the solid phase of PBT (A) [9] and PET (B) [10].

The polycondensation temperature influences the polycondensation speed as well as the size of the obtained molecular weights. Figure 2.4 shows examples for PBT and PET. A temperature just below the polyester's melt temperature is favorable. Since, however, that part of the polyester has already melted, simmering and baking of the pellets, temperatures up to 20 °C below the polyester's melt temperature are preferred. It is suitable to precrystallize the slowly crystallizing PET first in one or more steps. This is not necessary for PBT. To ensure reproducibility of conditions of polycondensation in the solid phase, it is necessary to keep a certain ratio of end groups. The amount of slowly reacting methyl groups should be small, while the ideal OH/COOH ratio is 2–10 for PBT [9] and 2 for PET [11].

2.3.2.5 Reaction Kinetics

Solid phase transesterifications [1, 5, 12–18], esterification [12, 19], and melt polycondensation, as well as polycondensation [1, 9–12, 14, 15, 19–28], have been investigated thoroughly. The mechanism of metal catalysis was also investigated [12, 29–31, 43]. Although there are not general statements, it is assumed that the transesterifications of dimethylterephthalate with 1,4-butanediol or ethylene glycol are first-degree following reactions that proceed via hydroxybutylterephthalic acid methylester, dihydroxybutylterephthalate, and oligomers as intermediates.

Catalysis by hydroxonium ions is assumed for the esterification of terephthalic acid with diols. The ions are made of the carboxyl group of terephthalic acid and of already formed esters [19].

$$R-COOH + R'-\overset{\overset{\displaystyle O}{\|}}{C}-OR'' \rightleftharpoons R-COO^{\ominus} + R'-\overset{\overset{\displaystyle \overset{\oplus}{O}H}{\|}}{C}-OR'' \qquad (2.4)$$

The reaction speed constant of the esterification of terephthalic acid with ethylene glycol increases in ester media.

Chain-building polytransesterification is the equilibrium reaction in melt polycondensation.

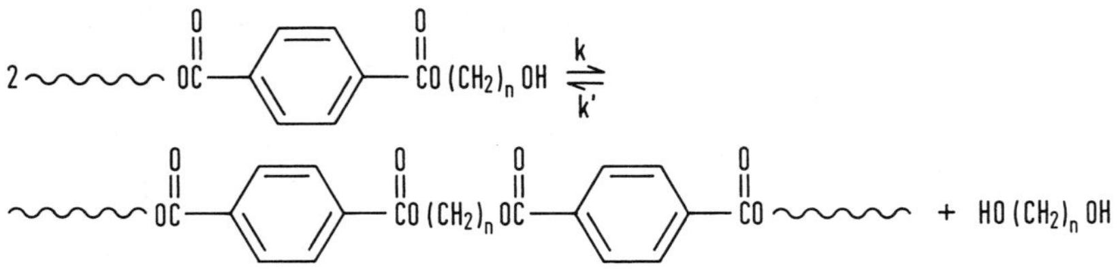

(2.5)

The equilibrium is shifted toward higher molecular weights by removing the glycol by-product from the melt. Different statements are made for the reaction degree. Additional factors that have an important influence on the reaction speed after temperature and concentration of the reactants are the layer thickness of the melt, the end group concentration, and the catalyst.

For polycondensation in the solid phase, the polyesterification must be seen as competitive reaction to the polytransesterification [11]:

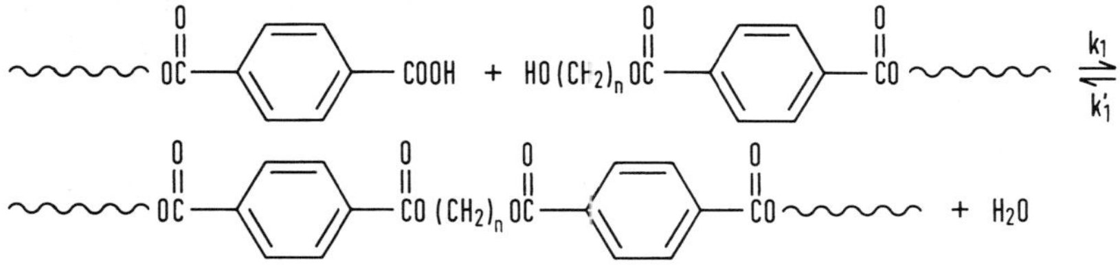

(2.6)

Both reactions use hydroxyalkylester end groups. Next to basic kinetic reflections, mathematical models for the technological polyester production for discontinuous [33], semicontinuous [34], and continuous processes [35, 36] have been developed.

2.3.2.6 Oligomers

PBT and PET contain low molecular weight linear and cyclic oligomers in amounts of 1–2 wt% [12, 37, 38]. The concentration of linear oligomers, which are intermediates of the polycondensation reaction, decreases very quickly during polycondensation. Cyclic oligomers are remade from already formed polyesters. The formation of oligomers from oligomer-free polyesters, from which oligomers were extracted, occurs after a short period of melting. Cyclic oligomers dominate. Therefore, oligomer-free products cannot be obtained under the conditions of operational scale production of PBT and PET.

2.3.2.7 Side Reactions

Several side reactions can happen during the production of PBT and PET. The main side reaction is the dehydration of the glycols. In case of 1,4-butanediol, an intramolecular dehydration to tetrahydrofuran occurs (reaction 2.7), and in case of ethylene glycol there is an intermolecular dehydration to diethylene glycol (reaction 2.8):

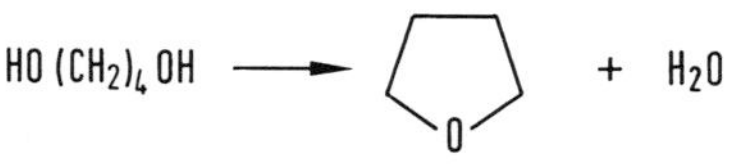

(2.7)

$$2\ HOCH_2CH_2OH \longrightarrow HOCH_2CH_2OCH_2CH_2OH + H_2O \qquad (2.8)$$

The dehydration is acid-catalyzed and is favored by a high carboxyl content. While tetrahydrofuran can be removed from the reaction mixture and does not affect the properties of PBT, diethylene glycol is condensed in PET during polycondensation. Therefore, commercial PET contains about 2.5 mol% diethylene glycol units, which reduces the melt temperature of PET, decreases the hydrolysis and light stability as well as the thermooxidative stability, and lowers the mechanical properties [39, 40]. The water formed by dehydration can cleave to produce ester bondages. Because of this, the number of carboxyl end groups increases.

Tetrahydrofuran is also formed by intramolecular separation from already produced hydroxybutylester end groups. Carboxyl end groups are formed at the same time [5, 39]:

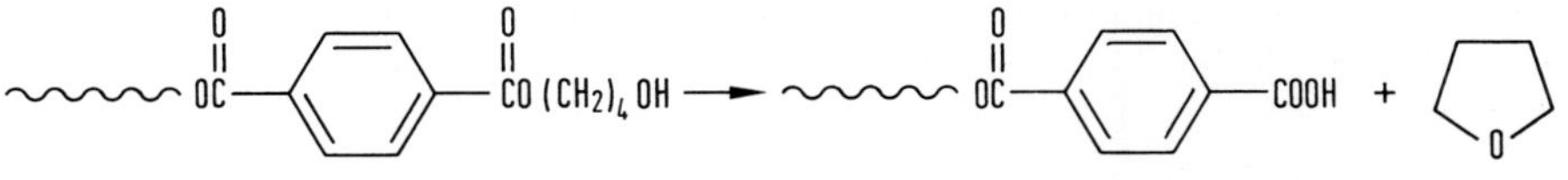

(2.9)

Other possibilities for the formation of diethylene glycol are discussed in the literature [12, 39, 40].

Another important side reaction during the production of PBT and PET is thermal degradation. Here the ester bond is cleaved, with the formation of a fragment with a carboxyl end group and another fragment with an olefinic double bond:

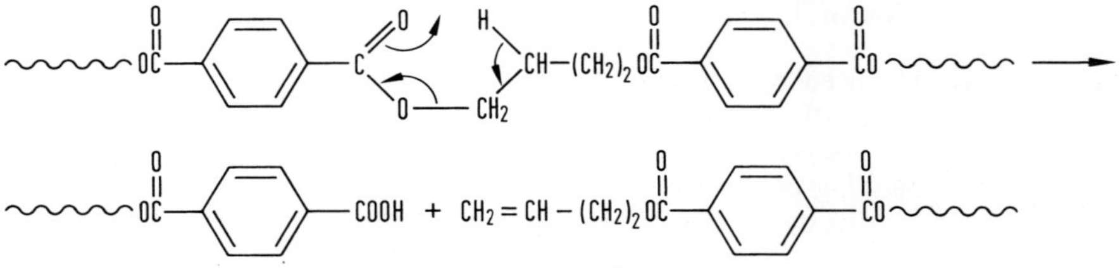

(2.10)

I

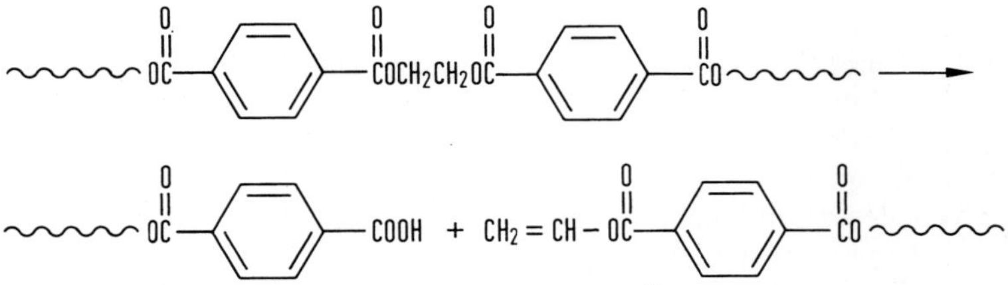

(2.11)

II

The 3-butene end group of PBT can form butadiene in another thermolysis step, or it can undergo transesterification to normal polyester under separation of 4-hydroxy-1-butene:

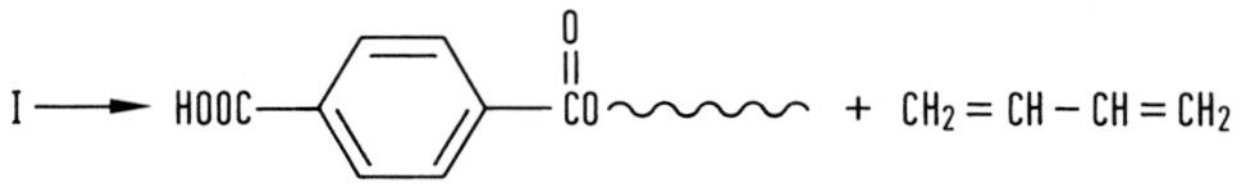

(2.12a)

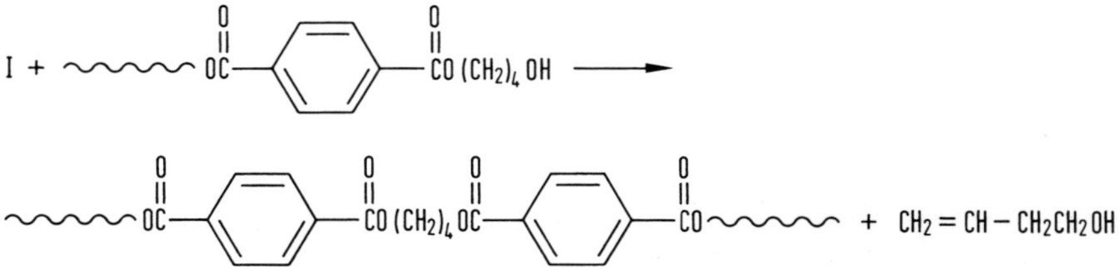

(2.12b)

In an analogous step, the vinyl ester end group of PET can also transesterify to produce acetaldehyde:

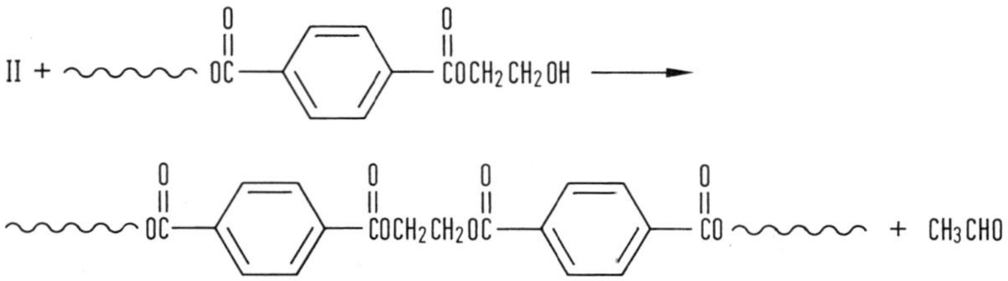

(2.13)

Thermooxidative degradation does not play a role in the production of polyesters because of the exclusion of oxygen. But hydrolytic degradation has a certain importance because water can be formed in one of the side reactions. It is significant that in kinetic terms, hydrosysis is 10^4 times quicker than thermolysis [1].

2.3.3 Technologies of Polyester Production

PBT and PET are produced by continuously and discontinuously working processes. Until 1963 nearly all the PET produced in the world was based on discontinuously working processes [41]. Discontinuously working plants can supply small amounts of a product, but often changing product types is easier and less expensive.

Large amounts of polyester having a constant composition can be produced more advantageously in continuously working plants, which guarantee homogeneous product quality and lower production costs. For PBT, the continuously working process is recommended for amounts of 12,000–14,000 tons/yr [42]. After Vickers–Zimmer AG had introduced the continuously working process [41, 43], several processes were developed. In general, all processes structure in the following steps (Figure 2.5):

- transesterification and elimination of methanol
- precondensation and elimination of excess or already formed glycol by-product
- polycondensation in one or more steps

To accommodate polycondensation kinetics in the advanced state, which is characterized by high melt viscosity of the polyester, the polycondensation reactor should satisfy the following demands:

- large surface volume
- small melt volume
- thin melt layer
- fast surface renovation
- good mixing
- small amount of back-mixing
- good heat transfer

It is also advantageous to adapt the polycondensation to the increasing melt viscosity during the process and to keep the residence time of high molecular weight polyesters short at the high temperatures used at the end of the polycondensation. This approach favors the production of polyesters with homogeneous molecular weight distribution and constant high quality.

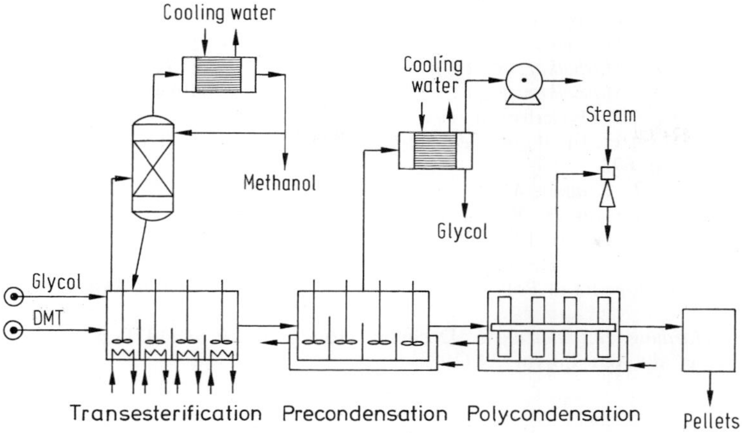

Figure 2.5. Continuous production of polyester from dimethylterephthalate [1].

References for Section 2.3

1. *Falkai, B. von*: Synthesefasern. Verlag Chemie, Weinheim 1981
2. *Hoffmann, G.; Mayer, D.*: Terephthalsäuredimethylester. In: Ullmanns Encyclopädie der technischen Chemie. 4. Ed. Vol. 22. Verlag Chemie, Weinheim 1982, p. 529
3. *Hoffmann, G.; Mayer, D.*: Terephthalsäure mit Isophthalšure. in [2] p. 519
4. *Gräfje, H.*, et al.: Butanediols, Butenediol, and Butynediol. In: Ullmann's Encyclopedia of Industrial Chemistry. 5. Ed. Vol. A 4, Verlag Chemie, Weinheim 1985, p. 455
5. *Tuček, E.; Dinse, H. D.*: Acta Polym. 31 (1980) p. 429
6. GB-PS 578079 and US-PS 2 465 319, Du Pont (both patents with priority of 1941, *Whinfield, J. R.; Dickson, J. T.*)
7. *Buyle Padias, A.; Hall, H. K.*: J. polym. Sci., Polym. Chem. Ed. 19 (1981) p. 1021
8. DE-PS 26 26 827 (1976) Bayer AG
9. *Dinse, H. D.; Tuček, E.*: Acta Polym. 31 (1980) p. 108
10. *Jabarin, S. A.; Lofgren, E. A.*: J. Appl. Polym. Sci. 32 (1986) p. 5315
11. *Schaaf, E.*, et al.: Acta Polym. 32 (1981) p. 250
12. *Ludewig, H.*: Polyesterfasern, Chemie und Technologie. Akademie-Verl., Berlin 1975, p. 125
13. *Dimov, K.*, et al.: Faserforsch. Textiltech. 24 (1973) p. 381
14. *Rafler, G.*, et al.: Faserforsch. Textiltech. 24 (1973) p. 235
15. *Bonatz, E.*, et al.: Faserforsch. Textiltech. 24 (1973) p. 313
16. *Adorova, I. V.*, et al.: Acta Polym. 32 (1981) p. 282
17. *Yamanis, J.; Adelman, M.*: J. Polym. Sci. 14 (1976) p. 1945 and 1961
18. *Datye, K. V.; Raje, H. M.*: J. Appl. Polym. Sci. 30 (1985) p. 205
19. *Baranova, T. L.*, et al.: Faserforsch. Textiltech. 29 (1978) p. 577
20. *Rafler, G.*, et al.: Acta Polym. 31 (1980) p. 732
21. *Bonatz, E.; Rafler, G.*: Acta Polym. 31 (1980) p. 402
22. *Bonatz, E.*, et al.: Faserforsch. Textiltech. 24 (1973) p. 309
23. *Pell, T. M.; Davis, T. G.*: J. Polym. Sci. 11 (1973) p. 1671
24. *Chang, S.*, et al.: J. Appl. Polym. Sci. 28 (1983) p. 3289
25. *Gostoli, C.*, et al.: J. Appl. Polym. Sci. 29 (1984) p. 2873
26. *Rafler, G.*, et al.: Acta Polym. 30 (1979) p. 253
27. *Rafler, G.*, et al.: Acta Polym. 34 (1983) p. 48
28. *Lin, C. C.; Baliga, S.*: J. Appl. Polym. Sci. 31 (1986) p. 2483
29. *Rafler, G.*, et al.: Faserforsch. Testiltech. 27 (1976) p. 579
30. *Wolf, K. H.; Herlinger, H.*: Angew. Makromol. Chem. 65 (1977) p. 133
31. *Dimov, K.; Terlemezyan, E.*: J. Polym. Sci. 10 (1972) p. 3133

32. *Kumar, A.*, et al.: J. Appl. Polym. Sci. 29 (1984) p. 1045
33. *Ravindranath, K.; Mashelkar, R. A.:* J. Appl. Polym. Sci. 26 (1981) p. 3179
34. *Ravindranath, K.; Mashelkar, R. A.:* J. Appl. Polym. Sci. 27 (1982) p. 471
35. *Ravindranath, K.; Mashelkar, R. A.:* J. Appl. Polym. Sci. 29 (1984) p. 437
36. *Griehl, W.; Jäger, P.:* Faserforsch. Textiltech. 26 (1975) p. 378
37. *Sattler, H.:* Polyester. In: *Winnacker, K.; Küchler, L.:* Chemische Technologie. Vol. 6. Hanser, Munich 1982, p. 675
38. *Borman, W. F. H.; Kramer, M.:* Coat. Plast. Prepr. 34 (1974) p. 77
39. *Penton, H. R.; Bouman, H.:* Polym. Prepr. 17 (1976) 2, p. 521
40. *Scheller, H.:* Chemiefasern 12 (1965) p. 923
41. *Clagett, D. H.:* Polyester Resins. In: *Mark, H. F.; Bikales, N. M.; Overberger, C. G.; Menges, G.* (Eds.): Encyclopedia of Polymer Science and Engineering. Vol. 6. Wiley, New York 1985, p. 111
42. *Hardung-Hardung, H.; Dietze, M.:* Dechema-Monographien 68 (1971) p. 509
43. *Rafler, G.*, et al.: Acta Polym. 39 (1988) p. 315

2.4 Modification of Polyester

The use of additives permits the properties of polybutylene terephthalate to be changed in many ways. Processing can be improved through the use of stabilizers, lubricants, release agents, and nucleating agents. The use of blowing agents makes possible the production of foamed polyesters. As important as these processing aids are, even more significant are functional additives that allow the final properties to be considerably varied. Examples include stabilization against various environmental influences, strengthening via filler and reinforcers, improved flame retardance, and coloring with dyes and pigments. In specific cases, polyesters may be modified with antistatic agents, optical brighteners, and softening agents.

2.4.1 Stabilization

2.4.1.1 Stabilization Against Thermal Degradation

Thermal degradation of polyesters occurs via cleavage of the ester bond. Polyethylene terephthalate is more stable than polybutylene terephthalate [1]. The ester cleavage is strongly influenced by the various polycondensation and transesterification catalysts [2]. The acceleration of the ester cleavage is probably due to the strong polarization of the carbonyl bond by coordination with the metal catalyst. The reaction is thought to go through a cyclic intermediate as follows [3]:

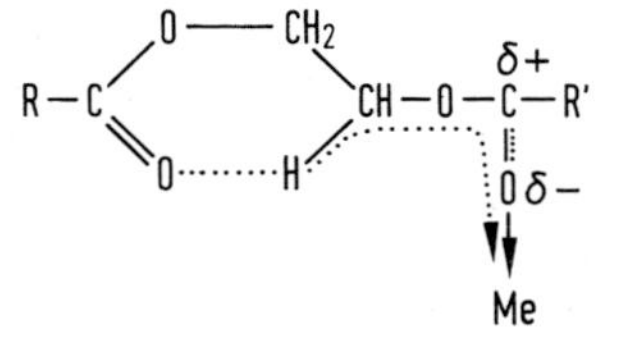

Because of the decreased inductive effect in polybutylene terephthalate, catalysts have less influence in thermal degradation of polybutylene terephthalate than of polyethylene terephthalate. Organic phosphites, phosphates, phosphonates, and phosphinates suppress degradation by forming an insoluble complex with the metal catalyst [4].

The kinetics of the thermal degradation of polyesters have been investigated [2, 5]. An assortment of commercially available stabilizers is shown in Table 2.1. In general, amounts of 0.02–0.2 wt % stabilizer are added. Other effective stabilizers known in the literature include 5-hydroxyisophthalic acid, which reduces the formation of acetaldehyde [6], and α-diethylphosphosuccinates such as ethyl-2-hydroxyethyl-α-diethylphosphosuccinate [7].

2.4.1.2 Stabilization Against Thermooxidative Degradation

Whereas the thermal degradation of polyester plays a key role during processing, thermooxidative degradation is of prime importance in the use of molded parts at high temperatures. There are numerous publications regarding the mechanism of thermooxidative degradation [2, 4]. It is generally accepted that the first reaction consists of the formation of a hydrogen peroxide on a methylene group. In polyethylene terephthalate the ethylene glycol units favor the formation of hydrogen peroxides. In the second step, homolysis of the hydrogen peroxide leads to OH or OOH radicals and O or CH radicals, respectively. Another route for the formation of hydrogen peroxide is reaction with polyester to yield a polyester O radical and a polyester CH radical. Further reactions of the radicals lead to the disproportionation of the ester bond to give a carboxylic acid, carbon dioxide, and unsaturated end groups. Additional radical reactions can lead to crosslinking.

Stabilizers against thermooxidative degradation function as antioxidants. They inhibit the radical mechanism by acting as radical sinks for the reactive carbon and peroxy radicals. Effective antioxidants belong to the following general classes:

- sterically hindered phenols and aromatic amines, which act as radical sinks
- organic sulfur compounds and organic phosphorus compounds, which tie up the oxygen radicals

Some commercially available stabilizers are shown in Table 2.2. The phosphorus compounds listed in Table 2.1 function as thermal and thermooxidative stabilizers. Review articles are available [8–10].

Certain combinations of stabilizers show synergistic effects. Combinations of thioethers and hindered phenols, phosphites with hindered phenols, and combinations of trimethylhexylborate and triphenylphosphites are especially effective [11, 12].

In addition to the commercially available stabilizers, other effective stabilizers against thermooxidative effects are claimed in the patent literature [13–17].

2.4.1.3 Stabilization Against Hydrolysis

The stability of a polyester against hydrolysis plays an especially important role in processing [18]. A very small amount of water (0.01%) can cause a decrease in viscosity and molecular weight through hydrolytic cleavage of the ester bond [2]. Hydrolytic sensitivity and water concentration are so important because hydrolysis proceeds 10^4 times faster than thermal degradation and 10^3 times faster than thermooxidative degradation [4]. For this reason, numerous patents describe the stabilization of polyesters against hydrolysis. Classes of compounds include epoxides, carbodiimides, phosphoric acids, and esters. The most effective protection against hydrolysis during processing is, however, a well-dried polyester.

Table 2.1. Organic Phosphorous Compounds as Stabilizers Against Thermal Oxidative Degradation

Chemical Name	Structural Formula	Commercial Products
Triphenyl phosphite	$P(OC_6H_5)_3$	Irgafos TPP (Ciba-Geigy) Weston TPP (Borg-Warner)
Tris(nonylphenyl)phosphite (TNPP)	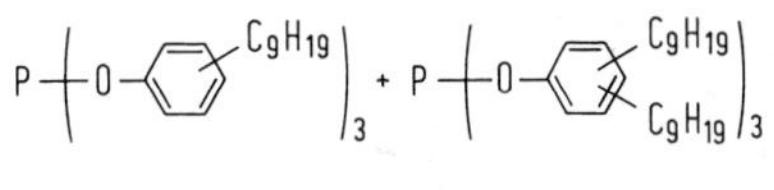	Irgafos TNPP (Ciba-Geigy) Weston TNPP (Borg-Warner)
Tris (2,4-di-*tert*-butylphenyl)phosphite	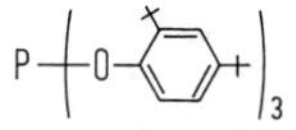	Irgafos 168 (Ciba-Geigy)
Bis(alkyl) pentacrythritol diphosphite	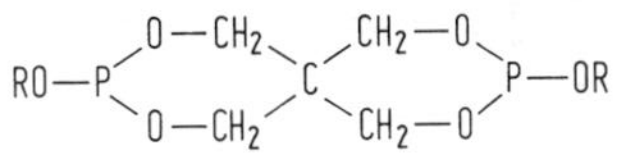 R = Stearyl R = 2,4-di-tert.-butylphenyl	Weston 618 (Borg-Warner) Ultranox 626 (Borg-Warner)
Tetrakis (2,4-di-*tert*-butylphenyl)-4,4′-diphenylene diphosphite	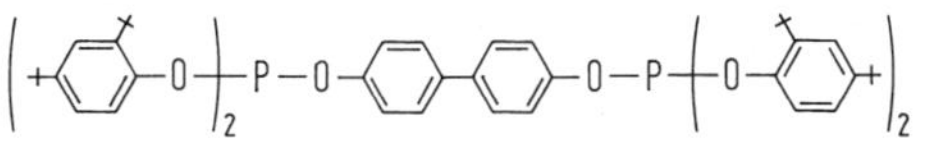	Sandostab P-EPQ (Sandoz) Irgafos E-EPQ (Ciba-Geigy)
3,5-Di-*tert*-butyl-4-hydroxybenzyl phosphonic acid diethyl ester	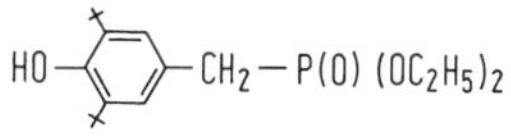	Irganox 1222 (Ciba-Geigy)

Table 2.2. Antioxidants as Stabilizers Against Thermoxidative Degradation

Chemical Name	Structural Formula	Commercial Product
Monophenols		
2,6-Di-*tert*-butyl-4-methylphenol	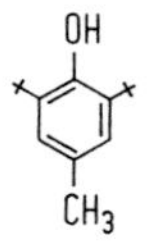	Vulkanox KB (Bayer)
2,6-Di-*tert*-butyl-4-*sec*-butylphenol	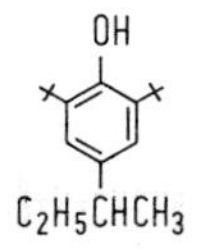	(Schenectady)
Diphenols		
4,4′-Methylene-bis(2,6-di-*tert*-butylphenol)	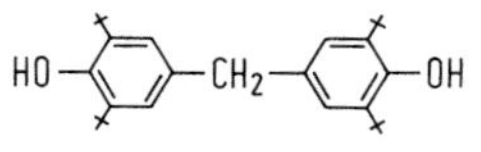	Lowinox 002 (Lowi) Ethanox 702(Ethyl)
2,2′-Ethylidene-bis(4,6-di-*tert*-butylphenol)	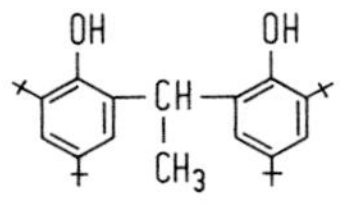	
4,4′-Butyliden-bis(6-*tert*-butyl-3-methylphenol)	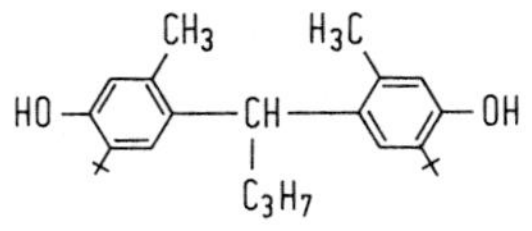	Santowhite Powder (Monsanto)
4,4′-Thiobis-(6-*tert*-butyl-3-methylphenol)	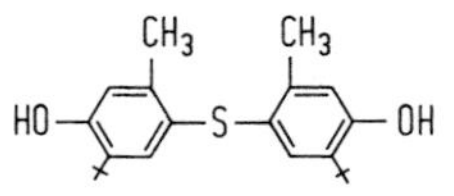	Santonox R (Monsanto) Lowinox 44S36 (Lowi)
1,6-Hexamethylen-bis(3,5-di-*tert*-butyl-4-hydroxyhydorcinnamate)	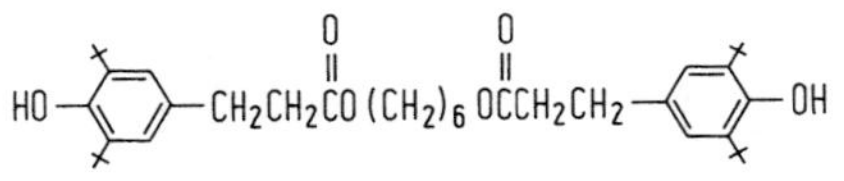	Irganox 259(Ciba-Geigy)
N,N′-Hexamethylene-bis(3,5-di-*tert*-butyl-4-hydroxyhydrocinnamide)	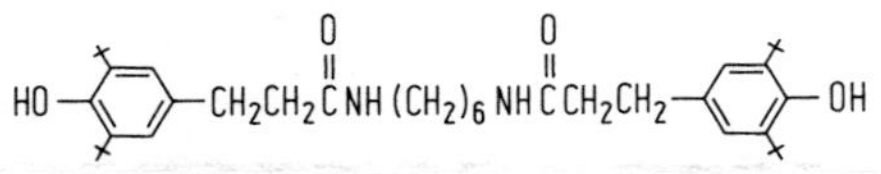	Irganox 1098 (Ciba-Geigy)
N,N′-Bis(3,5-di-*tert*-butyl-4-hydroxyhydrocinnamoyl)hydrazine	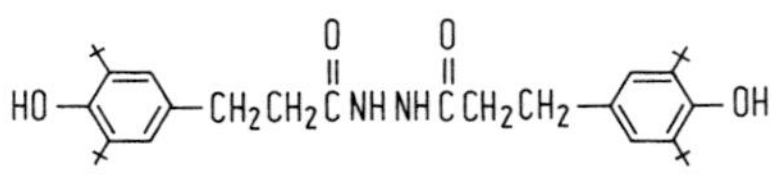	Irganox MD 1024 (Ciba-Geigy)
Polyphenols		
Bis[3,3-bis(4-hydroxy-3-*tert*-butylphenyl)-butyric acid]glykolester	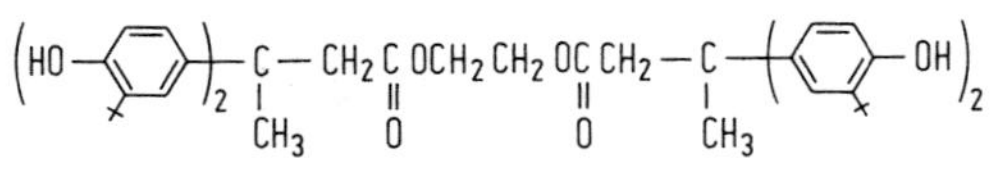	Hostanox 03 (Hoechst)
Tetrakis[methylene(3,5-di-*tert*-butyl-4-hydroxyhydrocinnamate)]methane	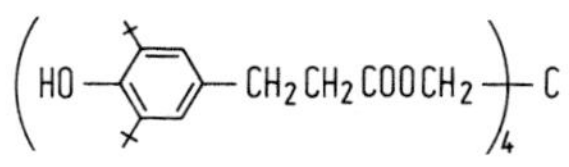	Irganox 1010 (Ciba-Geigy)
1,3,5-Trimethyl-2,4-6-tris(3′,5′-di-*tert*-butyl-4′-hydroxybenzyl)benzene	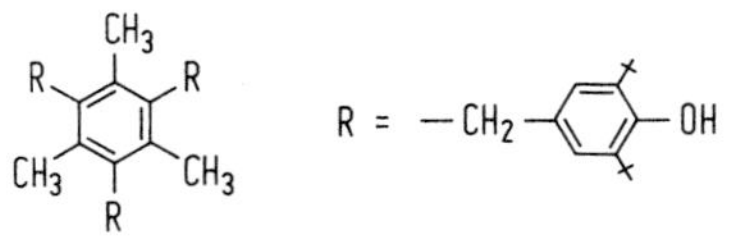	Irganox 1330 (Ciba-Geigy) Ethanox 330 (Ethyl)
1,3,5-Tris(3,5-di-*tert*-butyl-4-hydroxybenzyl)-isocyanurate	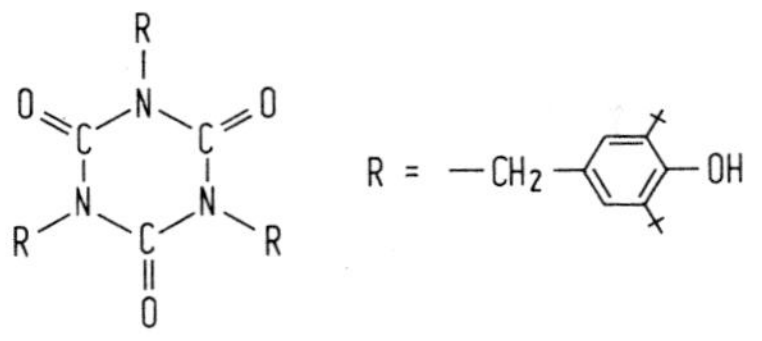	Good-rite 3114 (BF Goodrich)
1,3,5-Tris(4-*tert*-butyl-3-hydroxy-2,6-dimethylbenzyl)isocyanurate	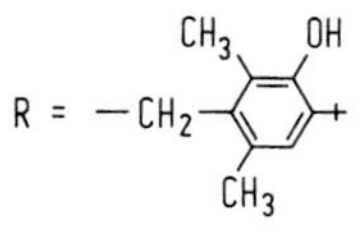	Cyanox 1790 (Am. Cyanamid)
Butylated product of *p*-cresol and dicyplopentadiene		(Goodyear Tire)
3:1-Condensate of 3-Methyl-6-*tert*-butylphenol with crotonic aldehyde		(ICI)
Hydrochinones		
2,5-Di-*tert*-amylhydrochinone	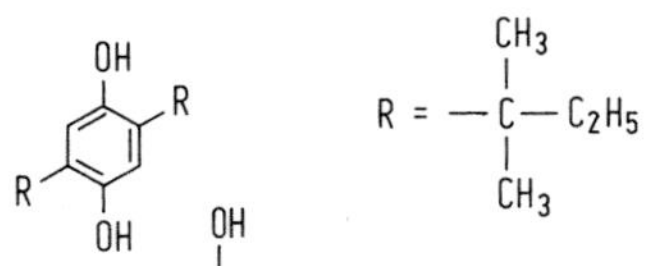	Santovar A (Monsanto)
tert-Butylhydrochinone	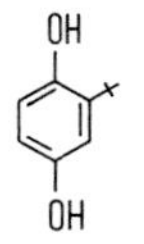	(Eastman)
Diarylamines		
N-Phenyl-α-naphthylamine	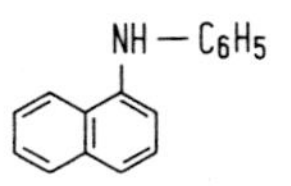	(Uniroyal)
4,4′-Bis(α,α-dimethylbenzyl) diphenylamine	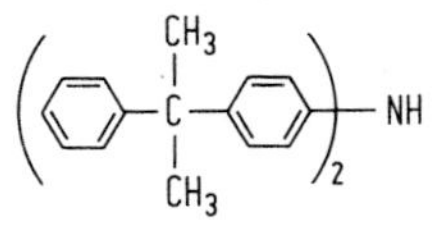	Naugard 445 (Uniroyal)
p-Oriented diphenylamine alkylated with styrene		(Goodyear Tire, Mobay)
Thiother		
β,β′-Thio-di-(propionic acid stearyl ester)	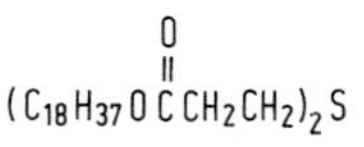	Irganox PS 802 (Ciba-Geigy) Carstab DSTDP (Cincinnati Milacron) Hostanox SE 2 (Hoechst)
β,β′-Thio-di-(propionic acid lauryl ester)	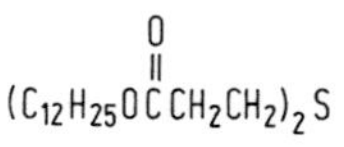	Irganox PS 800 (Ciba-Geigy)

Table 2.3. UV Stabilizers

Chemical Name	Structural Formula		Commercial Products
2,4-Dihydroxybenzophenone	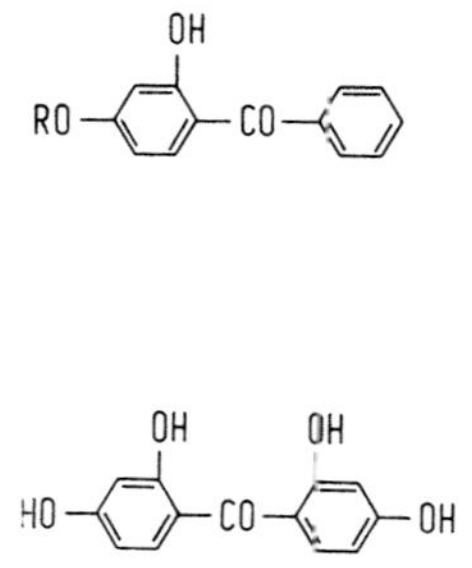	R = H	Syntase 100 (Neville-Synthèse Organics) Uvinul 400 (BASF)
2-Hydroxy-4-methoxybenzophenone		R = CH_3	Cyasorb UV 9 (Am. Cyanamid) UV-Absorber 325 (Mobay) Uvinul M-40 (BASF)
2-Hydroxy-4-*n*-octyloxybenzophenone		R = C_8H_{17}	Cyasorb UV 531 (Am. Cyanamid) Mark 1413 (Argus) Uvinul 408 (BASF)
2-Hydroxy-4-isooctyloxybenzophenone		R = i - C_8H_{17}	Carstab 701, 702 (Carstab)
2-Hydroxy-4-dodecyloxybenophene		R = $C_{12}H_{25}$	Eastman Inhibitor DOBP (Eastman) Syntase 1200 (Neville-Synthèse Organics)
2,2′,4,4′-Tetrahydroxybenzophenone			Uvinul D 50 (BASF)
2-(2′-Hydroxy-5′-methylphenyl)benzotriazole	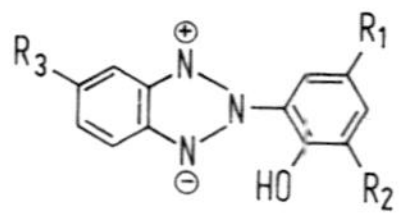	$R_1 = CH_3$ $R_2, R_3 = H$	Tinuvin P (Ciba-Geigy)
2-(2′-Hydroxy-5′-*tert*-octylphenyl)benzotriazole		R_1 = tert.-C_8H_{17} $R_2, R_3 = H$	Cyasorb UV 5411 (Am. Cyanamid)
2-[2′-Hydroxy-3′,5′-(di-*tert*-butyl)phenyl]-benzotriazole		R_1, R_2 = tert.-C_4H_9 $R_3 = H$	Tinuvin 320 (Ciba-Geigy)
2-[2′-Hydroxy-3′,5′-(di-*tert*-amyl)phenyl]-benzotriazole		R_1, R_2 = tert.-C_5H_{11} $R_3 = H$	Tinuvin 328 (Ciba-Geigy)
2-(2′-Hydroxy-3′-*tert*-butyl-5′-methylphenyl)-5-chlorobenzotriazole		$R_1 = CH_3$ R_2 = tert.-C_4H_9 $R_3 = Cl$	Tinuvin 326 (Ciba-Geigy)
2-(2′-Hydroxy-3′,5′-di-*tert*-butylphenyl)-5-chlorobenzotriazole		R_1, R_2 = tert.-C_4H_9 $R_3 = Cl$	Tinuvin 327 (Ciba-Geigy)
2-Ethylhexyl-2-cyano-3,3-diphenylacrylate	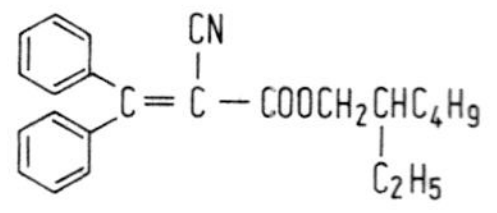		Uvinul N 539 (BASF)
2-Ethyl-2′-ethoxyaxanilide	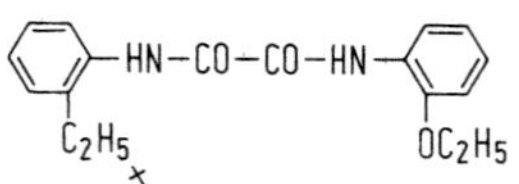		Sanduvor VSU (Sandoz)
n-Hexadecyl-3,5-di-*tert*-butyl-4-hydroxybenzoate	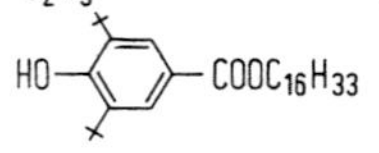		Cyasorb UV 2908 (Am. Cyanamid)

2.4.1.4 Stabilization Against Photooxidative Degradation

Polyethylene terephthalate and polybutylene terephthalate are relatively stable against photooxidation [4, 19]. The degradation mechanism, which has been intensively investigated, is principally a chain cleavage mechanism rather than a crosslinkage [4, 20, 21]. Photooxidative products are CO, CO_2, and carboxylic acids. UV stabilizers must be effective at wavelengths below 360 nm and especially under 320 nm.

Some commercially available stabilizers are shown in Table 2.3 [22]. Additional UV stabilizers are described in the patent literature [23–28].

2.4.2 Nucleation

The important properties of polyethylene terephthalate and polybutylene terephthalate are based in part on their semicrystallinity.

Heterogeneous nucleating agents function by reducing the surface free energy. In this way the critical crystal thickness is decreased and the crystallizability is increased [29]. Under certain processing conditions, polybutylene terephthalate crystallizes so fast that a heterogeneous nucleating agent is generally not necessary. Molding temperatures of 80 °C are sufficient, but the optimal isothermal crystallization temperature is 195 °C. The optimum isothermal crystallization temperature can be increased. For example, reinforced grades with 30% glass fibers have an optimum crystallization temperature of 203 °C. Further addition of flame-retardant packages with 7% antimony trioxide and 10% Dechloran 515 increases the temperature to 211 °C [30].

In comparison to polybutylene terephthalate and other semicrystalline polymers, polyethylene terephthalate crystallizes at a significantly slower rate. This property has significance for injection molding. The crystal growth rate of the spherulites is 10 µm/min, compared to 400 µm/min for polyoxymethylene and 5000 µm/min for high density polyethylene [31]. In addition, certain chemically inert impurities which function as seeds and lead to many small spherulites per unit volume, sometimes result in heterogeneous nucleation. The induction time of crystallization is shortened and the crystallization end point is achieved earlier under such conditions.

Besides depending on the crystallization temperature, the crystallizability of the material per unit volume depends on the size distribution and nucleation strength of the nucleating agent [32]. In practice, insoluble inorganic compounds such as metal oxides (titanium dioxide, magnesium dioxide, silicon dioxide, antimony trioxide), silicates (mica, kaolin, talc), or boron nitrides are employed, typically in particle sizes of 3 µm and concentrations of approximately 0.5 wt %. Talc is especially effective (see Figure 2.6).

Salts of mono- and polycarboxylic acids such as sodium stearate, as well as lignite waxes, are similarly effective nucleating agents. In the patent literature, other groups of compounds are recommended as nucleating agents [33]. Such additives make it possible to injection-mold PET at temperatures of 120–140 °C in economically reasonable cycle times. These ingredients can be added as the PET is produced or during compounding. It is also possible to add the nucleating agents as powder to the polymer pellets.

2.4.3 Lubricants and Release Agents

Semicrystalline polymers flow significantly better than typical amorphous polymers, and lubricants are not generally necessary. During the molding of large and complex parts, the use of such agents may be appropriate to improve the flow and molding behavior.

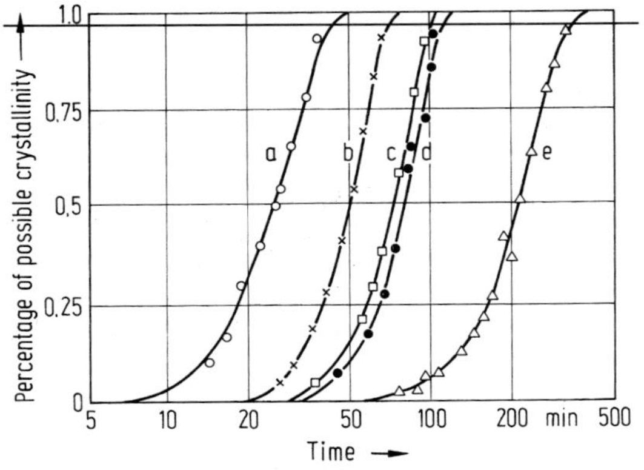

Figure 2.6. Crystallization isotherms of nucleated (0.5%) versus nonnucleated PET at 100 °C [32]: a, talc; b, kaolin; c, SiO_2; d,TiO_2; e, without nucleating agent.

Lubricants that also function as release agents are of special interest. Table 2.4 gives examples of effective lubricants and release agents.

Table 2.4. Processing and Molding Aids

Chemical Class	Examples of Commercial Products
Hydrocarbons	
Paraffine	Irgawax 367 (Ciba-Geigy)
	Loxiol G 22 (Henkel)
Polyethylene wax	Hoechst-Wachs PE 520 (Hoechst)
	Hoechst-Wachs PED 521 (Hoechst)
Fatty acid	
Stearyl alcohol	Irgawax 365 (Ciba-Geigy)
	Loxiol G 51 (Henkel)
Higher carboxylic acids	
Montan wax	Hoechst-Wachs S (Hoechst)
Metal salt of higher carbon	
Calcium stearate	Irgawax 331 (Ciba-Geigy)
Zinc stearate	Listab Zn (Metallgesellschaft)
Calcium montanate	Hostalub VP CA W2 (Hoechst)
Waxes	
Montan ester wax	Hoechst-Wachs E (Hoechst)
partially saponificated montan ester wax	Hoechst-Wachs OP (Hoechst)

2.4.4 Reinforcing

The reinforcement of polybutylene terephthalate and polyethylene terephthalate with glass fibers is significant for many applications. Thermal, mechanical, and electrical properties experience a significant change. The most important effects are increases in strength (tensile and flexural) and stiffness (elastic modulus). In addition, hardness, creep resistance, and notched impact strength are increased, whereas elongation is decreased. The heat distortion temperature is influenced by the increased stiffness but the glass transition temperature of the polyester is not altered by the reinforcement. From the practical point of view, it is important that the thermal expansion coefficient be appreciably decreased. The combination of increase in dielectric strength and decrease in leakage current strength is emphasized for electrical and electronic applications.

A disadvantage of glass fiber reinforcement is the inclination, especially of large thin-walled parts, to blemish. This tendency is attributed to the anisotropic distribution of the glass fibers in the part.

The effectiveness of glass fiber reinforcement is based essentially on the transfer of the load from the matrix to the fiber via the interface. The glass fiber must, therefore, strongly adhere to the matrix, and it must possess a sufficient aspect ratio [34, 35]. Long, short, and ground glass fibers are suitable. To protect the glass surface and to improve the interfacial adhesion between the fiber and the matrix, glass fibers are often treated with sizing agents.

The interfacial adhesion between polyester matrix and glass fiber is improved through use of adhesion promoters. In general, these materials are based on epoxysilanes with hydroxyzable alkoxysilane group (i.e., γ-glycidoxypropyltrimethoxysilane). The effectiveness of the adhesion promoter is based on the ability of the hydrolyzable alkoxysilane group to react with OH groups on the glass surface and on the other side of the molecule: that is, the ability of the epoxy group to react with end groups of the polyester.

In general, glass fibers for thermoplastic polyesters are produced from alkali-poor aluminum borosilicate glass (E-glass). For certain high performance applications, C-glass, D-glass, or S-glass can be used. Because of its high alkaline content, A-glass is not considered for thermoplastic polyesters. Table 2.5 summarizes different glass types [36–39]. Glass fiber producers include Bayer, Gevetex, Owens-Corning, PPG Industries, Saint Gobain Industires, Silenka, and Vetrotex [38, 39].

In addition to glass fibers, other high modulus, high temperature fibers have been investigated in polybutylene terephthalate and polyethylene terephthalate, and the first products have been marketed [40]. Compared to glass-reinforced material, carbon fiber reinforced polyester shows higher modulus, less thermal expansion, and less shrinkage. Noteworthy is the clearly higher electrical surface capacitance which leads to better antistatic properties. A disadvantage is the low impact strength [39, 41]. Although thus far applications have been limited, the shrinking price of carbon fibers may lead to wider use of these materials.

Amide fiber reinforced polyester (PBT) shows no significant improvement [42].

In contrast to reinforcement with specialty fibers, reinforcement with mineral fillers has achieved significance, originally because of the lower cost of the mineral filler compared to polyester resin and glass fibers. With the availability of fillers surface treated with sizing agents similar as glass fibers, there has been growth in the development of filled material with desired properties [43]. Most producers of polyester resins offer filled grades.

Mineral fillers are classified according to their shape: spherical, cubical, blocks, sheets, or fibrous [44]. The properties of the filled polyester depend largely on filler shape, filler size, and filler content.

The most important effect of fillers is the increase in stiffness. In many cases other mechanical, thermal, and electrical properties are changed. Furthermore, mineral fillers have a strong influence on the kinetics and thermodynamics of crystallization [45]. Materials used in the reinforcement of polyethylene terephthalate and polybutylene terephthalate have included glass microspheres, glimmer, talc, wollastonite, and microcrystalline silicon dioxide in the mineral form of novaculite. Table 2.6 lists fillers, a selection of producers, and important properties of filled polyesters.

Table 2.5. Composition Properties of Glass Used as Fibers to Add Strength [36–39]

Glass Type	SiO_2 (%)	Al_2O_3 (%)	Na_2O/K_2O (%)	CaO (%)	MgO (%)	B_2O_3 (%)	Thickness (g/cm^3)	Fiber Diameter (μm)	Tensile Strength (kN/mm^2)	Plastic Modulus (kN/mm^2)	Elongation (%)	Observation
A	72	0.6	14	10	2.5	–	2.48	10	2.4	67.6	1.5	High alkaline content
C	65	4	8	14	3	6	2.49		2.8	69.0		In compoarsion to E-glass higher resistance to acids
D	74	0.3	2.5	0.5	–	22	2.16		2.4	51.7		For high performance electronics
E	54	1.4	0.6	17.5	4.5	8	2.54	3–13	3.4	72.4	2.5	For electrical applications
S	65	25	–	–	10	–	2.49	9	4.6	86.2	5	For applications with higher tensile strength and high modulus of elasticity

Table 2.6. Mineral Fillers and Filled Polyesters [39, 40]

Filler	Form	Example	Properties of Filled Polyester
Microspheres	0.3–44 μm	Ballotini/Potters-Ballotini	High stiffness, surface luster lower warpage
Mica	Flakes with high aspect ratio	Suzorite/Marietta Resources International	High strength, high heat distortion temperature, good electrical properties, lower warpage, high stiffness
Talc (fine milled)	Thin flakes	Micro-Talc/Norwegian Talc Mikro-Talkum/J. Scheruhn Mikro-Talkum/Naintsch Mineralwerke	High stiffness
Wollastonite (calcium silicate)	Needles, aspect ratio 15:1	Wollastonite/Interpace Casiflux/Quarzwerke	High strength, high stiffness, high heat distortion temperature, good electrical properties
Novaculite	Microcrystalline SiO_2	Novacite/Malvern Minerals Novacup/Malvern Minerals	High stiffness, high heat distortion temperature, good electrical properties

Through combination of fillers and glass fibers, customized products for specific targeted applications can be produced. A range of such products is offered on the market.

Titanium dioxide filled products, in addition to improved stiffness and hardness, show 95% light reflection [46].

2.4.5 Flame Retardance

The steady growth in sales of polymers in professional devices and household appliances has led to a sharp increase in flammability regulations. This is especially true for applications of polybutylene terephthalate and polyethylene terephthalate, which are heavily used in electrical and electronic equipment. The development of flame retardants calls for specific criteria for flammability. For example, in judging the effectiveness of flame protection, the spread of the flame is very important; thus the tendency of thermoplastic polyesters to have burning drips becomes an issue. In addition, for many applications the challenge is to minimize the generation of smoke and toxic gases. For the electrical and electronics industries, which are among the main uses of flame-retardant polyesters, the challenge is to limit generation of corrosive gases during a fire.

In actuality, the pyrolytic degradation of the polymer occurs before ignition. The mechanism is characterized by the initial formation of H and OH radicals [47]:

Initiation
$$RH \longrightarrow R\bullet + H\bullet \quad (2.14)$$

Propagation
$$R\bullet + O_2 \longrightarrow ROO\bullet \quad (2.15a)$$
$$ROO\bullet + RH \longrightarrow ROOH + R\bullet \quad (2.15b)$$

Branching
$$ROOH \longrightarrow RO\bullet + \bullet OH \quad (2.16a)$$
$$RO\bullet + RH \longrightarrow ROH + R\bullet \quad (2.16b)$$
$$RH + \bullet OH \longrightarrow H_2O + R\bullet \quad (2.16c)$$

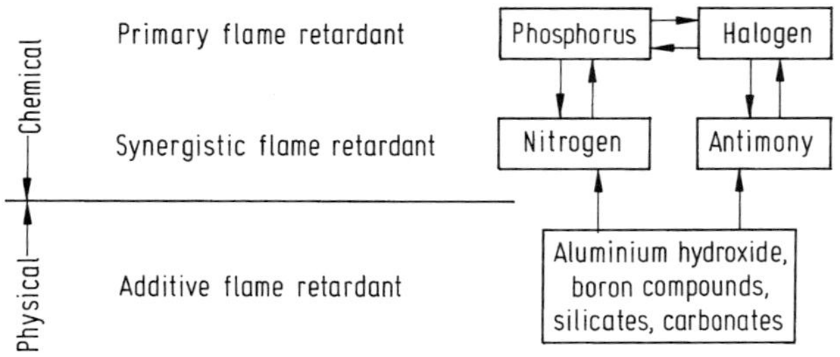

Figure 2.7. Classification of flame retardants [48].

This oxidative degradation leads to gaseous products and carbonaceous residue. Flammable gases mix with air and as the temperature rises, a spark can ignite the material. At higher temperatures (above the autoignition temperature), the polymer can ignite spontaneously.

Of the various possible flame retardants (see Figure 2.7) bromine compounds in synergistic combination with antimony trioxide are most effective [48]. Physical flame retardants are the exception. The effectiveness of organic bromine compounds is based on their ability to inhibit oxidative degradation by taking reactive H and O radicals out of the reaction [47, 49].

$$R'Br \longrightarrow R'\bullet + Br\bullet \tag{2.17}$$

$$RH + Br\bullet \longrightarrow R\bullet + HBr \tag{2.18}$$

$$\bullet OH + HBr \longrightarrow H_2O + Br\bullet \tag{2.19}$$

and

$$H\bullet + R'Br \longrightarrow R'H + Br\bullet \tag{2.20a}$$

or

$$H\bullet + R'Br \longrightarrow R'\bullet + HBr \tag{2.20b}$$

Especially effective organic bromine compounds are high melting, nonmigrating solids with a high bromine content. Examples are oligocarbonates based on tetrabromobisphenol A, brominated epoxy compounds, brominated diphenyl, brominated polystyrene, brominated polyacrylates, and derivatives of tetrabromophthalic acid [40, 43]. Table 2.7 shows examples of effective bromine-containing flame retardants [50, 51].

Because of the possibility of formation of polybrominated dioxins and dibenzofurans during incinerations or fires, polybrominated diphenylether retardants are no longer used in Germany [40]. Similarly, the development of polymers containing brominated diphenylether moieties has been stopped [52]. For this reason polyester producers are intensively investigating halogen-free flame retardants.

The effectiveness of brominated compounds is significantly improved by the use of antimony compounds. Although numerous patents claim synergism from different antimony compounds, the overwhelming choice of polyester producers is antimony trioxide. The synergistic activity of antimony trioxide is not clearly understood. It is generally accepted that important reactions that occur in the gas phase influence the radical chain mechanism [49]. In comparison to flame retardance with brominated antimony compounds, the role of phosphorus, phosphorus compounds, and other materials as flame retardants in commercial polyesters is of minor importance. Numerous studies have, however, been published [53–60]. By incorporation of reactive halo-

Table 2.7. Bromine Flame Retardants [50, 51]

Chemical Name	Structural Formula	Molecular Weight (g/mol)	Bromine Content (%)	Melting Temperature (°C)	Example
Poly(pentabromobenzyl)-acrylester	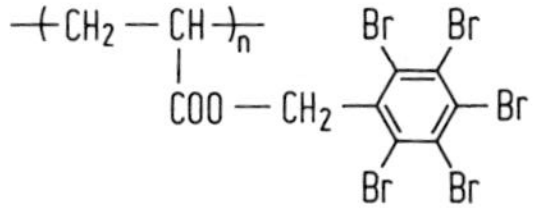	80,000	70–71	205–215	FR-1025 (Dead Sea Bromine)
Poly(tribromostyrene)	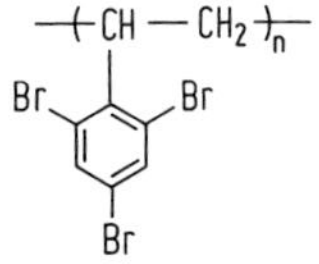	150,000	68	220–230	Pyro-check 68 PB (Ferro)
Tetrabromobisphenole oligocarbonate	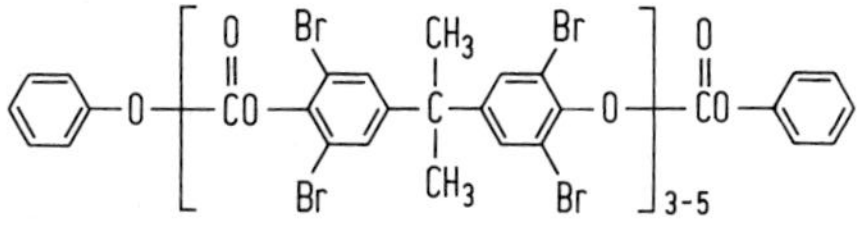	3,000 3,400	52 58	205–240 230–260	BC 52 (Great Lakes) BC 58 (Great Lakes)
Ethylene-bistetrabromophthalimide	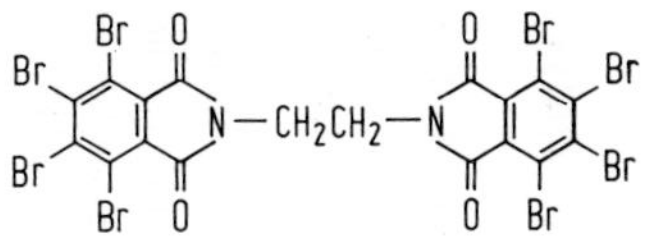	952	67	to 450	Saytex BT-93 (Saytech, Ethyl)
Brominated epoxy resins		3,400–4,000 1,800–2,200 40,000	50–52 50–52 51–53	115–125 100–110 145–155	F 2300 F 2310 } (Makhteshim) F 2400
Octabromodiphenyl + nonabromodiphenyl	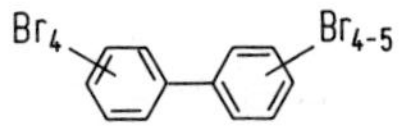	785.5	81–83	200–290	Bromkal 80-9D (Chemische Fabrik Kalk)
Decabromodiphenyl	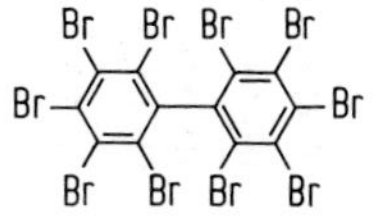	919	84		Adine 0106 (Akzo)

gen-containing flame retardants—for example, bis(2-hydroxyethyl)tetrabromobenzimidazole—copolyesters are obtained. In comparison to low molecular weight flame-retardant additives, these materials cannot migrate. The first products based on copolybutylene terephthalate are commercially available [61, 62].

Burn tests of polyesters treated to impart flame retardance evaluate combustibility, flash points, flame spread, and heat radiance. Depending on the application, they use either standard test bars or finished parts and follow national or international codes. Because the main applications for flame-retardant polyesters are in consumer or professional electronics, the standard tests for electronics are worth considering. The following tests apply for Germany [49].

Test Procedures for Polyester Materials	
Warm bar test	VDE 0304 part 3; DIN 53 459
Flammability test	VDE 0318 part 1a; DIN 40 802, part 1
Tests for Polyester Parts	
Glow wire test	VDE 0471 part 2
Glow contact test	VDE 0471 part 3
Arc resistance test	VDE 0471 part 4
Flame test	VDE 0471 part 5; DIN 50 051
Arc tracking test	VDE 0303 part 1; DIN 52 480
	VDE 0471 part 6

In the United States the testing and approval of Underwriters' Laboratories Inc. (UL) is important for judging the safety of electronic devices and components. The fire safety test procedures are collected in Regulation 94 (UL 94). This regulation has also achieved significance for the specification of polyesters in Germany [49].

Reviews of flame retardance, flame-retardant additives, and flammability tests are found in the literature [49, 63–66]. Abstracts of U.S. patents on flame-retardant polyesters through 1979 are found in the literature [67, 68].

2.4.6 Modification of Polyethylene terephthalate to Improve Processibility

The treatment of polyethylene terephthalate with nucleating agents makes it possible to injection-mold PET at temperatures of 120–140 °C in economically reasonable cycle times. Heating of the mold with oil is necessary. Polybutylene terephthalate requires shorter cycle times and can be molded at 80 °C. This processing advantage of PBT has kept the market for PET in injection molding small, despite the lower raw material and conversion costs of the latter materials. Plasticizers, especially in combination with hightly effective nucleating agents, have been found to limit mold temperatures to under 110 °C (in some cases under 100 °C) and at the same time to reduce cycle times. This discovery has won some market share for PET.

Progress in this regard has been reported in the patent literature over the past ten years by means of modification with:

- monomeric aliphatic esters like adipic acid-di(2-ethylhexy)-ester [69]
- monomeric aromatic esters like dioctylphthalate or neopentylglycol dibenzoate [70]
- oligomeric polyesters like polyhexane adipate or poly(triethylene glycol)phthalate [71]
- sulfonic acid esters like C_{12}–C_{18} alkylsulfonic acid-*p*-phenyloctylester [72]
- plasticizers like neopentylglycol dibenzoate in combination with sodium salts of dimer acids and the sodium ionomer of olefin/methacrylic acid polymers [73] or in combination with complexes of sodium acetate or sodium propionate and polyethylene glycol [74]

- octyl/decyl-terminated ethylene adipate–phthalate [75]
- the sodium salt of crosslinked microgels from styrene/methacrylic acid/vinyl acetate/divinylbenzene [76]
- the sodium salt of phenolsulfonic acid [7]
- plasticizers like dioctylphthalate, neopentylglycol dibenzoate, or toluenesulfonamide in combination with sodium salicylate [78] or with the sodium salt of oligoethylene terephthalate [79]
- phthalimidoester as a plasticizer in combination with sodium ionomers of olefin/methacrylic acid polymers [80]
- epoxidized alkyl acids, unsaturated triglycerides, and aliphatic carboxylic acids in combination with nucleating agents from sodium salts of carboxylic acids [81]
- diesters and diglycidyl ethers of polyethylene glycol in combination with nucleating agents [82]

Improvements in the processibility of PET led to the development, after Akzo (Arnite), DuPont (Rynite), and Miles Inc./Bayer (Petlon), of other PET products by companies such as BASF (Ultradur), Hüls (Vestodur), EMS Chemie (Grilpet), Rohm & Haas (Ropet), and Allied (Petra).

References for Section 2.4

1. *Rafler, G.*, et al.: Acta Polym. 32 (1981) p. 608
2. *Ludewig, H.*: Polyesterfasern. Akademie-Verl., Berlin 1975, p. 131
3. *Zimmermann, H.; Lohmann, P.*: Acta Polym. 31 (1980) p. 686
4. *Cooney, J. D.*, et al.: J. Appl Polym. Sci. 28 (1983) p. 2887
5. *Falkai, B. von:* Synthesefasern. Verlag Chemie, Weinheim 1981, p. 171
6. US-PS 4 093 593 (1977) Owens-Illinois, Inc.
7. US-PS 4 062 829 (1976) Emery Ind., Inc.
8. *Voigt, J.*: Kunststoffe, Zusätze. In: Ullmanns Encyklopädie der technischen Chemie. 4. Ed. Vol. 15. Verlag Chemie, Weinheim 1978, p. 253
9. *Klemchuk, P. P.*: Antioxydants. In: Ullmanns Encyclopedia of Industrial Chemistry. 5. Ed. Vol. A3. Verlag Chemie, Weinheim 1985, p. 91
10. *Dexter, M.*: Antioxydants. In: *Mark, H. F.; Bikales, N. M.; Overberger, C. G.; Menges, G.* (Eds.): Encyclopedia of Polymer Science and Engineering. Vol. 2. Wiley, New York 1985, p. 73
11. *Angelova, A.*, et al.: Angew. Makromol. Chem. 64 (1977) p. 75
12. *Woinova, S.*, et al.: Angew. Makromol. Chem. 64 (1977) p. 81
13. US-PS 4 064 103 (1976) BASF AG
14. US-PS 4 139 521 (1977) Allied Chem. Corp.
15. US-PS 4 254 018 (1978) Akzona Inc.
16. DE-OS 26 15 341 (1976) Bayer AG
17. EP-PS 48 878 (1980) Bayer AG
18. *Bartel, W.*: Kunststoffe 77 (1987) p. 384
19. *Trozzolo, A. M.*: Stabilization against oxidative Photodegradation. In: *Hawkins, W. L.* (Ed.): Polymer Stabilization. Wiley, New York 1972, p. 188
20. *Ranby, B.; Rabek, J. F.*: Photodegradation, Photooxidation and Photostabilization of Polymers. Wiley, London 1975, p. 231
21. *Gugumus, F.*: Photooxydation von linearen Polyestern. In: *Gächter, R.; Müller, H.* (Eds.): Taschenbuch der Kunststoff-Additive. 2. Ed. Hanser, Munich 1983, p. 129
22. *Dexter, M.*: UV-Stabilizers. In: *Mark, H. F.; Othmer, D. F.; Overberger, C. G.; Seaborg, G. T.* (Eds.): Encyclopedia of Chemical Technology. Vol. 23. Wiley, New York 1983, p. 615
23. US-PS 4 105 631 (1975) Eastman Kodak Co.
24. US-PS 4 076 687 (1976) Eastman Kodak Co.
25. US-PS 4 169 089 (1978) Argus Chem. Corp.
26. US-PS 4 116 928 (1976) Eastman Kodak Co.
27. US-PS 4 096 115 (1976) Eastman Kodak Co.

28. US-PS 4 069 227 (1976) Eastman Kodak Co.
29. *Wunderlich, B.:* Macromolecular Physics. Vol. 2. Academic Pr., New York 1976, p. 35
30. *Chang, E. P.*, et al.: Polym. Eng. Sci. 18 (1978) p. 933
31. *Domininghaus, H.:* Die Kunststoffe und ihre Eigenschaften. 2. Ed. VDI-Verl., Düsseldorf 1986, p. 493
32. *Groeninckx, G.*, et al.: J. Polym. Sci., Phys. Ed. 12 (1974) p. 303
33. *Jansen, J.:* Keimbildner für teilkristalline Polymere. In: [21], p. 687
34. *Ehrenstein, G. W.; Wurmb, R.:* Angew. Makromol. Chem. 60/61 (1977) p. 157
35. *Yamashiro, S.:* Kunststoffe 78 (1988) p. 231
36. *Mettes, D. G.:* Glass Fibers. In: *Lubin, G.* (Ed.): Handbook of Fiberglass and Advanced Plastics Composites. Van Nostrand Reinhold, New York 1969, p. 143
37. *Štěpek, J.; Jaoust, H.:* Additives for Plastics. Springer, New York 1983, p. 76
38. *Mohr, J. G.:* Fiberglass. In: *Katz, H. S.; Milewski, J. V.* (Eds.): Handbook of Fillers and Reinforcements for Plastics. Van Nostrand Reinhold, New York 1978, p. 467
39. *Bosshard, A. W.; Schlumpf, H. P.:* Füllstoffe und Verstärkungsmittel. In: [21], p. 411
40. *Katz, H. S.:* Carbon-Graphite Filaments In: [83], p. 562
41. *Caesar, H. M.:* Kunststoffe 77 (1987) p. 1004
42. *Keuerleber, R. H.; Schütt, K. P.:* Kunststoffe 76 (1986) p. 912
43. *Fischer, W.*, et al.: Kunststoffe 70 (1980) p. 650
44. *Ferrigno, T. H.:* Principles of Filler Selection and Use. In: [38], p. 11
45. [37], p. 83
46. DE-OS 27 19 429 (1977) Bayer AG
47. *Jolles, Z. E.; Jolles, G. I.:* Plast. Polym. 40 (1972) 12, p. 319
48. *Hoke, C. E.:* SPE J. 29 (1973) p. 36
49. *Troitzsch, J.:* Brandverhalten von Kunststoffen. Hanser, Munich 1982
50. *Troitzsch, J.:* Kunststoffe 74 (1984) p. 627
51. *Jenkner, H.:* Brandschutzausrüstung für Thermoplaste. In: [21], p. 549
52. Kunststoffe 76 (1986) p. 988
53. *Koch, P.J.*, et al.: J. Appl. Polym. Sci. 19 (1975) p. 227
54. *Granzow, A.; Cannelongo, J. F.:* J. Appl. Polym. Sci. 20 (1976) p. 689
55. *Granzow, A.*, et al.: J. Appl. Polym. Sci. 21 (1977) p. 1687
56. *Day, M.*, et al.: J. Appl. Polym. Sci. 27 (1982) p. 575
57. *Aseeva, R. M.; Zaikov, G. E.:* Flame retardant Effectiveness and Mechanism of Action. In: *Scott, G.* (Ed.): Developments in Polymer Stabilisation. Vol. 7. Elsevier, London 1984, p. 260
58. DE-OS 23 20 007 (1973) Bayer AG
59. US-PS 4 180 496 (1977) Nissan Chem. Ind. Ltd.
60. DE-OS 29 15 116 (1978) Pennwalt Corp.
61. *Breitenfellner, F.:* Kunststoffe 72 (1982) p. 284
62. *Breitenfellner, F.; Kainmüller, T.:* Swiss Plast. 8 (1986) 10, p. 29
63. *Cullis, C. F.; Hirschler, M. M.:* The Combustion of Organic Polymers. Clarendon Pr., Oxford 1981
64. *Mark, H. F.; Othmer, D. F.; Overberger, C. G.; Seaborg, G. T.* (Eds.): Encyclopedia of Chemical Technology. Vol. 10. Wiley, New York 1980. Darin: *Lyons, J. W.:* Flame Retardants, an overview. p. 348; *Avento, J. M., Touval, I.:* Antimony and other inorganic Compounds. p. 355; *Larsen, E. R.:* Halogenated Flame Retardants. p. 373; *Weil, E. D.:* Phosphorous Compounds. p. 396
65. *Landrock, A. H.:* Handbook of Plastics Flammability and Combustion Toxicology. Noyes Publ., Park Ridge 1983
66. *Gann, R. G.; Dipert, R. A.; Drews, M. J.:* Flammability. In: *Mark, H. F.; Bikales, N. M.; Overberger, C. G.; Menges, G.* (Eds.): Encyclopedia of Polymer Science and Engineering. Vol. 7. Wiley, New York 1987, p. 154
67. *Ranney, M. W.:* Fire Resistant and Flame retardant Polymers. Noyes Data Corp., Park Ridge 1974, p. 114
68. *Yehaskel, A.:* Fire and Flame retardant Polymers. Noyes Data Corp., Park Ridge 1979, p. 53
69. US-PS 4 435 546 (1977) Bayer AG
70. US-PS 4 223 125 (1976) Bayer AG
71. US-PS 4 223 113 (1976) Bayer AG
72. DE-PS 26 39 428 (1976) Bayer AG
73. EP-OS 50 851 (1980) Du Pont de Nemours

74. US-PS 4 558 085 (1983) Plastics Engineering Comp.
75. EP-OS 15 145 (1979) Rohm and Haas Comp.
76. EP-OS 112 167 (1982) Celanese Corp.
77. DE-OS 31 03 142 (1981) Asahi Kasei Kogyo K. K.
78. EP-OS 185 981 (1984) Allied Corp.
79. EP-OS 102 768 (1982) Du Pont de Nemours
80. EP-OS 214 112 (1985) Monsanto Comp.
81. EP-PS 73 042 (1981) Ethyl Corp.
82. EP-OS 237 843 (1986) Toyo Boseki Kabushiki Kaisha

2.5 Blends with Other Polymers

Although polybutylene terephthalate and polyethylene terephthalate show good toughness and ductility in unnotched tests, sharp notches effectively decrease the impact strength. This limitation has led to intensive and successful efforts to modify polyesters, especially PBT, with elastomers.

There have been additional successful developments, especially in filled products. Examples include the modification of thermoplastic polyesters with amorphous thermoplastics and the modification of PBT with PET. Modified elastomers and thermoplastics have significantly contributed to the successful growth of polybutylene terephthalate in particular [1].

2.5.1 Binary Mixtures with Elastomers

2.5.1.1 Blends with Butadiene Copolymers and Acrylate Copolymers

2.5.1.1.1 Unreinforced Products

Modification of thermoplastic polyesters with elastomers containing various copolymers results in significant increases in notched impact strength and multiaxial impact strength. Examples are styrene, acrylonitrile, and methyl methacrylate graft copolymers, crosslinked polybutadiene, and various polyacrylates. If the copolymers have good compatibility with molten polyesters and are well dispersed during processing, they can be added in larger amounts. Good mixing is essential to avoid delamination during injection molding, which can seriously compromise key physical properties.

Small amounts of elastomeric modifier strongly influence notched impact strength and especially multiaxial impact resistance [1–9]. Normally brittle thermoplastic polyesters, when combined with suitable elastomers (at levels of 15% elastomers), undergo a brittle-to-ductile transition (see Figure 2.8). Although the impact strength increases, mechanical properties that derive from the crystallinity of thermoplastic polyesters, particularly tensile strength, stiffness, and hardness, are compromised. For each application there is an optimum compromise.

The use of suitable elastomers to improve the notched and multiaxial impact strengths of engineering polyesters is effective not only at room temperature but also at low temperatures. Indeed after modification, the ductile transition can be suppressed to below −40 °C. Polybutadiene graft copolymers and polyacrylate graft copolymers are particularly effective. Defined bimodality of the elastomer particles as well as heating of the polyester/elastomer mixture almost to the melt temperature of the polyester increase the effectivity of the elastomers even more [10–12]. The addition of small amounts of

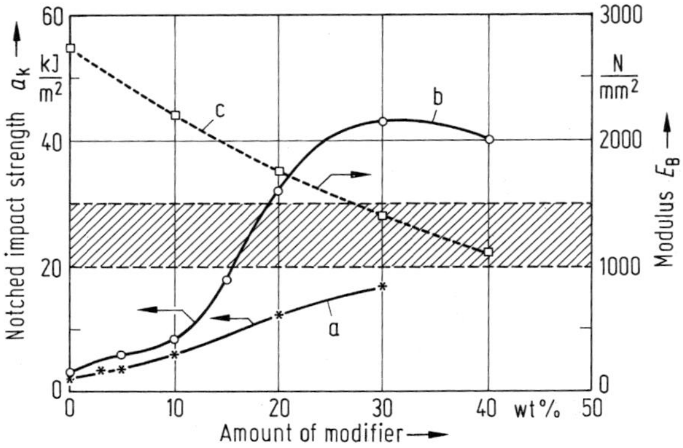

Figure 2.8. Notched impact strength a_k (Charpy) and modulus E_F of unreinforced, elastomer-modified PBT as a function of the amount of modifier [1]. The shaded region shows the brittle-to-ductile transition: a, for average viscosity PBT; b, for high viscosity PBT; c, E_F for both.

polytetrafluoroethylene to such polyester/elastomer blends effectively decreases the ductile/brittle transition to even lower temperatures, and at the same time tensile strength, stiffness, and elongation of the mixture are improved [13]. Although such products have ductile/brittle transitions at −20 °C and less, the crystallinity of polyesters imparts tensile strength and stiffness and makes possible extended use at high temperatures.

Large amounts of these blends are used to produce plastic bumper assemblies, which can be painted at the same time as the car body. Figure 2.9 shows the notched impact strength of a blend of polybutylene terephthalate and various modifiers as a function of temperature [14–16].

Polybutadiene graft rubbers in combination with thermoplastic polyesters impart superior low temperature toughness. The maximum use temperature of such blends is, however, 100 °C. For this reason, this blend is used where long-term temperature stability is not required and or where sharp temperature increases occur, as in the ovens

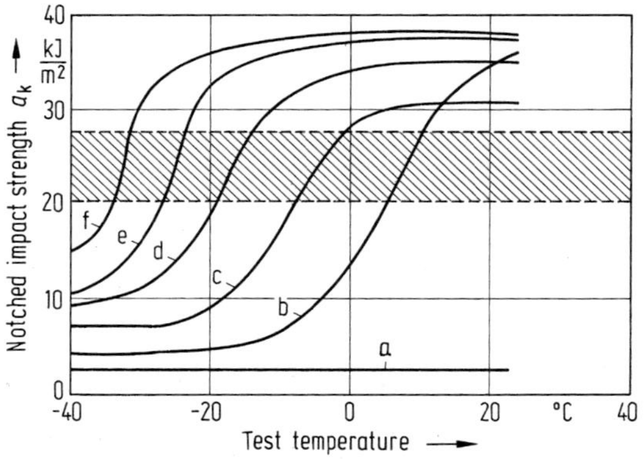

Figure 2.9. Notched impact strength of a blend of polybutylene terephthalate and various modifiers as a function of temperature [14]; the shaded region shows the ductile-to-brittle transition: a, not modified; b, polyacrylate graft copolymer; c, polyolefin graft copolymer; d, with PC and polybutadiene graft copolymer [15]; e, polybutadiene graft copolymer; f, modified polybutadiene graft copolymer.

used after car body painting. In contrast, polyacrylate-grafted rubbers do not subtract from the low temperature toughness of a product but offer good long-term temperature stability and thereby win applications in the electrical and electronics sectors. For example, the toughness of PBT that has been modified with polyacrylate-grafted rubber is unchanged after 1000 hours at 140 °C [1].

Besides the main effect of improved toughness, certain modifiers act as crystallization enhancers. For example, butadiene grafted with blends of PET and polymethyl methacrylate achieve high crystallinity in molded parts at temperatures as low as 80–120 °C [17].

2.5.1.1.2 Reinforced Products

As described in Section 2.4.4 the notched impact strength of polybutylene terephthalate and polyethylene terephthalate is increased through reinforcement with glass fibers. This notched impact strength level is also increased in reinforced products via elastomer modification. Examples of this effect with polybutadiene and polyacrylate graft copolymers are shown in Figure 2.10. Because of such elastomer modification, polyacrylate-modified, glass fiber reinforced polyesters are used in the electrical, electronics, industrial, and transportation markets, which take advantage of the higher continuous use temperatures and increased thermal stability during the processing of polyacrylate-modified material.

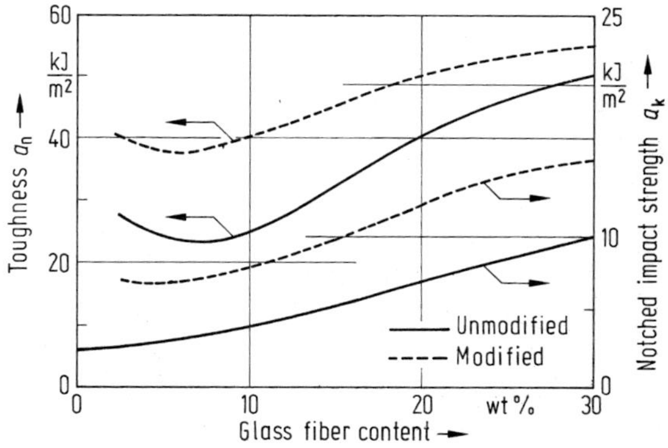

Figure 2.10. Toughness and notched impact strength (Charpy) as a function of glass fiber content for unmodified and elastomer-modified PBT [1].

It is noteworthy that the test specimens show only hinged breaks in comparison to normal brittle failure. The improvement in toughness is also effective at temperatures under 0 °C. With constant glass fiber reinforcement, the increase in impact strength with increasing elastomer content is minimal. Notched impact strength shows a somewhat greater linear increase. The damaging energy, however, exhibits an exponential increase with elastomer content; a fourfold increase in damage work is observed with only 9% elastomer, and with 15% elastomer a 15-fold increase is achieved (see Figure 2.11) [18]. Despite this superior improvement in toughness and increased tensile elongation, properties such as strength, modulus, and heat distortion temperature decrease only slowly with increasing elastomer content up to 18%. The behavior of different elastomer-modified, glass-reinforced polyesters in falling bolt tests is very informative in the damage course at low deformation speeds until first crack initiation or, in the case of

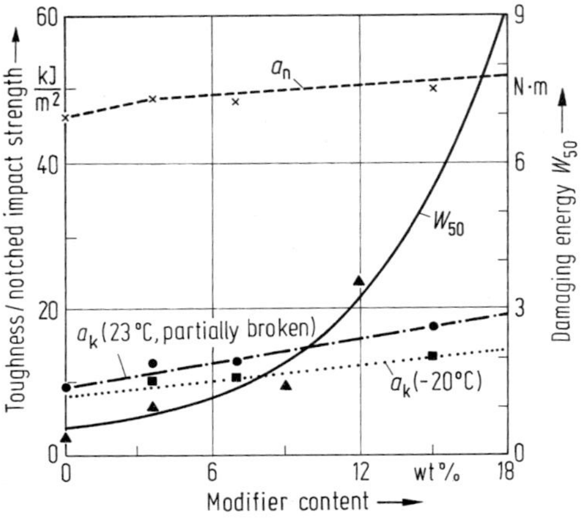

Figure 2.11. Toughness and damaging energy of elastomer modified (polyacrylate graft copolymer and polybutadiene graft copolymer) in glass reinforced PBT [1, 18].

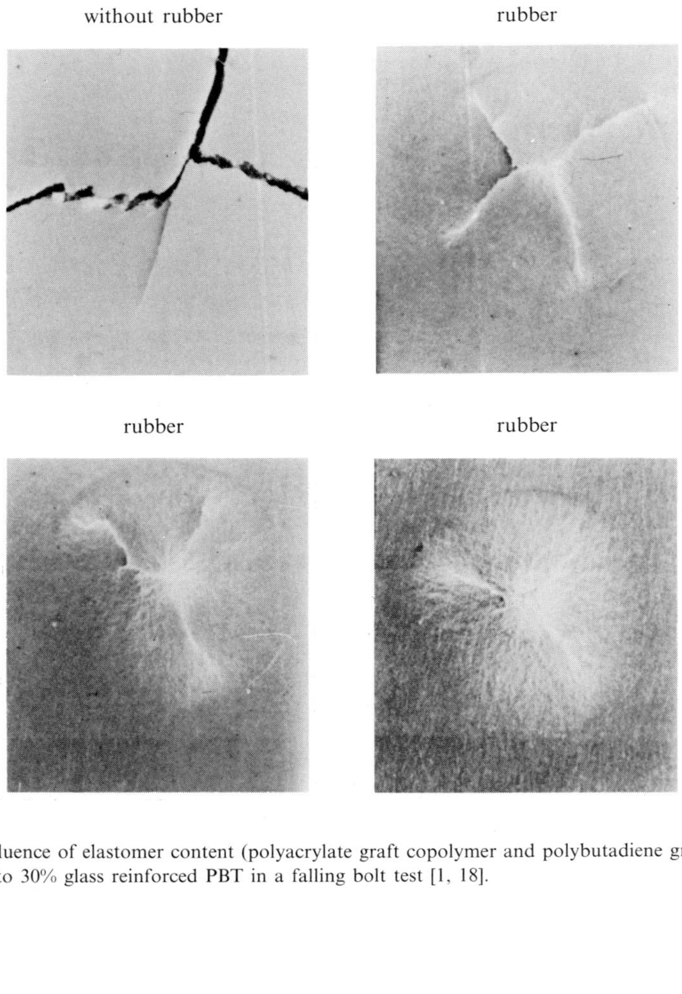

Figure 2.12. Influence of elastomer content (polyacrylate graft copolymer and polybutadiene graft copolymer) on the damage to 30% glass reinforced PBT in a falling bolt test [1, 18].

higher deformation speeds in the penetration test, until the test specimen has been completely destroyed.

For slow deformation speeds the first crack initiation occurs at later times as the elastomer content increases. At the same time, the character of the crack is changed; the damage to non-elastomer-modified, glass-reinforced polyesters proceeds from some rough cracks to ductile crack, followed by a tough break with microcracks (see Figure 2.12).

At higher deformation rates, test specimens with higher elastomer content exhibit energy absorption even in the course of complete destruction (see Figure 2.13). Micrographs of fracture surfaces show that with increasing elastomer content, the adhesion between matrix and glass fibers improves and the matrix undergoes ductile failure.

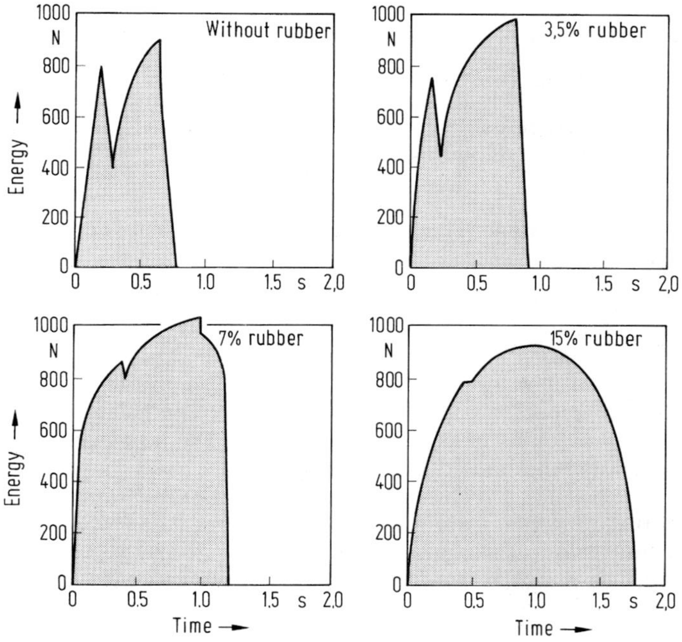

Figure 2.13. Influence of the elastomer content on the course of the time/energy curves of 30% glass reinforced PBT in high deformation rate falling dart tests [1, 18].

Polyacrylate-modified, glass-reinforced polyesters used in the electrical and electronics markets have special importance because the addition of flame-retardants decreases toughness [1].

2.5.1.2 Blends with Olefinic Elastomers (EPM, EPDM)

Whereas butadiene and acrylate copolymers are used routinely as impact modifiers for thermoplastic polyesters, references to ethylene–propylene and ethylene–propylene–diene monomer rubbers (EPM and EPDMs) are less common. This is mostly due to the poor compatibility of the components. For this reason recent developments have concentrated on achieving improvements in compatibility through the use of small amounts of copolymers with succinic acid anhydride groups [19] or through use of EPDM/EMA/MAA grafted rubbers.

2.5.2 Binary Blends with Other Thermoplastics

2.5.2.1 Blends of Polybutylene Terephthalate with Polyethylene Terephthalate

The injection molding of glass fiber reinforced PBT leads to long direction orientation and anisotropy in the molded part. An additional deficiency is the poor surface quality of the molded part. Upon blending with 45 wt % polyethylene terephthalate is miscible with PBT in every ratio [20], warpage is clearly diminished and surface quality improved [21, 22]. Because the processibility of polybutylene terephthalate in the blend is only slightly compromised, many polyester producers have incorporated PBT/PET blends into their product portfolios.

Besides the technical advantages of the products just mentioned, there is a cost savings to be realized from incorporating polyethylene terephthalate, especially if the PET is recycled (bottle scrap).

Because these blends are processed, the residence time and the molding time must be kept as short and low as possible to avoid transesterification due to large residues of catalyst remaining from the production of the PET. To avoid this transesterification, the catalyst can be deactivated by stabilizers.

2.5.2.2 Blends of Polyethylene Terephthalate or Polybutylene Terephthalate with Other Polyesters

The patent literature contains extensive reports of blends of polyethylene terephthalate or polybutylene terephthalate with other polyesters. Practical applications of blends with block copolyether esters stand out as having improved notched impact strength [23–25]. A number of these blends have been brought to market by polyester producers.

Improvement in the impact strength of polybutylene terephthalate is also observed upon addition of small amounts of poly(ε-caprolactone). The heat distortion temperature of this blend is not compromised [26].

Blends of polyethylene terephthalate and polybutylene terephthalate with polyethylene isophthalate exhibit improved barrier properties against oxygen and CO_2 [27]. The barrier properties also improve upon addition of small amounts of glycol acid esters [28] and particularly blends of polybutylene terephthalate with amorphous polyesters from 1,4-dimethylol-cyclohexane, ethylene glycol, and terephthalic acid [29].

The warpage of glass-reinforced polybutylene terephthalate is significantly diminished with fully aromatic polyesters [30]. Blends of polyethylene terephthalate with aromatic polyesters instituted before and after transesterification have been thoroughly described [31].

2.5.2.3 Blends with Amorphous Heat-Stable Thermoplastics

Assuming good compatibility, blends of thermoplastic polyesters with amorphous, heat-stable polymers are characterized by the chemical resistance, easy flow, and crystallinity of the polyester on the one hand and the toughness, decreased shrinkage, and high heat distortion temperature of the amorphous polymer on the other hand. This is especially true of polyester and polycarbonate blends for which there are numerous patents [32–34]. These blends are especially interesting for reinforced products because the warpage is significantly decreased [35]. A few polyester producers have developed special products for housings for household, automotive, and electrical/electronic appliances.

It is imperative to hinder transesterification in polyester/polycarbonate blends [36]. These blends have a distinctive color if, during the production of the polyester, a dibutyltin diacetate catalyst is used instead of a titanium catalyst [37]. Blends with polysiloxane-*co*-polycarbonate block copolymers are especially interesting because only a small amount of such copolymers significantly inproves the toughness of the polyester [38–40].

In a manner similar to polycarbonate, polysulfones influence the properties of polyethylene terephthalate and polybutylene terephthalate. Glass-reinforced blends are distinguished by decreased mold shrinkage, decreased warpage, and improved dimensional stability coupled with higher heat distortion temperature and better chemical resistance. Electrical connectors and cookware designed for microwave ovens are typical applications [22].

2.5.2.4 Blends with Polyolefins

Because of the poor compatibility of polyethylene with polyethylene terephthalate and polybutylene terephthalate, blends containing large amounts of polyethylene have not achieved commercial significance. Small amounts of linear low density polyethylene (LLDPE) improve the molding of PBT [41]. Polyolefin added in amounts under 10% improves the thermoformability of PET [42]. Copolymerization with acrylates such as ethylacrylate or butylacrylate increases the compatibility of the polyolefin with the polyester, and an improved toughness results [43, 44]. These blends experience a significant increase in the notched impact strength if they are briefly heated to just under the melting point of the polyester [45].

2.5.2.5 Other Blends

A large number of blends with various polymers are described in the open literature. None of these blends have until now been sold commercially. Blends with high heat polymers like poly(ether ketone) and polyphenylenesulfide [46] may achieve significance if the volume of these polymers increases and the prices fall accordingly.

2.5.3 Ternary Blends

Blends with polycarbonate, elastomers, and polybutylene terephthalate or polyethylene terephthalate are used in load-bearing parts in automobiles. In some blends, the polycarbonate is the main component and in some the polyester is the main component [47].

References for Section 2.5

1. *Rempel, D.:* Kunststoffe 76 (1986) p. 900
2. DE-PS 23 78 377 (1973) Montecatini Edison S.p.A.
3. DE-PS 24 44 584 (1974) BASF AG
4. DE-OS 27 26 256 (1976) Rohm and Haas Co.
5. EP-PS 22 216 (1979) Bayer AG
6. EP-PS 50 262 (1980) Bayer AG
7. EP-PS 50 265 (1980) Bayer AG
8. EP-PS 64 207 (1981) Bayer AG

9. EP-PS 63 263 (1981) Bayer AG
10. EP-PS 131 202 (1983) Bayer AG
11. EP-OS 139 084 (1983) Chemische Werke Hüls AG
12. EP-OS 142 675 (1983) Bayer AG
13. EP-OS 166 295 (1984) Bayer AG
14. *Rempel, D.:* Kunststoffe 74 (1984) p. 578
15. Pocan S 1506. Firmenschrift Bayer AG 1984
16. Pocan KL 1-7907. Firmenschrift Bayer AG 1983
17. EP-PS 63 263 (1981) Bayer AG
18. *Binsack, R.*, et al.: Kurzglasfaserverstärkte, teilkristalline Verbunde am Beispiel Polyamid un Polybutylenterephthalat. 8 Kunststoff-Kolloquium DKI, Darmstadt 1982
19. US-PS 4 558 096 (1983) Goodyear Tire & Rubber Co.
20. US-PS 3 953 394 (1971) General Electric
21. *Witt, W.:* Kunststoffe 73 (1983) p. 498
22. *Giles, H. F.:* Mod. Plast. Enc. 63 (1986) 10a, p. 110
23. DE-OS 23 63 512 (1973) Bayer AG
24. US-PS 3 907 926 (1973) Du Pont de Nemours
25. US-PS 4 469 851 (1983) GAF Corp.
26. DE-OS 23 31 826 (1973) General Electric
27. US-PS 4 551 368 (1982) Goodyear Tire & Rubber Co.
28. US-PS 4 565 851 (1983) Eastman Kodak
29. EP-OS 121 883 (1983) Eastman Kodak
30. DE-OS 20 02 814 (1979) General Electrical
31. *Kimura, M.*, et al.: J. Appl. Polym. Sci. 29 (1984) p. 1629
32. US-PS 3 218 372 (1961) Kunoshima
33. BE-PS 661 850 (1964) General Electric
34. DE-OS 24 17 002 (1974) BASF AG
35. DE-OS 27 08 381 (1976) General Electric
36. US-PS 4 560 722 (1984) General Electric
37. WO 87/01122 (1985) Eastman Kodak
38. *Bopp, R. C.*, et al.: Technical Information Series, General Electric 1987, p. 1
39. US-PS 4 155 898 (1978) General Electric
40. US-PS 4 161 498 (1978) General Electric
41. US-PS 4 564 658 (1982) General Electric
42. US-PS 4 572 852 (1982) Goodyear Tire & Rubber Co.
43. EP-PS 13 941 (1979) Bayer AG
44. *Braun, D.; Illing, W.:* Plastverarbeiter 38 (1987) 1, p. 32
45. EP-OS 139 084 (1983) Chemische Werke Hüls AG
46. EP-OS 246 016 (1986) Celanese Eng. Resins
47. *Witt, W.:* Kunststoffe 77 (1987) p. 1010

2.6 Production of Compounds

2.6.1 Preparation of Raw Materials

Polybutylene terephthalate and polyethylene terephthalate are rarely extruded or injection-molded without modification. As described in Sections 2.4 and 2.5, processing is improved via processing aids, and the basic properties of the polyesters are altered with functional additives. Generally these additives are introduced either during precompounding or directly to the melt during processing. They can also be added simultaneously with the base polyester. For a specific additive, there is an optimum method that is dependent on the additive level, on the thermal and mechanical stability of the additive, on miscibility and, finally, on the compounding machine. In certain cases, premixing is required. When the additive level is very low, master batches are used to improve accuracy and to facilitate addition. Because of the hydrophilic nature of the polyesters, both the polyester and the additives must be predried. A desiccant dispenser may be directly attached to the processing equipment for this purpose.

2.6.2 Compounding

The compounding of polyesters can be achieved with either a kneader or an extruder. Depending on the task, various static, transport, kneading, or mixing elements are employed. These, in turn, are chosen in accordance with the characteristics of the components—for example, the melt point of the polyester, the form and size of the additives, and the homogeneity of the mixture. The required energy is provided primarily by viscous shearing.

Polybutylene terephthalate and polyethylene terephthalate tend to degrade thermally. For polybutylene terephthalate, the permissible melt temperature depends on the residence time in the extruder and the extruder configuration (see Figure 2.14). For example, polybutylene terephthalate can be compounded with glass fiber in a single-screw extruder at 260 °C. Because the retention time is shorter, mixing in an oscillating kneader at higher temperatures (280–285 °C) is possible without disadvantageous degradation although there is intensive shearing.

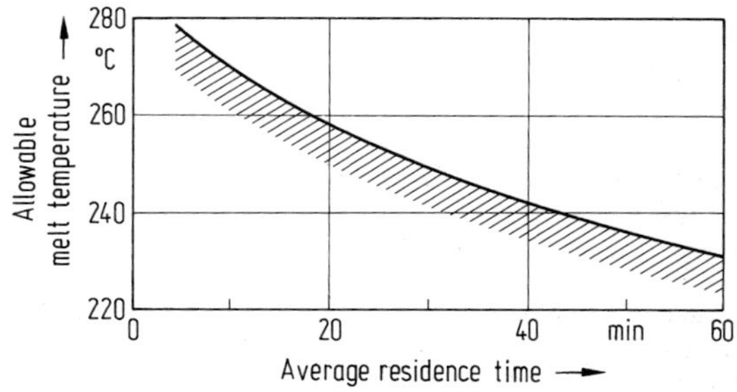

Figure 2.14. Allowable melt temperatures during extrusion and injection molding of PBT as a function of residence time [1, 2]. The areas under the curve have no influence on product quality.

Unfilled polybutylene terephthalate approaches Newtonian flow in the temperature range 240–280 °C. The melt temperature increases only slightly over the length of an extruder. In contrast, the melt temperature climbs quickly upon the addition of glass fibers (see Figure 2.15) because the melt viscosity increases rapidly. This characteristic behavior must be considered if thermal degradation is to be avoided in the production of reinforced grades. To maintain the melt temperature below certain limits, it is necessary to add the glass fibers to the melt in a manner that serves to limit the grinding of the fiber and the wearing down of the screws [2].

It is also possible to keep the melt temperature down by combining unmelted polyester pellets with the glass in a second extruder element. This arrangement allows for faster throughput and economical processing. On the other hand, it is possible to economically extrude high molecular weight polybutylene terephthalate (intrinsic viscosity < 1.3 dL/g) and glass fiber economically, without degrading the polymer [2].

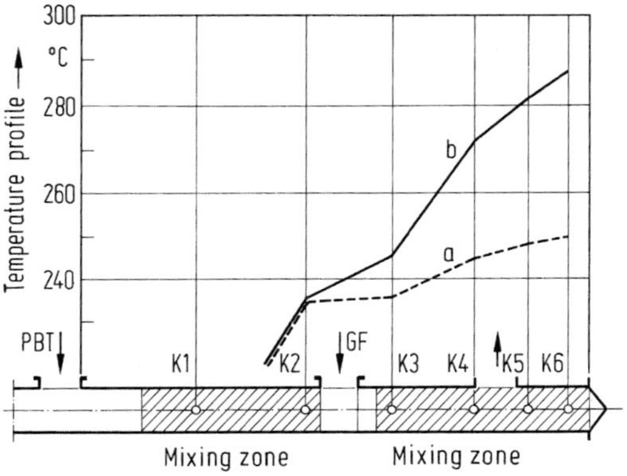

Figure 2.15. Temperature profile for a continuous extruder for unreinforced (a) and reinforced (b) polybutylene terephthalate [2]: (barrel diameter $D = 200$ mm; throughput $Q = 700$ kg/h).

2.6.3 Compounding Machines

There are a number of machine systems for compounding polybutylene terephthalate and polyethylene terephthalate. Examples are kneaders from Buss AG and twin-screw extruders (ZSK machines from Werner & Pfleiderer GmbH).

Figure 2.16 shows a scheme for the production of polyester compounded with filler, flame retardants, glass fibers, and second polymers or additional polyester in a two-stage Buss kneader with a horizontal discharging screw. The mixing section of the kneader can be vacuum vented. The compound melt is transported from the kneader into a single-screw discharge extruder, which builds a pressure appropriate for extrusion. The throughput depends on the type of the kneader and the product composition (Table 2.8).

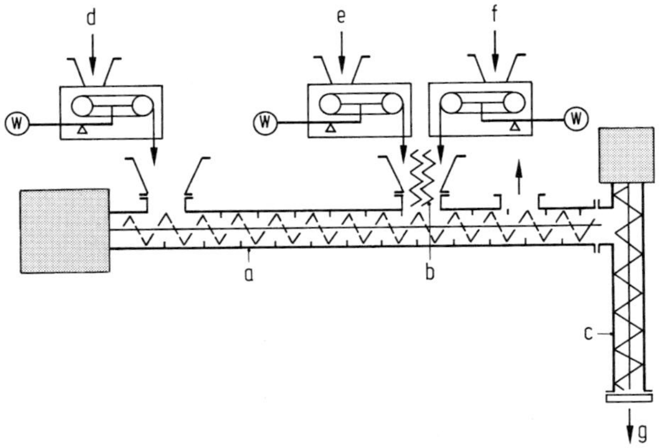

Figure 2.16 Schematic for a production facility using a two-stage buss kneader (model MDK-E) to extrude glass-reinforced polybutylene terephthalate [2]: a, continuous kneading ($L = 11$ D); b, double addition parts; c, exit extrusion, ($L = 5$ D); d, polybutylene terephthalate; e, flame retardant, filler secondary polymer, additional PBT; f, glass fiber; g, to granulation.

Table 2.8. Production Capacity of Buss Extruders for the Production of Glass-Reinforced PET and PBT

Model Number in MDK-E Series	Production Capacity (kg/h)
46	10–30
100	100–300
140	350–700
200	800–15,000

Figure 2.17 shows a scheme for a production facility for polybutylene terephthalate elastomer blends [3]. The facility is based on self-cleaning, twin-screw extruder from Werner & Pfleiderer. The components are added to the ZSK machine either volumetrically as a premix or gravimetrically in several separate component currents. The material proceeds through the following steps: plasticization, mixing, homogenization, dispersion, off-gassing, and hot melt granulation, or stranding (Figure 2.17).

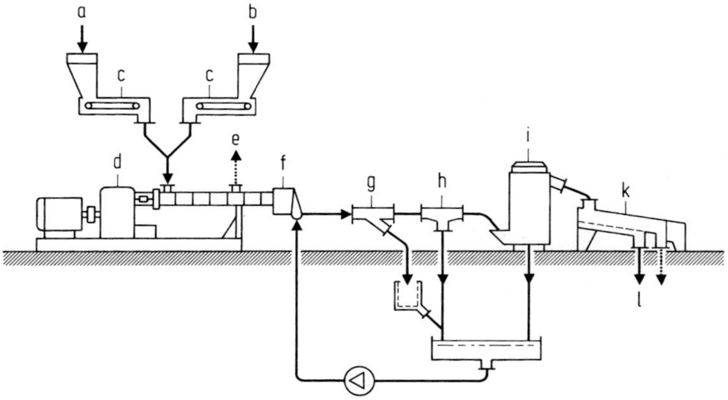

Figure 2.17. Schematic for a production facility for the blending of polybutylene terephthalate and elastomers or pigment master batches, based on a twin-screw extruder with hot melt granulation: a, polybutylene terephthalate; b, elastomers or pigment master batches; c, metering; d, twin-screw extruder; e, vacuum; f, water granulation; g, conveyor; h, water separation; i, centrifugal dryer; k, sorter; l, to packaging.

Table 2.9. Production Capacity for ZSK Machines (screw speed 30–300/min) for the Production of Blends from Polybutylene Terephthalate and Elastomers or Pigment Master Batches

ZSK Model no.	Propulsion Power (kW)	Throughput Range (kg/h)
25 M 8.6	5.4	10–30
40 M 33	20	60–120
58 M 96	50	200–350
70 M 175	108	400–700
92 M 385	240	800–14000
133 M 1160	728	2000–4000

Table 2.9 gives throughput data for production machines of various types. Addition of glass fibers or use of greater amounts of fillers can decrease the throughput by as much as 20% [3].

References for Section 2.6

1. *Brunner, J.:* Kunstst. Plast. 22 (1975) 9, p. 41
2. *Stade, K. J.:* Die Herstellung von glasfaserverstärktem Polybutylenterephthalat auf einem kontinuierlich arbeitenden Kneter. Company brochure Buss AG
3. Verfahren und Methoden: Technische Kunststoffe. Company brochure Werner & Pfleiderer GmbH 1987

2.7 Properties

The properties of polybutylene terephthalate and polyethylene terephthalate are very heavily determined by their semi-crystalline character and their low water absorption. In general they are distinguished by the following properties:

- high dimensional stability under heat
- high stiffness and hardness
- excellent tribological properties
- good bearing strength
- good electrical properties
- good resistance to chemicals
- good stress-cracking resistance
- excellent flow characteristics
- short processing cycles
- good mold release properties

In addition, reinforced products show high mechanical strength even at elevated temperatures. The pronounced decrease of toughness due to sharp notches can be counteracted even at low temperatures by mixing with impact modifiers. All copolyether polyesters are distinguished by their elastomeric properties, with high elasticity and high resilience depending on the ether/ester ratio.

2.7.1 Morphology and Crystallinity

Polybutylene terephthalate and polyethylene terephthalate are semicrystalline polyesters with crystalline and amorphous domains. The crystalline fraction determines the melting point; the amorphous fraction, the glass transition temperature. Instead of strict boundaries separating the crystalline and amorphous domains, there are only transition areas. Therefore, as with amorphous thermoplastic polymers, there are no exact melting points. Rather, there are temperature ranges that depend on processing as well as the thermal and mechanical history of the material.

The morphological structure of the thermoplastic polyesters is of special importance for their dynamic mechanical behavior.

2.7.1.1 Polybutylene Terephthalate

Polybutylene terephthalate has a crystallinity of about 40%. Quenching the melt (e.g., in ice water) results in an amorphous material. On the other hand, tempering enhances the crystalline fraction up to 60% [1]. The crystallization speed during cooling of the melt is very high. It increases steeply with decreasing isotherm crystallization temperature (Figure 2.18) and decreases after reaching a maximum. Polybutylene terephthalate crystallizes in two different structures, which can be reversibly transformed into each other. If untreated, the α type dominates, with a gauche-trans-gauche conformation of the butylene chain. If stretched, depending on stretching conditions, significant amounts of the β type with all-trans-conformation may be formed [2–6]. The amount of β phase increases with increasing stretching ratio and decreasing stretching temperature (Figure 2.19). Tempering leads to a transformation of the β phase into the thermodynamically more stable α phase. The speed of transformation increases with increasing tempering temperature and annealing time (Figure 2.20). The parameters of the unit cell for the α and β phases are given in Table 2.10 [2–4].

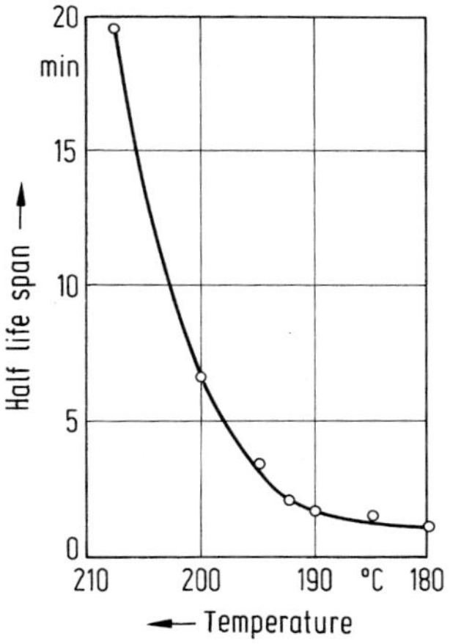

Figure 2.18. Dependence of the half-life period of crystallization of polybutylene terephthalate (Goodyear VFR 4716, $\bar{M}_w = 45{,}000$; $\bar{M}_w/M_n \approx 2$) on the isothermal crystallization temperature [1].

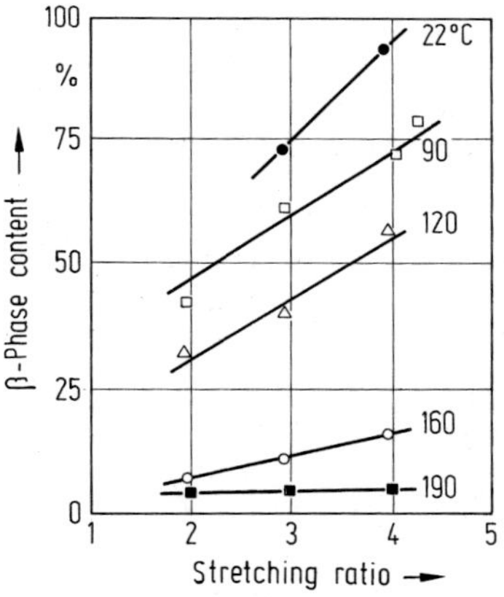

Figure 2.19. Content of β-phase polybutylene terephthalate as a function of stretching ratio and stretching temperature [3].

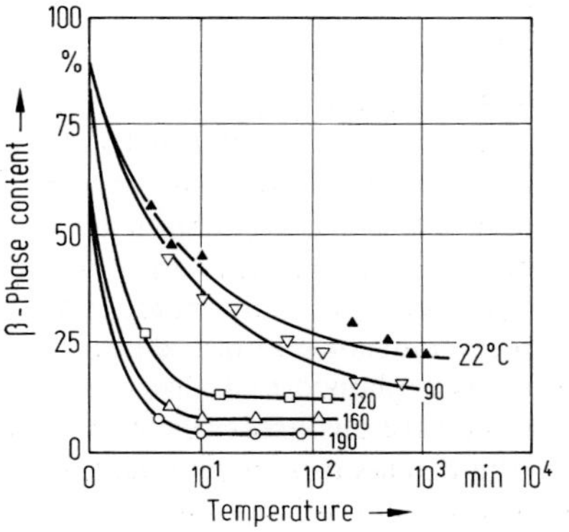

Figure 2.20. Content of β-phase polybutylene terephthalate with a stretching ratio of 4:1 at room temperature as a function of time and temperature of tempering [3].

Table 2.10. Unit Cell Parameters of Polybutylene Terephthalate and Polyethylene Terephthalate

Parameter and Units		PBT α Form[1]	PBT β Form[2]	PET[3]
a	nm	0.482	0.496	0.444
b	nm	0.593	0.580	0.591
c	nm	1.174	1.300	1.067
α	grad	100	102	100.1
β	grad	115.5	120.5	117.0
γ	grad	111	105	111.8
V	nm^3	0.261	0.267	0.2106
Crystal density	g/cm^3	1.403	1.37	1.515

[1] Crystallized at 220 °C [2].
[2] Stretched [4].
[3] Crystallized at 250 °C [2].

The density of polybutylene terephthalate crystals depends on the crystallization temperature. Because of the high mobility of the PBT chain, the density starts to rise at temperatures as low as 120 °C and reaches its maximum value at about 180 °C (Figure 2.21). The overall density for amorphous PBT is 1.26 g/cm^3 and, for well-crystallized, unfilled molded parts, 1.32 g/cm^3 [7].

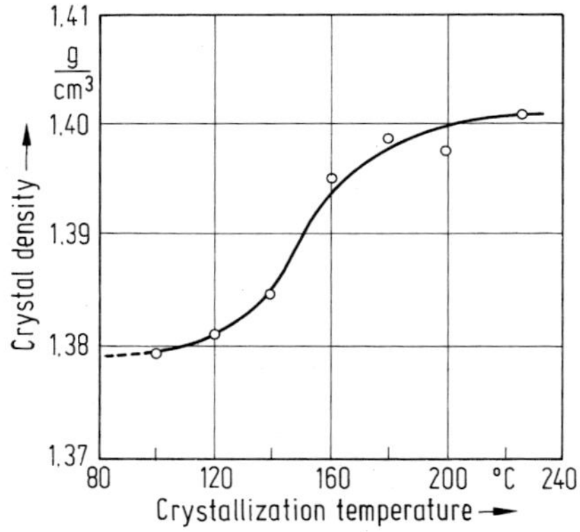

Figure 2.21. Crystal density of polybutylene terephthalate as a function of crystallization temperature.

The glass transition temperature depends on crystallinity, tempering, content and kind of fillers, and measuring method. It is in the temperature range of 30–50 °C. The glass transition temperature for completely amorphous polybutylene terephthalate has been determined by extrapolation to be 15 °C.

Dynamic mechanical relaxation experiments indicate that the temperatures for the α and β relaxations of nonoriented polybutylene terephthalate are independent of the degree of crystallization, but the peak heights decrease with increasing crystallinity. This conclusion is explained as follows.

In the amorphous areas, the α relaxation is characterized by overall chain motions, whereas the β relaxation is characterized by limited motions of the glycol and carboxyl groups. Both kinds of motion decrease with increasing crystallinity. Correspondingly,

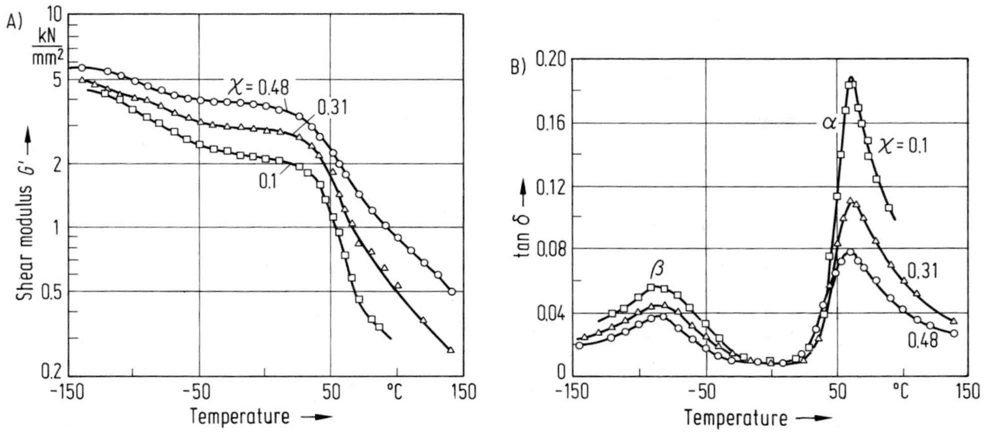

Figure 2.22. shear modulus G' (A) and mechanical loss factor tan δ (B) at 10 Hz for polybutylene terephthalate as a function of temperature and crystallinity χ [8].

the decrease in elastic modulus is less pronounced at both relaxations, especially at elevated temperatures (Figure 2.22) [8].

Orientation causes a significant temperature shift of the α relaxation to higher temperatures because of limited chain mobility. The elastic modulus is lower for temperatures below the α relaxation, but decreases less at higher temperatures (Figure 2.23).

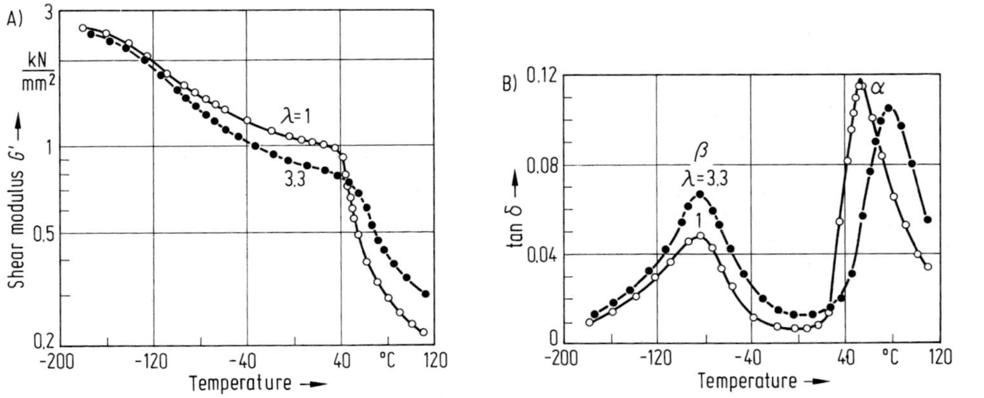

Figure 2.23. Shear modulus G' (A) and mechanical loss factor tan δ (B) at 1 Hz for isotropic ($\lambda = 1$) and oriented ($\lambda = 3.3$) polybutylene terephthalate as a function of temperature [8].

The melting temperature for polybutylene terephthalate is 222–232 °C; the melt enthalpy is 140 J/g [9].

2.7.1.2 Polyethylene Terephthalate

The crystal structure of polyethylene terephthalate corresponds to a triclinic unit cell with the parameters given in Table 2.10. Contrary to the case of polybutylene terephtha-

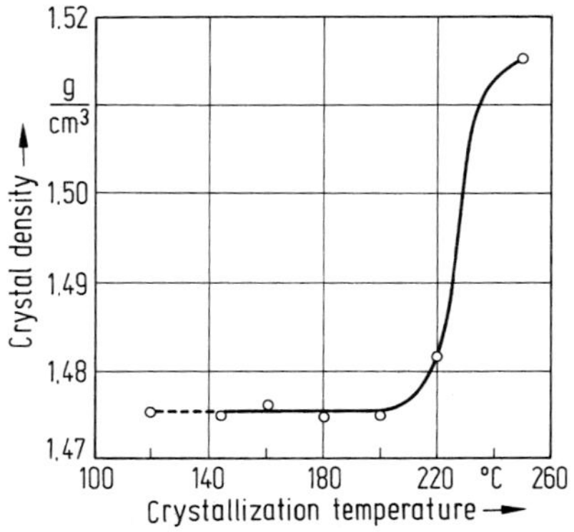

Figure 2.24. Crystal density of polyethylene terephthalate as a function of crystallization temperature [2].

late, the lower mobility of the PET chain prevents the crystal density from increasing with increasing crystallization temperature up to 200 °C. A steep increase is observed, however, above 200 °C (Figure 2.24).

The glass transition temperature of commercial polyethylene terephthalate depends on its purity, degree of crystallinity, and thermal history, as well as the method of determination. Correspondingly it varies between 67 and 140 °C [10].

Pure, well-tempered polyethylene terephthalate melts at 280 °C. Commercial products, however, melt between 250 and 265 °C because they contain impurities that lower their crystallinity. The major impurity, which is due to the formation of diethylene glycol and its condensation into the PET molecule during PET production, decreases both crystallinity and melting temperature. The decrease of the melting temperature can be estimated as follows:

melting temperature (°C) = 271 − 5.5 × (wt % of diethylene glycol)

The melting enthalpy is 166 J/g [9].

The degree of crystallinity depends on the crystallization temperature, and crystallization time and varies between 0% (amorphous) and 61% (Table 2.11) [11]. The degree of crystallization as a measure of molecular order decreases with increasing molecular weight, just as the melting enthalpy does [12]. The crystallization kinetics clearly differs from that of polybutylene terephthalate. The slow crystallization speed may be enhanced through nucleation (see Section 2.4.2).

Table 2.11. Degree of Crystallinity of Isothermally Crystallized Polyethylene Terephthalate [11]

Crystallization Temperature (°C)	Crystallization Time (h)	Degree of Crystallization
–	–	Amorphous
88	145	0.07
89	140	0.14
90	120	0.24
100	17	0.29
105	17	0.30
120	17	0.34
150	17	0.39
200	17	0.54
230	17	0.61

2.7.2 Mechanical Technological Properties

2.7.2.1 Short-Term Properties

Table 2.12 lists some properties of different polybutylene terephthalates, unmodified as well as impact-modified. These are average values for noncolored products, and the possibility of variation if pigments are applied must be taken into account for specific applications.

Glass fiber reinforced polybutylene terephthalate shows a significantly higher Izod impact strength, which can be again improved by modification with elastomers. Highly elastomer modified, nonreinforced polybutylene terephthalate retains its initial Izod impact strength down to temperatures of −20 °C through −40 °C (Pocan S 1506 and KU 1-7914).

Stress–strain diagrams for various unmodified, reinforced, and flame-retarded products are given in Figure 2.25 [13]. Comparisons at different temperatures of a polybutylene terephthalate reinforced with 30% glass fibers with an unmodified product indicate that there is more than the expected doubling of the tensile strength, especially at temperatures above 60 °C. This effect is attributed to very effective glass fiber coatings (Figure 2.26) [14].

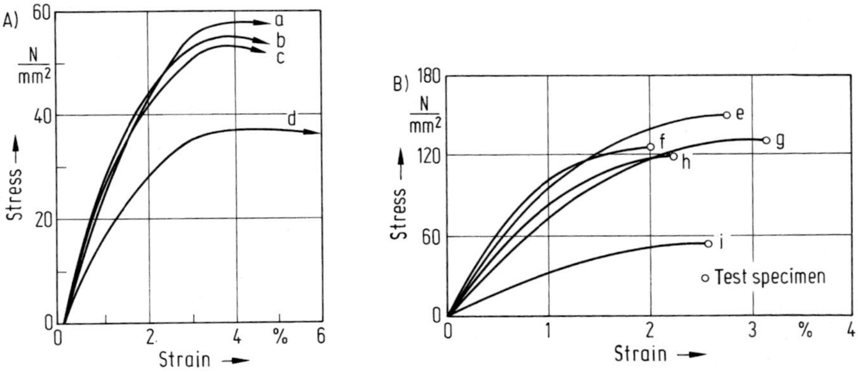

Figure 2.25. Stress–strain diagram at 23 °C of unmodified (A) and reinforced (B) polybutylene terephthalate [13]: a, high viscosity (BGA-type); b, medium viscosity; c, high viscosity; d, elastomer modified; e, 30% GF; f, 30% GF (V0); g, 20% GF; h, 20% GF (V0); i, 25% M. Reinforcements are glass fiber (GF) or mineral (M) type; V0, UL94V0.

Table 2.13 lists the mechanical short-term properties of various modified polyethylene terephthalates [15].

Typical stress–strain behavior for elastomeric copolyesters is shown in Figure 2.27 [16]. Any quantitative aspects of the curves depend on composition; qualitatively, there are three regions to consider. The high modulus in region I is due to quasi-elastic deformation that also affects the crystalline matrix. The deformation is reversible up to a strain of about 10%. At higher strains (region II) there is a pronounced irreversible change in the crystalline matrix, leading to a significant orientation of the cyrstallites. When the orientation is concluded, stress characteristics are observed which are typical for crosslinked elastomers (region III). The stress is mainly carried by the continuous elastomeric matrix and additionally, at high strains, by the oriented, crystallized hard segments.

Table 2.12. Mechanical Properties from Short-Term Tests of Polybutylene Terephthalate; Selected Pocan Products from Bayer Used as Examples [13]
GF, Glass fiber reinforced; M, mineral reinforced; V0, UL94V0

Property	Unit	Medium Viscosity	High Viscosity	20% GF B 3225	30% GF B3235	25% M B 7375	V0 20% GF B 4225	V0 30% GF B 4235	Elastomer-modified S 1506	Test Procedure	Specimen Dimensions (mm)
Tensile stress (50 mm/min)	N/mm²	57	55	–	–	55	–	–	35	ISO R 527 DIN 53 455 (tensile stress to ultimate tensile strain at break, 5 mm/min)	tensile dumb-bell No. 3 (DIN 53455) 4 mm thick
Tensile strain (50 mm/min)	%	3.7	3.7	–	–	2.2	–	–	4.2		
Ultimate tensile strain (at break) (50 mm/min)	%	>20	>50	–	–	–	–	–	>50		
Tensile strength (5 mm/min)	N/mm²	–	–	130	150	61	120	135	–		
Ultimate tensile strain (at break) (5 mm/min)	%	–	–	3.1	2.7	2.7	2.2	2.0	–		
Flexural strength	N/mm²	92	89	190	220	98	185	220	55	ISO 178 DIN 53452	80 × 10 × 4
Rim strain at highest force	%	6.0	6.1	4.5	4.0	5.0	3.6	3.3	5.8		
Tensile modulus of elasticity (1 mm/min)	N/mm²	2,700	2,600	8,000	10,500	3,500	9,000	11,500	1,700	ISO R 527 DIN 53457	tensile dumb-bell No. 3
Tensile creep modulus (1,000 h)	N/mm²	1,800	1,750	4,600	8,000	2,600	5,000	8,000	580	ISO 899 DIN 53444	see above
Flexural modulus of elasticity	N/mm²	2,300	2,300	6,000	8,000	3,100	7,000	9,000	1,600	DIN 53457	80 × 10 × 4
Impact strength 23 °C	kJ/m²	NB	NB	35	45	40	30	35	NB	ISO 179 DIN 53453	50 × 6 × 4
0 °C	kJ/m²	–	–	–	–	–	–	–	NB		
– 20 °C	kJ/m²	NB	NB	33	42	35	28	32	NB		
– 40 °C	kJ/m²	NB	NB	31	40	–	27	30	NB		

Izod impact strength	23 °C	kJ/m²	100	150	42	60	25	35	47	NB	ISO 180	80 × 10 × 4
	– 30 °C	kJ/m²	65	140	38	60	20	32	43	NB		
Notched impact strength	°C	kJ/m²	2	3	7	9	2	6	8	B	ISO 179 DIN 53453	50 × 6 × 4
		kJ/m²								NB		
	0 °C	kJ/m²	–	–	–	–	–	–	–	B		
	– 20 °C	kJ/m²	2	3	7	9	2	6	8	B		
	– 40 °C	kJ/m²	–	–	–	–	–	–	–	12		
Izod notched impact strength	23 °C	J/m	33	45	75	90	27	60	75	900	ASTM D 256	63.5 × 12.7 × 3.2
	0 °C	J/m	–	–	–	–	–	–	–	900		
	– 20 °C	J/m	30	40	70	85	25	55	70	750		
	– 40 °C	J/m	–	–	–	–	–	–	–	150		
Izod notched impact strength	23 °C	kJ/m²	7	8	11.5	13	6.5	10	12	70	ISO 180/1 A	80 × 10 × 4
	– 20 °C	kJ/m²	6.5	7.5	11	12.5	6.5	9	11.5	50		
	– 30 °C	kJ/m²	6.5	7.5	–	–	–	–	–	50		
	– 40 °C	kJ/m²	–	–	–	–	–	–	–	15		
Ball indentation hardness		N/mm²	125	120	175	190	145	200	220	70	ISO 2039 (Method B), DIN 53456	50 × 6 × 4

B, partially broken NB, not broken

Table 2.13. Mechanical Properties from Short-Term Tests of Polyethylene Terephthalate; Products of the Rynite Series (DuPont) [15] Used as Examples

Property	Unit	30% GF 530	V0 30% GF FR 530	45% GF 545	55% GF 555	35% M-GF 935	Elastomer-modified 30% GF 430	Elastomer-modified 35% GF SST 35	Test Procedure
Tensile strength									
−40 °C	N/mm^2	218	207	242	221	121	186	196	ASTM D 638,
23 °C	N/mm^2	158	152	193	196	97	124	103	ISO R 527,
93 °C	N/mm^2	83	86	92	96	44	45	62	DIN 53455
150 °C	N/mm^2	55	56	67	71	30	30	46	
Tensile strain at break									
−40 °C	%	3	2	2	1	2	3	3.5	ASTM D 638
23 °C	%	3	2	2	2	2	3	6	ISO R 527,
93 °C	%	6	5	5	4	5	5	10	DIN 53455
150 °C	%	7	7	6	5	5	5	10	
Shear strength									
−40 °C	N/mm^2	93	–	96	–	–	–	–	ASTM D 732
23 °C	N/mm^2	79	66	86	83	54	59	38	
93 °C	N/mm^2	61	–	64	–	–	–	–	
150 °C	N/mm^2	39	–	46	–	–	–	–	
Tensile modulus of elasticity									
−40 °C	kN/mm^2	11.0	–	17.2	–	–	–	–	ASTM D 638,
23 °C	kN/mm^2	9.9	–	14.6	–	–	–	–	ISO R 527,
93 °C	kN/mm^2	4.3	–	6.8	–	–	–	–	DIN 53457
150 °C	kN/mm^2	3.1	–	3.7	–	–	–	–	
Flexural modulus of elasticity									
−40 °C	kN/mm^2	10.3	11.0	15.2	20.7	11.7	9.7	9.0	ASTM D 790,
23 °C	kN/mm^2	9.0	10.3	13.8	17.9	9.6	7.6	6.9	ISO R 178
93 °C	kN/mm^2	3.6	4.3	5.5	9.2	3.4	2.4	2.5	
150 °C	kN/mm^2	2.7	2.9	4.0	5.7	2.2	–	1.9	

Flexural strength										
	– 40 °C	N/mm^2	289	280	344	358	181	297	276	ASTM D 790, ISO R 178
	23 °C	N/mm^2	230	220	283	310	148	195	145	
	93 °C	N/mm^2	114	107	141	165	62	69	69	
	150 °C	N/mm^2	76	69	96	110	43	–	–	
Bending creep rupture										
27.6 N/mm^2, 1,000 h	23 °C	%	0.52	–	0.33	–	–	–	–	ASTM D 2990
27.6 N/mm^2, 5,000 h	23 °C	%	0.56	0.46	0.32	0.19	0.48	0.63	–	
	60 °C	%	1.18	1.01	0.70	–	–	1.52	–	
	125 °C	%	1.65	1.86	1.14	0.81	2.0	1.96	–	
Ultimate bending stress		N/mm^2	260	240	340	–	170	–	–	DIN 53452
Izod impact strength										
	– 40 °C	J/m	96	80	123	–	–	123	160	ASTM D 256, ISO R 180
	23 °C	J/m	101	85	117	123	64	140	235	
	93 °C	J/m	139	–	139	–	–	–	–	
	125 °C	J/m	–	–	–	–	–	365	–	
	150 °C	J/m	176	–	155	–	–	–	–	
Charpy notched impact strength										
	23 °C	kJ/m^2	9	8	11	10	6	–	–	ISO R 179, DIN 53453
Compressive strength		N/mm^2	172	172	179	196	141	164	81	ASTM D 695
Deformation under load,										
27.6 N/mm^2	23 °C	%	0.4	0.5	0.4	–	–	0.7	–	ASTM D 621
	50 °C	%	1.6	1.2	1.2	–	–	2.8	–	

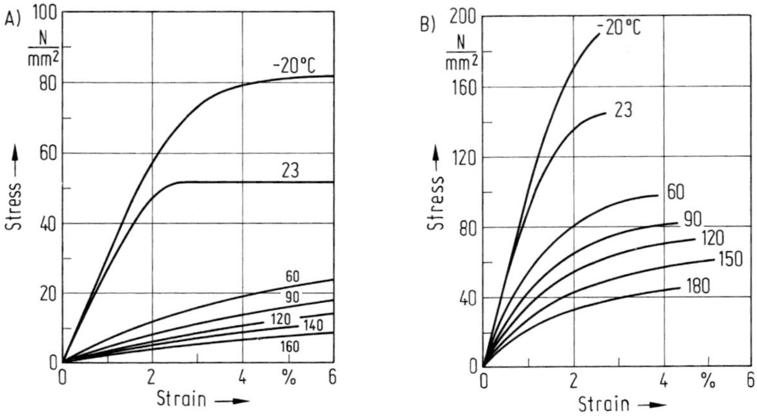

Figure 2.26. Stress–strain diagrams for unmodified (A) and 30% glass fiber reinforced (B) polybutylene terephthalate at various temperatures [14].

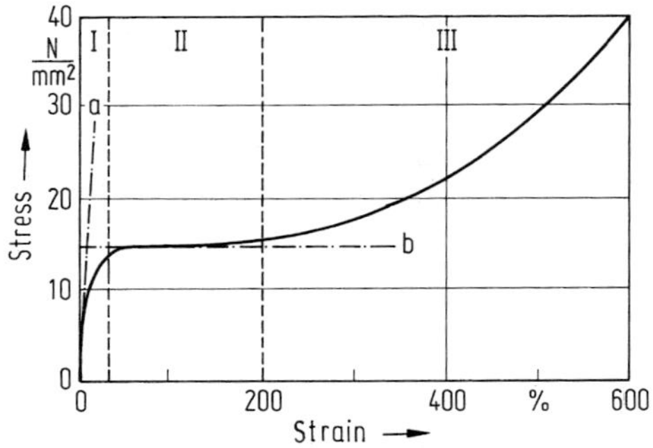

Figure 2.27. Stress–strain diagram of an elastomeric copolyetherester [16]: a, Young's modulus; b, yield stress.

2.7.2.2 Long-Term Properties

In addition to properties tested in short-term trials, the changes in properties under repeated stress are extremely important for the design of parts. The dependence of stress and strain on load and time can be estimated from isochronic stress–strain curves. Figure 2.28 shows such curves for various modified polybutylene terephthalates at different temperatures.

Figure 2.29 demonstrates the high creep rupture strength of unmodified polyethylene terephthalate [17]. For comparison to polybutylene terephthalate, Figure 2.30 shows the higher level of tensile strength for polyethylene terephthalate [18]. The creep modulus, $E_c(t)$, can be calculated from the slope of the isochronic stress–strain curves as the ratio of the constant stress and the time-dependent strain $\varepsilon(t)$:

$$E_c(t) = \sigma \cdot [\varepsilon(t)]^{-1} \quad (2.21)$$

Figure 2.31 shows the creep modulus as a function of time for polybutylene terephthalate in two forms—unmodified, high viscous, and 30% glass fiber reinforced—at various temperatures.

The flexural creep modulus of semicrystalline, unmodified polyethylene terephthalate is given in Figure 2.32. The variation of the creep modulus with time indicates a low creep tendency for semicrystalline polyesters.

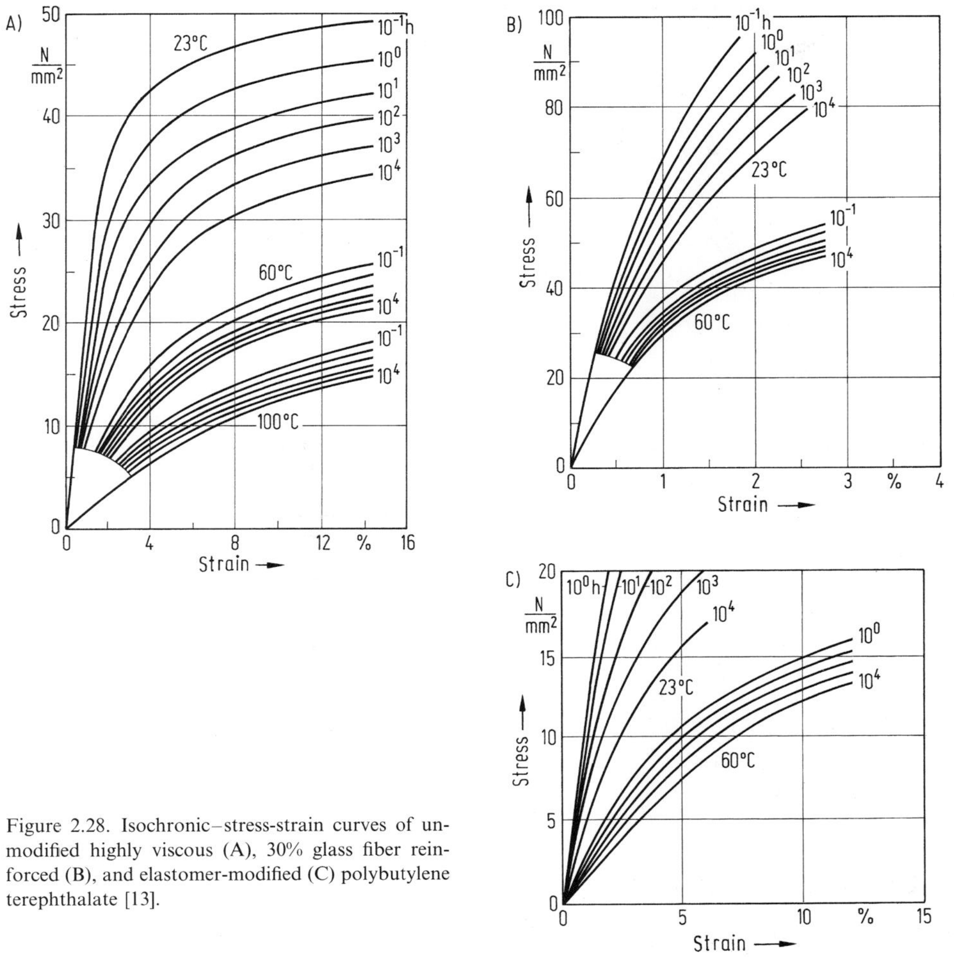

Figure 2.28. Isochronic–stress-strain curves of unmodified highly viscous (A), 30% glass fiber reinforced (B), and elastomer-modified (C) polybutylene terephthalate [13].

Figure 2.33 shows the small decrease in tensile strength of polybutylene terephthalate reinforced with 30% glass fibers under permanent stress, especially at elevated temperatures.

The dynamic long-term behavior of a material is determined in fatigue tests and plotted as Wöhler curves. The shape of these curves depends greatly on the conditions of measurement (e.g., frequency, shape of stress function, ratio of sample surface to sample cross section, thermal conductivity, temperature). Figure 2.34 shows the fatigue test data for both unmodified and 30% glass fiber reinforced polybutylene terephthalate (normal adjustment and V0 adjustment). In Figure 2.35, flexural fatigue strength at the alternating and swelling ranges are given for unmodified polyethylene terephthalate, as well as for unmodified and 30% glass fiber reinforced polybutylene terephthalate [19]. The diagrams show the high resistance to fatigue, especially for reinforced materials. Figure 2.36 is a similar stress–strain diagram (see also Figure 2.30) and shows that polyethylene terephthalate has a higher resistance to flexural fatigue than does polybutylene terephthalate.

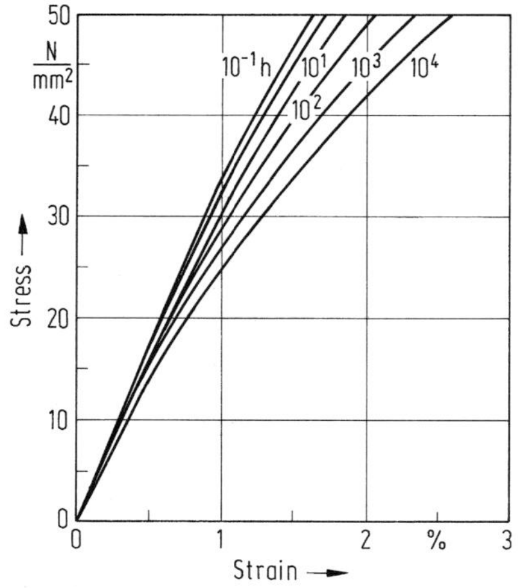

Figure 2.29. Isochronic stress–strain diagrams of semicrystalline unmodified polyethylene terephthalate [17].

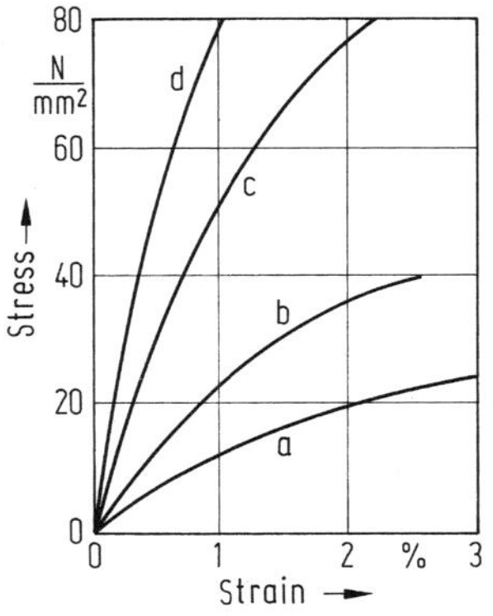

Figure 2.30. Isochronic stress–strain curves of polybutylene terephthalate and polyethylene terephthalate for 1000 hours at 23 °C [18]: a, unmodified PBT; b, unmodified PET; c, PBT with 30% glass fiber; d, PET with 30% glass fiber.

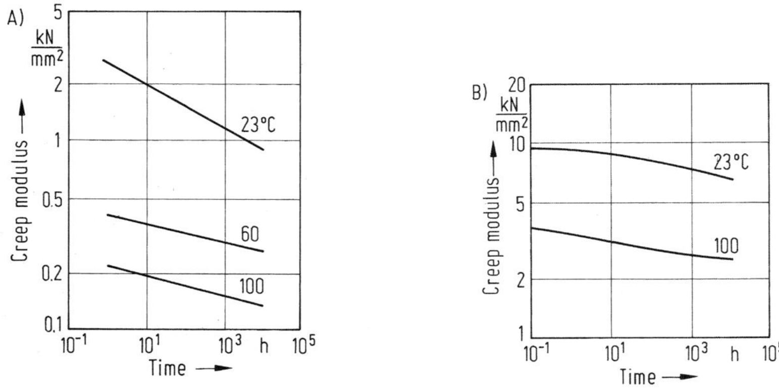

Figure 2.31. Creep modulus E_c of polybutylene terephthalate [13]: (A) unmodified, high viscosity; (B) 30% glass fiber reinforced.

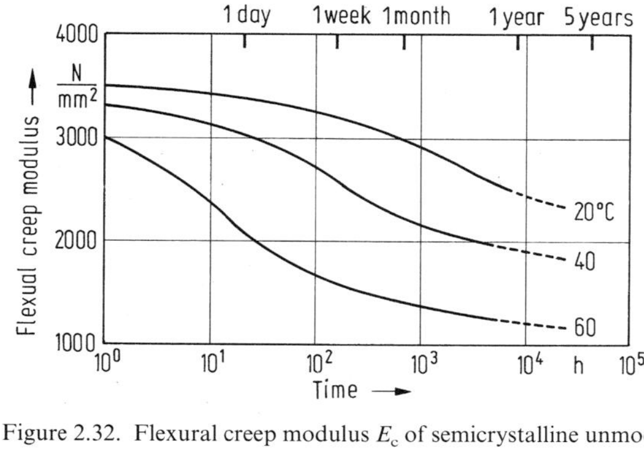

Figure 2.32. Flexural creep modulus E_c of semicrystalline unmodified polyethylene terephthalate [17].

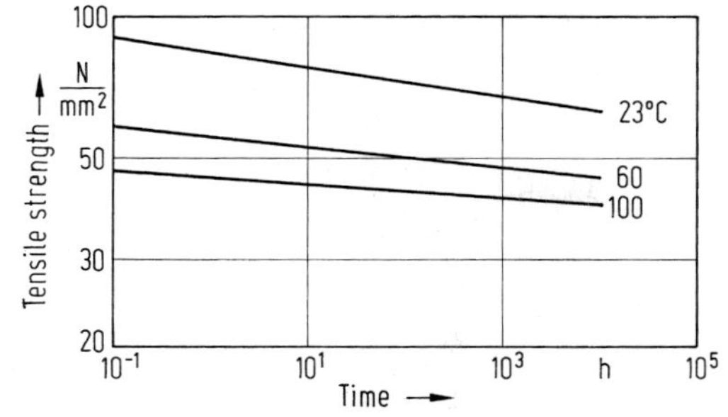

Figure 2.33. Tensile strength over time of 30% glass fiber reinforced polybutylene terephthalate [13].

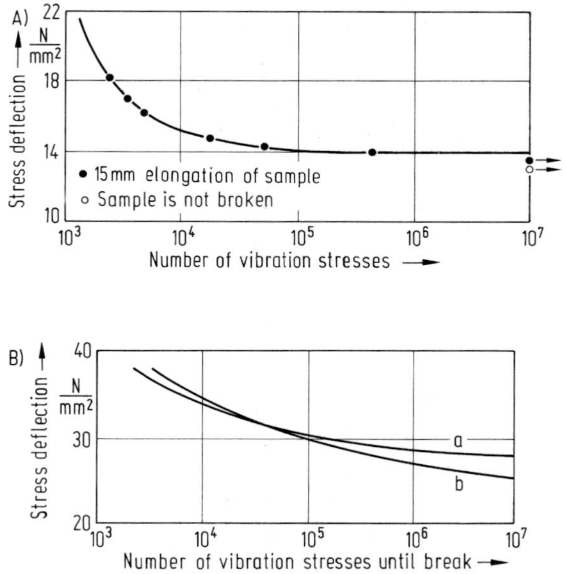

Figure 2.34. Tension fatigue test of polybutylene terephthalate. (A) unmodified, high viscosity; (B) 30% glass fiber reinforced [13]: a, normal adjustment; b, V0 adjustment.

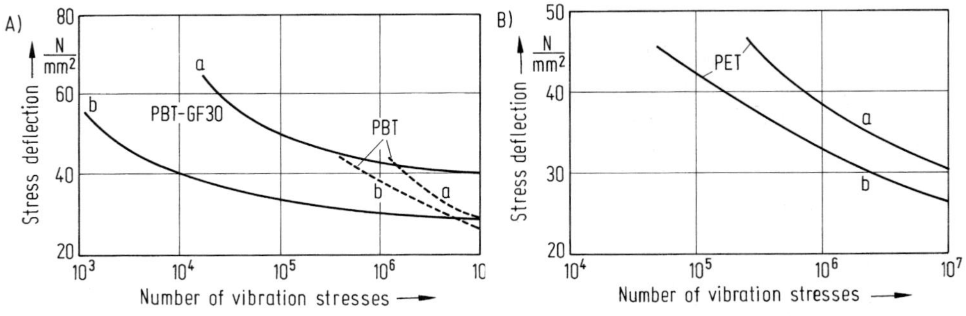

Figure 2.35. Results of flexural fatigue test of unmodified and 30% glass fiber reinforced polybutylene terephthalate (A) and unmodified semicrystalline polyethylene terephthalate (B) [19]: a, alternating range; b, swelling range.

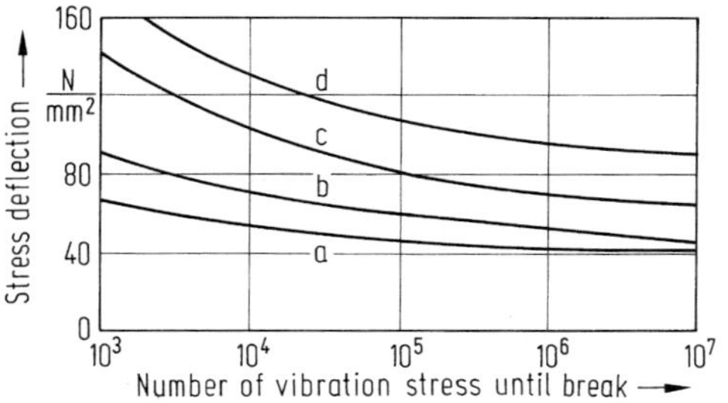

Figure 2.36. Results of flexural fatigue tests of polybutylene terephthalate and polyethylene terephthalate [18]: a, unmodified PBT; b, unmodified PET; c, PBT with 30% glass fibers; d, PET with 30% glass fibers.

2.7.3 Thermal Properties

2.7.3.1 Short-Term Properties

The behavior under temperature stress of polybutylene terephthalate and polyethylene terephthalate is determined by their semicrystallinity as described in Section 2.7.1. This is evident in the temperature dependence of the shear modulus and the mechanical loss factor (tan δ) or, in other dynamic mechanical experiments, in the softening of the amorphous segments and in the melting of the crystallites. Figure 2.37 shows diagrams for unmodified, glass fiber reinforced, and elastomer-modified polybutylene terephthalate (see also Figures 2.22 and 2.23).

Because of the high degree of crystallinity, molded parts can be heated briefly close to the melting temperature without significant damage or deformation. Therefore, the temperature for practical use is between 200 and 250 °C, depending on degree of reinforcement needed to accommodate short-term stress. The maximum temperature for short-term use of elastomer-modified polybutylene terephthalate is lower than that of the standard product, depending on the elastomer content (see Table 2.14). For practical applications, glass fiber reinforced products show a much smaller decrease in the maximum temperature than nonreinforced products; this effect is precisely due to the reinforcing effect of the glass fibers.

Maximum service temperatures for polyethylene terephthalate are higher than for polybutylene terephthalate because of the higher glass transition point and the higher melting point of the former (see Figure 2.38 and Table 2.15).

Enthalpy curves of reinforced polybutylene terephthalate [14], as well as the curve of unmodified polyethylene terephthalate [19], are shown in Figures 2.39 and 2.40.

The coefficient of linear thermal expansion of nonreinforced and of filled products (where the filler does not have a preferential direction, e.g., glass spheres) is dependent on the direction of molding. But because of the orientation of the glass fibers, reinforced products show a strong dependence on the direction of injection (see Table 2.14).

The low thermal conductivity of thermoplastic polyesters allows their use in applications with thermal insulation [20].

2.7.3.2 Long-Term Properties

The temperature for long-term use (i.e., the temperature at which relevant mechanical properties have decreased to half the initial levels after 25,000 hours) is 140–150 °C for polyethylene terephthalate in nonreinforced and reinforced standard products, 130–140 °C for polybutylene terephthalate (see Tables 2.14 and 2.15), and in special cases up to 160 °C [13]. For elastomer-modified products it is reduced to about 90 °C, depending on kind and content of the elastomer.

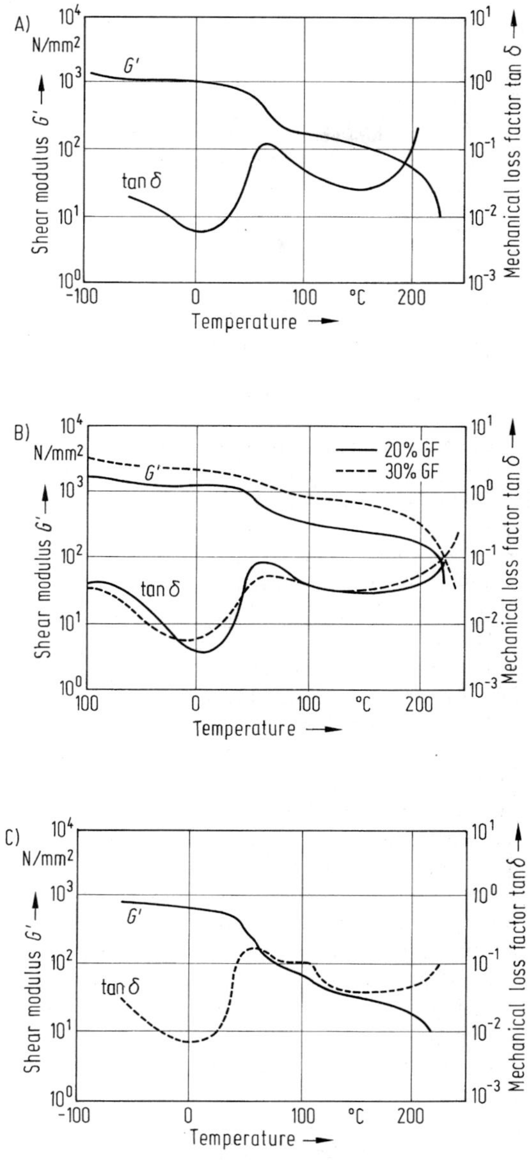

Figure 2.37. Dynamic mechanical testing of polybutylene terephthalate [13]: (A) unmodified, (B) reinforced, (C) elastomer modified.

"Thermal aging" does not simply mean thermal stress. The term includes the influence of oxygen at elevated temperatures. In addition to the basic properties of the polyesters, the kind of stabilization is important. For example, increasing crystallinity causes a decrease in the impact strength of polybutylene terephthalate reinforced with 30% glass fibers and with added stabilizers during the first hour from 50 kJ/m² to 45 kJ/m² at 140 °C or to 43 kJ/m² at 160 °C; after that only a minor decrease is observed over 1,000 hours (Figure 2.41). The ultimate tensile strength also increases as a result of crystallization and remains nearly constant at 160 °C up to about 1,000 hours. Modified, and with 30% glass fiber reinforcement, polyethylene terephthalate in standard products

Table 2.14. Thermal Properties of Polybutylene Terephthalate; Selected Products from the Pocan series (Bayer) [13] Used as Examples

Property	Unit	Medium Viscosity B 1305	High Viscosity B 1505	20% GF B 3225	30% GF B 3235	25% M B 7375	V0 20% GF B 4225	V0 30% GI B 4235	Elastomer-modified S 1506	Test Procedure	Specimen Dimensions (mm)
Temperature of deflection under load											
HDT/A (1.81 N/mm²)	°C	70	70	200	200	70	180	195	60	DIN 53461, ISO 75	110 × 10 × 4
HDT/B (0.45 N/mm²)	°C	170	170	210	215	185	205	210	105		
Vicat softening temperature											
VST A/50 (10 N, 50 K/h)	°C	–	–	220	220	215	220	220	–	DIN 53460, ISO 306	10 × 10 × 4
VST B/50 (50 N, 50K/h)	°C	–	–	215	215	185	205	210	124		
VST B/120 (50 N, 120 K/h)	°C	180	180	210	215	185	205	210	130		
Ball indentation (heat limit)	°C	190	190	206	210	200	195	205	180	VDE 0470	50 × 6 × 4
Temperature for practical use (without stress)											
Permanent	°C	140	140	140	140	140	140	140	ca. 90	UL 746 B is nearly equivalent to Vicat A	80 × 10 × 4
Short-term	°C	200	140	210	210	200	210	210	160		
Coefficient of linear thermal expansion											
(Longitudinal), 23°C 80°C	10^{-4} K^{-1}	1.2	1.2	0.4	0.3	1.1	0.3	0.25	1.3	DIN 53752	10 × 10 × 4
(Transverse), 23–80°C	10^{-4} K^{-1}	1.2	1.2	1.0	0.9	1.1	0.9	0.85	1.3		
Thermal conductivity	W/(m · K)	0.25	0.25	0.25	0.25	0.25	0.25	0.25	0.25	DIN 52612	80∅ × 3

Combustibility (UL 94)										UL 94	
Sample thickness 1		HB		HB	HB	HB					127 × 12,7 × 0,8
Sample thickness 2		HB	HB	HB	HB	HB	V0	V0	HB		127 × 12,7 × 1,6
Sample thickness 3		HB	HB	HB	HB	HB	V0	V0	HB		127 × 12,7 × 3,2
Maximum temperature by glow wire test Sample thickness	°C	750	750	750	750	750	960	960	750	DIN VDE 0471, Book 2/1, IEC 695-2-1	80∅ × 2
Oxygen index	%	25	25	21	23	22	32	31	22	ASTM D 2863-77	80 × 10 × 4
Melt flow index 206°C, 2.16 kg	ml/10 min	60	17	16	13	49	15	11	4	DIN 53735, ISO R 1133	Granulates
Melt flow index 260°C, 5 kg	ml/10 min	140	39	10	39	119	36	28	18		
Flowability, spiral flow 260°C	mm	600	400	370	340	550	330	290		380	2 × 8,7
Melting temperature	°C	225	225	225	225	225	225	225	225	DSC, ISO 3146, Method C	
Injection molding tool temperature	°C	260	260	260	260	260	260	260	250		
Injection molding material temperature	°C	80	80	80	80	80	80	80	80		
Flow front velocity	mm/s	180	180	540	540	540	540	540	180		
Mold processing shrinkage (longitudinal/transverse) (tool temperature 80 °C)	%	1.9/1.9	1.8/1.8	0.35/1.5	0.3/1.4	1.7/1.7	0.4/1.5	0.3/1.4	1.8/1.8	DIN 16901	150 × 90 × 3
Overall shrinkage (longitudinal/transverse) after tempering 1 h, 150 °C	%	2.3/2.3	2.2/2.2	0.4/1.7	0.35/1.6	2.2/2.2	0.4/1.6	0.3/1.5	2.5/2.5[1]		

[1] Tempering for 1 hour at 130 °C.

Table 2.15. Thermal Porperties of Polyethylene Terephthalate of Selected Products from the Rynite Series (DuPont) [15]

Properties	Unit	30% GF 530	V0 30% GF FR 530	45% GF 545	55% GF 555	35% M-GF 935	Elastomer-modified 30% GF 430	Elastomer-modified 35% GF SST 35	Test Procedure
Melting temperature	°C	254	254	254	255	252	250	250	ASTM D 789, ISO R 1218, DIN 53736
Temperature of deflection under load (1.8 N/mm²)	°C	224	224	226	229	215	210	220	ASTM D 648, ISO 75, DIN 53461
Coefficient of linear thermal expansion									ASTM D 696, DIN 52328 (VDE 0304/1)
−30–49 °C	10^{-5} K^{-1}	3.2	2.5	2.4	1.4	2.7	3.4	4.3	
50–79 °C	10^{-5} K^{-1}	2.7	2.0	2.4	1.1	1.8	2.0	2.4	
80–160 °C	10^{-5} K^{-1}	2.2	1.1	1.6	0.7	1.8	3.2	4.3	
Thermal conductivity	W/(m · K)	0.29	0.32	0.25	0.33	0.26	0.29	–	ASTM C 177
Vicat softening temperature	°C	–	216	–	–	–	–	–	DIN 53460
Temperature index									UL
Electrical	°C	140	150	140	140	140	140	–	
Mechanical with impact	°C	140	150	140	140	140	140	–	
Mechanical without impact	°C	140	150	140	140	140	140	–	
Combustibility (sample thickness 0.8 mm)		HB	V0	HB	HB	HB	HB	–	UL 94
Oxygen index	%	20	33	20	–	–	21	–	ASTM D 2863
Glowing wire ignition	S	132	300 +	146	151	–	173	–	UL 746 A

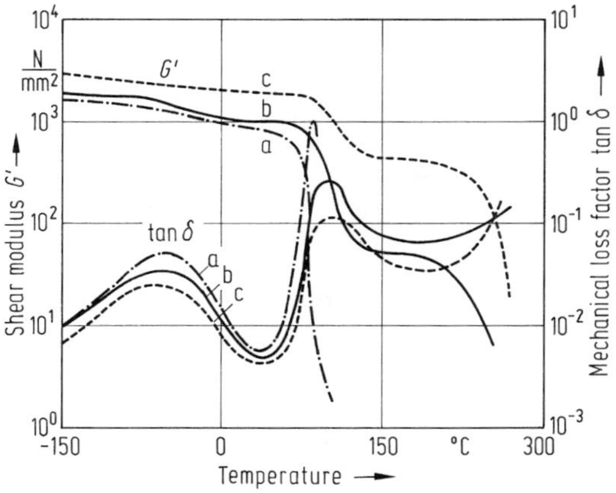

Figure 2.38. Torsional pendulum testing of nonreinforced (a, amorphous; b, semicrystalline) and 20% glass fiber reinforced (c) polyethylene terephthalate [17].

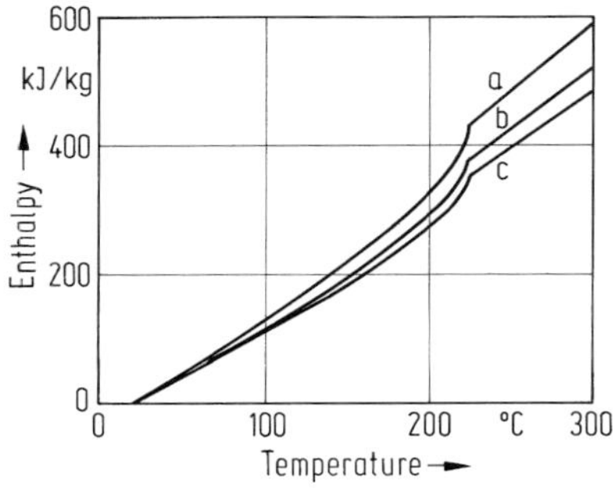

Figure 2.39. Enthalpy (H) of polybutylene terephthalate as a function of temperature ($H = 0$ for 20 °C) [14]: a, unmodified; b, with 20% glass fibers; c, with 30% glass fibers.

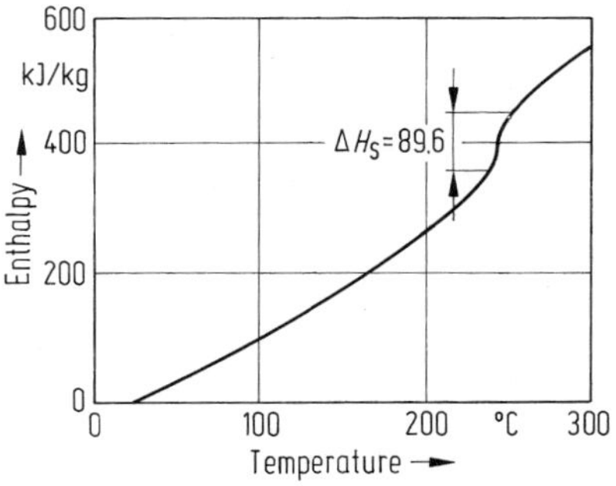

Figure 2.40. Enthalpy (H) of nonreinforced polyethylene terephthalate as a function of temperature ($H = 0$ for 20 °C) [19].

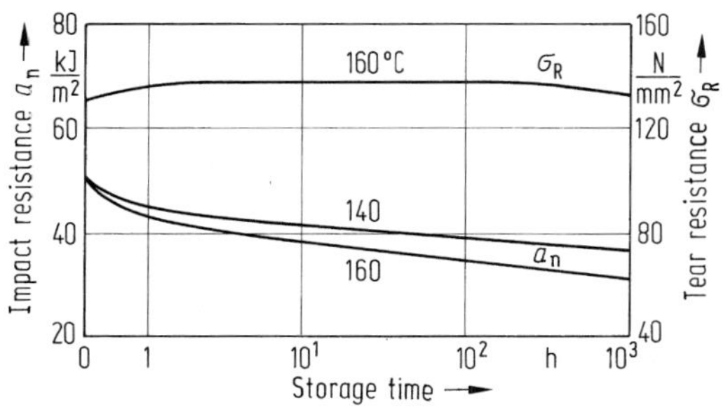

Figure 2.41. Aging in hot air of polybutylene terephthalate reinforced with 30% glass fibers [13].

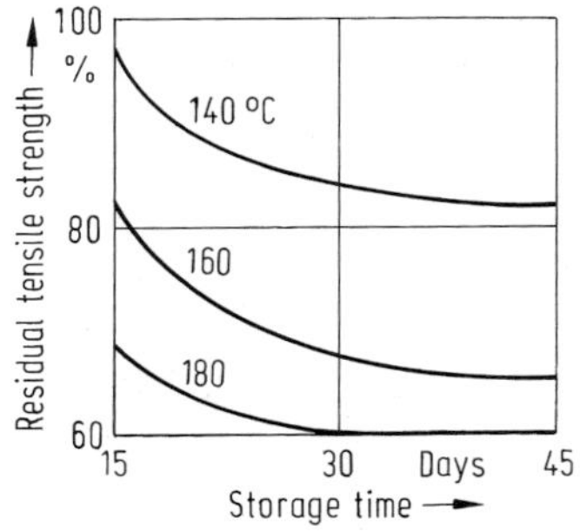

Figure 2.42. Aging in hot air of polyethylene terephthalate reinforced with 30% glass fibers for standard and V0 products [21].

Table 2.16 Electrical and General Properties of Polybutylene Terephthalate; Selected Products from the Pocan Series (Bayer) Used as Examples [13]
GF = glass fiber reinforced, M = reinforced with minerals, V0 = UL94V0

Properties	Units	Medium Viscosity B 1305	High Viscosity B 1505	20% GF B 3225	30% GF B 3235	25% M B 7375	V0 20% GF B 4225	V0 30% GF B 4235	Elastomer-modified S 1506	Test Procedure	Specimen/ Dimension (mm)
Dielectric constant:											
50 Hz		3.2	3.2	3.6	3.8	4.4	3.6	4.0	3.1	DIN VDE/0303, Part 4, IEC 250	80 ∅ × 1
1 MHz		3.1	3.1	3.5	3.5	4.2	3.5	3.9	3.0		
Dielectric loss factor × 10^4											
50 Hz		30	20	17	22	26	19	17	19		
1 MHz		190	220	190	188	196	175	159	167		
Dielectric strength										DIN VDE/0303, Part 2, IEC 243	80 ∅ × 1
K 20/P 50	kV/mm	30	27	34	34	26	29	28	28		
Dried	kV/mm	28	27	32	33	26	29	29	26		
4 days, 23 °C, 80%											
Comparison of tracking	V										
CTI[a)]		600	600	475	425	475	225	200	600	DIN VDE/0303, Part 1, IEC 112	⩾15 × 15 × 4
CTI 100-tear drop		575	575	450	400	450	200	175	575		
volume resistivity	V										
CTI M		500	500	200	200	400	175	175	600		
CTI M 100-tear drop		475	475	175	175	375	150	150	575		
Volume resistivity	Ω · cm									DIN VDE/0303, Part 3, IEC 93	80 ∅ × 1
Dried		10^{16}	10^{16}	10^{16}	10^{16}	10^{15}	10^{16}	10^{16}	10^{16}		
4 days, 23 °C, 80%		10^{16}	10^{16}	10^{16}	10^{16}	10^{15}	10^{16}	10^{16}	10^{15}		

Surface resistivity Dried 4 days, 23 °C, 80%	Ω	10^{16} 10^{15}	10^{15} 10^{15}	10^{16} 10^{16}	10^{15} 10^{15}	10^{13} 10^{13}	10^{16} 10^{15}	10^{16} 10^{16}	10^{16} 10^{15}	10^{16} 10^{15}	
Electrolytic corrosion effect		AN 1.2	AN 1.2	AN 1.2	AN 1.2	AN 1.2	AN 1	AN 1	AN 1.2	DIN 53489, IEC 426 ASTM D 495	30 × 10 × 4
Arc resistance	s	180	180	120	120	120	60	60			
Density	g/cm³	1.30	1.30	1.46	1.55	1.56	1.57	1.65	1.22	DIN 53479, ISO/R 1183	50 × 6 × 4
Water absorption (23 °C, saturated)	%	0.2	0.2	0.2	0.2	0.2	0.2	0.2	0.2	DIN 53495, ISO 62	50 × 50 × 3

[a] CTI: Comparative tracking index

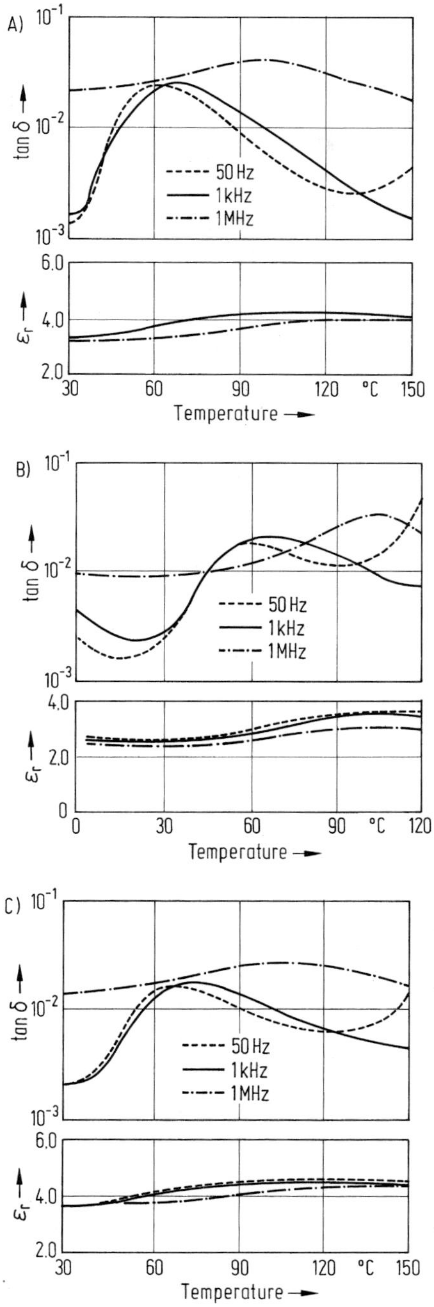

Figure 2.43. Dielectric loss factor tan δ and dielectric constant ε_R of polybutylene terephthalate at 50 Hz, 1 kHz, and 1 MHz: (A) high viscosity, nonreinforced, (B) reinforced with 30% glass fibers, (C) with 30% glass fibers, V0 adjustment [13].

and V0 products retains 82% of the initial tensile strength after aging in hot air at 140 °C for 45 days. This level is reduced to 60% after aging at 180 °C (Figure 2.42).

Since at low temperatures an extremely long time is necessary to produce property changes, the application possibilities for thermoplastic polyesters are limited only by the highest use temperatures of the materials.

2.7.4 Electrical Properties

Polybutylene terephthalate and polyethylene terephthalate have excellent electrical and dielectric properties. Therefore they are valuable engineering plastics for electrical and electronic parts (see Tables 2.16 and 2.17). As a special advantage, these properties are independent of the ambient humidity and constant over a large range of ambient temperatures. Since thermoplastic polyesters do not start electrolytic corrosion and show good tracking resistance, they are of special importance in assuring the reliability of electronic parts such as relays.

Figure 2.43 demonstrates the independence of the dielectric constant from frequency and temperature through examples of polybutylene terephthalate modified in different ways. In addition, polyethylene terephthalate shows good tracking resistance and high, nearly constant, dielectric strength over a wide range of temperatures (Figure 2.44).

2.7.5 Combustibility

Nonreinforced thermoplastic polyesters burn slowly, with dripping; there is a yellow-orange flame that produces carbon black. Glass fiber reinforced materials exhibit a higher burning speed because the glass fibers reduce dripping and support the creation of a flame front [14, 17]. The oxygen index depends on the type of polyester and kind of product and lies between 20 and 25%. The classification according to UL94 is HB (horizontal burning).

Thermoplastic polyesters are excellent engineering plastics for the utility and electronics industries, for electrical tools, and for household equipment. For safety reasons, many of these applications call for self-extinguishing materials. Therefore, most polyester suppliers have developed a variety of self-extinguishing products which, depending on polyester type and product adjustment, have an oxygen index of 30–35% and are classified as V0 according to UL 94. Some products even achieve a classification of 5-V [13, 20].

2.7.6 Resistance to Hydrolysis and Chemicals

The semicrystalline character of polybutylene terephthalate and polyethylene terephthalate is the basis for these materials' excellent resistance to chemicals. Organic solvents, especially aliphatic hydrocarbons, gasoline, alcohols, ethers, long-chain esters, fats, oils, perchlorinated hydrocarbons, and fluorinated hydrocarbons as used in aerosal sprays do not affect thermoplastic polyesters at room temperature. They are also resistant to weak bases and acids, water, and aqueous solutions of neutral and acidic salts at room temperature. On the other hand, they are attacked by strong acids, oxidizing acids, strong bases, ketones, and phenols. Above 60 °C, thermoplastic polyesters are increasingly sensitive to monofunctional alcohols, aromatics, and ketones. But they remain resistant to mineral and transformer oils, gasoline, and brake fluids (see Table 2.18).

Table 2.17. Electrical Properties of Polyethylene Terephthalate; with Products from the Rynite Series (DuPont) [15] Used as Examples

Properties	Unit	30% GF 530	V0 30% GF FR 530	45% GF 545	55% GF 555	35% M-GF 935	Elastomer-modified 30% GF 430	Elastomer-modified 35% GF SST 35	Test Procedure
Volume resistivity	Ω · cm	10^{15}	10^{15}	10^{15}	–	10^{15}	10^{15}	–	ASTM D 257
Surface resistivity	Ω	10^{14}	10^{14}	10^{14}	–	10^{14}	10^{13}	–	ASTM D 257
Dielectric constant									
100 Hz		–	–	–	–	–	3.5	–	ASTM D 150,
1 kHz		3.6	3.6	4.0	–	3.8	3.4	–	DIN 53483
1 MHz		3.5	3.6	3.9	–	3.7	3.4	–	
Dielectric strength	kV/mm								ASTM D 149
Short term									
0.8 mm disk		45.3	40.9	–	–	–	–	–	
1.6 mm disk		29.6	25.6	–	–	–	27	–	
3.2 mm disk		21.7	16.9	21.3	19.7	22.5	21.7	21	
Step by step									
3.2 mm disk		17.3	13.8	17.3	–	20.9	18.7	–	
Dielectric loss factor × 10^4									ASTM D 150,
100 Hz		–	–	–	–	–	50	–	ISO R 1325,
1 kHz		50	70	50	–	80	60	–	IEC 250
1 MHz		120	100	110	–	100	160	–	DIN 53483
Arc resistance	s	125	117	126	126	131	79	–	ASTM D 495
High voltage arc resistance	m/s	–	$1.1 \cdot 10^{-3}$	–	–	–	–	–	–
Tracking resistance	V								IEC 112
CTI 1 mm		–	200	–	–	–	–	–	–
2 mm		–	225	–	–	–	–	–	–
3 mm		–	225	–	–	–	–	–	–
CTI M 1 mm		–	200	–	–	–	–	–	–
2 mm		–	200	–	–	–	–	–	–
3 mm		–	175	–	–	–	–	–	–

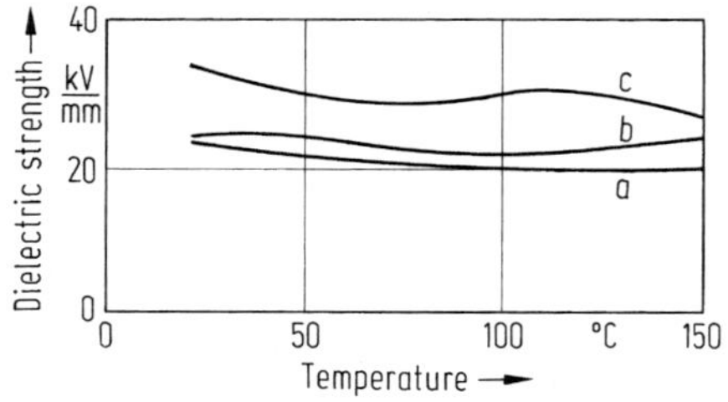

Figure 2.44. Dielectric strength of polybutylene terephthalate [13]: a, unmodified, high viscosity; b, with 30% glass fibers, V0 nonreinforced adjustment; c, with 30% glass fibers.

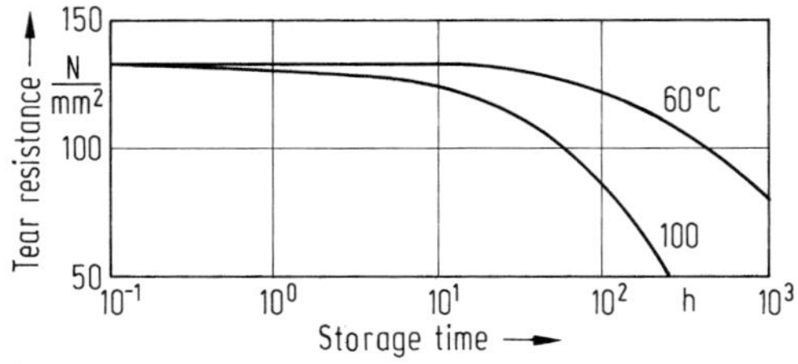

Figure 2.45. Ultimate tensile strength of polybutylene terephthalate reinforced with 30% glass fibers after immersion in water at 60 and 100 °C [13].

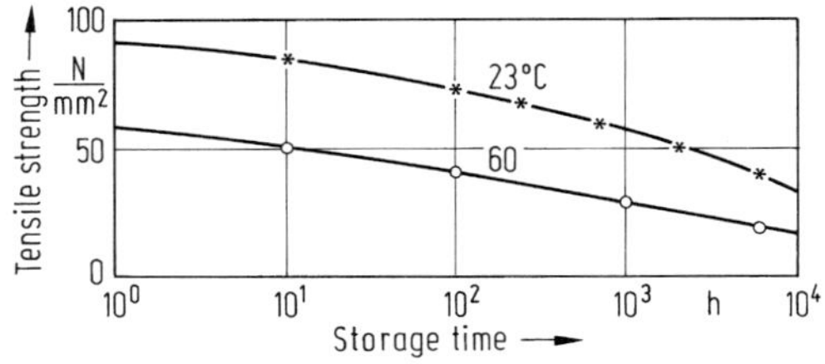

Figure 2.46. Creep rupture strength of polybutylene terephthalate reinforced with 30% glass fibers in water at 23 and 60 °C [13].

Water and aqueous solutions lead to hydrolysis at elevated temperatures, and this causes a decrease of mechanical strength (Figures 2.45 and 2.46).

2.7.7 Resistance to Weathering and Irradiation

Both polybutylene terephthalate and polyethylene terephthalate show good resistance to weathering, as required for outdoor applications, and this advantage can be increased by the addition of carbon black. The impact strength of polybutylene terephthalate reinforced with 30% glass fibers is reduced by less than 50% after 5 years of natural weathering in a Central/Eastern European climate or after 3 years in the Xenon Weather-O-Meter (Figure 2.47).

Polyesters are resistant to high energy radiation. polyethylene terephthalate is degraded at doses of 20 kJ/(kg ·h) only if the absorbed energy exceeds 100 kJ/kg. At an absorbed energy of 1000 kJ/kg, 80% of the initial tensile strength and tensile strength at break is retained [17].

2.7.8 Tribological Behavior

For polybutylene terephthalate there is only a slight difference between adhesion and sliding with respect to frictional resistance. Table 2.19 contains the static and dynamic

Table 2.18. Resistance to Chemicals for Polybutylene Terephthalate and Polyethylene Terphthalate [19]

Chemical	Concentration (%)	PET		PBT	
		20 °C	60 °C	20 °C	60 °C
Acetic acid	10	r	rr	r	r
Acetone	–	r	rr/n	n	n
Ammonia	10	r	–	r	n
Ammonia	Conc.	r/rr	–	n	n
Benzene	–	r	n	n	n
Bleach, industrial	–	r	r	–	–
Butanol	–	r/rr	rr	r	rr
Butyl acetate	–	r	r	r	n
Calcium chloride	10	r	r	r	r
Carbon tetrachloride	–	r	r	r	n
Chlophen A 60	–	r	r	–	–
Chloride of lime	–	r	r	–	–
Chlorobenzene	–	r	n	n	n
Chloroform	–	n	n	n	n
Citric acid	10	r	r	r	r
Dibutylphthalate	–	r	r	r	r
Diesel fuel	–	r	r	r	r
Dioxane	–	r	n	–	–
Ethanol	–	r	rr	r	rr
Ether	–	r	r	r	r
Ethyl acetate	–	r	rr/n	rr	n
Ethyl chloride	–	r	n	–	–
Ethyl glycol	–	r	r	–	–
Formic acid	10	r	r/rr	r	r
Gasoline	–	r	r	r	rr/n
Glacial acetic acid	–	r	n	rr/n	n
Glycerin	–	r	r	r	r
Glycol	–	r	rr	r	r
Heating oil	–	r	r	r	r
Hydrochloric acid	10	r/rr	rr	r	r/rr
Hydrogen peroxide	35	r	r	r	n
Isopropanol	–	r/rr	rr	r	rr
Lubricating oil	–	r	-	r	–
Methanol	–	r	rr	–	–
Methyl glycol	–	r	rr/n	–	–
Methylene chloride	–	n	n	n	n
Methylethyl ketone	–	r	n	–	–
Mineral oil	–	r	–	n	–
Motor oil	–	r	r	r	r
n-Hexane	–	r	r	r	r
Nitric acid	10	r/rr	n	r	r/n
Olive oil	–	r	r	r	r
Perchlorethylene	–	r	r	r	n
Petroleum	–	r	r	r	r
Phosphoric acid	25	r	r	r	r
Photographic fixing bath liquid	–	r	–	r	–
Potassium dichromate	10	r	r	r	r
Potassium hydroxide	10	rr/n	n	–	–
Potassium permanganate	10	r	r	r	r
Sodium bisulfite	10	r	r	r	r
Sodium carbonate	10	r	r	r	r
Sodium hydroxide	10	n	n	–	–
Sulfuric acid	10	r	r	r	r

Table 2.18 (Continued)

Chemical	Concentration (%)	PET 20 °C	PET 60 °C	PBT 20 °C	PBT 60 °C
Tetrahydrofuran	–	r	n	–	–
Tetralin	–	r	r	–	–
Toluene	–	r	r/rr	r/rr	n
Transformer oil	–	r	–	r	–
Trichloroethylene	–	r	n	n	n
Triethylene glycol	–	r	–	–	–
Xylene	–	r	r	r	–

r = resistant; weight increase up to 3% or weight loss up to 0.5% or reduction of tensile strength up to 15%
rr = resistant, but restrictions may apply; weight increase 3 to 8%, weight loss 0.5–3%, or reduction of tensile strength by 15–30%
n = nonresistant; weight increase of more than 8%, weight loss of more than 3%, or reduction of tensile strength by more than 30%
– = not tested

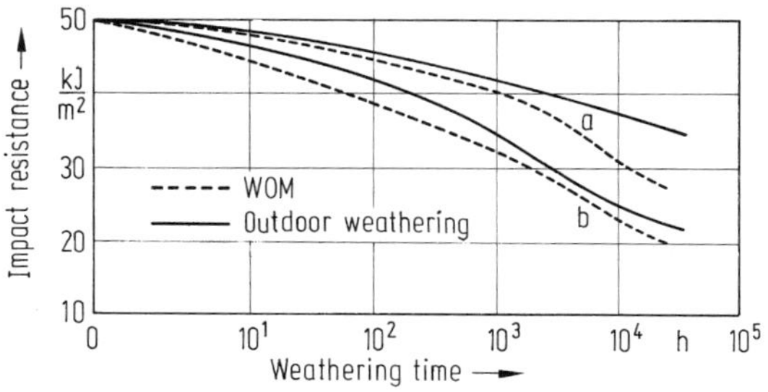

Figure 2.47. Resistance to weathering of polybutylene terephthalate reinforced with 30% glass fibers. Weathering took place in a Xenon Weather-O-Meter and in Europe climates (B) [13]: a, black samples; b, uncolored samples.

coefficients of friction for some polyethylene terephthalate products. For modified, glass fiber reinforced polybutylene terephthalate, the coefficient of friction (after ASTM D 1894) versus itself and versus steel lies between 0.17 and 0.20 [15].

Table 2.19. Coefficient of Friction of Polybutylene Terephthalate [13] (measured according to DIN 53375, against lapped steel)

PBT Product	23 °C		80 °C		120 °C		140 °C	
	μ_{stat}	μ_{dyn}	μ_{stat}	μ_{dyn}	μ_{stat}	μ_{dyn}	μ_{stat}	μ_{dyn}
Nonreinforced, high viscosity	0.125	0.101	0.078	0.061	0.073	0.047	0.089	0.046
30% glass fibers, standard adjustment	0.119	0.113	0.080	0.061	0.060	0.045	0.090	0.053
30% glass fibers, V0 adjustment	0.122	0.113	0.080	0.066	0.066	0.046	0.067	0.043

2.7.9 Abrasion Resistance

The abrasion was 63 mg for unmodified polyethylene terephthalate and 82 mg for a polyethylene terephthalate with 30% glass fibers under the same conditions after 500 cycles on the Taber abraser (DIN 53754). Applying ASTM D 1044, an abrasion of 15 mg was found after 1,000 cycles for polyethylene terephthalate reinforced with 30% glass fibers [21].

2.7.10 Water Absorption

Semicrystalline polyesters absorb only small amounts of water even if stored in water. Four-day absorption values are 0.1 to 0.2%; for glass fiber reinforced products, the values from DIN 53472 are even less than 0.1%. About half a year of storage underwater is necessary to obtain saturation. Under these conditions, unmodified polybutylene terephthalate absorbed 0.45% and unmodified polyethylene terephthalate absorbed 0.6%. Glass fiber reinforced polybutylene terephthalate absorbs less than 0.2% even after this long period of time (see Figures 2.48 and 2.49).

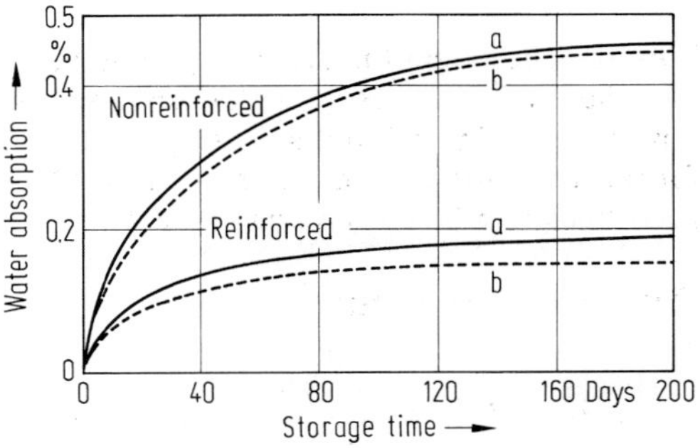

Figure 2.48. Water absorption of nonreinforced and reinforced polybutylene terephthalate under various conditions of storage [14]: a, storage in water at 23 °C; b, storage in standard climate (25 °C, 50% relative humidity).

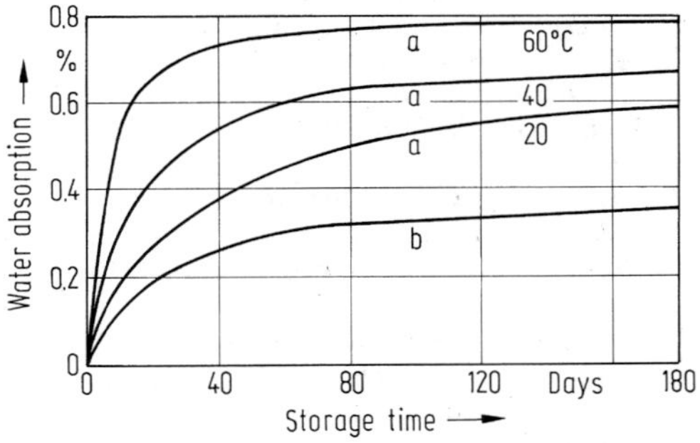

Figure 2.49. Water absorption of semictystalline, nonreinforced polyethylene terephthalate [17]: a, storage in water at different temperatures; b, storage in standard climate (25 °C, 50% relative humidity).

2.7.11 Permeation Behavior

Polyethylene terephthalate shows a higher water vapor permeability than polyolefins, but a lower one than is found for polycarbonates, polyamides, and polyoxymethylene. The permeability for carbon dioxide is similar to that of PVC-U (not chlorinated PVC),

Table 2.20. Permeation of Polyethylene Terephthalate [22]

Medium	Unit	Not oriented	Oriented
Oxygen	$\frac{cm^3 \cdot mm}{m^2 \cdot d \cdot bar}$	4–4.5	1.8–3.5
Carbon dioxide	$\frac{cm^3 \cdot mm}{m^2 \cdot d \cdot bar}$	10–11	5–8
Water vapor	$\frac{g \cdot mm}{m^2 \cdot d}$	3	0.9–2

but considerably lower than that of polyolefins. the permeability for oxygen is considerably lower than that of polyolefins (polypropylene) and about 30% lower than that of PVC-U (see Table 2.20) [22].

Films of polyethylene terephthalate are nearly impermeable to volatile aromatic ingredients. Therefore, combinations of polyethylene and polyethylene terephthalate films are used for packaging.

The permeability of polybutylene terephthalate is very similar to that of polyethylene terephthalate [14].

2.7.13 Flow Behavior

Polyethylene terephthalate and polybutylene terephthalate melts flow very well. Even complicated molded parts with deep flows can be easily produced. The flow length in the flow spiral at 260 °C and a cross section of 2 mm × 6.7 mm is between 330 and 600 mm for most commercial and experimental PBT products, including elastomer-modified and glass fiber reinforced materials (see Table 2.14).

References for Section 2.7

1. *Chang, E. P.*, et al.: Polym. Eng. Sci. 18 (1978) p. 933
2. *Bornschlegl, E.; Bonart, R.*: Colloid Polym. Sci. 258 (1980) p. 319
3. *Lu, F. M.; Spruiell, J. E.*: J. Appl. Polym. Sci. 31 (1986) p. 1595
4. *Hall, I. H.; Pass M. G.*: Polymer 17 (1976) p. 807
5. *Yokouchi, M.*, et al.: Macromolecules 9 (1976) p. 266
6. *Tashiro, K.*, et al.: Macromolecules 13 (1980) p. 137
7. *Jaquiss, D. B. G.; Borman, W. F. H; Campbell, R. W.*: Polyesters, Thermoplastic. In: *Mark, H. F.; Othmer, D. F.; Overberger, C. G.; Seaborg, G. T.* (Eds.): Encyclopedia of Chemical Technology. Vol. 18. Wiley, New York 1982, p. 549
8. *Leung, W. P.; Choy, C. L.*: J. Appl. Polym. Sci. 27 (1982) p. 2693
9. *Illers, K. H.*: Colloid Polym. Sci. 258 (1980) p. 117
10. *Jadhav, J. Y.; Kantor, S. W.*: Polyesters, Thermoplastic. In: *Mark, H. F.; Bikales, N. M.; Overberger, C. G.; Menges, G.* (Eds.): Encyclopedia of Polymer Science and Engineering. Vol. 12. Wiley, New York 1988, p. 217
11. *Groeninckx, G.*, et al.: J. Polym. Sci., Polym Phys. Ed. 14 (1976) p. 591
12. *Rao, M. V. S.; Dweltz, N. E.*: J. Appl. Polym. Sci. 31 (1986) p. 1239
13. Pocan—thermoplastischer Polyester, company brochure Bayer AG 1987
14. *Domininghaus, H.*: Polybutylenterephthalat. In: Die Kunststoffe und ihre Eigenschaften. 2. Ed. VDI-Verl., Düsseldorf 1986, p. 507
15. Rynite, thermoplastische Polyester. Company brochure Du Pont 1986
16. *Schroeder, H.; Cella, R. J.*: Polyesters, Elastomeric. In: [10], p. 75
17. *Domininghaus, H.*: Polyethyleneterphthalat. In: [14], p. 492
18. *Breitenfellner, F.*: Kunststoffe 66 (1976) 10, p. 610
19. *Carlowitz, B.*: Polyethylenterephthalat un Polybutylenterephthalat. In: Thermoplastische Kunststoffe. Zechner & Hüthig, Speyer 1980, p. 105
20. *Irish, J. R.*: Polybutylene Terephthalate. In: *Margolis, J. M.* (Ed.): Engineering Thermoplastics. Dekker, New York 1985, p. 19
21. *Bier, P.; Ong, G. N.*: Kunststoffe 71 (1981) p. 573
22. *Domininghaus, H.*: Polyethylenterephthalat als Barriere-Kunststoff. In: [14], p. 525

2.8 Analysis

Thermoplastic polyesters are investigated with respect to product quality by the polyester manufacturers, and they are checked before and after processing with respect to specifications. The most important factors are the chemical analysis of the composition and the physical analysis, with emphasis on molecular weight, morphology, and thermal transition temperatures. Additives can be determined by either chemical or physical methods.

2.8.1 Chemical Composition

The chemical composition of a polymer is usually known from the manufacturer. If that is not the case, if there are copolyesters with variable compositions, or if side reactions that occur during polycondensation are to be investigated, it is necessary to determine the composition independent of the manufacturing process.

2.8.1.1 Diols and Dicarbonic Acids

Polyesters can be easily hydrolyzed in boiling ethanolic potassium hydroxide. Potassium terephthalate is of low solubility and precipitates. It is determined gravimetrically [1]. The dicarbonic acids produced by hydrolysis can be extracted after acidification of the hydrolysate and transformed into the corresponding dimethylesters via reaction with diazomethane. Then they can be separated by gas chromatography and determined quantitatively [2].

A quick test for the dicarbonic acids and the diols works as follows: hydrolysis of the polyesters with potassium hydroxide in 2-ethoxyethanol, acidification of the hydrolysate, solution of the dicarbonic acids in pyridine, and silylation with *N,O*-bis(trimethylsilyl) trifluoroacetamide, followed by gas chromatographic separation of the trimethylsilylethers of the diols and the trimethylsilylesters of the dicarbonic acids [3]. A gas chromatographic determination of the dimethylesters of dicarbonic acids is also possible after methanolysis of the polyesters under pressure at 220 °C [4] or after methanolysis with sodium methanolate/acetic acid methylester in methanol and subsequent extraction with benzene/diphenyl ether [5].

After polyesters have been degraded in hydrazine, the dicarbonic acids can be determined as hydrazides by polarography [6] and the diols can be separated by gas chromatography [2, 7]. The content of diethylene glycol units as a typical defect structure in polyethylene terephthalate can be determined by gas chromatography after various pretreatments: ethanolysis under pressure at 25 °C [8]; hydrolysis with methanolic sodium hydroxide [9], potassium hydroxide in 2-ethoxyethanol [3], or barium hydroxide [10]; pressure hydrolysis [11]; or reaction with hydrazine [7, 12–14]. Small concentrations of diethylene glycol units can be detected by proton NMR spectroscopy (400 MHz) of the polyester [15].

In copolyesters the average length of the sequences and the degree of order are determined by high resolution proton NMR [16] or by ^{13}C NMR spectroscopy [17].

2.8.1.2 Oligomers

During both the polycondensation of and the melting of polyesters, linear and cyclic oligomers are formed (see Section 2.3.2.6). There are different methods of determining the oligomer content in polyesters after extraction or selective precipitation from solution. The methods for polybutylene terephthalate are compiled in Table 2.21 [18–23]; those for polyethylene terephthalate are found in Table 2.22 [24–31].

Table 2.21. Analysis of Oligomers in Polybutylene Terephthalate
GPC, gel permeation chromatography; HPLC, high performance liquid chromatography; TLC, thin-layer chromatography

Isolation of Oligomers	Analysis	References
Extraction with methylene chloride or methylene chloride/chloroform	GPC	18
Synthesized oligomers	HPLC (adsorption with tetrahydrofuran/heptane/acetic acid, reversed phase chromatography with methanol/water)	19
Precipitation from solution in *o*-dichlorobenzene by cooling and use of chloroform or precipitation from trifluoroacetic acid	HPLC (adsorption with hexane/dioxane) or TLC (with hexane/chloroform 10 : 5 and chloroform/hexane 10 : 3)	20
	^{13}C-NMR analysis (solid state)	21
Synthetic oligomers	HPLC (reversed phase chromatography with methanol/water) or TLC	22
	Identification by field desorption mass spectroscopy	23

Table 2.22. Analysis of Oligomers in Polyethylene Terephthalate

Isolation of Oligomers	Analysis	References
Extraction with xylene or dioxane or precipitation from cooling in dimethylformamide by cooling	TLC column chromatography	24
Extraction with dioxane	TLC	25
Esterification of terephthalic acid with ethylene glycol	GPC of the transesterification product	26
Extraction with chloroform or precipitation from solution in trifluoracetic acid with chloroform/acetone	TLC, HPLC (adsorption with chloroform/hexane/isopropanol)	27
Extraction with chloroform	GPC, including preparative GPC, identification by melting point, IR or mass spectroscopy	28
Extraction with chloroform or precipitation from dimethylformamide solution by cooling	TLC	29
Precipitation or extraction	HPLC (adsorption with methylene chloride/heptane and methylene chloride/dioxane)	30
Extraction with chloroform	HPLC (adsorption with hexane/dioxane)	31

2.8.1.3 End Groups

If transesterification is incomplete, methoxycarbonyl end groups may remain in the polyester. These can be determined by gas chromatography after reaction with hydrazine [32]. There are various methods for the determination of hydroxy end groups:

- acylation with 3,5-dinitrobenzoyl chloride in hot nitrobenzene, hydrolysis of the unreacted benzoyl chloride, and determination of the residual dinitrobenzoic acid so produced [33]
- acylation with benzoyl chloride in quinoline and nitrobenzene, saponification of the benzoic acid ester, and determination of the benzoic acid by UV spectrosocpy [34]
- acylation with either sulfobenzoic acid cyclic anhydride or succinic anhydride, followed by acidimetric titration of the consumption of that reagent [35, 36]
- acylation with dichloroacetic acid anhydride and determination of the chlorine content by x-ray fluorescence [37]

There are also various methods to determine carboxyl end groups. Usually alkalimetric titrations are predominant. Examples include titration of polyester solutions in hot mixtures of these compounds:

- benzyl alcohol/chloroform with sodium hydroxide in benzyl alcohol [38]
- nitrobenzene/isopropanol with potassium hydroxide in isopropanol [39]
- aniline with methanolic sodium hydroxide [40, 41]
- *o*-cresol/chloroform with alcoholic potassium hydroxide [12, 42–44] or with tetrabutylammonium hydroxide in aniline [45]
- phenol/chloroform with potassium hydroxide in benzyl alcohol [46]
- 2,6-dimethylphenol/chloroform with ethanolic potassium hydroxide [47]

The content of carboxylic end groups can also be determined by UV spectroscopy at 240 nm after reaction with hydrazine [48]. IR spectroscopy methods have also been described [49–51].

Vinyl ester end groups which are produced by thermal polyester cleavage can be determined Coulometrically after bromination of the polyester solution in dichloroacetic acid that contains potassium bromide, water, and mercury(II) chloride [52].

2.8.1.4 Other Analyses

Acetaldehyde, which is formed during the thermal decomposition of polyethylene terephthalate, can be determined by gas chromatography after programmed evaporation [46, 53].

2.8.2 Molecular Weight

The molecular weight, which highly influences the processibility of polyester molding parts and the mechanical properties of parts made from these items, is usually determined indirectly by solution viscosity using the Mark–Houwink relation, which describes the dependence of the intrinsic viscosity [η] on the molecular weight M.

$$[\eta] = K \cdot M^{\alpha} \tag{2.22}$$

For polybutylene terephthalate and polyethylene terephthalate, Mark–Houwink plots have been published for various solvents (see Table 2.23) [36, 46, 49, 54–58]. In these cases, the molecular weights have been determined by the end groups (number average, $\bar{M}_n$) or by light scattering (weight average, $\bar{M}_w$).

Since absolute molecular weights (weight average, $\bar{M}_w$) could be determined for fractionated polyester samples by light scattering in 1,1,1,3,3,3-hexafluoro-2-hydroxypropane, the Mark–Houwink relations based on these data are very accurate. Plots of such Mark–Houwink relations based on absolute molecular weights are given in Figure 2.50.

Table 2.23. Values for the constants K and α in the Mark–Houwink Relation

Polyester	Solvent	Temperature (°C)	$K \times 10^4$	α	Molecular Weight Average	Reference
Polybutylene terephthalate	Phenol/tetrachloroethane (3 : 2)	30	1.666	0.871	$\bar{M}_n$	[54]
	Phenol/*o*-dichlorobenzene (1 : 1)	25	4.37	0.72	$\bar{M}_w$	[55]
	m-cresol	100	1.09	0.847	$\bar{M}_n$	[54]
Polyethylene terephthalate	*o*-chlorophenol	25	3.0 ± 1.2	0.77 ± 0.09	$\bar{M}_n$	[49]
	Trifluoroacetic acid	30	4.33	0.68	M_w	[56]
	Phenol/tetrachloroethane (3 : 5)	30	2.29	0.73	$\bar{M}_w$	[46]
	Phenol/tetrachloroethane (1 : 1)	25	2.1 ± 0.3	0.82 ± 0.03	$\bar{M}_n$	[36]
	Phenol/tetrachloroethane (1 : 1)	20	7.55	0.685	$\bar{M}_n$	[57]
	Phenol/tetrachloroethane (3 : 2)	25	3.72	0.73	$\bar{M}_n$	[58]
	Phenol/*o*-Dichlorobenzene (1 : 1)	25	4.69	0.68	$\bar{M}_w$	[55]

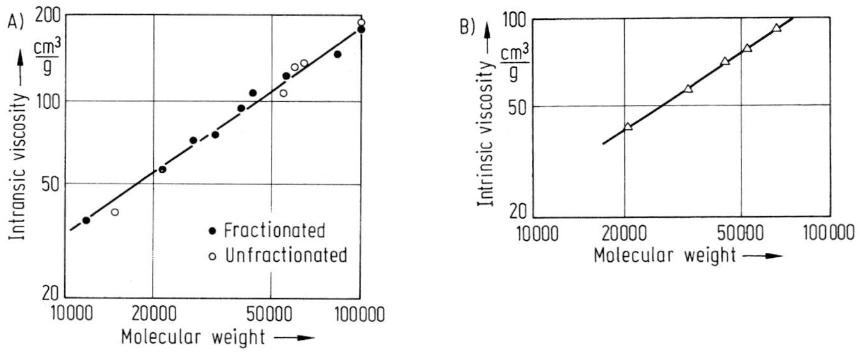

Figure 2.50. Mark–Houwink relationship for polybutylene terephthalate (A) and polyethylene terephthalate (B) (1 : 1 phenol/*o*-dichlorobenzene, 25 °C) [55].

A similar relationship exists for the melt viscosity and molecular weight. Equation 2.23 holds true for polybutylene terephthalate at a shear stress $\tau < 10^{-1}$ N/mm² at 250 °C [54]:

$$\eta_S = 1.85 \cdot 10^{-16} \cdot \bar{M}_n^{4,8}. \tag{2.23}$$

The corresponding relationship for polyethylene terephthalate at a shear stress $\tau < 10^{-1}$ N/mm² at 280 °C [59] is given in equation 2.24:

$$\eta_S = 1.77 \cdot 10^{-10} \cdot \bar{M}_n^{3,5}. \tag{2.24}$$

By knowing the relation between melt index and intrinsic viscosity, one can establish a relation between melt index and molecular weight [60].

Besides the molecular weight, the molecular weight distribution is an important figure of merit, characterizing the quality of polyesters during production as well as during processing. It is generally obtained by gel permeation chromatography (GPC).

Table 2.24. Determination of Molecular Weight Distributions by GPC

Polyester	Column	Elution	Temperature (°C)	Reference
Polybutylene terephthalate	Polystyrene gel	Hexafluorisopropanol	20	61
	Polystyrene gel	*m*-Cresol	105	62
	Polystyrene gel	*m*-Cresol	100	54
Polyethylene terephthalate	Polystyrene gel	*m*-Cresol	120	58
		Nitrobenzene/tetrachloroethane (0.5 : 99.5)	20	63, 64
	Polystyrene gel	*o*-Chlorophenol/chloroform (25 : 75)	25	65
		Methylene chloride/hexafluoro isopropanol (70 : 30)		66
		Hexafluorisopropanol	23	67
		Hexafluorisopropanol/ Pentafluorophenol	23	67
	Silica gel	0.01 M sodium trifluoroacetate solution in hexafluoroisopropanol		68

Table 2.24 summarizes some GPC methods, with data on columns and eluants for polybutylene terephthalate and polyethylene terephthalate [54, 61–68].

2.8.3 Additives

Different additives require the application of specific methods for qualitative and quantitative determinations. Inorganic additives are determined from the residues of ashing. Atomic absorption spectroscopy (AAS) as well as the related techniques of atomic emission (AE) spectroscopy and atomic fluorescence (AF) spectroscopy are well suited for detecting traces of metals in organic and inorganic additives. AAS may be best suited for quantitative trace analysis of metals down to the ppm or even ppb level [69].

References for Section 2.8

1. *Schröder, E.; Franz, J.; Hagen, E.:* Ausgewählte Methoden zur Plastenanalytik. Akademie-Verl., Berlin 1976
2. *Valk, G.; Stein G.:* Die alkalische Hydrolyse physikalish und chemisch modifizierter Polyesterfasern. Forschungsber. Nordrhein-Westfalen Nr. 2612. Westdeutscher Verl., Opladen 1977
3. *Allen, B. J.*, et al.: Anal. Chem. 49 (1977) p. 741
4. *Tuček, E.; Dinse, H. D.:* Acta Polym. 34 (1983) p. 41
5. *Jankowski, S. J.; Garner, P.:* Anal. Chem. 37 (1965) p. 1709
6. *Berger, W.; Ewert, K. H.:* Faserforsch. Textiltech. 24 (1973) p. 377
7. *Haken, J. K.:* Progr. Org. Coat 7 (1979) p. 209
8. *Janssen, R.*, et al.: Makromol. Chem. 77 (1964) p. 153
9. *Gaskill, D. R.*, et al.: Anal. Chem. 39 (1967) p. 106
10. *Glamann, H.; Woitzik, J.:* Faserforsch. Textiltech. 25 (1974) p. 86
11. *Ponder, L. H.:* Anal. Chem. 40 (1968) p. 229
12. *Hovenkamp, S. G.; Munting, J. P.:* J. Polym. Sci., Polym. Chem. Ed. 8 (1970) p. 679

13. *Dinse, H.; Tuček, E.:* Faserforsch. Textiltech. 21 (1970) p. 205
14. *Stein, G.; Dugal, S.:* Melliand Textilber. 55 (1974) p. 713
15. *Kricheldorf, H. R.*, et al.: Polym. Bull. 4 (1981) p. 547
16. *Yamadera, R.; Murano, M.:* J. Polym. Sci., Polym. Chem. Ed. 5 (1967) p. 2259
17. *Kricheldorf, H. R.:* Makromol. Chem. 179 (1978) p. 2133
18. *Burzin, K.*, et al.: Makromol. Chem. 74 (1978) p. 93
19. *Hässlin, H. W.*, et al.: Makromol. Chem. 181 (1980) p. 301
20. *East, G. C.; Grishab, A. M.:* Polym. Commun. 23 (1982) p. 323
21. *Havens, J. R.; König, J. L.:* Polym. Commun. 24 (1983) p. 194
22. *Müller, F. J.*, et al.: Makromol. Chem. 184 (1983) p. 2487
23. *Pleskova, A. P.; Fajdel, G. I.:* Acta Polym. 37 (1986) p. 43
24. *Peebles, L. H.*, et al.: J. Polym. Sci., Polym. Chem. Ed. 7 (1969) p. 479
25. *Valk, G.*, et al.: Melliand Textilber. 51 (1970) p. 504
26. *Minárik, M.*, et al.: Angew. Makromol. Chem. 64 (1977) p. 147
27. *Hudgins, W. R.*, et al.: J. Appl. Polym. Sci., Appl. Polym. Symp. 34 (1978) p. 145
28. *Shiono, S.:* J. Polym. Sci., Polym. Chem. Ed. 17 (1979) p. 4123
29. *Ha, W. S.; Choun, Y. K.:* J. Polym. Sci. Polym. Chem. Ed. 17 (1979) p. 2103
30. *Wick, G.; Zeitler, H.:* Angew. Makromol. Chem. 112 (1983) p. 59
31. *Cimecioglu, A. L.*, et al.: J. Appl. Polym. Sci. 32 (1986) p. 4719
32. *Vink, D.; Wijk, R. van:* Z. Anal. Chem. 264 (1973) p. 293
33. *Zimmermann, H.; Tyronadt, A.:* Faserforsch. Textiltech. 18 (1967) p. 487
34. *Challa, G.:* Makromol. Chem. 38 (1960) p. 105
35. *Zimmermann, H; Kolbig, C.:* Faserforsch. Textiltech. 18 (1967) p. 536
36. *Conix, A.:* Makromol. Chem 26 (1958) p. 226
37. *Houwelingen, G. D. B. van*, et al.: Z. Anal. Chem. 293 (1978) p. 396
38. *Pohl, H. A.:* Anal. Chem. 26 (1954) p. 1614
39. *Hendrix, W.:* Z. Ges. Textilind. 66 (1964) p. 937
40. *Urbanski, J.*, et al.: Handbook of Analysis of Synthetic Polymers and Polymers and Plastics. Horwood, Chichester 1977
41. *Majewska, J.; Warzywada, J.:* Melliand Textilber. 43 (1962) p. 480
42. *Van Lingen, R. L. M.:* Z. Anal. Chem. 247 (1969) p. 232
43. *Dinse, H. D.; Tuček, E.:* Acta Polym. 31 (1980) p. 108
44. *Rafler, G.*, et al.: Acta Polym. 30 (1980) p. 599
45. *Houwelingen, G. D. B. van:* Analyst 106 (1981) p. 1057
46. *Chang, S.*, et al.: J. Appl. Polym. Sci. 28 (1983) p. 3289
47. *Garmon, R. G.*, in: *Slade, P. E.:* Polymer Molecular Weights: End Group Determinations. P. 1. Dekker, New York 1975
48. *Nissen, D.*, et al.: J. Appl. Polym. Sci. 18 (1974) p. 1953
49. *Ward, I. M.:* Trans. Faraday Soc. 53 (1957) p. 1406
50. *Patterson, D.; Ward, I. M.:* Trans. Faraday Soc. 53 (1957) p. 291
51. *Addleman, R. L.; Zichy, V. J. I.:* Polymer 13 (1972) p. 391
52. *Aalbers, J. G. M.; Houwelingen, G. D. B. van:* Z. Anal. Chem. 314 (1983) p. 472
53. *Dong, M.*, et al.: J. Chromatogr. Sci. 18 (1980) p. 242
54. *Borman, W. F. H.:* J. Appl. Polym. Sci. 22 (1978) p. 2119
55. *Horbach, A.*, et al.: Angew. Makromol. Chem. 98 (1981) p. 35
56. *Wallach, M. L.:* Makromol. Chem. 103 (1967) p. 19
57. *Koepp, H. M.; Werner, H.:* Makromol. Chem. 32 (1959) p. 79
58. *Hergenrother, W. L.; Nelson, C. J.:* J. Appl. Polym. Sci., Polym. Chem. Ed. 12 (1974) p. 2905
59. *Gregory, D. R.:* J. Appl Polym. Sci. 16 (1972) p. 1479
60. *Garske, D.; Kaufmann, W.:* Kunststoffe 68 (1978) p. 87
61. *Slagowski, E. L.*, et al.: J. Appl. Polym. Sci. 21 (1977) p. 2293
62. *Müller, F. J.*, et al.: Angew. Makromol. Chem. 83 (1979) p. 171
63. *Paschke, E. E.*, et al.: J. Appl. Polym. Sci., Polym. Chem. Ed. 15 (1977) p. 983
64. *Ponder, L. H.:* Am. Chem. Soc., Div. Polym. Chem., Preprints 21 (1980) 1, p. 169
65. *Jabarin, S. A.; Balduff, D. C.:* J. Liquid Chromatogr. 5 (1982) p. 1825
66. *Overton, J. R.; Browning, H. L.:* Org. Coat. Appl. Polym. Sci. Proc. 4 (1983) p. 940
67. *Berkowitz, S.:* J. Appl. Polym. Sci. 29 (1984) p. 4353
68. *Naoki, M.*, et al.: J. Appl. Polym. Sci., Polym. Phys. Ed. 23 (1985) p. 2567
69. *Möhler, H.:* Kunststoffe 77 (1987) 4, p. 395

2.9 Toxicology

Polyethylene terephthalate and polybutylene terephthalate are physiologically inert [1, 2]. In Germany, there are no reservations or restrictions against the manufacture and use of polyterephthalate diol ester products with respect to federal law regulating food and general supplies (Section, 5, Chapter 1, Number 1). Recommendation XVII of the committee of the national health administration (BGA) communicates requirements for purity [3]. The U.S. FDA has regulated the use of polyesters for food packaging in the United States [4–7].

Oligomers of polyethylene terephthalate do not show any mutagenic activity [8]. Polybutylene terephthalate implants did not show any inflammatory reaction even over an extended period of time. Body fluids slowly degraded the polyester [9].

Because there are no restrictions stemming from toxicological properties, polybutylene terephthalate is used extensively for household supplies that may come into contact with food, whereas polyethylene terephthalate is the dominant form used in materials for the packaging of food.

Most thermoplastic polyesters can be processed under normal precautions. Local exhaust may be necessary while processing some polyesters, however, because volatile components of certain flame retardants may be respiratory tract irritants. Disposable gloves should be used for frequent handling of still-warm parts made of such products—for example, in finishing or packaging operations [10].

Local exhaust is always needed if superheating (e.g., during an interruption of production with filled cylinder) or burning (e.g., cleaning of tools in an oven) occurs [10].

References for Section 2.9

1. *Domininghaus, H.:* Polyethylenterephthalat. In: Die Kunststoffe und ihre Eigenschaften. 2nd ed. VDI-Verl., Düsseldorf 1986, p. 493; and Polybutylenterephthalat, p. 507
2. *Brochhagen, F.:* Toxikologische Relevanz von Polykondensaten und deren Ausgangsprodukten. In: *Bartl, H.; Falbe, J.* (Eds.): Methoden der organischen Chemie (Houben-Weyl). Vol. E 20, Part 3. Thieme, Stuttgart 1987, p. 2324
3. BGBl. 27 (1984) p. 289 and BGBl. 30 (1987) p. 3
4. Code of Federal Regulations, Title 21, Subpart F, Regulation 121. 1958, p. 2524
5. Code of Federal Regulations, 21 CFR 177.1630
6. Code of Federal Regulations, 21 CFR 177.1660
7. Code of Federal Regulations, 21 CFR 177.1315
8. *Nowak, E. N.*, et al.: SPE Ann. Tech. Conf. Preprints 38 (1980) p. 543
9. *Montgomery, R. R.:* In: Patty's Industrial Hygiene and Toxicology. 3rd ed. Vol. II c. Wiley, New York 1982, p. 4431
10. Arnite, Thermoplastischer Polyester. Company brochure Akzo Plastics 1987

2.10 Processing

Thermoplastic polyester is easily processed in many ways; it can be injection-molded into molded parts; extruded into profiles, sheets, film, monofilaments, and fibers; or blow-molded for hollow parts.

Polybutylene terephthalate is used primarily for injection molding, whereas polyethylene terephthalate is used primarily for extrusion, and in the last few years for blow molding.

2.10.1 Injection Molding Processing

2.10.1.1 Raw Material Preparation

Thermoplastic polyester absorbs only small quantities of water, even when immersed (see Section 2.7.10). At processing temperatures, however, this behavior changes [1] (see also Section 2.4.1.3). A water content of less than 0.05% for PBT and less than 0.02% for PET is recommended for processing [2–7]. Therefore, most polyester producers sell their products as dried granules packaged in moisture-proof bags.

To avoid condensation of humidity on the cold surface of the granules, the bags should be stored in the processing area for 12–24 hours and opened just prior to processing. If exposure to ambient air will be longer (about 8 hours for PBT, or 5–6 hours for PET), predrying before processing is a necessity. Predrying in general is recommended for the manufacture of small precision parts [3]. A drying time of 120 °C and approximately 3 cm layer thickness of material for 2–3 hours in dry air or desiccant dryers, or 3–4 hours in forced-air dryers is recommended [4]. Longer drying times, up to 12 hours, are necessary for certain applications.

The hopper of the injection molding machine should be heated to prevent the hot resin exiting the dryer from cooling and allowing condensation. The hopper should be closed during processing to avoid the absorption of humidity.

It is possible to use a dry air or desiccant dryer as a hopper dryer. Alternatively, a separate large container with connecting tubing to the injection molding machine may be used.

Today's technology is advanced to the point that ventilation units on injection molding machines are no longer adequate substitutes for predrying.

Polyester resin is usually sold in tinted form. However the processor can tint the colorless material in house by pigment coating in drum mixers. Examples include the addition of 0.2% Panier oil with a dye concentrate (master batches) or liquid pigment dispersion [2].

2.10.1.2 Injection Molding Machines

2.10.1.2.1 Screw Injection Molding Machines

Polybutylene terephthalate and polyethylene terephthalate can be processed on all injection molding machines equipped with plastifying screws.

Cylinder, Screws, and Residence Time

Screws with an L/D ratio between 17 and 23 and a flight/depth ratio of 1:2 allow good homogenization/mixing and careful processing [2]. Especially suitable are standard three-zone screws. To prevent backflow of the melt during injection and the application of pressure, a nonreturn flow valve is needed.

To minimize thermal degradation, the machine output and cylinder diameter should be selected so that the melt volume is 1.2 to a maximum of 4 times the shot volume (i.e. the volume of the parts and sprues combined); such a choice will keep the residence time in the cylinder as short as possible. The residence time should not exceed 5–10 minutes, depending on the product composition [2, 8, 9].

Nozzles

Processing can be done with open nozzles and pressure discharge controls. Short nozzles with an opening of 3 mm or more minimize heat due to friction, as well as pressure loss.

This is especially true for glass fiber reinforced, filled, and/or flame retardant products. Shutoff nozzles, especially those with complicated flow channels, require an effective and accurate heating element. After a long period of molding glass fiber reinforced parts, cylinder and nozzles should be purged with plain resin to avoid deposits of glass fibers upon reheating.

With the exception of hot runner systems, the nozzle should be detached from the mold after the injection–holding pressure period to avoid rapid cooling and freeze-up [2, 3].

Cylinder Temperature

The best melt temperature for unreinforced polybutylene terephthalate is 240–260 °C. Reinforced resin should be processed at a temperature about 10 °C higher. Flame-retardant grades, should, if at all possible, be processed below 260 °C. Upper temperature limits should not be exceeded, even when using hot runner systems. To ensure exact temperature settings, it is recommended that the actual melt temperature be measured at the nozzle exit and the temperature settings adjusted accordingly.

Lower cylinder temperatures are selected when reduced melt throughput and longer cycle times are used to minimize heat loss [2–4, 8–10].

The melt temperature for polyethylene terephthalate is higher because of the higher crystalline melting temperatures of this material. The temperature limits are between 265 and 295 °C. The faster crystallizing, modified resins can be molded at lower temperatures: 260–280 °C [2–5, 7].

The melt temperatures for elastomer-modified polybutylene terephthalate should be 240–260 °C, with a maximum residence time in the cylinder and the hot runner system of 6 minutes [11].

Injection Speed

Polybutylene terephthalate and most polyethylene terephthalate resins have a very fast speed of solidification and crystallization. On one hand, this allows very fast processing with very short cycle times; but on the other hand, the injection speed, especially for thin-walled parts, must be very high indeed, to avoid premature cooling and melt freeze/solidification during the injection process. This rapid solidification and crystallization improves the surface, especially of glass fiber reinforced parts, considerably.

A high injection speed requires thorough ventilation of the mold, since pressurized air can cause burns. The ventilation channels must be enlarged and/or the number increased. The injection speed for thick-walled parts and unreinforced resin can be reduced to avoid air inclusion [2–4, 8, 9].

Injection Pressure and Back Pressure

The injection pressure is largely dependent on the wall thickness of the part, which in turn determines the length and behavior of the flow. The injection pressure must be set to ensure uniform filling of the mold. Normally this calls for 1300 bar; 1500–1800 bar must be used for elastomer-modified or large-surface PBT parts.

To avoid flashing of parts, or damage to the mold, switching from injection pressure to back pressure must be accurate.

The secondary pressure is about 40–70% less than the injection pressure, and it should be held until the melt has solidified. Insufficient secondary pressure time results in sink marks and voids. Longer secondary pressure time is needed for larger wall thicknesses and larger gates [2–5, 9, 11]. A secondary pressure that is too long or too high increases stress in the molded part and reduces transparency and impact strength [2].

Back Pressure and Screw Speed

Back pressure, which enhances a homogeneous melt, should be set to ensure a uniform melt, free of air bubbles, and a uniform part weight. Generally 3–6 bar is adequate. The screw rpm should be set to achieve a plastifying time that is just within the cooling time.

Back pressure and rotational screw speed should be set as low as possible to avoid heat due to friction or, with glass fiber reinforced products, excessive fiber shortening [2, 3, 9].

Clamping Force

Thermoplastic polyesters have low viscosity with good flow characteristics. A strong clamping force is needed to avoid flashing of parts.

2.10.1.2.2 Injection Molding Machines with Ventilation Units

Polybutylene terephthalate can be processed on injection molding machines with ventilation units. However, the design and the length of screw, that are necessary, require a longer melt residence time in the cylinder. Ventilation units are not recommended for polyethylene terephthalate [2]. It is important to have an optimum ratio of shot volume to part weight to prevent excessive thermal degradation [2, 3].

2.10.1.2.3 Piston Injection Molding Machines

Piston injection molding machines are unsuitable for processing. To offset the heat transfer from the cylinder wall to the melt, a higher cylinder temperature is necessary, which can result in excessive degradation of the polyester. Also, because of lack of homogenization, optimized molded part properties are not achievable [2, 3, 10].

2.10.1.2.4 Mold Temperature

Maximum property qualities of PBT and PET parts, with reduced postmold shrinkage, are achieved only after sufficient crystallinity has been reached. "Sufficient crystallinity" here refers to the degree of crystallinity and the morphological structure.

Fast-crystallizing PBT can be processed at mold temperatures of 30–130 °C. Thin-walled parts require higher mold temperatures. Generally, temperatures of 60–80 °C are adequate to produce parts with minimal postmold shrinkage. Higher mold temperatures are required for difficult-to-fill parts or for precision parts, particularly those made out of unreinforced material [2, 3, 8, 10] for use at elevated temperatures.

PET generally should be molded at a temperature of 130 °C or higher to obtain parts with high rigidity and a smooth/shiny surface that can be demolded without deformation. If mold temperatures are below 130 °C, PET parts do not reach complete crystallization. The partially amorphous morphology that results can lead to demolding problems and part deformation [2].

Amorphous, totally transparent parts with a maximum wall thickness of 5 mm can be made from certain PET resins by rapid cooling of the melt at mold temperatures of 10–20 °C [2].

Specially modified PET resins can also be molded into partially crystallized parts at 120 °C mold temperature [5, 7, 12].

2.10.1.2.5 Interruptions During Processing and Resin Change

If there are multiple interruptions in cycles during processing prior to restarting, a melt remaining in the plastification unit must be removed and the cylinder purged with new resin. If processing is interrupted for an extended period, melt remaining in dead spaces of hot runner molds can be removed only by prolonged purging. Prepurging with high density polypropylene or polyethylene followed by temperature reduction of the hot runner molds is recommended. Purging with polypropylene or polyethylene is also recommended when changing to polyester from a thermoplastic resin with a considerably higher processing temperature (> 300 °C), especially flame-retardant grades [2, 3, 8–10].

2.10.1.3 Molds

A good mold design is a must for perfect injection molding and perfectly molded parts. A molded part with uniform wall thickness ensures uniform mold shrinkage and prevents distortion. Sharp-edged notches and level changeovers should be avoided because they will affect the strength of the part [2, 8].

2.10.1.3.1 Runners and Gates

Careful consideration must be given to the location of the gate to avoid distortion or warpage due to anisotropic shrinkage. The preferred location for the gate is at the largest wall thickness of the part. This ensures a long and uniform flow pattern. Also, glass fiber reinforced resin fills best into a straight-lined mold cavity. Therefore, plaques should be molded with a centrally located sprue, but with a diaphragm gate. The runners should be round or rounded off with a trapezoid cross section.

Molding with a sprue is of great advantage for rotation–symmetrical and thick-walled, high quality, dimensionally stable parts because the back pressure is on during the entire cooling/solidification cycle. Molding with restricted gates at the parting line is advantageous for large parts and for parts with wall thicknesses exceeding 1.5 mm. This allows for effective back pressure and fast filling of the cavity. Smaller parts, having a wall thickness of less than 2 mm, mold well with a tunnel runner and a pinpoint gate. Central injection of flat or cylindrical shaped forms in a multicavity mold can also be done with a pinpoint gate.

Rotation–symmetrical parts with strict, limited tolerances and high demands on mechanical strength are produced by uniform, symmetrical filling of the mold cavity, without interfering mold seams, using a diaphragm gate or a fan gate. Custom-shaped molds can also be injected on or in a mole. Dimensional accuracy is obtained through uniform back pressure. Smaller rotation–symmetrical parts are injected by using multicavity molds with ring gates. Seam formation can be prevented by reducing the size of the cross section [2, 3, 9, 13].

2.10.1.3.2 Mold Ventilation

Thorough ventilation of the mold is very important. Without ventilation the mold-filling process would be slowed and parts could sustain localized burns. The high speed needed for injecting high viscosity polyesters and especially glass fiber reinforced resins call for a tight seal on the mold separating lines. Therefore, ventilation must take place via air channels located at the end of the flow path, and air must be able to escape along seam joints. The vent channels should be 1.5–5 mm wide and 0.01–0.2 mm deep, depending on mold volume, length of flow path, and product type.

If it is not possible for air to escape at the parting line at the seam, injector pins with greater tolerances can be inserted at such locations. The air can then escape through the gap between the pin and the hole [2, 3, 8, 13].

2.10.1.3.3 Mold Temperature Controls

Molded parts with optimum properties can be achieved only by using exacting mold temperatures and cycle times. To ensure these exact temperatures and times, effective mold temperature controllers are necessary. This requirement must be taken into consideration when designing the mold. It is important that a uniform temperature be retained throughout the entire mold to ensure uniform cooling of the molded part. This can be accomplished by means of an adequate number of well-designed cooling channels around, and as close as possible to, the cavity. High mold temperatures of 130 °C can be reached with closed-loop circulating oil or water temperature controllers. To keep the mold at a constant temperature, especially at high temperatures, pressure-resistant insulating plates 10–12 mm thick should be used to isolate the mold from the heat of the injection molding machine.

Larger inserts must be prewarmed to the mold temperature before molding to prevent cracks in the plastic part due to inadequate weld deposition at lines and increased postmold shrinkage when secondary processing is carried out [2, 3, 13].

2.10.1.3.4 Demolding

Trouble-free demolding, especially for parts made from unreinforced polyethylene terephthalate, is accomplished with an adequate number of ejectors or ejector plates with accurately dimensioned ejector surfaces [2].

2.10.1.3.5 Shrinkage, Postmold Shrinkage, and Deformation

Shrinkage (S) expressed as a percentage is calculated by determining the difference between a characteristic length (l_W) of the mold cavity and the corresponding length (l_F) of the cooled molded part at 20 °C ($\pm 2°$) after 24 hours (± 4 h):

$$S = \frac{l_W - l_F}{l_W} \cdot 100 \tag{2.25}$$

Only shrinkage due processing is measured. Postmold shrinkage (PS), expressed as a percentage is calculated by determining the difference between the length (l_{F1}) of the cooled part at 20 °C ($\pm 2°$) and the length (l_{F2}) after a specified storage time and a specified temperature [10].

$$PS = \frac{l_{F1} - l_{F2}}{l_{F1}} \cdot 100 \tag{2.26}$$

Shrinkage of PET that can be processed into amorphous parts is primarily caused by thermal contraction of the molded part during cooling to room temperature. For this reason, a shrinkage rate of 0.2–0.4% is not very high.

Because a high degree of crystallinity is obtainable with polyethylene terephthalate and polybutylene terephthalate, shrinkage of these materials is considerable. Shrinkage depends on the following factors:

Molded parts:

- wall thickness
- length of flow path
- type of runner system

Processing:
- gate size
- melt and mold temperatures
- injection speed
- back pressure and holding time

Polyesters:
- composition
- glass fiber orientation

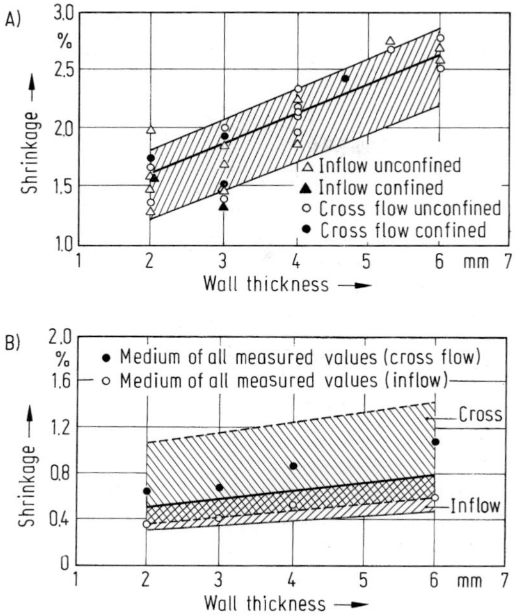

Figure 2.51. Shrinkage of (A) unreinforced, high viscosity and (B) 30% glass fiber reinforced polybutylene terephthalate versus wall thickness (mold temperature approximately 80 °C) [4].

Thicker walls, narrow runners, and long flow paths cause increased shrinkage (see Figure 2.51). Shrinkage can be largely reduced by injecting at the thickest area of the part.

The temperature of the mold also influences shrinkage. Generally, shrinkage increases with high mold temperatures (see Figure 2.52). Shrinkage can be reduced by increasing back pressure and holding time (see Figure 2.53). Glass fiber reinforced polyesters shrink less than unreinforced resins (compare Figure 2.51B with 2.51A and 2.52B with 2.52A).

There is a considerable difference in shrinkage in the direction of flow and across the direction of flow. This is due to the fiber orientation. Shrinkage is impeded to a considerably greater degree in the direction of flow than across the direction of flow (see Figure 2.51B). Very flat parts are subject to warpage caused by differences in shrinkage. Therefore, special thought must be given to the design of the mold and to the processing parameters.

With unreinforced resin, there is almost equal shrinkage in direction of flow and across the direction of flow (see Figure 2.51A). Shrinkage differences are almost nil with glass sphere or mineral filled resins.

The minimal postmold shrinkage apparent with polybutylene terephthalate can be decreased further by increasing the mold temperature (see Figure 2.54). Therefore, precision parts of unreinforced polybutylene terephthalate should be processed at a mold temperature of 120–140 °C and glass fiber reinforced PBT at 80–90 °C.

Elastomer-modified PBT used for large surface parts behaves much like unreinforced PBT (see Figure 2.55).

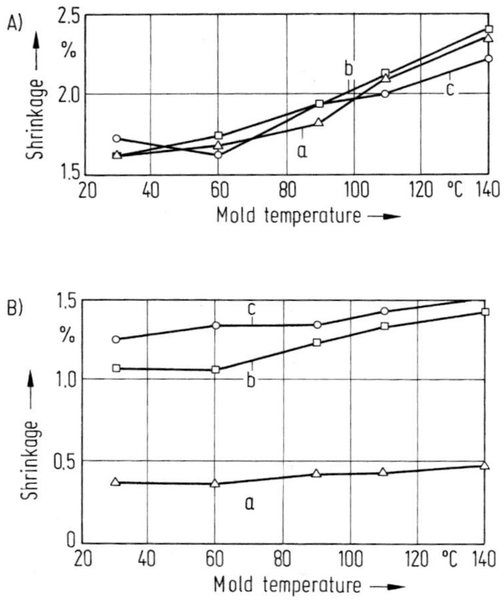

Figure 2.52. Shrinkage of (A) unreinforced, high viscosity and (B) 30% glass fiber reinforced polybutylene terephthalate versus mold temperature measured on plaques (148 mm × 105 mm × 3 mm) with a film gate [4]: a, shrinkage in flow direction; b, across flow at gate; c, across flow at end of flow path.

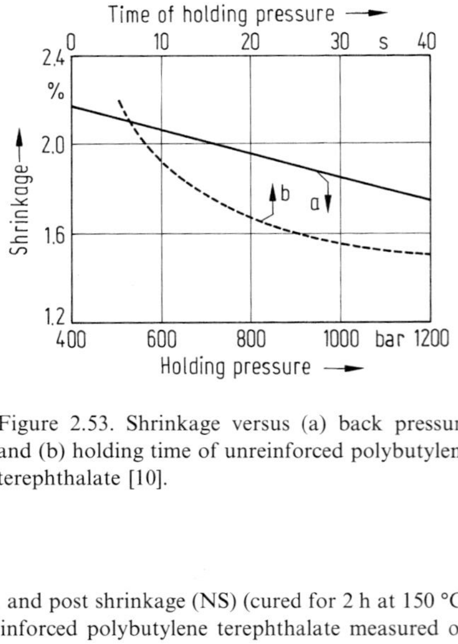

Figure 2.53. Shrinkage versus (a) back pressure and (b) holding time of unreinforced polybutylene terephthalate [10].

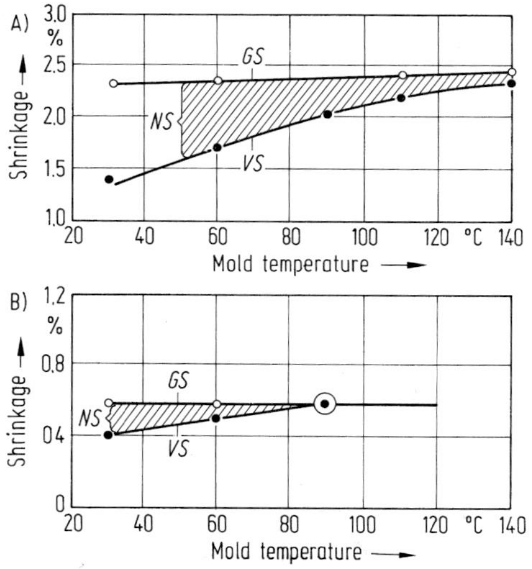

Figure 2.54. Total shrinkage (GS), processing shrinkage (VS), and post shrinkage (NS) (cured for 2 h at 150 °C) of (A) unreinforced, high viscous and (B) 30% glass fiber reinforced polybutylene terephthalate measured on 3 mm plaques [4].

2.10.1.3.6 Processing with Hot Runner Molds

The processing temperatures outlined in Section 2.10.1.2.1 are the limits for molding polybutylene terephthalate and polyethylene terephthalate with hot runner molds. A maximum temperature of 260 °C should not be exceeded in the hot runner distribution and nozzle area for PBT. A minimum temperature of 230 °C is necessary at the mold

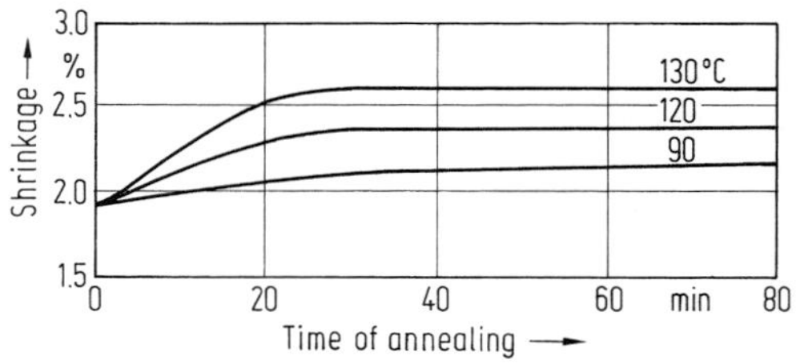

Figure 2.55. Processing shrinkage (mold temperature 80 °C) and postmold shrinkage of elastomer-modified PBT, measured at 3 mm thick plaques [11].

gate to prevent freeze-up at the nozzle. Nozzle and mold must therefore be thermally separated by a large gap, and contact surface must be minimal. The heating capacity within the nozzle should be designed in a way that will not cause high melt temperatures. The temperature within the individual runners must be regulated by very accurate temperature controllers.

Connecting hot runner nozzles in series is not suitable owing to the narrow processing temperature range and the thermal sensitivity of polyester. Therefore, the nozzles should be placed so that the length of the flow path to each gate is the same, ensuring equal pressure loss and equal residence time. External heating of the hot runner nozzles with an unrestricted flow path is advantageous.

If a short sprue is used, or a large gate diameter (about 1.5–2 mm), the parts can be injected directly. Indirect injection using a distribution channel and a hot runner nozzle must be done with an adequately sized hot runner gate diameter of about 2–3 mm [3].

2.10.1.3.7 Cycle Time

Cycle time depends mainly on the time needed for injection, for holding pressure, and for cooling. The thermal sensitivity of polyester limits the extent to which cycle time can be dependent on machine utilization capacity and melt temperature (see Figure 2.56).

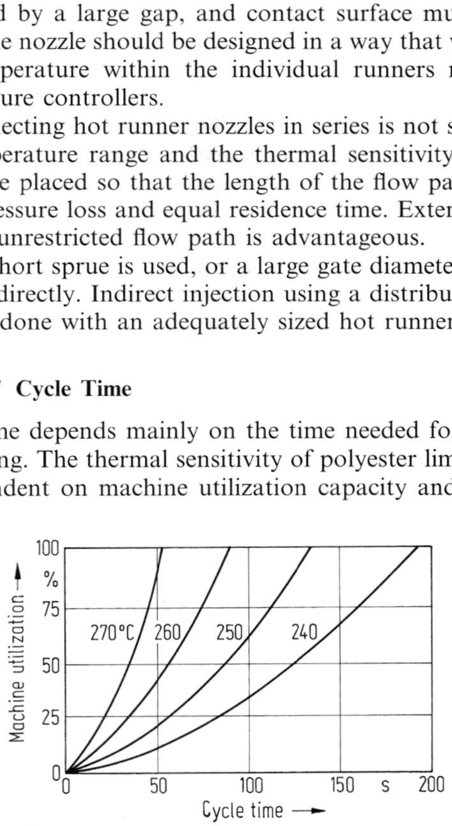

Figure 2.56. Allowable cycle time of polybutylene terephthalate versus machine utilization and melt temperature [3].

Total cycle time for a 11.4 g screw guide of unreinforced medium viscosity PBT (Vestodur 2000, Hüls) is 27 seconds; for a 100 g container lid of 30% glass fiber reinforced PBT (Vestodur SGF 30), the cycle time is 30 seconds. An example of total cycle time for a 6 kg large-surface part having a wall thickness of 3.5 mm elastomer-modified PBT (Pocan S 1506, Bayer) is 90 seconds [11].

2.10.2 Extrusion

Thermoplastic polyesters are easily extruded into various forms, such as sheeting, pipes, films, and fibers. The recommended melt temperature for polybutylene terephthalate is 240–260 °C and for polyethylene terephthalate, 260–275 °C.

Generally, to ensure adequate melt stability, higher molecular weight polyesters are used for extrusion than for injection molding.

Commercially speaking, extrusion is of secondary importance to injection molding for polybutylene terephthalate, whereas extrusion followed by blow molding dominates for polyethylene terephthalate.

Extruded polybutylene terephthalate remains partially crystalline. Films up to 0.2 mm thick, quickly cooled and adequately sheared, retain spherolites so small that the film is highly transparent and tough. Chill-roll temperatures of 60–80 °C are recommended for the production of film and sheet. When making pipes, interior air pressure can be employed to obtain the desired shape. It is effective to cool the extruded piping and profiles with 10–20 °C water to retain their dimensions [14].

Elastomer-modified resin is suitable for making flexible piping (e.g., hoses) [8]. Low viscosity PBT is suitable for extrusion laminating of paper and cardboard for packaging instant food mixes and prepared pastries. To obtain good adhesion, it is important to warm the carrier material to 120–130 °C just prior to lamination [8].

High quality PET films are produced by biaxial stretching of the extruded film. Neither film nor fiber processing methods are described here. Descriptions of the processing of polyester film [15, 16] and of polyester fibers [17–20] can be found in the literature.

2.10.3 Blow Molding

PET is increasingly being processed into hollow parts using blow molding. There are three major blow molding processes: extrusion, injection, and injection stretch blow molding. The latter method dominates because of distinct advantages in making beverage containers [14].

Injection stretch blow molding incorporates a biaxial stretching step between the preform molding step and the blow molding step while the PET is in the thermoelastic stage. This stretching process is an easy way to obtain high orientation of the polyester molecules, thereby improving the physical properties of the material. The preform is made by either injection or extrusion. With multiple molds 10,000 bottles per hour can be produced. Efficient stretch blow molding machines are computer controlled and can produce 0.1–13 liter bottles [21].

To obtain constant part quality and uniform processing, the specification for 2-liter PET bottles calls for an upper viscosity limit of 0.72 dL/g ($\overline{M}_n$ about 24,000). The specifications for 1-liter and smaller bottles generally call for higher molecular weight PET, to avoid haze and loss of toughness due to crystallization, with an upper viscosity limit of 0.8–1.0 dL/g [14].

2.10.4 Processing of Recycled Materials

For successful recycling, the original parts should consist of virgin material that had been processed under optimum conditions. For minimal loss of tensile and impact strength, the percentage of regrind should not exceed 20–30% for unreinforced, 20% for glass fiber reinforced, and 10–15% for flame-retardant PBT. It is recommended that the percentage of recycled material for each application be checked to determine loss of properties for individual parts. Recycled material should not be used for parts under mechanical stress. A higher percentage of regrind is possible in parts for which high heat stability is the only requirement [3, 4, 13].

References for Section 2.10

1. *Bartel, W.:* Kunststoffe 77 (1987) 4, p. 384
2. Arnite, Thermoplastischer Polyester, Spritzgießen. Company brochure Akzo Plastics 1987
3. Crastin, Thermoplastischer Polyester in der Spritzgießverarbeitung. Company brochure Ciba-Geigy 1987
4. Pocan—Thermoplastischer Polyester. Company brochure Bayer AG 1987
5. Petlon, Thermoplastic PET Polyester Resin. Company brochures Mobay 1984 to 1986
6. Petra. Company brochure Allied Chemical Plastics
7. Rynite, Thermoplastic Polyester Resin. Company brochures Du Pont 1981 to 1986
8. Polybutylenterephthalat Vestodur. Company brochure Hüls AG 1985
9. Hostadur B. Company brochure Hoechst AG 1987
10. *Brunner, J.:* Kunstst. Plast. 22 (1975) 9, p. 37
11. Pocan S 1506, Schlagzähmodifiziertes Polybutylenterephthalat. Company brochure Bayer AG 1985
12. *Bier, P.; Ong, G. N.:* Kunststoffe 71 (1981) 9, p. 573
13. *Brunner, J.:* Kunstst. Plast. 22 (1975) 10, p. 9
14. *Jadhav, J. Y.:* Polyesters, Thermoplastic. In: *Mark, H. F.; Bikales, N. M.; Overberger, C. G.; Menges G.* (Eds.): Encyclopedia of Polymer Science and Engineering. Vol. 12. Wiley, New York 1988, p. 217
15. *Werner, E.; Janocha, S.; Hopper, M. J.; Mackenzie, K. J.:* Polyesters, Films. In: [14], p. 193
16. *Janocha, S.:* Herstellung von Polyethylenterephthalatfolien. In: *Harnisch, H.; Steiner, R.; Winnacker, K.* (Eds.): Chemische Technologie. 4. Ed. Vol. 7. Hanser, Munich 1986, p. 520
17. *Ludewig, H.:* Polyesterfasern. Akademie-Verl., Berlin 1975
18. *Falkei, B. von:* Synthesefasern. Verlag Chemie, Weinheim 1981
19. *Davis, G. W.; Hill, E. S.:* Polyester Fibers. In: *Mark, H. F.; Othmer, D. F.; Overberger, C. G.; Seaborg, G. T.* (Eds.): Encyclopedia of Chemical Technology. Vol. 18. Wiley, New York 1982, p. 531
20. *Davis, G. W.; Talbot, J. R.:* Polyesters, Fibers. In: [14], p. 118
21. *Mank, M.; Hofmann, W.:* Kunststoffe 78 (1988) p. 391

2.11 Secondary Processing

2.11.1 Machining

Reinforced and unreinforced parts made from thermoplastic polyester generally require secondary operations, such as joining, boring, sawing, planing, milling, cutting, stamping, drilling, forming, and polishing. The tendency of polyester to "smear" is reduced by the high softening temperature of the material, which is due to its high crystallinity. However, the temperature at the secondary processing location should not be allowed to exceed 200 °C.

Sharpened conventional tools, such as those used for working with steel, nonferrous metals, or wood, are adequate. Tools with hardened cutting edges (HSS or carbide) are recommended for flame-retardant grades. The combination of a cutting speed of up to

60 m/min and a feed of 0.6 mm/revolution without cooling is recommended for drilling up to 10 cm deep. A cutting speed of 500 m/min and a feed of 0.5 mm/revolution is recommended for milling. For lathing, it is possible to use a cutting speed of up to 400 m/min, a feed up to 0.25 mm/revolution, and a cutting depth up to 10 mm.

The saw blade must have enough free cut to avoid smearing while sawing. A cutting speed of about 2500 m/min is possible [1–4].

2.11.2 Welding

Unreinforced polyester parts respond well to different welding methods, including ultrasonic welding, friction welding, hot wire welding, and heated gas welding. Although weld seam strength is reduced by increased glass fiber and flame-retardant content, this loss can be compensated by broader seam design. Ultasonic welding is superior because joining time is short and weld seam strength good. These characteristics are advantageous not only for welding, but also when it is necessary to incorporate metal parts (e.g., gear housings). Parts to be ultrasonically welded must fit precisely (see Figure 2.57). No extra pressure may be exerted on the welded surface. Postmolding and riveting can also be done with ultrasonic or hot stamp welding. Tape welding is also possible (see Figure 2.58). Initial cost is low but the joining time is longer compared to ultrasonic welding.

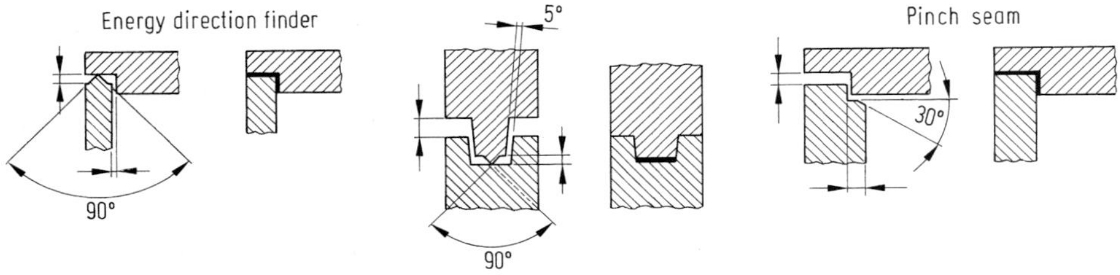

Figure 2.57. Sketches of joint design for polybutylene terephthalate ultrasonic welding [3].

Heated gas welding with additives is suitable for joining semifinished parts. To ensure a perfect weld, the parts must be clean (e.g., nongreasy) on the surface to be joined, but they do not need to be dried [1–5].

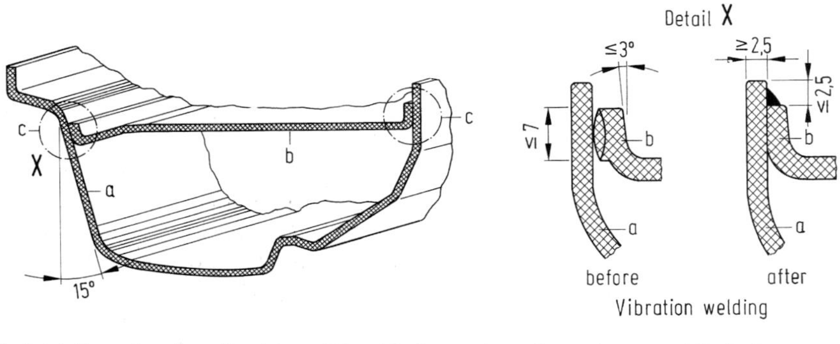

Figure 2.58. Joint formation for vibration welding (similar to hot wire and tape welding) shown on an automobile bumper made from elastomer-modified polybutylene terephthalate [3]: a, bumper; b, reinforcing plate; c, joint area.

2.11.3 Adhesion

Gluing thermoplastic polyester with a solvent-based adhesive is difficult because polyester has good chemical resistance. Solvents that might be useful are health hazards and are not recommended [3].

Two-component adhesives that are suitable include epoxy, silicone, and polyurethane. Cyanacrylate and hot melt adhesives are also suitable. The area to be joined should be lightly roughened and clean (e.g., nongreasy). The adhesion strength obtained, however, will be below the specified product strength [1–5].

2.11.4 Coating

Thermoplastic polyester can be coated with practically all known coating systems. Polyester's solvent rsistance is so good, however, that many of these coatings adhere only poorly to the surface. For this reason a primer should be used with all standard coating systems, as is done with metal coating systems in the automotive industry.

Conventional metal coatings are relatively brittle and therefore have little elasticity compared to polyester. This can lead to cracks at impact and stress conditions that would transfer onto the polyester part, resulting in premature failure. The application of an elastic primer as buffer between the more elastic polyester part surface and the hard coating greatly reduces premature failure.

Highly elastic coating systems, such as Bayer's 2K polyurethane coating systems, are especially suitable for retention of properties from parts made from elastomer-modified polybutylene terephthalate.

The high heat resistance of thermoplastic polyester, even that of elastomer-modified polybutylene terephthalate, makes possible on-line top-coating of automobile bodies using primers and coating systems for car bodies typically seen in the automotive industry.

Unreinforced polyester parts have a high surface quality that requires no preparation like sanding or filling. However cleaning the molded parts is a must—for example, in ultrasonic baths, or by washing in diluted alcohol [1–5].

2.11.5 Metallizing

Parts (e.g., automobile reflectors) can be metallized by vacuum coating. To guarantee a highly polished surface on reinforced and filled products, an intermediate coating is necessary. The same vacuum coating method can be used to apply a coating protecting the metallizeed surface (e.g., a silicone or cellulose base) [1–4].

2.11.6 Printing

Thermoplastic polyester products are easily imprinted with either flexogravure or silk screening. The flexogravure process requires a polyamide binder with or without nitrocellulose. The latest printing processes, such as transfer or thermodiffusion, are also suitable [1, 3–5]. Laser writing is an economical and flexible method of labeling.

2.11.7 Thermoforming

Polybutylene terephthalate and polyethylene terephthalate film can be thermoformed. The deformation temperature for polybutylene terephthalate is 230–240 °C. The narrowness of this processing window is due to the sharp melt point. Transparent polybutylene terephthalate films, however, are more easily shaped by combining prestretching and compressed air molding at 90–100 °C [4, 5].

References for Section 2.11

1. Arnite, Thermoplastischer Polyester. Company brochure Akzo Plastics 1987
2. Hostadur B. Company brochure Hoechst AG 1987
3. Pocan—Thermoplastischer Polyester. Company brochure Bayer AG 1987
4. Polybutylenterephthalat Vestodur. Company brochure Hüls AG 1985
5. *Carlowitz, B.:* Kunststoff-Tabellen. 3. Ed. Hanser, Munich 1986
6. *Foltin, E.; Radeck, A.:* Kunststoffe 78 (1988) p. 688

2.12 Applications

The range of applications of thermoplastic polymers is extraordinarily broad because these materials offer a good combination of mechanical, physical, and electrical properties, in conjunction with high deflection temperature, good heat aging resistance, weather resistance, and easy processing (Section 2.7). Typical applications are described in Sections 2.12.1–2.12.8.

2.12.1 Electrical Engineering, Electronics

Along with heat resistance, dimensional stability and low moisture absorption are very important in electrical engineering and electronics applications. These properties are necessary to achieve the miniaturization of electronic parts, which must be extremely heat resistant. Many glass fiber reinforced PBTs and PETs (especially flame-retardant and corrosion-resistant materials) can be used in these applications. If thin walls are present, connectors must show high dimensional stability to be suitable for the contacts that will be metallized at 260 °C. In these applications, flame-retardant PBT with glass fiber reinforcement is the best choice (Figure 2.59).

Higher and longer lasting temperature-induced stress shows up in parts that have been subjected to the new surface mount technologies (SMTs) such as vapor phase soldering (VPS) and infrared brazing. Heat-stable products such as flame-retardant, glass fiber reinforced PET (Figure 2.60) are necessary here, as well.

In addition to high temperature resistance and high dimensional stability, ground fault switches require very low distortion. This is achieved by using flame-retardant, glass fiber reinforced PBT that also contains glass beads (Figure 2.61).

Because of its high heat resistance and good chemical resistance, flame-retardant, glass fiber reinforced PBT is very suitable, even for picture tubes (Figure 2.62).

Electrical contacts must work perfectly and safely if used out of doors, under the hood of a car (to start the engine at low temperatures), in contact with oil, fuel, or water, and at points where sparks or short-circuits can be expected to form. A flame-retardant, glass fiber reinforced PBT is best suited for devices such as those shown in Figure 2.63.

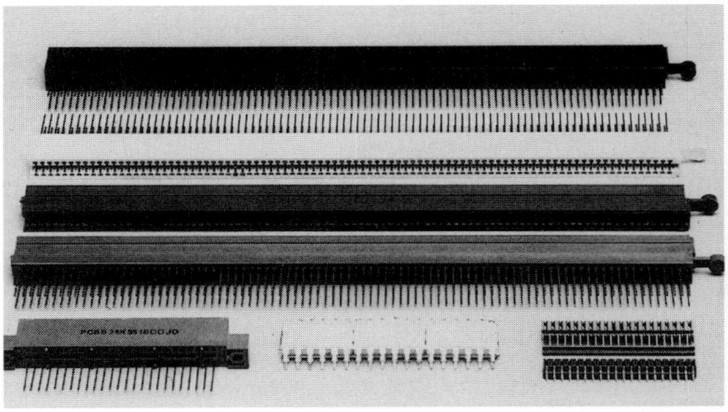

Figure 2.59. Connector strips made of flame-retardant, glass-reinforced PBT. (Photo: Bayer)

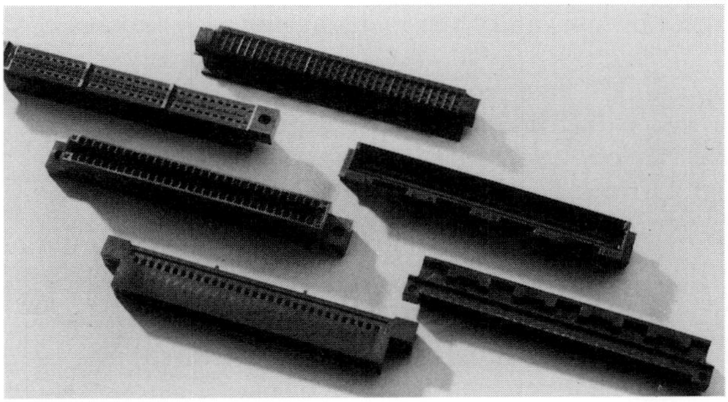

Figure 2.60. Connector housing made of flame-retardant, glass-reinforced PET. (Photo: General Electric Plastics)

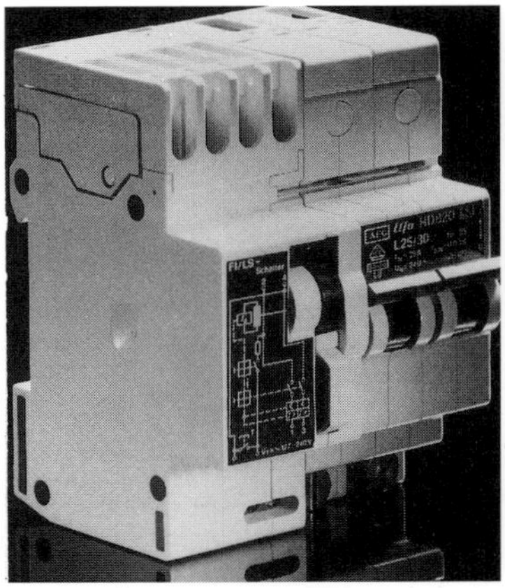

Figure 2.61. Circuit breaker made of flame-retardant, glass-reinforced, glass pellet filled PBT. (Photo: Bayer)

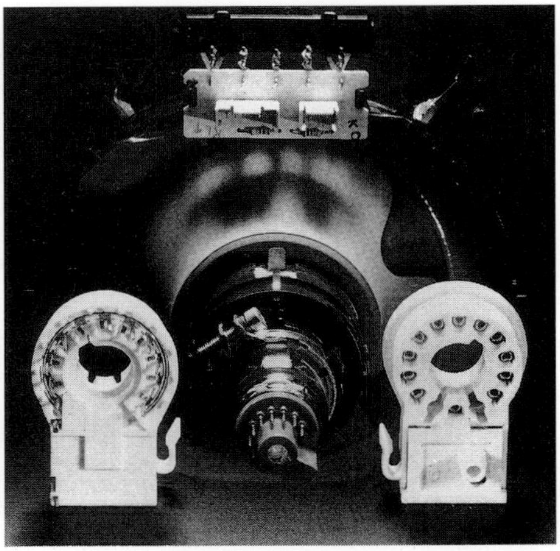

Figure 2.62. Picture tube base made of flame-retardant, glass-reinforced PBT. (Photo: Bayer)

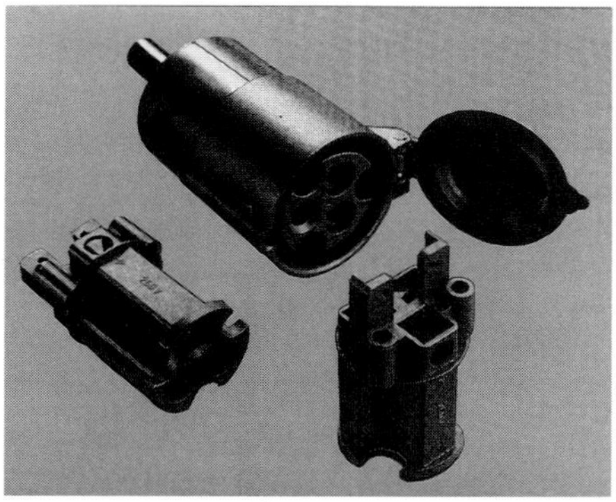

Figure 2.63. Connectors made of flame-retardant, glass-reinforced PBT. (Photo: Hoechst)

For relays it is very important that during long oven times and at high service temperatures, no substances diffuse out of the material, for such leakage could lead to contact corrosion.

Other important applications that require the foregoing properties include brush holders, electric engine parts, instrument chassis, condenser covers, LED displays, ventilators for electronic devices, crankshaft housings, series connectors, automatic cutouts, bobbins, bus bar carriers, and telephone distribution boxes.

2.12.2 Automotive Industry

Over the past years the use of plastics in cars has steadily increased. The roughly 40 kg of plastic used per car in 1972 had increased to 75 kg of plastic per car in 1985 (Figure 2.64). The average weight of plastic per car has risen worldwide to 100 kg in 1993. A quarter of overall plastic consumption is in the automotive industry, where thermoplastic polyester has a tremendous impact.

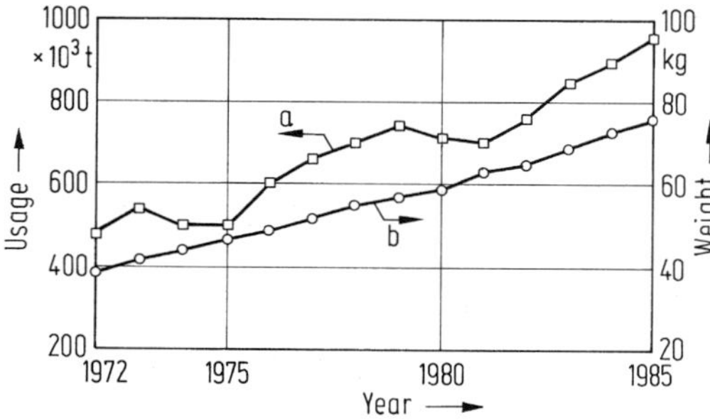

Figure 2.64. Use of plastics by Western automobile factories [1]: a, plastic use for automobiles; b, average for automobile.

Depending on the function, specific requirements must be met. For example, glass fiber reinforced PBT is used for electronic ignition devices and spark plugs because of its high impact resistance at high temperatures, tear resistance under pressure, resistance to fuel and other engine fluids, and high electrical capacitance (Figure 2.65).

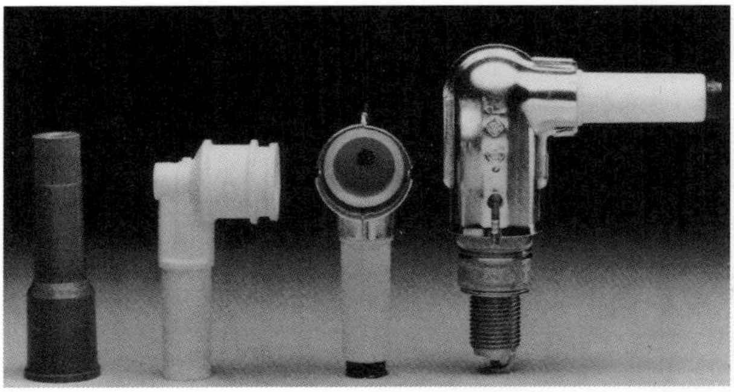

Figure 2.65. Spark plugs made of glass-reinforced PBT. (Photo: Bayer)

High temperature resistance in combination with dimensional stability and direct metallizability are the reasons for the production of reflectors using unreinforced PBT (Figure 2.66).

Figure 2.66. Reflectors made of unreinforced PBT. (Photo: Bayer)

High stiffness as well as weather and chemical resistance under dynamic stress are important reasons for using glass fiber reinforced (perhaps thermoplastic-modified) PBT for windshield wipers and windshield wiper sockets (Figure 2.67).

Elastomer modification of unreinforced and reinforced PBT opens up opportunities to combine the polymeric properties with high toughness even at winter temperatures. For that reason, outside mirrors, which are made of elastomer-modified, glass-reinforced PBT, are dynamically stress resistant, weather resistant, and impact resistant (Figure 2.68).

Bumpers, radiator grilles, air dams, and spoilers, with or without color, are perfect applications for elastomer-modified PBT. These accessories can be mounted on the car and painted at 155 °C. Bumpers painted in this way can withstand an impact of 4 km/h even at −20 °C following the ECE rules (Figure 2.69). Another advantage of this product is that the design engineer can specify the exact shape on the basis of wind tunnel test results.

Sufficient stiffness in bumpers can be achieved by reducing the weight—for example, by changing the design to obtain boxlike bumpers or to have a U-like profile on the back, which can be fastened to the actual bumper (Figures 2.70 and 2.71).

Figure 2.67. Built-in windshield wiper holder made of glass-reinforced PBT. (Photo: Bayer)

Figure 2.68. Mirror housing made of elastomer-modified PBT. (Photo: Bayer)

Figure 2.69. On-line finish coating of front bumpers made of elastomer-modified PBT. (Photo: Bayer)

Figure 2.70. Bumper made of box-like profile elastomer-modified PBT. (Photo: Bayer)

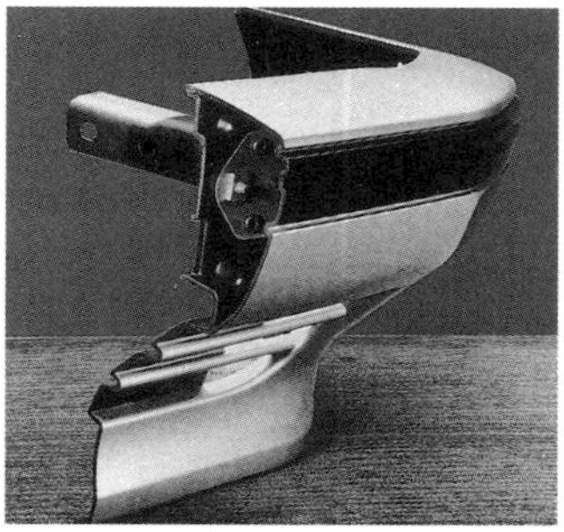

Figure 2.71. Bumper made of elastomer-modified PBT with welded-in U-profile. (Photo: Bayer)

Developmental blends (elastomer-modified PBT/polycarbonate blend, Pocan KU 1-7914) are being used for the lining of bumpers in the new BMW 7-Series. After extrusion, these large one-piece bumpers are painted with commercial polyurethane coatings. This bumper profile is then mounted on an aluminum beam in the front and on an SMC (sheet molding compound) beam in the back and can withstand an impact of 8 km/h without permanent damage. If a crush zone is added, an impact of 15 km/h can be absorbed without damage to the engine beams [2].

Other examples of applications in the automotive field are pedals, grilles, air filter housings, fuel lines, ashtrays, safety belt parts, electric mirrors, door handles, lock housings, and carburetor parts, as well as various motorcycle parts.

Because of their creep resistance, high impact resistance, low compression set, and flexibility even at low temperatures, elastic copolyether esters can be used as bumper supports (Figure 2.72) and blow-molded transmission boots as well as steering and shock systems (Figure 2.73).

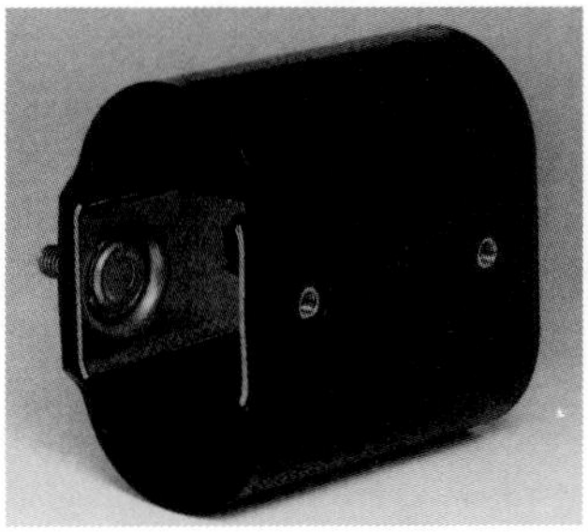

Figure 2.72. Bumper bracket made of copolyether ester between plastic bumper and chassis. (Photo: DuPont)

Figure 2.73. Blow-molded bellows made of copolyether ester for automotive transmission, steering, and shock absorption systems. (Photo: DuPont)

2.12.3 Lighting Technology

The generation of light is always accompanied by the development of heat. The high heat resistance that can be obtained, especially in glass fiber reinforced products, in conjunction with stiffness and dimensional stability, leads to many possible applications for thermoplastic polymers. Depending on the light source, temperatures exceeding 140 °C can be withstood (Figure 2.74).

Lightbulb sockets made of glass fiber reinforced PBT and PET do not change dimensions while in use. This property is important because the sockets hold electrical units that are sensitive to dimensional changes. Lightbulb sockets are tested by being exposed to a temperature 35 K above their long-term heat distortion temperature for 168 hours [3].

Other applications include adapters, flashlight housings, spotlights, and scuba lights.

2.12.4 Data Technology

Materials used in the field of data technology must show excellent sliding capability and surface polish in combination with high precision—for example, in the special function parts of computer keyboards, telephones, fax machines, and word processors.

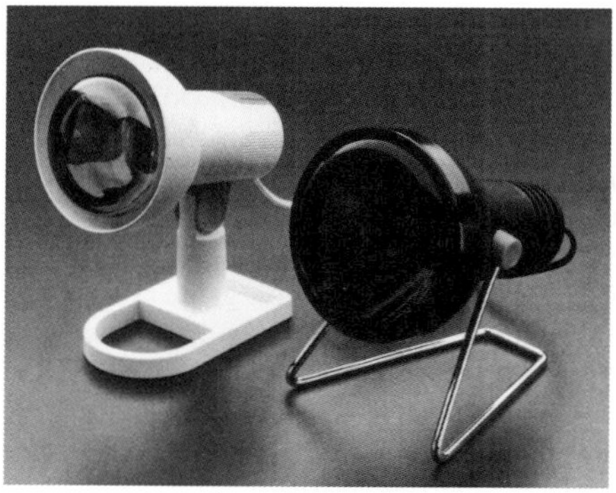

Figure 2.74. Light housings made of glass-reinforced PBT. (Photo: Hoechst)

Figure 2.75. Lightbulb sockets made of glass-reinforced PBT or PET. (Photo: Bayer)

Computer keyboards are becoming more and more important. The labeling on the keys is achieved by a faster and cheaper laser process and pressure transfer printing and thermal diffusion procedures. In transfer printing, letters printed on a foil are transported to, for example, a computer keyboard and pressed in at 180 °C until they reach a depth of 100 μm (Figure 2.76). Scales, signs, toys, and other items are being printed in the same manner.

Figure 2.76. Printed computer keys made of unreinforced PBT. (Photo: Bayer)

Jackets of light cables are a modern application of unreinforced, modified PBT, since they are stiff, resistant to chemicals, and easy to extrude. Further applications are push-fit fittings, telephone handset cradles, and printer ribbon guides.

2.12.5 Home Appliances

Thermoplastic polymers are used in home appliances mainly because of their high heat resistance. However, many other properties, such as electric and thermal insulation, stiffness, and moisture resistance, are very important, as well as resistance to deterioration from contact with chemicals and foods. Housings usually require a smooth, shiny surface that can also be textured. Examples for glass fiber reinforced PBT are irons (Figure 2.77), waffle irons (Figure 2.78), fryer ovens (Figure 2.79), and oven door handles (Figure 2.80). Other appliances are crepe makers, coffee makers, stove top covers, toasters, vacuum cleaner parts, hair curlers, hair dryers, silverware handles, and scissors.

Figure 2.77. Iron housing made of flame-retardant glass-reinforced PBT. (Photo: Bayer)

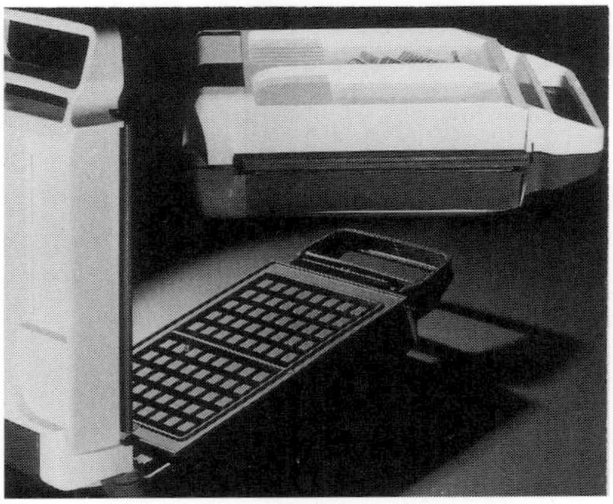

Figure 2.78. Waffle iron housing made of glass-reinforced, thermal-modified PBT. (Photo: Bayer)

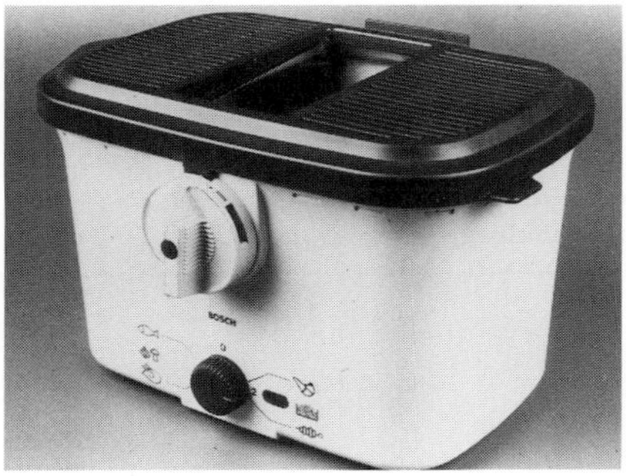

Figure 2.79. Fryer housing and buttons of glass-reinforced mineral-filled PBT. (Photo: General Electric Plastics)

Figure 2.80. Oven door handle made of PBT. (Photo: Akzo Plastics)

2.12.6 Other Applications

Besides the applications listed so far, thermoplastic polymers can be found in other fields that require certain properties. Examples are hobbies and medicine. Some examples are listed in Table 2.25.

Table 2.25. Sample Applications of Thermoplastic Polyester

Applications	Relevant Properties
Housing for glue guns made of glass-reinforced PBT (Figure 2.81)	Resistance to high temperatures and chemicals; high surface quality
Lawnmower housing made of glass-reinforced PET (Figure 2.82)	High stiffness, viscosity, and chemical resistance
Pumps (or drill) housings made of glass-reinforced PET (Figure 2.83)	High impact resistance
Anchor rope drums made of unreinforced PET (Figure 2.84)	Impact and saltwater resistance
Winter sporting goods made of elastomer-modified PBT, e.g., boots, boot clasps, ski binders, ski pole bases (Figure 2.85)	High stiffness; impact resistant down to −30 °C.
Housing for machine water pumps made of unreinforced PBT	Dynamic load capacity, chemical resistance, high impact resistance and little water absorption
Portable pumps with housing for engine and pump out of glass-reinforced PET	Good flow characteristics
Renal calculus remover with housing made of elastomer-modified PBT (Figure 2.86)	Simple surrounding of the metal parts
Inhalation counter made of unreinforced PBT	Good elasticity as well as good sliding properties and good durability
X-ray-transport electrodes made of carbon fiber reinforced polyester (Figure 2.87)	No disturbance of the x-ray picture, as with regular metal electrodes
Beer taps made of unreinforced PBT (Figure 2.88)	FDA quality; resistant to high pressures and chemicals
Safety locks made of reinforced PBT	Dimensional stability, stiffness, chemical resistance
Spools for textile industry made of glass-reinforced thermoplastic-modified PBT	Tear resistance, high stiffness, durability, resistance against textile dyes
Housings for hair dryers made of glass-reinforced, thermoplastic-modified PBT	High temperature resistance, stiffness, deformation resistance
Sinks made of unreinforced PBT	Scratch resistance, durability; resistant to pressure and chemicals

2.12.7 Bottles and Containers

Because of the good barrier properties of PET, the dominant application of this material as an industrial thermoplastic (besides films and fibers) is as bottles for carbonated drinks [4]. In 1985, 400,000 tons of PET was used for bottle production worldwide [5]. The U.S. entire market of 2-liter plastic bottles is almost completely covered by PET, while the market of 3-liter bottles made out of PET is constantly increasing. The use of PET for water bottles became possible when the content of acetaldehyde was lowered to 1 ppm [6]. Coloring has been restricted to the use of green, brown, and yellow pigments, which do not have any adverse health effects.

For alcoholic drinks up to 25 vol %, certain copolymers are suitable (e.g., Kodar PETG 6793, Eastman Kodak). Pasteurization and filling with hot liquids are possible when multilayer bottles made of PET and a more heat-resistant polymer are used. PET

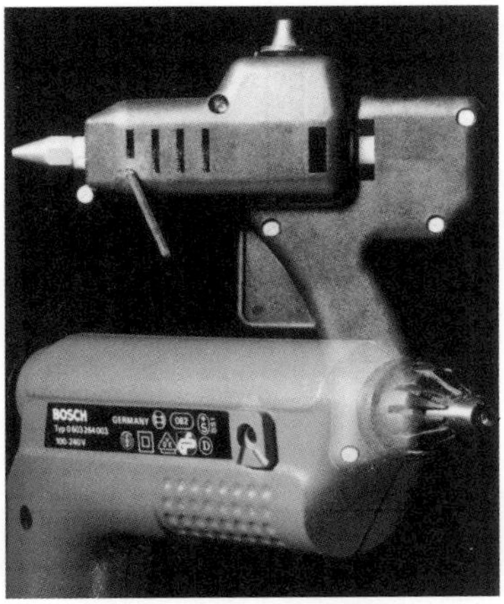

Figure 2.81. Housing for glue gun made of glass-reinforced PBT. (Photo: Bayer)

Figure 2.83. Pump made of glass-reinforced PET for use with drill. (Photo: Akzo Plastics)

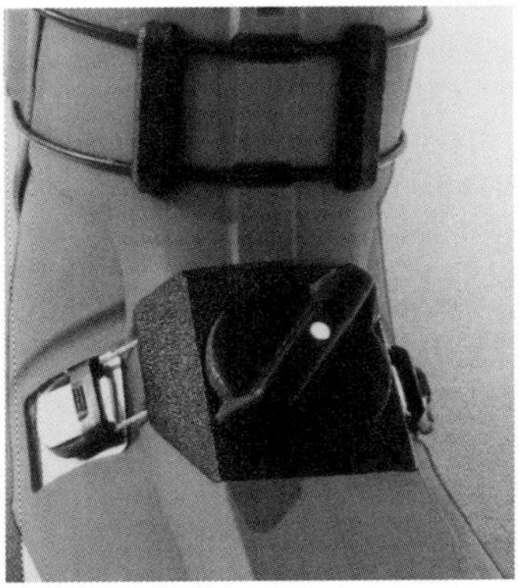

Figure 2.85. Skiing boot clasp made of unreinforced elastomer-modified PBT. (Photo: Bayer)

Figure 2.82. Lawnmower housing made of glass-reinforced PET. (Photo: DuPont)

Figure 2.84. Anchor rope drum made of unreinforced PET. (Photo: Akzo Plastics)

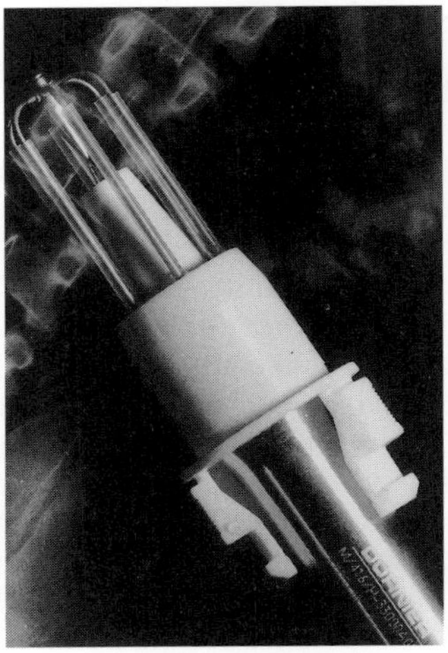

Figure 2.86. Renal calculus remover made of elastomer-modified PBT. (Photo: Bayer)

laminated with polyvinylidene chloride is used for beer bottles. Wine bottles, made out of PET and *m*-xylylene diaminoadipate for its good barrier properties, are being produced using the coinjection molding process. Five-layered bottles with polyvinyl alcohol used as a barrier polymer are outstanding because of their good barrier property. Noncrystalline PET copolymers are being used for bottles to hold cosmetics and medicines. Cooking oils, shampoos, and detergents are packaged in thin, highly transparent bottles made out of PET. Another application for PET is a 20-liter bottle for lacquers containing solvents [4–8].

In Germany, a deposit has been imposed on PET bottles to make sure that they will be returned and recycled.

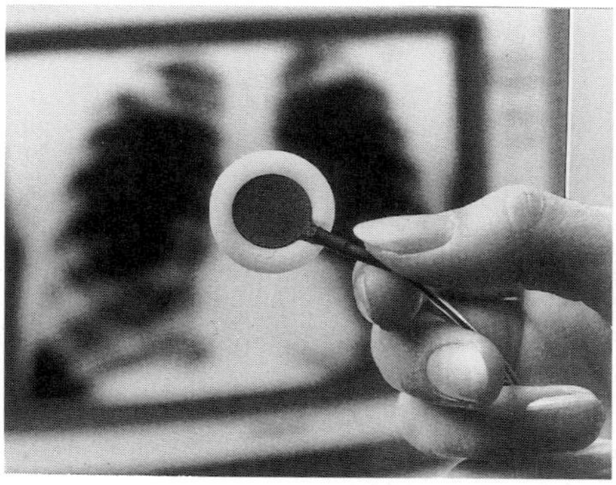

Figure 2.87. X-ray-transparent electrode made of carbon fiber reinforced polyester. (Photo: Akzo Plastics)

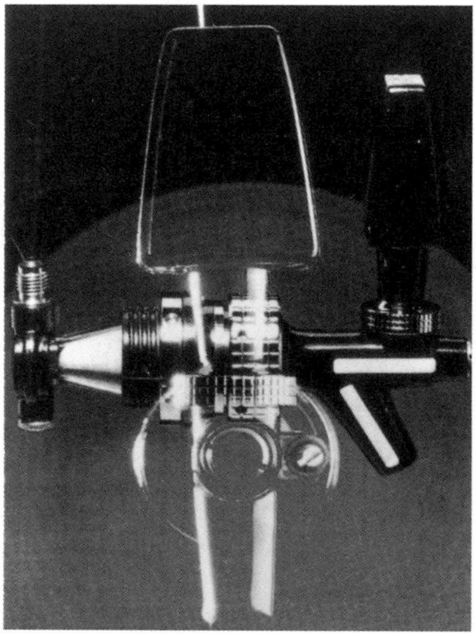

Figure 2.88. Beer tap made of unreinforced PBT. (Photo: Bayer)

2.12.8 Films

Polyethylene terephthalate is the predominant plastic used in films. Polymer films are being considered as engineering films because of their excellent thermal, mechanical, optical, and electrical properties. Films made of PET have more than 50 specific applications, ranging from regular films for photos to x-rays to photoresist films for printed circuits. These are the main uses, which call for PET's optical clarity, dimensional stability, and resistance to chemicals and time (Figure 2.89).

Because of the high mechanical resistance, dimensional stability, durability, and quality of PET films, magnetic recording on audio tapes, computer diskettes, and video tapes is another good application. For this purpose, the film is coated with a magnetically active material, typically iron or chrome oxide. In the electronics field, PET films are being used as insulators and heat barriers because of their high dielectric constant, high thermal dimensional stability, and high melting point. Many applications are found in reprography, especially for microfilms, technical reproduction, and drawing films. Thermal dimensional stability, dimensional stability against moisture, and stability under conditions of long storage are the significant properties here.

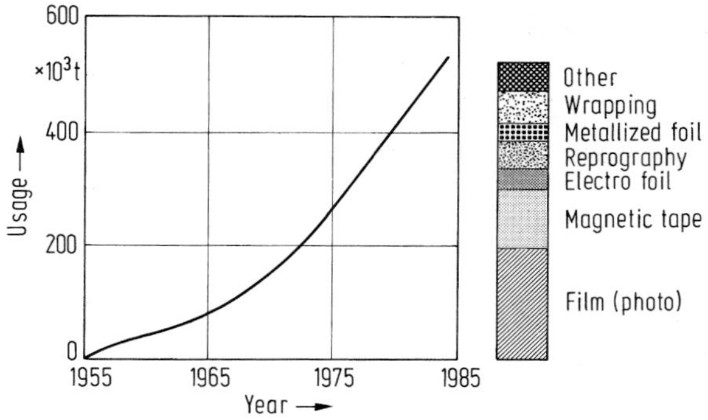

Figure 2.89. Worldwide use of PET films [9].

The polyester film production has risen worldwide to 1.2 million tons in 1995.

In the food market, PETs are used because they are stable and clear, have good barrier properties, are easy to print on, have good metallizability, and are coextrudable. For the same reason, PET films are being used for pharmaceutical and medical packaging.

The increasing use of PET films with a width of 3–6 μm is due to the high mechanical stiffness, dimensional stability, heat resistance, time resistance, and cleanliness of these foils. They are being used for donor films for the pressure transfer procedure mentioned in Section 2.12.4. There are many other applications as well, including covers, solar panels, transfer foils, typewriter ribbons, tapes, reflector foils, thermal shrink-wrap, and skin for sausages.

Other films being used are:

- copolymers based on 1,4-dimethylolcyclohexane (for heat-shrink wrappers)
- copolymers based on isophthalic acid (as the outer layer of coextruded polymer films for heat sealing)
- copolymers based on naphthalene dicarboxylic acid (Q-film, Teijin: for limited thermal shrinking)

References for Section 2.12

1. *Lehnert, G.:* Kunststoffe 78 (1988) p. 248
2. Kunststoffe 78 (1988) p. 623
3. Pocan—Thermoplastischer Polyester. Company brochure Bayer AG 1987
4. *Nitschke, C.:* Mod. Plast. Enc. (1986/1987) p. 45
5. *Domininghaus, H.:* Polyethylenterephthalat als Barriere-Kunststoff. In: Die Kunststoffe und ihre Eigenschaften. 2. Ed. VDI-Verl., Düsseldorf 1986, p. 525
6. *Caesar, H. M.:* Kunststoffe 77 (1987) p. 1004
7. Kunststoffe 77 (1987) p. 44
8. Kunstst. J. 21 (1987) 12, p. 95
9. *Seifried, W.:* Kunststoffe 75 (1985) p. 773

2.13 Recycling of Polyester Wastes

When thermoplastic polymers are processed by normal injection molding and extrusion, there will be waste that to a certain degree can be recycled (see Section 2.10.4). Since PET is so widely used in fabrics, foils, and films, it can be recycled and reused in many different ways. When the used parts are recycled, it is necessary to separate the PET from the remaining components. The different ways of recycling PET, for producing either new polymers or different polymers, are enumerated below.

1. Regenerating the terephthalic acid as the dimethyl ester through pressure hydrolysis, acid hydrolysis, or saponification with caustic in the presence of amines or under pressure, or by pressurized methanolysis in presence of phosphoric acid, benzene–naphthalenesulfonic acid, zinc acetate, or dimethyl terephthalate. An industrial method for the continuous depolymerization of PET wastes, as well as a cleaning process involving the dimethyl terephthalates, have been described [1]. The profitability of these methods is not clear, and for this reason the ICI recycling plant in Rozenburg, Netherlands, was closed in 1984 [2].
2. Decomposition of PET with ethylene glycol to monomeric terephthalic acid diglycolester. Catalytic acceleration of this decomposition can be achieved by using zinc, cobalt, manganese, or iron salts. There are many ways of cleaning the diglycolesters in preparation for new polycondensation [1].
3. Decomposition of PET with glycol and reaction of the resulting polyols with aliphatic dicarbonic acids to polymer polyols, which can react with organic polyisocyanates to form polyurethanes [3]. These procedures have been developed for industrial uses [4].
4. Decomposition of PET with propylene glycol and reaction of the decomposition product with maleic anhydride to form unsaturated polymers [5].

The isolation of PET to a powdered product can be done by dissolving the PET in solvent under pressure followed by precipitation [6]. Washed PET film pieces can also be regranulated. Many ways of isolating PET for subsequent reuse have been described [7], some with quite detailed technical data [8, 9]. Processing after isolation is possible for wood substitutes [10], for rayon for pillow fillings or for floor tiles [11], and for pallets made out of glass fiber reinforced polymers [12].

Some of the polymer producers, including Agfa, DuPont, Goodyear, and Eastman Kodak, have developed their own recycling systems, which mainly work up photographic films to retrieve their silver content. The recovery of PET is of little importance, if it is not burned anyway [13].

Different machine manufacturers offer recycling technology as well as recycling equipment. One U.S. company recycles 7.2 million PET bottles annually; the isolation of PET from high density polyethylene is the key process [14]. In Germany, a countrywide 65% recycling quota of PET bottles is being aspired for 1994. The corresponding PET recycling plants can be used economically at a rate of 6000 tons/year [15]. A plant built in the Netherlands has a capacity of 10,000 tons/year.

Besides these special ways of recycling polymers, there are also general procedures: for recycling plastic wastes [16], wrappings [17], or plastic parts in cars [18], for example.

References for Section 2.13

1. *Ludwig, H.:* Polyesterfasern, Chemie und Technologie. 2. Ed. Akademie-Verl., Berlin 1975
2. *Bartlett, K.:* Innovation. ICI-Veröff. 10, (1986/1987) p. 12–15

3. EP-OS 248 570 (1986) ICI
4. *Carlstrom, W. L.*, et al.: Mod. Plast. 62 (1985) 5, p. 100
5. *Vaidya, U. R.; Nadkarni, V. M.*: J. Appl. Polym. Sci. 34 (1987) p. 235
6. DE-PS 26 32 031 (1975) Forschungsinstitut für Textiltechnologie
7. *Leaversuch, R. D.*: Mod. Plast. Int. 17 (1987) 4, p. 40
8. Kunststoffe 77 (1987) p. 1140
9. Kunstst. J. 21 (1987) 12, p. 142
10. Mod. Plast. Int. 17 (1987) 6, p. 22
11. *Heater, P. L.*: Mod. Plast. Enc. (1988) p. 44
12. Kunststoffe 78 (1988) p. 154
13. Company brochure Eastman Kodak (1979)
14. *Jadhav, J. Y.; Kantor, S. W.*: Polyesters, Thermoplastic. In: *Mark, H. F.; Bikales, N. M.; Overberger, C. G.; Menges, G.* (Eds.): Encyclopedia of Polymer Science and Engineering. Vol. 12. Wiley, New York 1988, p. 217
15. *Brandt, S.; Landers, B.*: Kunststoffe 79 (1989) p. 312
16. *Schalles, H.*: Kunststoffe 77 (1987) 11, p. 1127
17. *Franke, M.*: Kunstst. J. 22 (1988) 5, p. 64
18. *Menges, G.*, et al.: Kunststoffe 78 (1988) p. 573
19. Kautschuk + Gummi-Kunstst. 43 (1990) p. 269

2.14 Market Development

The market was slow immediately after the introduction of PET in 1966 and PBT in 1970 for injection molding, since the multitude of present applications had not been discovered. From about 1975, use of these materials has steadily increased, as a result of the development by the automotive industry, the electronic industry, and the machinery industry of more and more possible applications. An average increase of the use of PBT of 8% is expected for the future. Table 2.26 shows the market development from 1983 up to projections for 1995 [1].

Table 2.26. Market Development of PBT Molded Items [1]

Application	Use (× 1000 tons)					Average Annual Growth Rate (%)
	1983	1984	1985	1990	1995	
Western Europe						
Electronic technologies	5.0	5.6	6.4	12.0	16.0	10.2
Automotive	4.0	4.7	5.3	10.6	21.0	14.8
Appliances	2.0	2.2	2.5	4.0	6.0	9.6
Other	4.0	4.2	4.6	6.4	9.0	7.0
Total	15.0	16.7	18.8	33.0	52.0	10.9
United States						
Electronic technologies	8.8	10.6	12.9	16.8	20.5	7.3
Automotive	10.7	13.0	14.2	17.5	16.0	3.4
Appliances	5.0	6.0	6.7	8.4	13.0	8.4
Alloys and blends	5.0	9.0	11.0	13.5	19.0	11.2
Other	9.0	10.1	10.9	13.8	17.5	5.7
Total	38.5	48.7	55.7	70.0	86.0	6.9

Table 2.26 (Continued)

Application	Use (× 1000 tons) 1983	1984	1985	1990	1995	Average Annual Growth Rate (%)
Japan						
Electronic technologies	14.0	16.0	17.0	19.0	20.0	3.0
Automotive	6.0	7.0	8.0	10.0	14.0	7.3
Alloys and blends	–	1.0	2.0	5.0	10.0	21.2
Other	5.0	6.0	7.0	9.0	11.0	6.8
Total	25.0	29.0	33.0	43.0	55.0	7.0
Total Western Europe, U.S.A. and Japan	78.5	94.4	107.5	146.0	193.0	7.8

The faster growing use of PBT in Western Europe in comparison to the United States and Japan is related to the steadily growing need for PBT in the electronics, mechanical, and automotive markets. The increase of PET, at 15%, is approximately twice as high as the increase for PBT. This difference is due to two factors: the small initial use and the strong increase in use in the field of automotive electonics. Since the

Table 2.27. Market Development of PET Molded Items [1]

Market/Application	Use (× 1000 tons) 1983	1984	1985	1990	1995	Average Annual Growth Rate (%)
Western Europe						
Electronic technologies	3.0	3.2	3.6	8.0	14.0	13.7
Automotive	0.5	0.6	0.6	1.0	2.5	14.4
Appliances	0.5	0.5	0.6	0.7	1.0	6.9
Other	0.5	0.6	0.6	1.8	3.5	17.0
Total	4.5	4.9	5.4	11.5	21.0	13.7
United States						
Electronic technologies	0.4	0.6	0.9	7.5	14.5	34.9
Automotive	2.0	2.5	3.3	5.4	7.9	12.1
Appliances	0.4	0.5	0.6	1.3	2.1	14.8
Other	1.1	1.2	1.7	3.0	6.0	15.2
Total	3.9	4.8	6.5	17.2	30.5	18.7
Japan						
Electronic technologies	5.1	6.2	7.1	11.4	15.8	9.9
Automotive	0.3	0.4	0.7	2.0	4.2	24.6
Appliances	–	–	0.2	1.5	3.0	
Other	1.1	1.5	1.8	4.1	6.5	16.0
Total	6.5	8.1	9.8	19.0	29.5	13.4
Total Western Europe, U.S.A., and Japan	14.9	17.8	21.7	47.7	81.0	15.2

Table 2.28. Annual Use of Thermoplastic PET in 1987 (× 1000 tons) [2]

Application	Western Europe	U.S.A.	Japan	Total use
Forms	5.5	8.5	9.5	23.5
Food trays	4.5	11.0	< 1.0	16.5
Bottles	135.0	325.0	58.5	518.5
Films	145.0	265.0	160.0	570.0
Other extrusion applications	12.0	21.0	5.0	38.0
Total	302.0	630.5	234.0	1166.5

cost for the raw materials of PET is lower, and since PET is more heat resistant and has better mechanical properties, it is increasingly being used for PBT applications. Table 2.27 shows the market development of PET for the same period covered in Table 2.26.

Although the use of PET as an injection molding material is steadily increasing, the absolute portion of total consumption of thermoplastic PET is low, since most of it is being used for bottles and foils (Table 2.28; see also Sections 2.12.7 and 2.12.8).

From the development so far and the estimated development in the future, it can be seen that thermoplastic polyesters have to be associated with the most vital group of the engineering thermoplastics.

References for Section 2.14

1. Engineering Plastics, A Strategic Assessment 1983–1995. Vol. 3. Chem. Systems, New York 1985
2. *Caesar, H. M.:* Kunststoffe 77 (1987), p. 1004

2.15 Trading Products

Many companies in Western Europe, the United States, and Japan produce PBT and PET by means of straight injection molding. Table 2.29 lists some of these companies and the names of their products. In general, producers have offered a wide variety of unreinforced, glass fiber reinforced, and flame-retardant products, each designed for specific applications. In addition, mineral- and glass-filled products, used in combination, as well as thermoplastic- and elastomer-modified products are offered. Some producers also have carbon fiber reinforced and copolymeric products.

Table 2.29. Producer and Name of Polyester Used for Injection Molding

Producer	Product Name	
	PBT	PET
Western Europe		
Ato Chem. SA	Orgater	Orgater
BASF AG	Ultradur	Ultralen
Bayer AG	Pocan	Petlon
Ciba-Geigy AG	Crastin	Crastin
DSM	Arnite T	Arnite T
EniChem	Pibiter	

Table 2.29. (Continued)

Producer	Product Name PBT	PET
Hoechst AG	Celanex	Impet
Hüls AG	Vestodur	Vestodur
Rhone-Poulenc Chemie SA	Techster T	Techster E
United States		
Allied Corp.		Petra
Bayer Corp.		Petlon
BASF Corp.	Ultradur	
Hoechst Celanese Corp.	Celanex	Impet
E.I. Du Pont de Nemours & Co., Inc.		Rynite
General Electric Co.	Valox	Valox
Japan		
Dainippon Ink & Chemicals, Inc.	Planac	
EPL. Ltd.	Valox	
Mitsubishi Chemical Industries, Ltd.	Novadur	
Mitsubishi Rayon Co. Ltd.	Tufpet, Shinko-Lac	Dianite
Polyplastics Co., Ltd.	Duranex	
Teijin Ltd.	Teijin PBT	FR-PET
Toray Industries, Inc.	Toray PBT	
Toyobo Co., Ltd.		Vylopet

Table 2.30. Producer and Product Names of Polyethylene Terephthalate Used for Bottles and Films

Producer	Product Name Bottle	Film
Western Europe		
Akzo Plastics BV	Arnite D	Arnite A[1]
Bayer AG		AGFA-Polyesterfilm
Hoechst AG	Polyclear	Hostaphen
ICI	Melinar	Melinex
Rhone-Poulenc Chemie SA		Terphan
Montedipe	Vivipack	Vivifilm[1]
United States		
Bemis Co., Inc.		Esterphane
E.I. Du Pont de Nemours & Co., Inc.		Mylar, Cronar
Eastman Kodak Co.	Kodapak	Tenite, Estar
Goodyear Tire and Rubber Co.	Cleartuf	
Minnesota Mining and Manufacturing Co.		Scotchpac, Scotchpar
Rohm & Haas Co.	Corodel	
Japan		
Diafoil		Diafoil
Teijin Ltd.		Tetoron
Toray Industries, Inc.		Lumirror
Toyobo Co., Ltd.		Espet

[1] Granulates/Pellets

Table 2.31. Producer and Product Name of Copolyester Elastomers

Producer	Product Name
Western Europe	
Akzo Plastics BV	Arnitel
United States	
E.I. Du Pont de Nemours & Co., Inc.	Hytrel
Eastman Kodak Co.	Ecdel
GAF Corp.	Gaflex
General Electric Co.	Lomod
Hoechst Celanese Corp.	Riteflex
Japan	
Toyobo Co., Ltd.	Pelprene

3 Polycarbonates

U. Grigo, K. Kircher, and P. R. Müller

3.1 Introduction

3.1.1 Definition

By general definition, polycarbonates are polyesters of carbonic acid. In other words, they are high molecular weight compounds containing carbonate groups as characteristic structures [1]. To increase the precision of this relatively wide definition, this chapter concentrates essentially on linear thermoplastic polymers of carbonic acid and aliphatic or aromatic dihydroxy compounds. The emphasis is on the aromatic polycarbonates.

3.1.2 Historical Development

Polycarbonates were produced for the first time by Einhorn in 1898 [2]. Phosgene was reacted with an aqueous solution of alkali salts of hydroquinone or resorcinol in the presence of toluene in a kind of phase transfer interfacial reaction. Bischoff and Hedenström got similar products upon reacting resorcinol or hydroquinone with diphenyl carbonate in the melt [1]. These products were insoluble in the usual solvents and, therefore, difficult to investigate. This is most likely the main reason for the lack of any other publications about aromatic polycarbonates for the next 50 years. In 1930 Carothers and Natta investigated high molecular weight polycondensates, including aliphatic polycarbonates [3]. However, these aliphatic polycarbonates were unusable as engineering resins because of their relatively low melting or softening points [3].

In 1953 Schnell and coworkers (Bayer) produced usable aromatic polycarbonates with plastic properties from 2,2-bis(4-hydroxyphenyl)propane (bisphenol A) [4, 5]. At the same time, Fox (General Electric) synthesized bisphenol A polycarbonate independently [2, 6]. In both cases the aromatic diol bisphenol A was used as the basis of investigations. This was technically seen as a serendipitous choice. Later several other bisphenols were investigated, but none of them possessed the favorable combination of properties, production costs, and availability offered by bisphenol A.

After 1953, different processes were developed for the production of bisphenol A polycarbonate. Today the phase transfer interfacial reaction is almost the only process that is used. Phosgene is fed into a mixture of an aqueous alkali bisphenol A solution and a halogenated hydrocarbon solvent (e.g., methylene chloride: see also Section 3.2.2) [7]. The first bisphenol A polycarbonates were sold by Bayer, especially in Europe, and by General Electric, mainly in the United States. After 1970, both companies widened their international activities. Since 1975, Bayer has produced and sold bisphenol A polycarbonate in the United States and General Electric in Europe. Other companies, including several Japanese firms, also produce polycarbonate today. The newest member of this group is Dow Chemical, which entered the market in 1985.

In the last three decades, bisphenol A polycarbonate has quantitatively occupied the second place, after polyamide, among all engineering plastics [5]. The reason is doubtless the unique combination of technological properties such as outstanding toughness, dimensional stability, temperature resistance, and transparency. Remarkable growth rates can be expected for bisphenol A polycarbonate in the future.

References for Section 3.1

1. *Peilstöcker, G.*, in: Kunststoff-Handbuch. Vol. VIII: Polycarbonate. Hanser, Munich 1973
2. *Einhorn, A.:* Liebig's Ann. Chem. 300 (1898) p. 135
3. *Carothers, W. H.*; *van Natta, F. J.*: J. Am. Chem. Soc. 52 (1930) p. 314
4. *Bottenbruch, L.*, in: *Becker, G. W.; Braun, D.* (Eds.): Encyclopedia of Polymer Science and Technology. Wiley, New York, Vol. 10, 1965, p. 710–764
5. DE-PS 971 790 (1953) Bayer AG (*Schnell, H.*; *Bottenbruch, L.*; *Krimm, H.*)
6. *Fox, D. W.:* ANTEC-Paper, Chicago 1985
7. *Clagett, D. C.; Schafer, S. J.:* Polym. Eng. Sci. 25 (1985) p. 458

3.2 Chemical Structure

In general, polycarbonates are linear thermoplastic polyesters of carbonic acid and aliphatic or aromatic dihydroxy compounds:

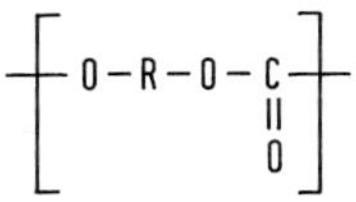

R = aliphatic or aromatic

In the case of 2,2-bis(4-hydroxyphenyl)propane polycarbonate (bisphenol A polycarbonate), the most important member of this class, R is the following structural unit:

R = 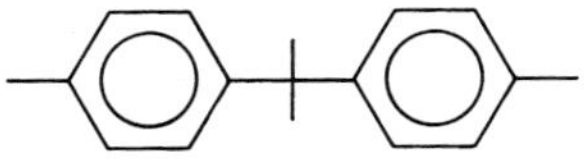

3.2.1 Aliphatic Polycarbonates [1]

Economically, aliphatic polycarbonates are much less important than aromatic polycarbonates and are therefore discussed only briefly here.

3.2.1.1 Preparation

Current methods for the preparation of aliphatic polycarbonates include transesterification of diols with lower dialkyl carbonates made of short-chain alcohols, dioxolanones (e.g. cyclic ethylene carbonates), or diphenyl carbonates in the presence of catalysts such as alkali metal, tin, and titanium compounds. Sensitive diols, such as 2,2′-dihydroxydiethyl sulfide, cannot be converted in this way. As in many other cases, however, polycondensation with phosgene or bischloroformates with aqueous alkali hydroxide is successful.

Based on 1,2-ethylene carbonate, which is made from carbon dioxide and ethylene oxide, other phosgene-free ways to produce aliphatic polycarbonates have been found: 1) starting with carbon dioxide via urea route; 2) transesterification with dialkyl carbonates; 3) starting with carbon monoxide.

Recently, the cationic, or radical, ring-opening polymerization of cyclic carbonic acid orthoesters leading to polyether polycarbonates has been described, as well as the ring-opening polymerization of six-membered cyclic carbonates [1]. Branched polycarbonates with outstanding optical properties are obtained by radical polymerization of diethylene glycol bisallyl carbonate [2].

3.2.1.2 Properties and Uses

The molecular weight of linear, aliphatic polycarbonates is process dependent; it tends to lie between 500 and 5,000, which is adequate for most applications.

High molecular weight polycarbonates ($\leq$ 30,000) are obtained by transesterification, whereas those with molecular weight exceeding 50,000 are obtained by the polymerization of carbonates possessing six-membered rings.

As expected, the starting monomers influence the properties of the polymers. In general, the melting range is lower than the melting range of aromatic polycarbonates. Thermal stability is also lower because the aliphatic carbonate group may undergo ester pyrolysis; therefore, pure aliphatic polycarbonates are rarely used as thermoplastics. When aliphatic polycarbonates contain terminal OH functional groups, they are widely used as the polyol for the synthesis of polyurethanes and polyurethane–urea elastomers.

In addition, certain aliphatic polycarbonates are mentioned in the literature as possible softeners [3], as well as stabilizers [4]. Other uses may be found in medicine or in the optical industry, for example, using branched, aliphatic polycarbonates as raw materials [5].

3.2.2 Aromatic Polycarbonates

Aromatic polycarbonates are prepared by the reaction of bisphenols with carbonic acid derivatives. 2,2-Bis(4-hydroxyphenyl)propane (bisphenol A) is usually the principal component, with phosgene and its derivatives or diphenyl carbonate as the carbonic acid derivative. Lately phosgene-free synthesis, starting with alkyl carbonates or carbon monoxide/bisphenol/heavy metal catalyst combinations, have been found more often in the literature. These preparation processes are described in Section 3.3.

3.2.2.1 Homopolycarbonates

Technically usable polycarbonates can be produced of many bisphenol-type monomers. Of special importance as polycarbonate raw materials are alkylidene and cycloalkylidene bisphenols. Table 3.1 offers an overall look at homo- or copolycarbonates based on modified bisphenols [6–42]. In addition to many other polycarbonates, bisphenol A polycarbonate offers a well-balanced combination of the extraordinary properties that have made it the most important industrial thermoplastic. This colorless, transparent, impact-resistant, dimensionally stable material is usable in a very wide temperature range, is flame retardant, and has good electrical properties. Section 3.7 describes the properties of bisphenol A polycarbonate thoroughly.

In the course of polycarbonate research, numerous new aromatic polycarbonates have been prepared from special monomers [1, 6, 7, 43–46] to overcome some of the inadequacies of bisphenol A polycarbonate (reduction of melt viscosity, hydrolytic

Table 3.1. Aromatic Polycarbonates Made from Bisphenols

Monomers R	Monomers R'	Bisphenol: Melt temperature °C	Polycarbonate: Glass transition temperature °C	Polycarbonate: Melting range °C	Polycarbonate: Special properties[1)]	Literature
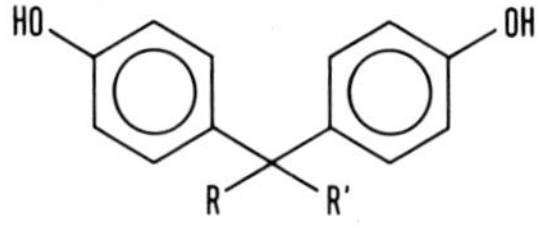						
—H	—H	163	147	223 to 225	—	[6 to 8]
—H	—CH_3	123	130	185 to 195	—	[6 to 8]
—H	—n-C_3H_7	129	123	150 to 170	—	[6 to 8]
—H	—i-C_3H_7	155	149	170 to 180	—	[6 to 8]
—H	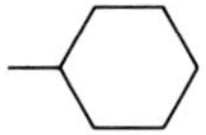	223 to 224	190	—	—	[9, 10]
		221 to 222	198	—	—	
—H		207 to 208	207	—	—	[9]
—H	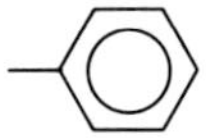	161	159	200 to 215	—	[6 to 9]
—CH_3	—CH_3	157	149	220 to 230	*LOI* = 26.5%	[6 to 9]
—CH_3	—C_2H_5	125	134	205 to 220	—	[6 to 8]
—CH_3	—n-C_3H_7	149	137	200 to 220	—	[6 to 8]
—CH_3	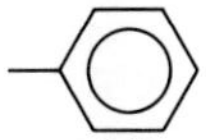	188	176	210 to 230	—	[6 to 8]
—C_2H_5	—C_2H_5	206 to 208	149	175 to 195	—	[11]
—n-C_3H_7	—n-C_3H_7	155	148	190 to 200	—	[6 to 8]
		—	—	175 to 200	—	
—i-C_3H_7	—i-C_3H_7	160 to 161	142	200 to 220	—	[11]
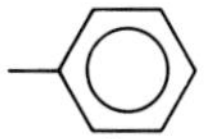	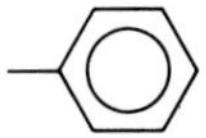	286 to 295	221	210 to 240	high refraction index —	[6 to 8, 12]
—H	—CF_3	—	159	—	flame retardant	[13]
—H	—CCl_3	202 decomposition	220	decomposition	—	[14]

1) *LOI*: oxygen index

Table 3.1. Aromatic Polycarbonates Made from Bisphenols (continued)

Monomers			Bisphenol	Polycarbonate			Literature
			Melt temperature °C	Glass transition temperature °C	Melting range °C	Special properties	
R		R'					
	$=CCl_2$	(Bisphenol C)	214 to 215 decomposition	168	230 to 260	flame retardant, fibers	[14, 15]
$-CF_2Cl$		$-CF_2Cl$	–	180	–	–	[16, 17]
$-CF_3$		$-CF_3$	161	149	–	flame retardant	[13, 18 to 20]
$-CH_3$		$-CN$	170	200	–	–	[21]
$-CN$		$-CN$	147	186	–	–	[21]
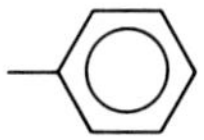		$-CN$	208	138	210 to 225	–	[6 to 8]
$-CH_3$		$-(CH_2)_2COOCH_3$	133 to 134	–	≈ 126	–	[22, 23]
$-CH_3$		$-PO(OCH_3)_2$	217	–	–	flame retardant	[24]
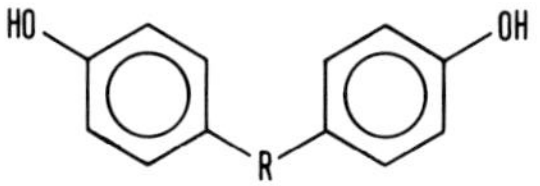							
			157	167	240 to 250	–	[6 to 8]
			190	179	250 to 260	high heat distortion temperature	[6 to 9]
			200	224	–	high heat distortion temperature	[9]
			221 to 223	256	–	high heat distortion temperature	[9]

Monomers	Bisphenol Melt temperature °C	Polycarbonate Glass transition temperature °C	Melting range °C	Special properties 1)	Literature
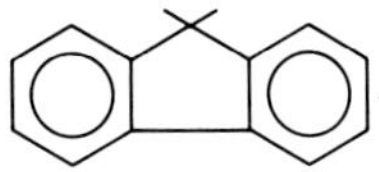	224	356 275	≈ 390		[12, 25, 26]
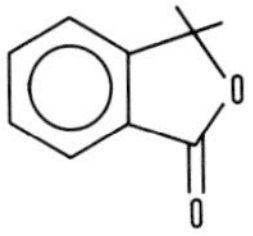	262 to 263	270	–	*LOI* = 38%	[27, 28]
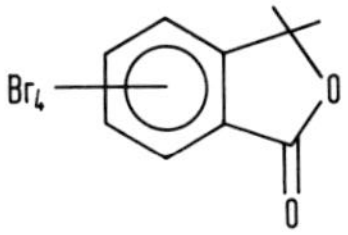	316	290	–	*LOI* = 65 to 75%	[27]
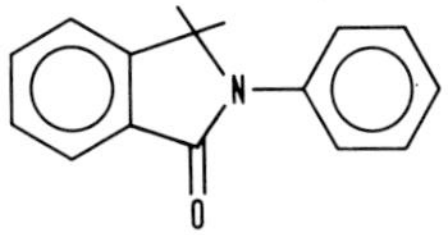	280	250	–	*LOI* = 38%	[27]
$-CH_2-CH_2-$	199	–	290 to 300	crystalline, non-soluble	[6 to 8]
$-(CH_2)_{10}-$	140	–	135 to 150	–	[6 to 8]
	141	120	210 to 230	–	[6 to 8]
	145	162	230 to 250	–	[6 to 8]
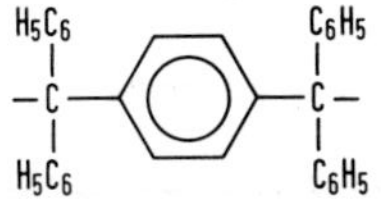	304	228	–	–	[29]

1) *LOI*: limiting oxygen index

Table 3.1. Aromatic Polycarbonates Made from Bisphenols (continued)

Monomers	Bisphenol	Polycarbonate			Literature
	Melt temperature °C	Glass transition temperature °C	Melting range °C	Special properties	
$-C(=O)-$	213 to 215	–	–	photo-sensitive	[30]
$-CH{=}CH-CO-$	197	–	–	photo-crosslinking	[6 to 8, 31]
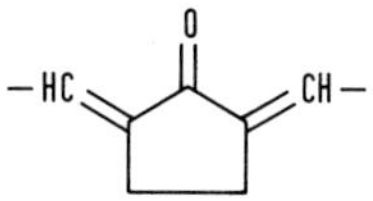	319 to 321	–	–	photo-crosslinking	[6 to 8, 31]
$-O-$	162	145	230 to 235	–	[6 to 8]
$-S-$	152	113	220 to 240	–	[6 to 8, 32]
$-SO_2-$	249	–	200 to 210	–	[6, 7, 33]
Bisphenol structure					
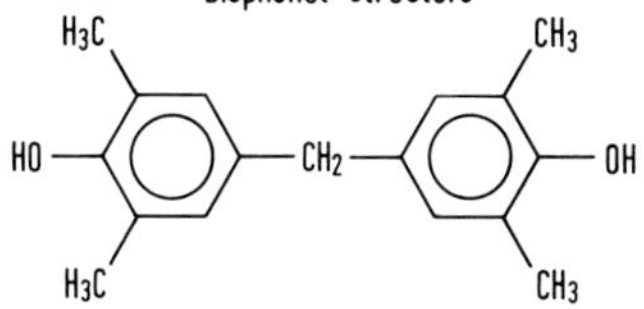	175	185	–	flame retardant	[34, 35]
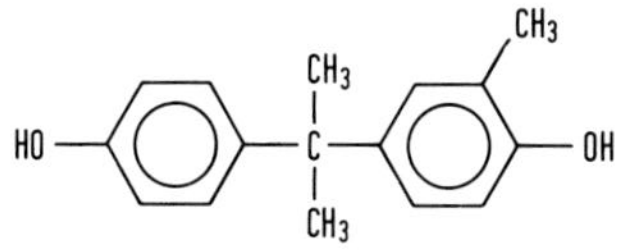	113	140	210 to 220	–	[6 to 8]
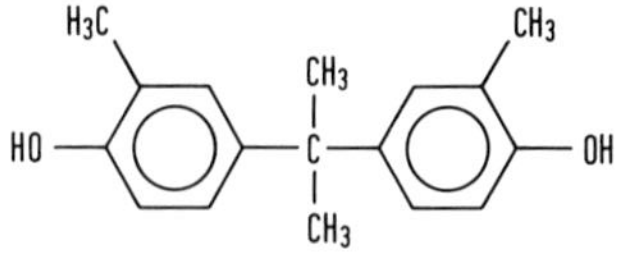	137	128	–	–	[34]

Monomers	Bisphenol	Polycarbonate			Literature
	Melt temperature °C	Glass transition temperature °C	Melting range °C	Special properties	
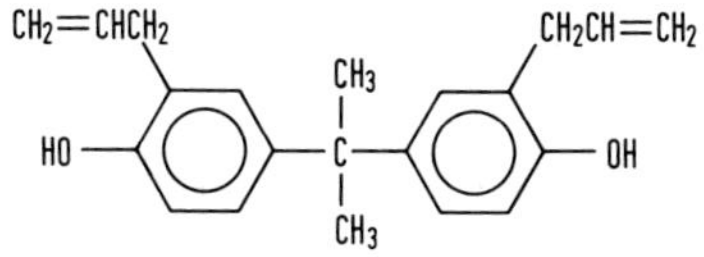	–	–	–	cross-linking	[35]
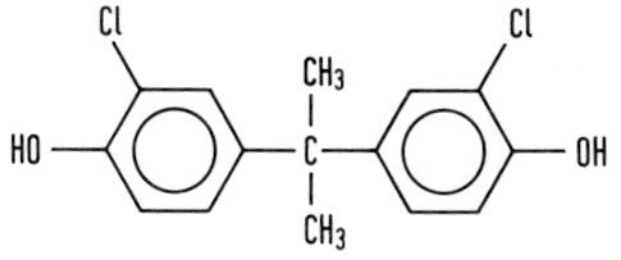	91	147	190 to 210	non-permeable to gas	[6 to 8]
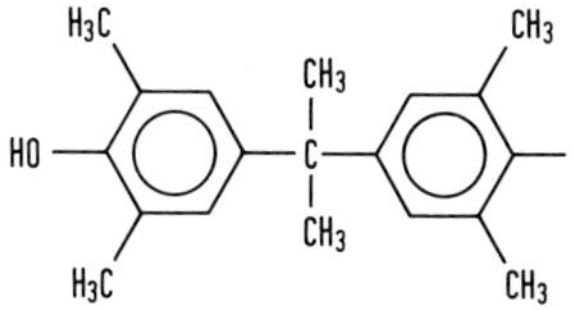	165	207	–	high hydrolytic stability	[34]
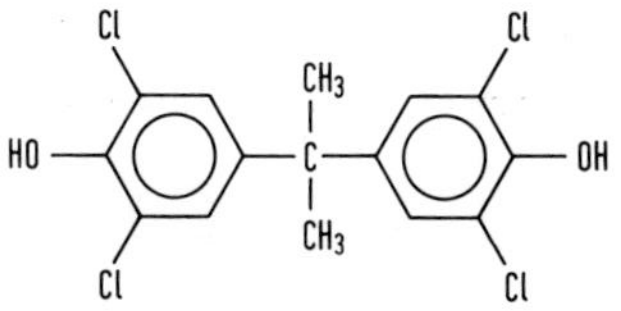	133	230	250 to 260	flame retardant, fibers	[6 to 8, 17]
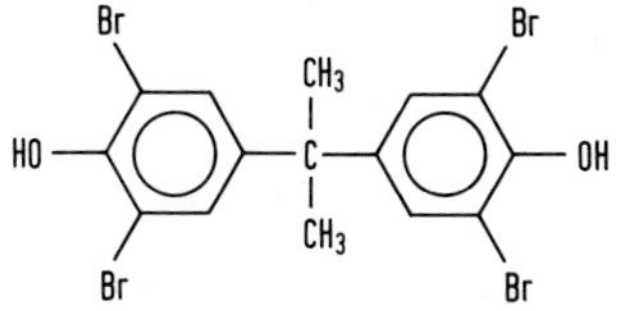	178 to 180	265	–	flame retardant	[6 to 8, 17]

Table 3.1. Aromatic Polycarbonates Made from Bisphenols (continued)

Monomers	Bisphenol: Melt temperature °C	Polycarbonate: Glass transition temperature °C	Polycarbonate: Melting range °C	Polycarbonate: Special properties	Literature
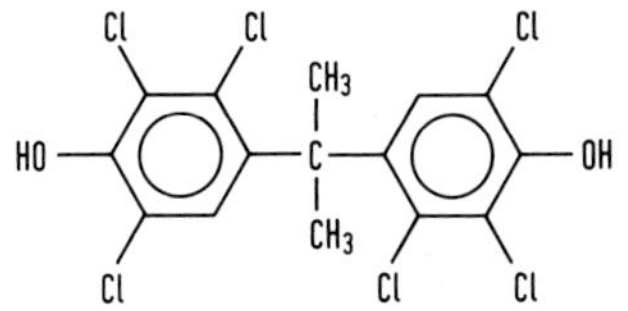	238 to 340	303	–	flame retardant	[36]
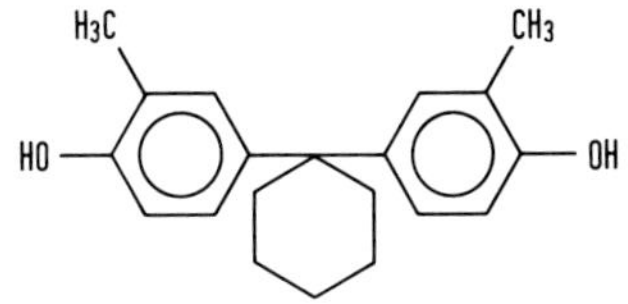	189	–	–	non-permeable to gas	[37]
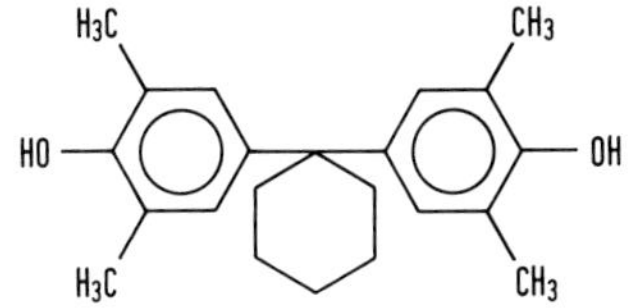	202	213	–	–	[34]
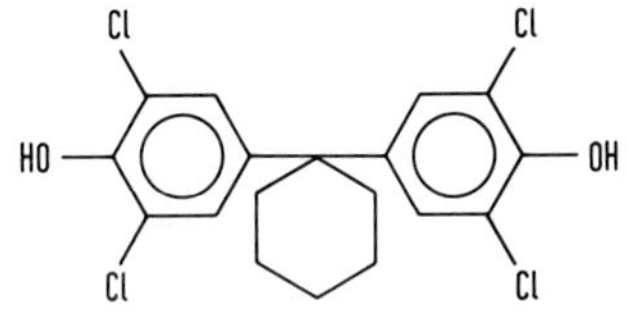	183 to 184	290	–	flame retardant, fibers, high heat distortion temperature	[9]
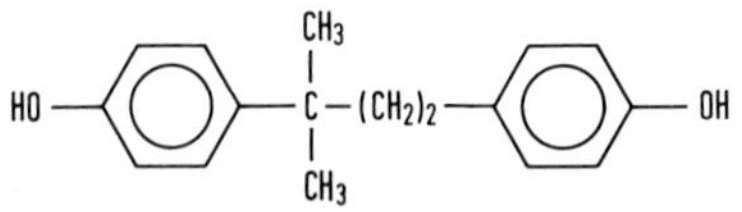	–	200	–	–	[34]

Monomers	Bisphenol	Polycarbonate			Literature
	Melt temperature °C	Glass transition temperature °C	Melting range °C	Special properties	
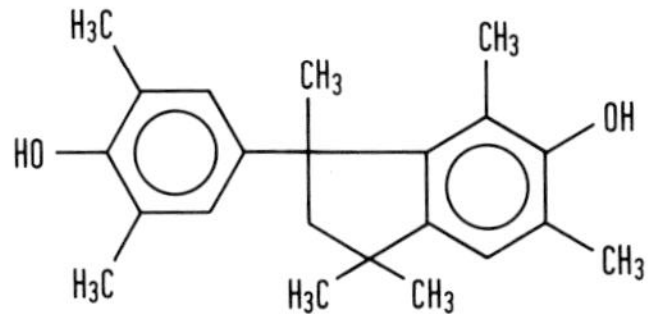	160	265	–	high heat distortion temperature	[38]
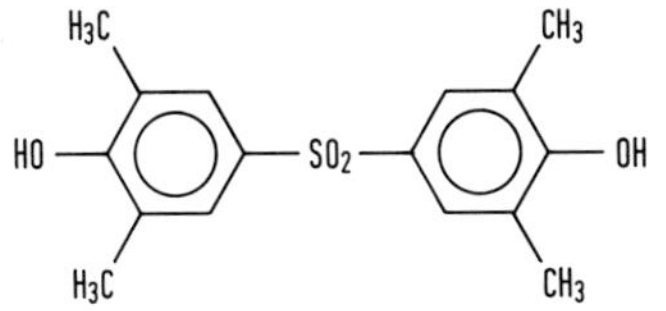	–	250	–	–	[39]
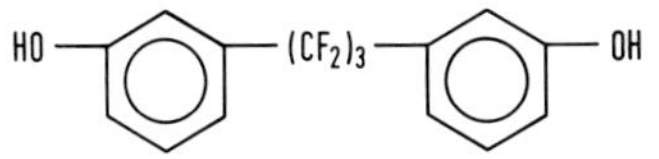	–	46	–	elastomer, glass adherent	[40]
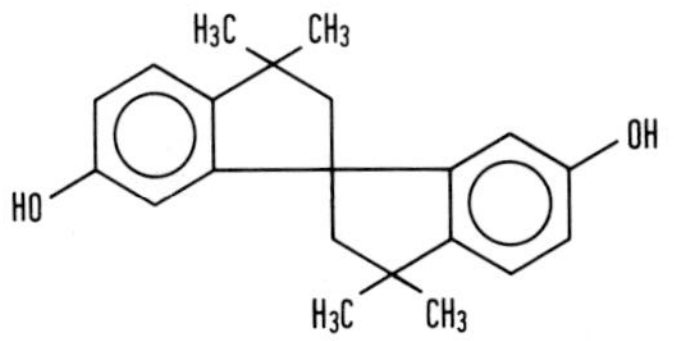	–	230	–	low birefringence, high heat distortion temperature	[41, 42]

instability to bases, and stress cracking on contact with some organic solvents) or to improve still further its good properties (increase of glass transition temperature and use temperature; flame retardance). Compared to bisphenol A polycarbonate, the overall properties of polycarbonates made of alternative bisphenols do not justify, in general, their high production costs. As a result, none of them entered the market as new engineering resins until now. The one exception is described in more detail in Section 3.2.3.4 [1]. Nevertheless, several bisphenols are used as homopolymer additives in blends (i.e., oligomers of tetrabromobisphenol A as flame retardants) or as comonomers to modify bisphenol A polycarbonate (see Section 3.2.2.2).

Another interesting polycarbonate is 3,3′,5,5′-tetramethylbisphenol A (TMBPA) polycarbonate [1]. Compared to bisphenol A polycarbonate ($T_g = 150\ °C$), TMBPA polycarbonate has a much higher glass transition temperature ($T_g = 203\ °C$) combined with a surprisingly low melt viscosity. It has an increased hydrolytic stability, a decreased inclination to crystallize, a better solubility in organic solvents, and a lower refractive index than bisphenol A polycarbonate. However, because of reduced aging stability, lower impact strength, more complicated synthesis, and higher production costs, an industrial use for TMBPA polycarbonate has not yet been found. But TMBPA polycarbonate in blended form is technically interesting because of its excellent compatibility with several thermoplastics, including bisphenol A polycarbonate, polystyrene, styrene–maleic anhydride copolymer, special aliphatic polyesters, and polyvinyl chloride [1, 47, 48, 49].

3.2.2.1.1 End Groups

Because of their extremely high melt viscosity, aromatic polycarbonates are processible only up to certain molecular weights; therefore, monofunctional compounds must be used as molecular weight regulators. The end groups of bisphenol A polycarbonate cannot be neglected because of the relatively low molecular weight of commercialized grades ($M_w \sim 20{,}000–30{,}000$) but must be seen as a kind of comonomer, since the polymer contains about 4 mole % of them.

Very different compounds can be used as molecular weight regulators: in general, phenols, alcohols, carboxylic and sulfonic acid chlorides, amines and amides connected to carbonate esters, and sulfonate or urethane groups. Only phenols and aromatic acid chlorides lead to thermally stable products and have practical importance.

The use of highly branched *p*-alkylphenol instead of phenol or *t*-butylphenol has become important because low viscosity polycarbonates having a high melt flow index with good impact strength, which would be rather brittle with phenyl end groups, can be produced in this way [50, 51] (see Sections 3.7 and 3.10).

3.2.2.1.2 Branched Polycarbonates

Incorporation of tri- or multifunctional monomers in polycarbonates produces branched polycarbonates at low concentrations and crosslinked materials at high concentrations. The difference between branched and crosslinked polycarbonates can be seen in their respective total solubilities.

Branched polycarbonates differ in their rheological behavior from the linear polymers. They have a strong structural viscosity—that is, because of their non-Newtonian flow behavior, they show a sufficiently high melt viscosity during extrusion, but with increasing shearing rate, it drops to the required melt viscosity range for injection molding (Figure 3.1). Branching agents, such as tri- and tetrafunctional phenols and carbonic acids [52, 53], as well as bisphenols with carbonic acid side chains [54], have been used for a long time.

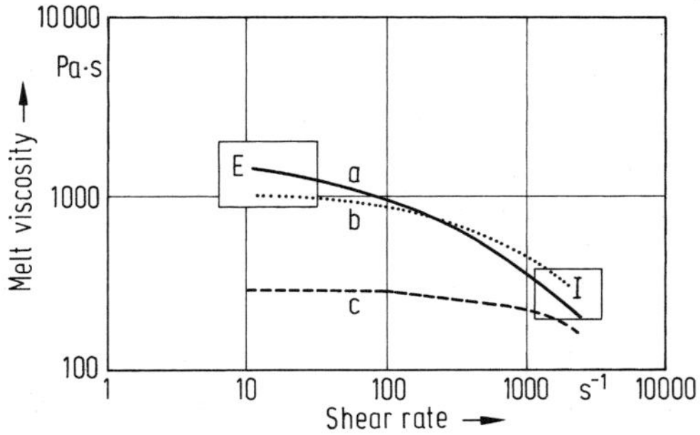

Figure 3.1. Flow curves of branched bisphenol A polycarbonates = a, Makrolon 1143; b, bisphenol A polycarbonate ®Makrolon 3200 (high viscosity); c, bisphenol A polycarbonate ®Makrolon 2600 (low viscosity). Curves show behavior from extrusion region E to injection molding region I.

In addition to thermal stability of the branching points, it is important that the functional groups of the branching agent are not too close together and that they have similar chemical reactivity at polycondensation. Some good examples are 1,4-bis(4′,4″-dihydroxytriphenylmethyl)benzene (I) as well as trisphenol TC (II).

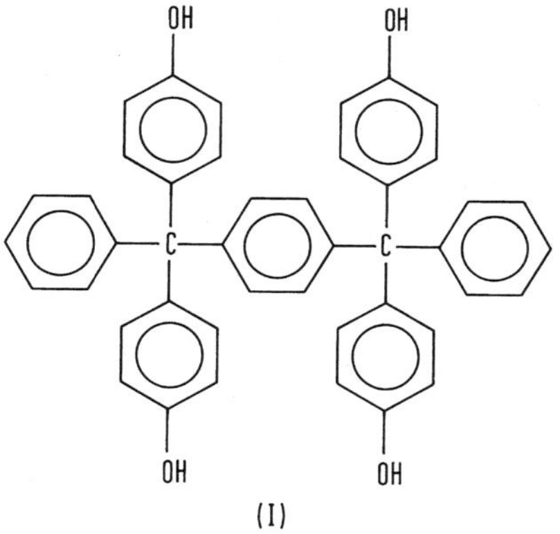

(I)

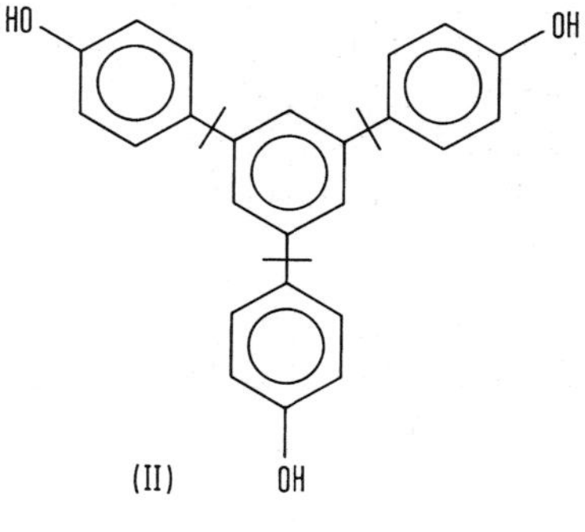

(II)

Nitrogen-containing branching agents are also described: cyanic chloride [55] and 3,3-bis(4-hydroxyphenyl)-2-oxo-2,3-dihydroindole [56].

3.2.2.2 Bisphenol A Copolycarbonates

The most important bisphenol A copolycarbonates are briefly described. Differentiation is made between copolycarbonates that substitute other bisphenols for part of the bisphenol A and copolycarbonates that replace a part of the carbonic acid component with other acids, especially terephthalic and/or isophthalic acid.

3.2.2.2.1 Bisphenol A Copolycarbonates with Modified Bisphenols

Copolycarbonates made of bisphenol A and tetrabromobisphenol A are technically important because only a small mole percentage of the comonomer enhances the flame-retardation properties while having little effect on the standard characteristics of bisphenol A polycarbonate. Tetrachlorobisphenol A causes similar properties in bisphenol A polycarbonate but is a less effective flame retardant. Therefore, no flame-resistant copolycarbonate based on this cobisphenol has been available [1].

Poly- and copolycarbonates based on 2,2-bis(4-hydroxyphenyl)-1,1-dichloroethylene have an impact strength similar to that of bisphenol A polycarbonate, as well as high transparency and good flame retardance. They produce a low amount of fumes and give decomposition products of low toxicity [57].

Notched impact strength at low temperatures [58, 59] or the critical thickness (transition from ductile to brittle behavior in parts >5–6 mm thick) of bisphenol A polycarbonate is increased by copolycondensation with 4,4′-dihydroxydiphenyl sulfide [60]. Compared to impact-modified polycarbonates, these materials offer better handling stability and high transparency [58].

Compared to bisphenol A polycarbonate, copolycarbonates based on sulfone bisphenols (III) are interesting because of their increased glass transition temperature and

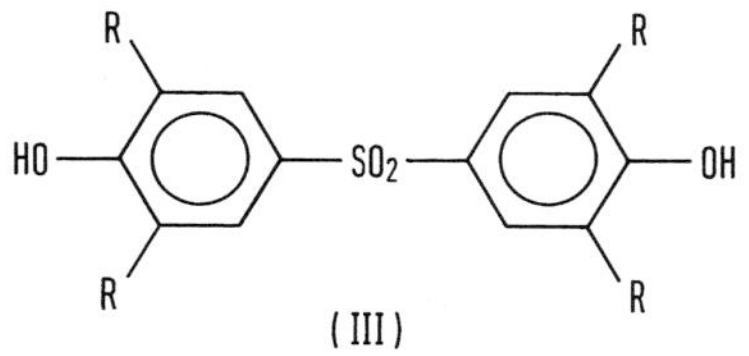

(III)

($R = H$ or CH_3)

thermostability and because of their improved stress-crack resistance to organic solvents [61, 62]. For the unsubstituted sulfone bisphenol ($R = H$), lower hydrolytic resistance is a big disadvantage. Bisphenol A copolycarbonate with phenolphthalein [63] or small amounts (0.5–12.0 mol %) of monoalkyl-bis(hydroxy-4-phenyl)alkyl phosphonate monoalkali salt [64] (IV) have better flame retardance.

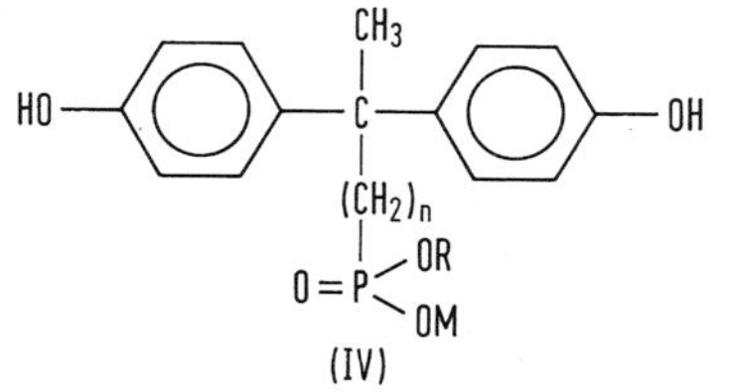

(IV)

$R = CH_3, C_2H_5$
$M = Na, K$
$n = 0, 1, 2$

3.2.2.2.2 Bisphenol A Polycarbonates with Modified Acid Units

The acid unit—carbonic acid—in bisphenol A polycarbonates can be partly replaced by aromatic dicarboxylic acids or their chlorides (i.e., terephthalic acid or isophthalic acid dichloride, or mixtures of these). Aromatic polyester carbonates are obtained.

These substitutions in bisphenol A polycarbonate cause an increase in the glass transition temperature. Because of their economical importance, polyester carbonates up to pure aromatic polyesters are thoroughly described in the chapter on polyarylates in Volume 3.

Partial replacement of the carbonate structure by phosphonate units to obtain aromatic polyphosphonatocarbonates is described in the literature [1] (equation 3.1).

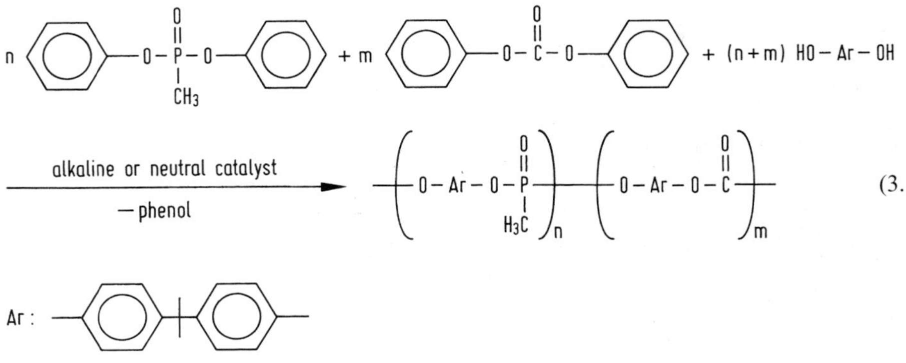

(3.1)

Schmidt and coworkers described a technically usable and economical method to produce polyphosphonatocarbonates [65]: starting with aromatic phosphoric acid esters, diphenyl carbonate, and diphenols, as well as small amounts of branching agents, high molecular weight aromatic polyphosphonatocarbonates are obtained in a melt polycondensation. Condensation of phosphoric acid structures in the polycarbonate chain yields a halogen-free, highly flame-retardant thermoplastic material. All other properties are almost identical to those of bisphenol A polycarbonate. Compared to pure polyphosphonates, polyphosphonatocarbonates are much more hydrolytically stable, as well as more economical to produce. Polyphosphonatocarbonates offer the synergistic advantages of both homopolycarbonates and polyphosphonate.

3.2.2.2.3 Block Copolycarbonates

By incorporating various amounts of chemically different segments—for example, flexible or hydrophilic blocks in the polymer chain—it is possible to modify the property profile of a wide range of polycarbonates. By choosing a suitable segment, it is possible to improve the excellent properties of thermoplastic polycarbonates or to change them considerably.

The synthesis of block copolycarbonates is best carried out in a homogeneous phase such as in an inert solvent or by interfacial polymerization. When melt transesterification is used, phase separation occurs most of the time because of the low compatibility of chemically different oligomers. Therefore, this process is used only in specific cases.

The molecular architecture of segmented polycarbonates is determined by the components of the polycarbonate. Accordingly, triblock (A—B—A) or multiblock structures ($[A{-}B]_n$) are preferred. They are prepared predominantly from prefabricated mono- or bifunctional terminated oligomers. By bonding prepolymers that react alternately (e.g., by reaction of equimolar amounts of hydroxyl-terminated oligomers and chloroformate-terminated oligomers), it is possible to obtain block copolycarbonates of exactly alternating sequence. On the other hand, the phosgenation of different hydroxyl-terminated species (oligomers, bisphenols) leads to irregular random coupling. A completely random structure is obtained only when the hydroxyl-terminated components have the same reactivity and fairly long sequences of the same structure are formed. Phosgenation of bisphenols of different acidity gives a block structure in limited yield [66].

Aromatic poly(polycarbonate-*b*-polycarbonate)s are prepared easily by a two-stage interfacial process in which 3,3′,5,5′-tetramethylbisphenol A, 3,3′,5,5′-tetrabromobisphenol A, and 3,3′,5,5′-tetracholorobisphenol A [67, 68] are used together with bisphenol A. The block copolymers have higher softening points than randomly coupled copolymers of similar composition [69].

For the preparation of aromatic–aliphatic polycarbonate block copolymers by the interfacial method, aliphatic hydroxyl groups must first be converted to more reactive terminal groups.

Thus, the bischlorocarbonates of polyoxyethylene glycol or polyoxytetramethylene glycol, for example, are used for the synthesis of poly(polyether-*b*-polycarbonate)s [70]. Under special conditions, the polyether diols may also be incorporated indirectly by the interfacial method [71]. By incorporation of more than 20–30 wt % polyether blocks (soft segments) in bisphenol A polycarbonate, thermoplastic elastomers with a remarkable combination of high tensile strength and elasticity are obtained [72].

These block copolymers have been developed for use as oil-resistant elastomers and biocompatible membrane materials. The polar hydrophilic character of the flexible polyether segments is utilized in the separation of polar from nonpolar gases, microfiltration, ultrafiltration, elctrophoresis, dialysis (blood purification), and reverse osmosis (desalination of seawater). Permeation properties are much better than those of cellulose hydrate membranes [1].

Poly(polyester-*b*-polycarbonate)s with aliphatic polyester segments based on dimeric fatty acit/hexane 1,6-diol soft blocks can be produced by interfacial polycondensation [73]. By incorporation of sufficient segment length, phase-separated materials with good properties are obtained because of the different solubility parameters of both compounds: 40 $(J/cm^3)^{1/2}$ for polycarbonate and approximately 32.5 $(J/cm^3)^{1/2}$ for the elastomer component. Polycarbonate elastomers with a high content of condensed dimeric fatty acid ester as soft segment can be used as packing materials for biological liquids. Segmented polycarbonates with less than 10 wt % condensed dimeric fatty acid ester have better impact strength at low temperatures, better mold release after thermoplastic molding, and better stress-crack resistance in organic solvents than comparable homopolycarbonates.

Poly(polysiloxane-*b*-polycarbonate)s, in particular those based on polydimethylsiloxane and bisphenol A, are the most interesting block copolymers investigated so far [1]. In general, they are prepared by cophosgenation of bisphenols with the corresponding siloxane oligomers possessing special terminal groups. Exactly alternating poly(polysiloxane-*b*-polycarbonate)s are produced by the reaction of hydroxyl-terminated oligocarbonates with chlorine-terminated polysiloxanes in the presence of tertiary amines, or with dimethylamino-terminated polysiloxanes with thermal elimination of dimethylamine.

These block copolymers give clear transparent elastomeric products with good toughness at low temperatures. Their mechanical and optical properties are due to their two-phase nature and the formation of domains (microphase separation), which act both as physical crosslinking points and as filler. The wettability and friction behaviors of blends with homopolycarbonates are explained in terms of an enrichment of the siloxane segments at the surface, which results from the low surface energy; only polytetrafluoroethylene is superior in this respect. Because water adsorption decreases with increasing siloxane content, certain block copolymers are suitable for coating aluminum or concrete to prevent ice formation. In laminates and fillers, the surface activity of the cocondensate is utilized. Partially crystalline products can be used as selective membranes of high heat distortion temperature.

In blends with polyalkylene terephthalates, poly(polysiloxane-*b*-polycarbonate)s act as impact modifiers: they couple high impact strength with high heat distortion temperature and low warpage in calcium carbonate reinforced or flame-retardant formulations of low molecular weight polyalkylene terephthalates.

The hydrolytic stability of the Si—O—C bond in small molecules is lower than that of the Si—C bond. This property is counteracted by the hydrophobic nature of these block copolymers, the steric hindrance of the long segments, and the small number of bonds. A synergistic effect leads to a maximum limiting oxygen index (LOI) at a siloxane content of approximate 20 wt % without dependence on the block length and the molecular weight. This phenomenon is attributed to the formation by solid phase reaction (pyrolysis) of a carbon layer, which results in a greater barrier action. Therefore, the flame retardance of such block copolymers is more favorable than the flame retardance of homopolycarbonates. The block condensates are thermally stable above 350 °C in air and above 400 °C under nitrogen.

3.2.3 Development Trends for Polycarbonates

Bisphenol A polycarbonate is one of the most important engineering thermoplastics. This lasting success is due to the special combination of properties (rated in comparison to other engineering thermoplastics), including

- excellent toughness
- very good dimensional stability
- transparency
- thermal stability
- good electrical properties

Nevertheless, the properties and the application potential of aromatic polycarbonates are not completely utilized. As Table 3.1 showed, the use of modified bisphenols makes possible a wide range of structural variations, Some structurally modified polycarbonates were presented in Section 3.2.2. A few of these—for example, polyester carbonates or polydimethylsiloxane block copolycarbonates—are already in commercial use.

Sections 3.2.3.1–3.2.3.4 give examples from the recent literature to show the development of aromatic polycarbonates by purposeful structural modification. Aside from structural variation, the properties of bisphenol A polycarbonate can be changed by adding suitable polymers tailor-made for a special use (see Section 3.5, below). ABS and polyalkylene terephthalates are preferably used as blend partners for bisphenol A polycarbonates. Polycarbonate ABS and polycarbonate polyalkylene terephthalate blends are commercially important, but they are not discussed further in this section.

3.2.3.1 Polycarbonates for Laser Optical Data Storage

Polycarbonates with extremely low melt viscosity are suitable as support materials for compact audio disks. Because of the excellent flow behavior of these polymers, a very precise picture of the matrix structure and a low amount of stress birefringence can be obtained (see also Sections 3.2.2.1.1 and 3.7.2.3). But for laser optical data storage—for example, with magneto-optical or phase change systems for information storage—the decrease in birefringence obtained by using low viscosity polycarbonates is insufficient. One possible way to decrease the birefringence of a polycarbonate to almost zero is to mix polycarbonate or structurally modified polycarbonate (e.g., tetramethylbisphenol A polycarbonate) with blend partners such as certain styrene copolymers having the opposite birefringence sign. Some examples are described in Section 3.5.

Another possible way to decrease or even bring to zero the birefringence of aromatic polycarbonates is the use of monomers of opposite birefringence. The relatively high birefringence of bisphenol A polycarbonate can be attributed to the alignment of the phenyl rings "in series." Incorporation of bulky polarizable side groups decreases the birefringence of polycarbonates visibly. Examples illustrating this principle are bisphenol A and Z polycarbonates [74].

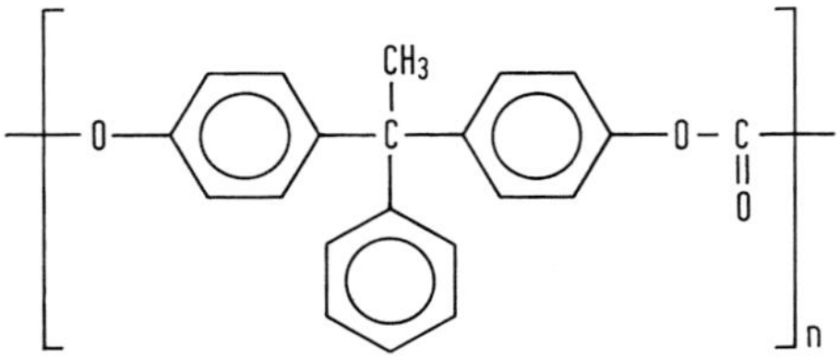

Bisphenol A polycarbonate: $T_g = 176\ °C$

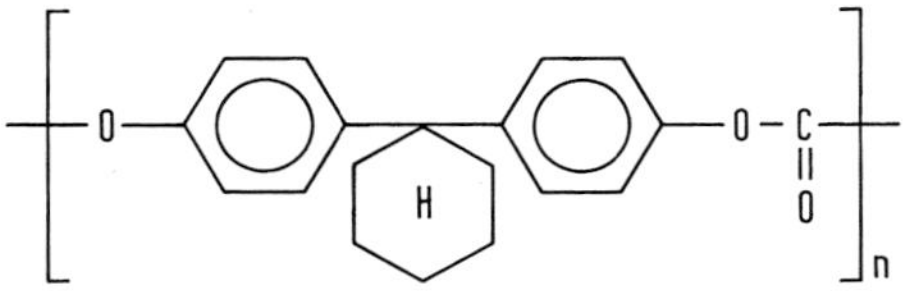

Bisphenol Z polycarbonate: $T_g = 179\ °C$

Intramolecular compensation for birefringence is also possible if the two phenyl rings of the bisphenol are fixed perpendicular to each other. 6,6′-Dihydroxy-3,3,3′3′-tetramethyl-1,1-spiro(bisindane)polycarbonate has this special structure [75]:

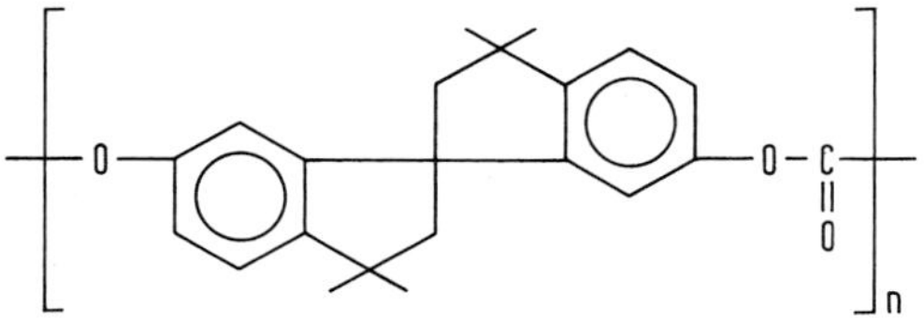

$T_g = 231\ °C$

The homopolymer is virtually birefringence-free, while the birefringence of the copolycarbonate with bisphenol A is dependent on the spiro(bisindane) content [42].

So-called structurally modified polycarbonates have reduced birefringence by virtue of an increase in glass transition temperatures or thermal stability [43]. Laser optical data storage disks based on structurally modified polycarbonates are not commercially available today.

3.2.3.2 Polycarbonate-Based Optical Fibers

Optical fibers based on polycarbonate are tougher than those made of glass. Therefore, larger fiber diameters with small radius of curvature are possible [76]. Furthermore, the polymeric fibers are connected with the source of light more easily. Light loss is higher than with optical fibers made of glass, however.

Optical fibers based on polymers offer many advantages when used to transmit light signals over short distances: for example, in aircrafts or cars. Here easy handling is more important than optical damping.

Polymers used in high purity form for optical fibers are polymethyl methacrylate (PMMA $T_g = 105$ °C), polystyrene (PS, $T_g = 100$ °C), and polycarbonate (PC, $T_g = 150$ °C). Compared to PMMA and PS, polycarbonate has a higher ductibility, allowing a much smaller ratio of curvature, as well as higher thermal stability, which is necessary for fibers that are used in engines. A disadvantage of polycarbonate fibers is the higher light damping.

Regarding optical fibers based on polycarbonate, it is important to produce optically pure polycarbonate fibers and to choose the right "cladding." The "cladding" is a transparent polymer covering the transmitting fiber that decreases or prevents the loss of light due to reflection. Therefore, the cladding polymer must have a lower index of refraction than polycarbonate. Polymethyl methacrylate, polymethylpentene, and copolycarbonates based on 2,2-bis(4-hydroxyphenyl)hexafluoropropane and bisphenol A are used [77].

3.2.3.3 Polycarbonates Made for Membranes

Polymers can be used as membranes to separate components that differ in their permeabilities. Transportation in polymer matrices was investigated intensively in the 1980s [78–80] and different models of transportation were developed.

Since they are soluble in special organic solvents, bisphenol A polycarbonate can be used to produce films and membranes having high heat distortion temperature and ductility. The flexibility of polycarbonate chains, which produces excellent toughness, also decreases packing density (e.g., compared to polyethylene terephthalate), hence increases gas permeability. Incorporation of soft segments, such as polyether segments or polydimethylsiloxane segments, further increases chain flexibility.

Membranes based on such block copolycarbonates are now commercially available [81] and are also used in form of membranes or hollow fibers for blood dialysis [82]. Polyether carbonate hollow fiber modules can be made with 8,640 single fibers, each of 230 μm diameter, with an effective membrane surface of 1 m^2 and having a high diffuse permeability because of their fine-pored structure. Hydrophilic polyether carbonate is much higher in separation capacity than classical membranes made of regenerated cellulose.

Similar to fibers made for dialysis, hollow fiber membranes with pores 100 times larger can be produced for plasma separation. As shown by this example, complex problems of membrane technology can be solved with the help of tailor-made polycarbonates. Compared to the film market, where bisphenol A polycarbonate is established as both a casting and an extruded medium, membrane technology has just begun to use polycarbonates. Tailor-made polycarbonates seem to be potential candidates for product development in this area because of their good solubility, which facilitates the production of membranes. Another major advantage is their broad potential for structural variation, which means property variation.

3.2.3.4 Polycarbonates with Higher Heat Distortion Temperature

The heat distortion temperature of bisphenol A polycarbonate ($T_g = 150$ °C) is already high. However, it is not high enough for some uses. In electronics, the trend to miniaturize such items as automotive front lights is well established. Therefore, structural units made of plastic should have high thermal stability. The heat distortion temperature of bisphenol A polycarbonate blends does not allow on-line coating, however, and this is the procedure most automobile producers use to coat visible plastic parts.

These few examples show that the need for polycarbonates with higher heat distortion temperatures is rising constantly.

One approach to increasing the heat distortion temperature of bisphenol A polycarbonate is replacement of the carbonate structure by aromatic dicarboxylic acids (e.g., terephthalic and/or isophthalic). These commercially available polyester carbonates have very interesting properties, but because of the aromatic ester structural units, they tend to yellow easily under UV radiation. Aromatic polyester carbonates are described more precisely in the chapter on polyarylates in Volume 3.

Another possibility for increasing the heat distortion temperature of aromatic polycarbonates is the purposeful variation of the bisphenol structure. As indicated in Table 3.1, there are several routes to this goal:

- tetrasubstitution of the (3,3′,6,6′) phenyl position ortho to the phenolic —OH [e.g., tetramethylbisphenol A, tetrahalogenated bisphenol A, tetramethylbis(4-hydroxyphenyl)sulfone]
- ensuring that the carbon bridging the two phenyl rings is part of a cycloaliphatic ring system [e.g., 1,1-bis(4-hydroxyphenyl)cyclohexane, 3,3,3′,3′-tetramethyl-1,1-spiro-(bisindane) bisphenol, phenolphthalein] or has a side-standing phenyl ring, like 1,1-bis(4-hydroxyphenyl)phenylethane.

In 1988 Bayer AG succeeded in increasing the glass transition temperature into the range of 240 °C with a polycarbonate based on 1,3,3′-trimethylcyclohexanone bisphenol (BPTMC) [83].

At "K 89" (Plastics fair at Düsseldorf/Germany in 1989) the first samples of copolycarbonates made of bisphenol TMC and bisphenol A were introduced under the trade name Apec HT [41].

The extremely short development time can be explained by the properties as well as the good availability of the bisphenol TMC polycarbonates. The starting monomer, bisphenol TMC, can be prepared in high yields ($>80\%$) by the hydrogenation of isophorone followed by the reaction of the dihydroisophorone with phenol [84, 87]:

Like bisphenol A polycarbonate, bisphenol TMC polycarbonate can be produced in an interfacial process. Because the reactivities of the phenolic hydroxy groups of bisphenol TMC and bisphenol A are nearly the same, copolycarbonates can be prepared in all mixture ratios. Therefore, it is possible to cover a deformation temperature range between 150 and 238 °C [84].

Based on this new polycarbonate technology, several copolycondensates with different deformation temperatures (glass transition or Vicat B temperatures of 160, 172, 185, 195, and 205 °C) were developed. These products, obtainable under the trade name ®Apec HT, are new high performance thermoplastics that exhibit a previously unattainable combination of transparency, high impact strength, good flowability, and thermal stability.

The most important properties of these new products are compared to Makrolon 2800, a bisphenol A polycarbonate, in Table 3.2 [85, 86].

The glass transition temperature of bisphenol TMC polycarbonate ($T_g = 238$ °C) is much higher than that of bisphenol A polycarbonate ($T_g = 149$ °C) because of the very limited mobility of the bisphenol TMC unit. To explain this, one must consider two parallel effects: the limited conformational mobility of the cyclohexane ring and the limited mobility (rotation) of the phenyl ring due to 1,3-diaxial interactions. Because the mobility of the carbonate group is obviously not affected, bisphenol TMC polycarbonate and copolycarbonates with bisphenol A are a little less tough than bisphenol A

Table 3.2. Properties of selected Bisphenol TMC-Copolycarbonates in comparison to Standard-Bisphenol A-Polycarbonate

Properties	Unit	Method	Standard BPA-PC	Apec HT KU 1-9331	KU 1-9351	KU 1-9371
Mechanical						
Tensile strength	N/mm^2	ISO/R 527, DIN 53 455	72	60	60	60
Elongation at break	%	ISO/R 527, DIN 53 455	120	70	70	50
Yield strength	N/mm^2	ISO/R 527, DIN 53 455	63	65	65	65
Elongation at yield	%	ISO/R 527, DIN 53 455	6	7	7	7
Tensile modulus	N/mm^2	DIN 53 457-t	2400	2200	2200	2200
Impact strength 23 °C	kJ/m^2	ISO 180/1 C	no failure	no failure	no failure	no failure
Impact strength −40 °C	kJ/m^2	ISO 180/1 C	no failure	no failure	no failure	no failure
Notched impact strength 23 °C	kJ/m^2	ISO 180/1 A	30	12	8	5
Thermal						
HDT Method A (1.81 N/mm^2)	°C	ISO 75, DIN 53 461	131	140	162	179
Method B (0.45 N/mm^2)	°C	ISO 75, DIN 53 461	138	152	174	195
Softening temperature VST/B 120	°C	ISO 306, DIN 53 460	148	160	185	205
Thermal expansion coefficient α	$10^{-6} \cdot K^{-1}$	DIN 53 752-3	70	75	75	75
Electrical						
Relative permittivity ε_r 50 Hz		IEC 250, VDE 0303, part 4	3.0	3.0	2.9	2.8
1 kHz		IEC 250, VDE 0303, part 4	3.0	3.0	2.9	2.8
1 Mhz		IEC 250, VDE 0303, part 4	2.9	3	2.9	2.8
Dissipation factor tan $\delta \cdot 10^3$ 50 Hz		IEC 250, VDE 0303, part 4	1	1.6	1.4	1.3
1 kHz		IEC 250, VDE 0303, part 4	1	0.8	1.0	0.8
1 Mhz		IEC 250, VDE 0303, part 4	10	8.7	10.2	8.9
Dielectric strength E_d	kV/mm	IEC 243, DIN VDE 0303, part 2	30	35	35	35
Optical						
Refractive index n_D		DIN 53 491	1.584	1.581	1.572	1.565
Light transmission, wall thickness: 1 mm	%	DIN 5036, part 3	89	90	90	90
Density	g/cm^3	ISO/R 1183, DIN 53 479	1.20	1.18	1.15	1.14

polycarbonate; but in general the toughness of these bisphenolic polycarbonates is still relatively high [84, 87].

Because of their lower melt viscosity, bisphenol TMC copolycarbonates can be processed more easily than aromatic polyester carbonates having a similar thermal stability and better, at lower melt temperatures, than other amorphous thermoplastics such as polyether imide, polysulfone, and polyether sulfones.

Because it has properties similar to bisphenol A polycarbonate at higher deformation temperatures, bisphenol TMC copolycarbonates (®Apec HT grades, Bayer AG) can be used for high thermal stress applications including auto lighting (reflectors, interior indicators caps, inner and outer lenses), household appliances, and in medical technology (hot steam sterilization at 134 °C) [88].

Other possible uses can be seen in blends and, because of the decreased optical birefringence compared to bisphenol A polycarbonates, in rewritable data storage disks (MOR) and in optical applications (lenses for projectors or CD players). Furthermore, the good compatibility of bisphenol TMC copolycarbonates with bisphenol A polycarbonate allows these materials to be recycled together.

References for Section 3.2

1. *Freitag, D.; Grigo, U.; Müller, P. R.; Nouvertné, W.*, in: Encyclopedia of Polymer Science and Technology, Vol. 11, Chapter: Polycarbonate. Wiley, New York, 1988
2. US-PS 3 497 478 (1970) Atlantic Richfield
3. US-PS 3 305 605 (1964) Union Carbide
4. US-PS 4 255 301 (1978) Argus Chem. Corp.
5. US-PS 4 260 564 (1979) PPG
6. *Schnell, H.*: Polym. Rev. 9 (1964)
7. *Schnell, H.*: Angew. Chem. 68 (1956) p. 633
8. Ind. Eng. Chem. 51 (1959) p. 157; Plast. Inst. Trans. 28 (1960) p. 143
9. *Jackson, W. J. Jr.; Caldwell, J. R.*: Ind. Eng. Chem., Prod. Res. Dev. 2 (1963) p. 246
10. DE-SO 3 343 898 (1984) General Electric
11. *Fox, D.*: Powder Coating to Recycling (Rubber). In: *Grayson, M.* (Ed.): Kirk-Othmer-Encyclopedia of Chemical Technology. 3. Ed., Vol. 18. Wiley, New York 1982, p. 479–494
12. *Morgan, P. W.*: Macromolecules 3 (1970) p. 536
13. WO-PS 82/02402 (1982) General Electric
14. *Sobieczewski, Z.; Wielgosz, Z.*: Plaste Kautsch. 15 (1968) p. 176
15. *Factor, A.; Orlando, C. M.*: J. Polym. Sci., Polym. Chem. Ed. 18 (1980) p. 579; *Mac Laury, M. R.*, et al.: J. Polym. Sci., Polym. Chem. Ed. 18 (1980) p. 2501
16. *Massa, D. J.; Flick, J. R.*: Polym. Prepr, Am. Chem. Soc., Div. Polym. Chem. 14 (1973) p. 1249
17. DE-OS 2 462 470 (1977) Bayer AG; *Gilbert, E. C.; Brindell, G. D.*: Am. Chem. Soc., Div. Org. Coat. Plast. Chem. Pap. 34 (1974) p. 136
18. *Gitina, R. M.*, Russ., et al.: Chem. Rev. 40 (1971) p. 679
19. EP 29 111 (1981) Mobay Chem. Corp.
20. *Said, E.*, et al.: Vysokomol. Soedin 7 (1965) p. 729
21. *Banucci, E. G.*: J. Polym. Sci., Polym. Chem. Ed. 11 (1973) p. 2947
22. *Fischer, R. P.*, et al.: J. Appl. Polym. Sci. 10 (1966) p. 245
23. *White, R. E.; Garlund, Z. G.*: J. Polym. Sci., Polym. Chem. Ed. 18 (1970) p. 1419
24. *Iliopulos, M. I.; Weider, H.*: Angew, Chem. 77 (1965) p. 618
25. *Kambour, R. P.*, et al.: J. Appl. Polym. Sci. 20 (1976) p. 3275
26. *Kambour, R. P.*, et al.: J. Polym. Sci., Polym. Lett. Ed. 16 (1978) p. 327
27. *Lind, M. S.; Pearce, E. M.*: J. Polym. Chem., Polym. Chem. Ed. 19 (1981) p. 2659; *Morgan, P. W.*: J. Polym. Sci. 2 (1964) p. 437
28. DE-OS 2 637 003 (1977) Dow Chemical Co.
29. DE-OS 2 425 291 (1975) Bayer AG
30. DE-OS 2 746 141 (1979) Bayer AG

31. *Borden, D. G.:* Am. Chem. Soc., Div. Org. Coat. Plast. Chem. Pap. 35 (1975) p. 96
32. DE-OS 2 721 595 (1977) Mobay Chem. Corp.; EP 40 863 (1981), Mobay Chem. Corp.; EP 84 578 (1983), Mobay Chem. Corp.
33. EP 547 (1979), Mobay Chem. Corp., EP 75 772 (1983) Bayer AG
34. *Serini, V.*, et al.: Angew. Makromol. Chem. 55 (1976) S. 175
35. EP 143 386 (1985) *A. Noguchi; K. Honna; G. S. Kolesnikov*, et al.: Vysokomol. Soedin. Ser. A 12 (1970) p. 1739
36. DE-OS 2 254 919 (1974) Bayer AG
37. EP 16 167 (1980) General Electric
38. DE-OS 2 615 038 (1977) Bayer AG; DE-OS 2 901 668 (1980) Bayer AG
39. *Fox, D. W.*, et al.: Techn. Pap. Reg. Techn. Cof. Soc. Plast. Eng. Baltimore-Washington Sect. (1985) p. 951
40. *Critchley, J. P.*, et al.: Chem. Ind. 61 (1969) p. 934
41. Mod. Plast. Int. (1989) 12
42. EP 287 887 (1988) General Electric
43. Mod. Plast. Int. (1988) 12, p. 16
44. *Schnell, H.:* Polym. Rev. 9 (1964) p. 1
45. *Moyers, C. G.*, et al.: CEP (1986) 5, p. 42–46
46. *Fox, D. W.:* Encycl. Chem. Techn. 18 (1982) p. 479
47. *Fernandes, A. C.*, et al.: Polymer 27 (1986) p. 1788; Polymer 27 (1986) p. 1799
48. *Humme, G.*, et al.: Makromol. Chem. 58/59 (1977) p. 85
49. *Casper, R.; Morbitzer, L.:* Makromol. Chem. 58/59 (1977) p. 1
50. DE-OS 2 842 005 (1980) Bayer AG
51. Makrolon FCR Polycarbonate Resin, company brochure Mobay Chem. Corp., USA 1985
52. DE-OS 2 254 918 (1974) Bayer AG
DE-OS 1 595 641 (1970) Bayer AG
DE-OS 1 570 533 (1970) Bayer Ag
53. DE-OS 2 113 347 (1972) Bayer AG
54. DE-OS 1 595 762 (1965) General Electric
55. DE-OS 2 718 466 (1978) Bayer AG
56. DE-OS 5 000 92 (1976) Bayer AG
57. *Factor, A.; Orlando, C. M.:* J. Polym. Sci., Polym. Chem. Ed. 18 (1980) p. 579
58. *Witman, M. W.*, et al.: SPE Techn. Pap. 26 (1980) p. 470
59. *Cama, F. J.; Morbitzer, L.:* SPE Ann. Techn. Conf. 28 (1982) p. 96–98
60. EP 40 863 (1981) Mobay Chem. Corp.; EP 84 578 (1983) Mobay Chem. Corp.
61. *Fox, D. W.*, et al.: ANTEC 1985, p. 951–953
62. EP 75 772 (1983) Bayer AG
EP 547 (1979) Mobay Chem. Corp.
63. *Liu, M. S.; Pearce, E. M.:* J. Polym. Sci., Polym. Chem. Ed. 19 (1981) p. 251
64. *Poisson, P.; Sturtz, G.:* Polym. Bull. 1 (1979) 897
65. DE-OS 2 925 208 (1979) Bayer AG
DE-OS 3 001 863 (1980) Bayer AG
66. EP 0 140 341 (1985) General Electric
67. DE-OS 2 211 956 (1973) Bayer AG
68. *Kolesnikow, G. S.*, et al.: Vysokomol. Soedin., Ser. A 12 (1970) 6, p. 1924
69. EP 61 060 (1982) Mitsubishi Chem. Ind., Ltd.
70. *Merrill, S. H.:* Polym. Sci. 55 (1961) p. 343
71. DE-OS 3 408 803 (1984) Bayer AG
72. *Goldberg, E. P.:* J. Polym. Sci., Polym. Lett. Ed. 4 (1964) p. 707
73. *Dhein, R.*, et al.: Angew. Makromol. Chemie 139 (1986) p. 157
74. *Buning, G. H. W.*, et al.: Proc. Int. Symp. on Optical Memory, Poster WC 23, Tokyo 1987
75. *Stueben, K.-C.:* Polym. Sci., Polym. Chem. Ed. 3 (1965) p. 3209
76. *Emslie, C.:* Mat. Sci. 23 (1988) p. 2281
77. EP 0 282 062 (1988) Idemitsu Petrochemical Co. Ltd.
78. *Alger, M. M.*, et al.: Polym. Sci., Poly. Phys. Ed. 27 (1989), p. 97
79. *Barbari, T. A.*, et al.: Polym. Sci., Poly. Phys. Ed. 26 (1988) p. 709 and p. 729
80. *Muruganandam, N.*, et al.: Polym. Sci., Poly. Phys. 25 (1987) p. 1999
81. *Kesting, E. R.:* Synthetic Polymer Membranes, Wiley, New York, 1985, p. 145/146
82. *Nerger, D.*, et al.: Angew. Makromol. Chem. 166/167 (1989) p. 57

83. EP 0 359 953 (1988) Bayer AG
84. *Freitag, D.; Fengler, G.; Morbitzer, L.:* Angew. Chem. 103 (1991) p. 1626
85. *Kämpf, G.; Freitag, D.; Fengler, G.:* Kunststoffe 82 (1992) p. 5
86. Apec HT. "Das neue Polycarbonat mit hoher Wärmeformbeständigkeit". Bayer AG, 1992
87. *Freitag, D.; Westeppe, U.:* Makromol. Chem., Rapid Commun. 12 (1991) p. 95
88. *Grigo, U.:* Swiss Plastics 16 (1994) p. 5

3.3 Preparation of Aromatic Polycarbonates

Aromatic polycarbonates are derived from the reaction of bisphenols with carbonic acid derivatives. The arrangement of the condensed monomers in polymers, which can be linear, branched, or macrocyclic, can be controlled by the reaction conditions. Chain termination of linear and branched polycarbonate grades is effected with monofunctional monomers (e.g., phenol), which are able to control the molecular weight and influence the properties of the polymers. Using polyfunctional compounds (e.g., multifunctional phenols), branches can be obtained. Specially terminated and branched polycarbonates were described earlier (Sections 3.2.2.1.1, 3.2.2.1.2).

3.3.1 Starting Materials

3.3.1.1 Carbonic Acid Components

Relatively few carbonic acid derivatives are suitable for use in the production of polycarbonates. Because the free acid is not stable, only derivatives of carbonic acid can be used as reaction partners with bisphenols: phosgene, its esters, and acid chloride esters (e.g., chloroformic acid esters) [1].

For a long time phosgene has been obtained by reaction of carbon monoxide and chlorine at activated carbon contacts (e.g., as starting material for dyes and isocyanates) [1]. Phosgene, $COCl_2$, is a colorless liquid that boils at 7.5 °C. At room temperature, it is a colorless gas with a characteristic odor. Phosgene is soluble in organic solvents and can be easily removed from these by heating. Because phosgene is toxic, special safety precautions must be observed [2, 3].

Instead of phosgene, chloroformic acid trichloromethylester ($Cl_3C—O—CO—Cl$, diphosgene) or ditrichloromethyl carbonate ($Cl_3C—O—CO—OCCl_3$, triphosgene) can be used. These decompose easily to 2 or 3 mol of phosgene, respectively [4].

Also usable for polycarbonate synthesis are diaryl carbonates and chloroformic acid esters of bisphenols, but these materials are not very important because phosgene as the acid component in the synthesis—the interfacial process—is commercially used today.

The bischloroformic acid ester of bisphenol A occurs as an intermediate during the interfacial reaction of phosgene and bisphenol A. The ester can also be produced independently and can be used for special polycarbonate synthesis—for example, for the production of cyclic oligomer carbonates (see Section 3.3.5).

Diaryl carbonates can be produced from phosgene and phenol, or obtained from the transesterification of dialkyl carbonates and phenol [5, 6]. Dialkyl carbonates are obtained from phosgene and alcohol, from the reaction of ethylene glycol carbonate with alcohols, or from the oxidative carbonylation of alcohols [7]. Under pressure, ethylene oxide and carbon dioxide form ethylene glycol carbonate [8], but this stage can be bypassed by reacting the alcohol directly with carbon dioxide in the presence of ethylene oxide as a water acceptor [9].

Dimethyl carbonate can also be produced by oxidative carbonylation directly from methanol and carbon monoxide in the presence of catalysts that contain palladium or copper [6, 10, 11]. This process was developed by Anic using homogeneously dissolved catalysts. Today the process is used with heterogeneous catalysis in the gas phase, which greatly simplifies the purification of the obtained dimethyl carbonate. In a similar way, diphenyl carbonate can be produced by oxidative catalytic reaction of phenol and carbon monoxide, but yields are still quite low [12, 13], as indicated by Figure 3.2.

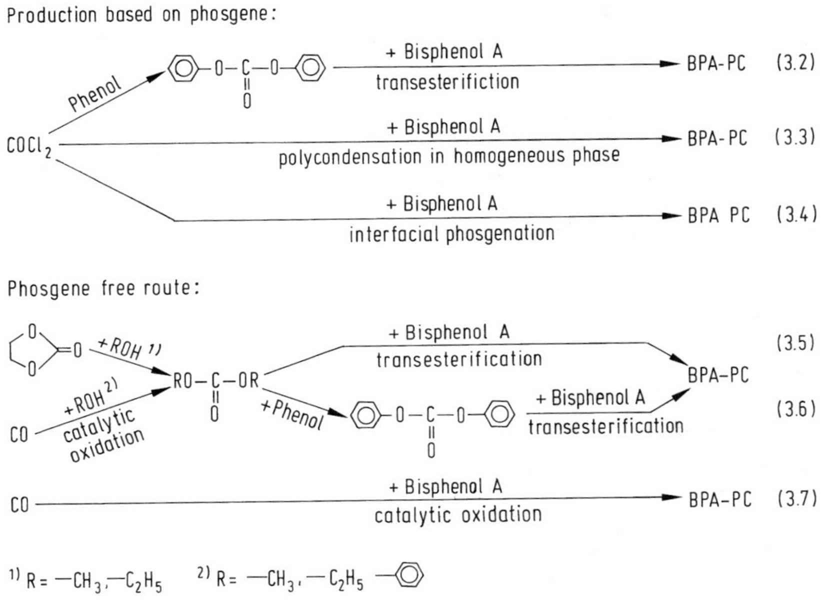

Figure 3.2. Two ways of producing aromatic polycarbonates—in this case, bisphenol A polycarbonate (BPA-PC).

The processes described by reactions 3.5–3.7 are being investigated quite intensively in the chemical industry as possible alternatives to the interfacial polycondensation approach that has been used until now.

3.3.1.2 Bisphenol Components

Several bifunctional phenols can be used to produce polycarbonates with good properties. Dihydroxydiphenylalkanes are called bisphenols [1]. The most important bisphenol is 2,2-bis(4-hydroxyphenyl)propane, or bisphenol A, the condensation product of 1 mol of acetone and 2 mol of phenol.

3.3.1.2.1 Physical Properties of Bisphenols [14, 15]

Bisphenols are colorless and odorless and most of them are solid substances. The most important bisphenols melt between 100 and 200 °C (bisphenol A, 156–157 °C). Bisphenols are virtually insoluble in water; their solubility in organic solvents is determined by the substituents. Bisphenol A is soluble only in polar media, such as ethers and alcohols, while large aliphatic substituents effect miscibility with aliphatic or aromatic hydrocarbons. The alkali metal salts of the bisphenols are soluble in water, although their solubility decreases with increasing degree of substitution. The boiling points of the bisphenols are very high and depend on the size of the molecule and the polarity. For this reason, and since many bisphenols decompose, these compounds are seldom distilled.

3.3.1.2.2 Chemical Properties of Bisphenols [14, 15]

The chemistry of the bisphenols is determined by the phenolic OH groups, the aromatic nuclei, and the bridge segment. Hence, they primarily undergo the same reactions as similarly substituted monophenols (esterification and etherification at the OH group, substitution and hydrogenation of the aromatic nucleus). Because of their bifunctionality, they are very useful in the synthesis of high molecular weight polyesters and polyethers as well as epoxide resins.

3.3.1.2.3 Synthesis of Alkylidene Bisphenols [14, 15, 16]

Bisphenols are prepared by the condensation of ketones or aldehydes with phenols, by the reaction of diols, dihalides, or similar compounds with phenols, or by the condensation of hydroxymethylphenols with phenols.

There are several possibilities for producing the most important bisphenol, bisphenol A. Today it is prepared by the reaction of acetone with 2 mol of phenol, with hydrogen chloride or a sulfonated crosslinked polystyrene as a catalyst in a fixed and fluidized bed. The less corrosive ion exchangers are preferred to hydrogen chloride. The yield is usually 80–95%.

The workup of the bisphenol reaction mixtures depends on the structure of the products and on the purity requirements. For polycondensation, bisphenols must be extremely pure [14]. Ultrapure bisphenol A (>99.9%) can be obtained by a special crystallization technique [16].

The recent literature describes some bisphenols that contain bifunctional units like oxygen, sulfur, $—SO_2—$, $—(CH_2)_n—$, $>C{=}CR_2$, and other groups, instead of the alkylidene bridge. These bisphenols are prepared by special, multistage synthesis. An overview was given in Table 3.1.

3.3.1.2.4 Toxicology [14, 15]

Bisphenols are much less toxic than phenol itself. Some bisphenols are even used as antioxidants in cosmetics and foodstuffs.

Orally administered, bisphenol A has no acute toxicity for rats and mice. In the mouse, it has an LD_{50} of 4 g/kg. Exposure to a vapor of bisphenol A causes irritation to skin, eyes, and the respiratory system. Compared to phenol, penetration through the skin is insignificant.

3.3.2 Production

Three different processes have been developed to produce linear polycarbonates:
- transesterification or melt polycondensation
- polycondensation in the homogeneous phase
- interfacial polycondensation

Furthermore, there are some experimental routes to the production of aromatic polycarbonates in a phosgene-free synthesis. Some principal possibilities for producing aromatic polycarbonates were summarized in Figure 3.2.

Today only interfacial polycondensation is important for production on an industrial scale. Sections 3.3.2.1–3.3.2.4 describe these processes more thoroughly, with emphasis on interfacial polycondensation.

3.3.2.1 Transesterification

In the transesterification process, aromatic esters of carbonic acid are condensed with bisphenols (bisphenol A) in the presence of basic catalysts [17, 18] to give high molecular weight polycarbonates:

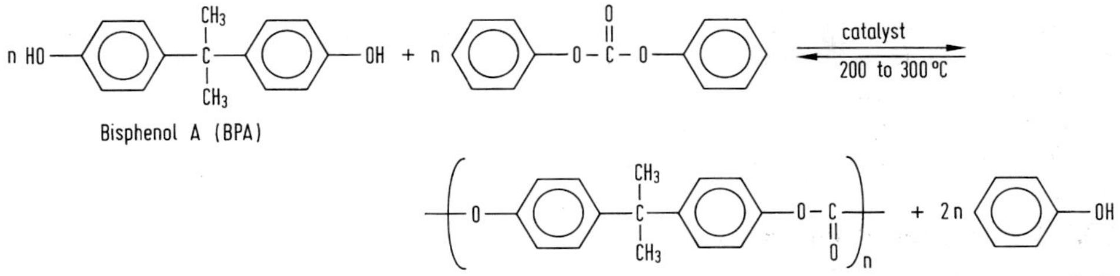

(3.8)

The equilibrium on the right-side of reaction 3.8 is necessary to obtain high molecular weights and is achieved by distilling the phenol under reduced pressure at relatively high temperatures [1, 19, 20].

Aromatic polycarbonates with a molecular weight as high as 50,000 can also be obtained by reaction with dialkyl carbonates, preferably diethyl carbonate, with bisphenol A in the presence of an excess of phenol and special tin-containing transesterification catalysts [21]. The intermediate in both processes is obviously diphenyl carbonate, the carbonate-donating agent.

New patents of Asahi [22, 23] describe the synthesis of bisphenol A polycarbonate from dimethyl carbonate and bisphenol A. In the first step, BPA dimethyl carbonate is prepared and transformed into high molecular weight polycarbonate by elimination of dimethyl carbonate (reaction 3.5 in Figure 3.2).

Polycarbonate of higher quality with respect to color and molecular weight is obtained by processing diphenyl carbonate from dialkyl carbonates by reaction with phenol, followed by purification, and then reacting with bisphenol A in the known transesterification process (reaction 3.6 in Figure 3.2).

In spite of the advantages—the absence of solvents and associated recovery steps, easy work-up of the polymer—melt transesterification has certain disadvantages. A complicated apparatus is required to sustain the long times at high temperatures and low pressures as well as to accommodate high melt viscosity of the polymer [1, 2].

The high temperatures may cause side reactions, such as discoloration, and the defective polycarbonates cannot be used in technically more sophisticated processes. Another disadvantage is that the production of higher molecular weight polycarbonates is limited.

A well-known variation of "classical melt transesterification" for other polycondensates (e.g., polyamides, polyesters) can be used to produce bisphenol A polycarbonate by solid phase polycondensation. First an oligomer precondensate is formed from diphenyl carbonate, at low temperature and in short reaction time. Then the oligocarbonate is transformed to the semicrystalline state by tempering or treatment with solvents (e.g., acetone) and later polycondensed at 220 °C in a solid phase reaction [24].

A phosgene-free synthesis can be achieved by starting the transesterification from such dialkyl carbonates as diethyl or dimethyl carbonate. The increasing number of patent documents describing phosgene-free processes to produce dialkyl carbonate and diphenyl carbonate, as well as aromatic polycarbonates, suggests that despite their disadvantages, these processes are becoming more attractive. More processes, starting from bisphenols and carbon monoxide, are described in Section 3.3.4.

3.3.2.2 Homogeneous Polycondensation

Polycondensation in a homogeneous phase was used for a brief period [17, 19, 25, 26]. Bisphenols like bisphenol A are phosgenated in a homogeneous organic medium containing a mixture of chlorinated hydrocarbons with pyridine, which acts as an acid acceptor and condensation catalyst.

This reaction is easy to carry out, since phosgene does not hydrolyze in an anhydrous medium. Pyridine is expensive, however, and the need to recover it from pyridine hydrochloride increases costs further. Pyridine must be removed quantitatively during work-up of the polycarbonate, since the presence of even trace amounts during thermoplastic processing may lead to discoloration of the polymer.

3.3.2.3 Interfacial Polycondensation

Interfacial polymerization, first developed in the 1950s at the Uerdingen plant of the Bayer AG, today is employed by most polycarbonate manufacturers [26].

In the first stage, the disodium salt of bisphenol A in aqueous alkaline solution (or suspension) is phosgenated in the presence of an inert organic solvent. The latter serves initially as a solvent for the phosgene that is introduced, but in the course of the reaction it also acts as a medium for the arylchlorocarbonates and oligocarbonates that are formed. Common solvents are chlorinated hydrocarbons (e.g., methylene chloride, 1,2-dichloroethane, chlorobenzene, chloroform) or anisole.

In the second stage, which takes place in organic media, the oligocarbonates are condensed to high molecular weight polycarbonates over catalysts, which can be added before or during phosgenation. Tertiary amines such as triethylamine or tri-*n*-propylamine are particularly effective [17, 19, 20, 27].

The reaction proceeds in a complicated manner that must be explained in simple terms [27, 28]. With an amine as catalyst, the reaction resembles a hydrolytic polycondensation.

Step. 1. Preparation of reactive monomers and oligomers of types A, B, and C (precondensation)

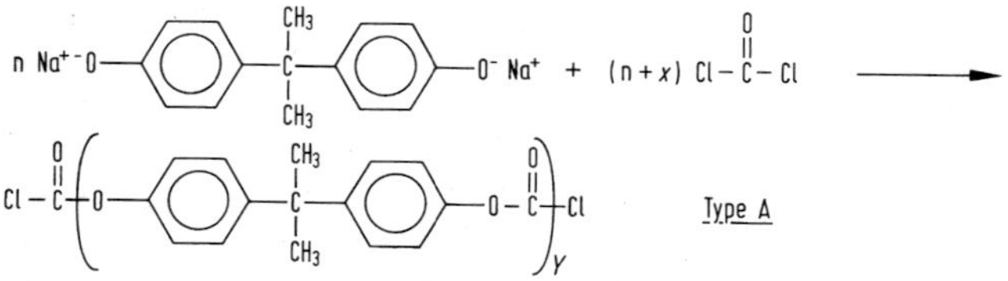

(3.9)

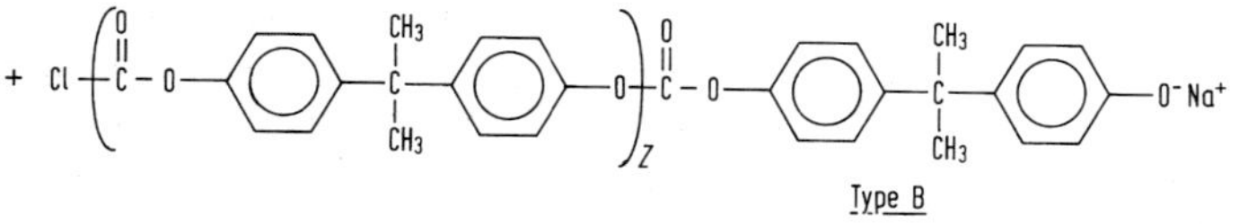

(3.10)

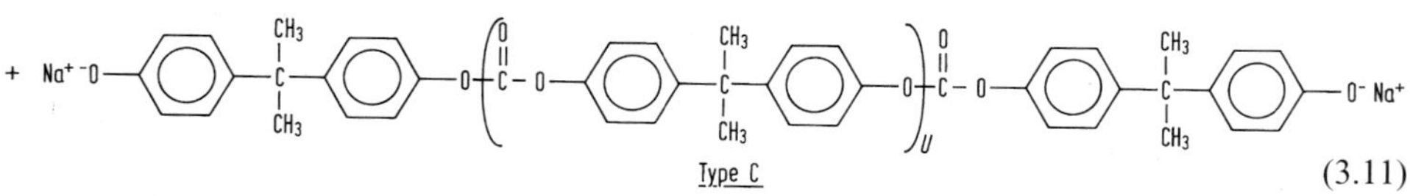

(3.11)

where $Y = 1-12$, $Z = 0-12$, $U = 0-12$, and $X = 0-0.2\,n$.

These reaction sequences are compound specific and occur within the two phases as well as at the phase boundary. The phosgene (20% excess) dissolved in the organic phase reacts—preferably at the phase boundary—with bisphenol A that is present as the disodium salt in the aqueous alkaline phase, giving the mono- or dichloroformate. The reaction occurs rapidly, and phosgene losses due to hydrolysis are small. Next the bisphenol A mono- or dichloroformate is condensed in a reaction that proceeds more slowly than the formation. The reaction of chloroformates with terminal phenolate groups produces reactive oligomers of types A, B, and C.

Step 2. Polycondensation with tertiary amines as catalysts

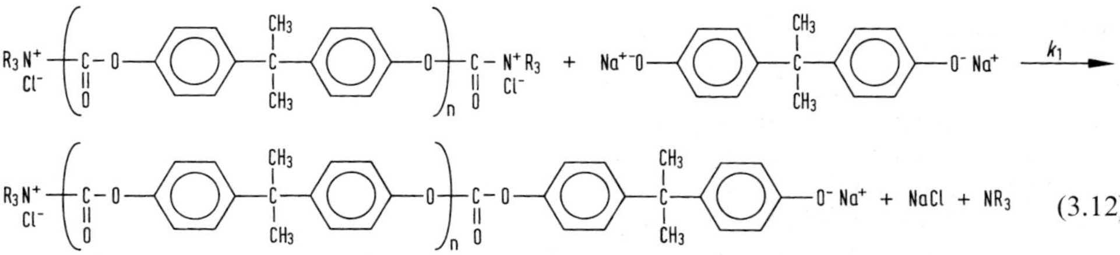

(3.12)

Competitive reaction

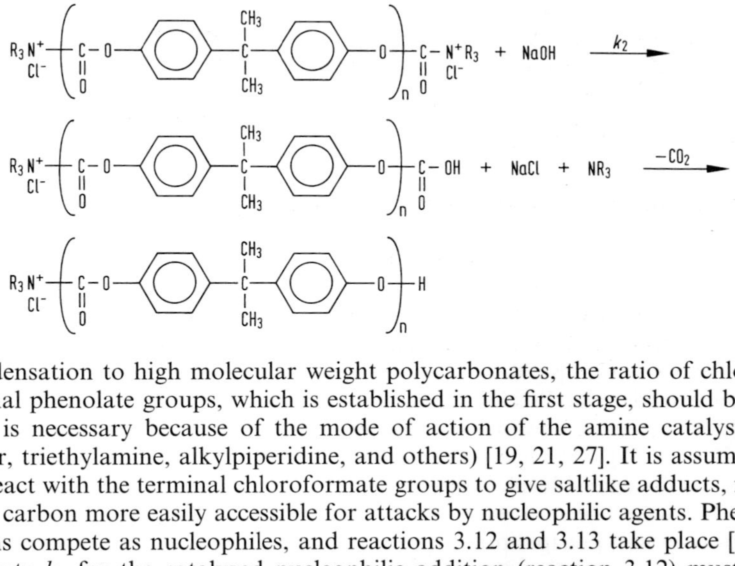

(3.13)

For condensation to high molecular weight polycarbonates, the ratio of chloroformate to terminal phenolate groups, which is established in the first stage, should be 1; a ratio of unity is necessary because of the mode of action of the amine catalysts used (in particular, triethylamine, alkylpiperidine, and others) [19, 21, 27]. It is assumed that the amines react with the terminal chloroformate groups to give saltlike adducts, making the carbonyl carbon more easily accessible for attacks by nucleophilic agents. Phenolate and OH^- ions compete as nucleophiles, and reactions 3.12 and 3.13 take place [17, 27, 29].

The rate k_1 for the catalyzed nucleophilic addition (reaction 3.12) must be higher than that for the hydrolysis of the terminal chloroformate groups, k_2 (3.13). This ensures that when an excess of terminal chloroformate groups is present, all terminal phenolate groups formed by hydrolysis are immediately covered. If the ratio of chloroformate to terminal phenolate groups falls to 1 (e.g., $k_2 > k_1$), condensation is terminated because of excess terminal phenolate groups, and the degree of polymerization is low.

Efficient mixing of the emulsion and attention to the alkali content of the aqueous phase are important for the success of a polycondensation reaction. In the synthesis of bisphenol A polycarbonate, a constant pH of 12 is desirable [25, 30]. However, synthesis conditions at a lower pH level have also been recommended [20, 25, 31]. As a rule, the temperature is between 10 and 35 °C, although a so-called hot phosgenation process has been described [32]. The desired molecular weight of bisphenol A polycarbonate, namely, between 10,000 and 200,000, can be obtained by adding the proper amount of molecular weight regulators (see Section 3.2.2.1.1).

The use of phase transfer catalysts, usually onium salts of group V elements (nitrogen, phosphorus, arsenic), causes formation to occur differently. These catalysts are useful for sterically hindered bisphenols (tetramethyl- or tetrabromobisphenol A) [33, 34], where they act as carriers between the aqueous and organic phases. The cation of the onium salts reacts with the phenolate anion to form an organophilic ion pair that migrates into the organic phase. There the phenolate moiety reacts with phosgene or oligocarbonates containing terminal chloroformate groups in the homogeneous phase. The reaction rate is determined by the diffusion of the adducts into the organic phase and the back-diffusion of the re-formed onium salt into the aqueous phase.

3.3.2.4 Preparation of Copolycarbonates

Copolycarbonates based on different bisphenols can be prepared by three processes, which result in statistical, alternating, or block copolycarbonates [1, 3, 19].

When the reactivities of the bisphenol components are comparable, copolycarbonates with a statistical distribution of the bisphenol units are obtained by interfacial polycondensation. Copolycarbonates with strongly alternating structure can be prepared by the reaction of the solution of the bischloroformic acid ester of one bisphenol with the sodium bisphenolate solution of the other bisphenol [1].

There are similar rules for the preparation of homogeneous block copolycondensation. In these cases, it is also more suitable to produce the bischloroformic acid ester of low molecular polycarbonates first and react it with low molecular polycarbonates of another composition in a second step [1].

3.3.3 Work-up of the Polycarbonate Solution

At the end of the reaction, the organic phase with the dissolved high molecular weight polycarbonate is separated from the aqueous phase, which contains alkali and some inorganic salts ($NaCl$, Na_2CO_3). Then the organic phase is washed with water (and eventually with acid), neutralized, and made electrolyte-free. There are several processes for the subsequent work-up [20, 26]. One is spray-drying, followed by work-up with a devolatilizing extruder [8]. Precipitation of polycarbonate from the organic phase by addition of polycarbonate nonsolvents (toluene, cyclohexane, pentane, acetone) is rarely used today [17, 19, 35]. Modified processes in which polycarbonate is precipitated from solutions of methylene chloride with a certain amount of nonsolvents, which are removed by hot water, seem to have more technical importance [36].

In the spray-drying/precipitation process, polycarbonate is received as a powder that can be extruded and granulated after drying. After multistep evaporation, the melted polycarbonate is obtained directly in strands that can be granulated. The granulate can be used without further processing. Since, however, for many uses special formulations are needed (e.g., addition of glass fibers, dye pigments, stabilizers, flame retardants), so that a second compounding extrusion is necessary (see Section 3.4).

The interfacial polycondensation process dominates the technical production scene because of the outstanding range of polycarbonate types produced. This variety compensates for the disadvantages of the process with respect to transesterification—for example, the use of solvents and the associated recovery, as well as the complicated work-up of the polycarbonate solution.

3.3.4 Synthesis of Aromatic Polycarbonates Starting with Bisphenols and Carbon Monoxide

The reaction of aromatic dihydroxydiaryl alkanes and carbon monoxide to give high molecular weight polycarbonates by a catalytic method is still in the experimental stage [37]. The process involves the catalytic oxidation of bifunctional phenols with the aid of metal complexes of group VIII-B, mainly palladium compounds, and the simultaneous incorporation of carbon monoxide to make polycarbonates. Air or several metal complexes, also used as catalysts, serve as oxidizers; hindered tertiary amines are used as auxiliary bases. This route is inefficient, however, as demonstrated by the large excess of catalyst and base required, and the low molecular weights achieved (1700–11,000). More difficulties can be expected during work-up of the polycarbonate solution, especially when the additives are separated and recovered.

A catalytic reaction between phenol and carbon monoxide (see Section 3.3.1.1) gives diphenyl carbonate as the starting material for the melt polycondensation. Purification of the diphenyl carbonate before the melt polycondensation should result in a polycarbonate of a better quality.

Predications about the future use of these processes in the technical production of polycarbonates are quite speculative.

3.3.5 Cyclic Aromatic Polycarbonates

While synthesizing ethylene carbonate in 1883, Nemirowski obtained a stable crystalline product that was later identified as a cyclic five-ring compound [38]. In 1898 Einhorn prepared a cyclic carbonate from catechol and phosgene [39]. Next to carbonates made of substituted catechols, this simple aromatic five-ring carbonate was the only cyclic aromatic carbonate known in the literature for a long time. These cyclic carbonates are very stable compounds, and ring-opening polymerization will not transform them into high molecular weight linear carbonates.

In 1930 Carothers et al. described the six-ring trimethylene carbonate and the cyclic dimer tetramethylene carbonate [40]. Later, the polymerization of cyclic six-ring esters and cyclic aliphatic carbonates was investigated [41].

Starting from the bischlorocarbonic acid ester of bisphenol A, Schnell and Bottenbruch (Bayer AG) succeeded in 1962 in synthesizing cyclic tetramer carbonate I ($n = 3$);

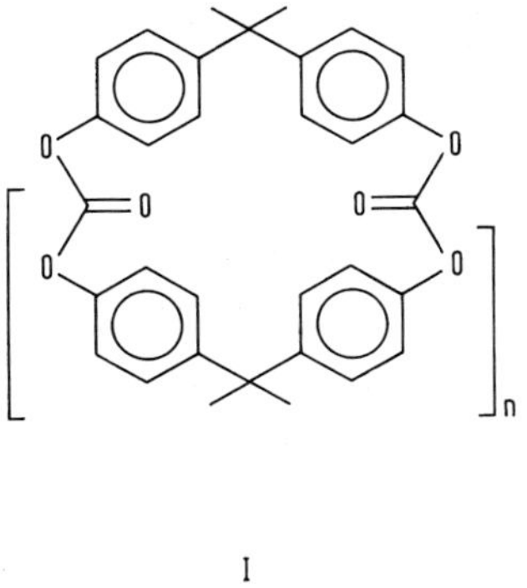

I

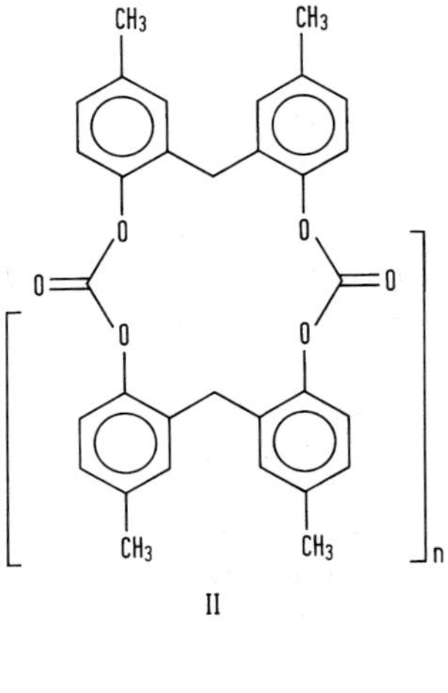

II

ring-opening polymerization with different catalysts led to a high molecular weight linear polycarbonate [42]. Later investigations by Moody and Prochaska at General Electric [43, 44] led to the dimer and trimer cyclics of bisphenol A (structure I, $n = 1, 2$) and to cyclic compounds of structure II ($n = 1, 2$).

Ring-opening polymerization of these compounds was described with metal oxides (e.g., antimony oxide, copper oxide), alkali metals, or alkali hydroxides (e.g., potassium hydroxide) at 300 °C [42–45].

At the beginning of the 80's, patents were published describing the synthesis and use of cyclic aromatic carbonates of structure I made of bischlorocarbonic acid esters [46]. Cyclic carbonates with a definite ring size (dimer, trimer, tetramer, . . .) were not synthesized, however, but rather an oligomer mixture of the cyclics with $n = 2–20$ (structure I). These mixtures also contained up to 30% high molecular weight polycarbonate (linear or cyclic, $n = 30$) [47]. Several methods have been described for the separation of the high molecular weight compounds, such as solvent extraction or precipitation reactions [48].

Polymerization can also be carried out with catalysts like aluminates, titanates, alcoholates, organotin compounds, and borates at temperatures from 250 to 350 °C and reaction times from minutes to hours.

Under these conditions, no molded parts can be obtained (RIM process, polymerization during molding), but the interesting possibility of producing composites by polymerization of prepared mixtures of cyclic carbonates and fillers (e.g., fibers, carbon) has been discussed [49].

The reaction of cyclic carbonate oligomer mixtures with the sodium salt of a polyphenylene ether leads to block copolymers [50]. Like these cyclic aromatic carbonate oligomers, macrocyclic polycarbonate ($M_w > 15{,}000$) has been known in the literature since 1979. Horbach, Vernaleken, and Weirauch obtained macrocyclic polycarbonates with molecular weights up to 100,000 by reacting the bischlorocarbonic acid ester of bisphenol A with bisphenol A under certain phase transfer reaction conditions [51]. The macrocyclic character of these polycarbonates is proved by terminal group analysis (terminal group deficiency $> 95\%$) as well as by ring-opening reactions.

References for Section 3.3

1. *Becker, G. W.; Braun, D.; Peilstöcker, G.; Krimm, H.* (Eds.), in: Polycarbonate, Kunststoff-Handbuch, Vol. 8: Hanser, Munich 1973
2. *Müller v. Blumencron, H.-O.; Gelbke, H. P.*, in: Ullmann's Encyclopädie der Techn. Chemie. Vol. 18. Verlag Chemie Weinheim 1979, p. 275–281
3. *Kirk-Othmer:* Encyclopedia of Chemical Technology. 3. Ed., Vol. 17. Wiley, New York, 1982, p. 416–425
4. a) Company information „MS Info 88-3" der Fa. Merck, Darmstadt 1988
   b) *Eckbert, H.; Forster, B.:* Ang. Chem. 99 (1987) p. 922
5. EP 879 (1977) Bayer AG
   EP 880 (1977) Bayer AG
6. *Buysch, H.-J.*, in: Ullmann's Encyclopedia of Industrial Chemistry. Verlag Chemie, Weinheim 1986, p. 197
7. EP 1 082 (1977) Bayer AG
   EP 1 083 (1977) Bayer AG
   EP 1 777 (1978) Bayer AG
8. *Peppel, W. J.:* Ind. Eng. Chem. 50 (1958) p. 767
9. DE-PS 2 748 718 (1977) Bayer AG
10. Chem. Eng. 94 (1987) 9, p. 28
11. Chem. Eng. 90 (1983) 9, p. 15
12. *Matthews, R., Hallgren, I. E.:* J. Organomet. Commun. 175, (1979) p. 135

13. *Hallgren, I. E.*, et al.: J. Organomet. Chem. 204 (1981) p. 135
14. *Buysch H. J.*, et al. in: Ullmann's Encylopädie der Technischen Chemie. 4. Ed., Vol. 18. Verlag Chemie, Weinheim 1981, p. 215
15. *Kirk-Othmer:* Encyclopedia of Chemical Technology. 3. Ed., Vol. 17. Wiley, New York 1982, p. 373–383
16. *Moyers, C. G.:* CEP (1986) 5, p. 42–45
17. *Bottenbruch, L.*, in: Encyclopedia Polymers, Science and Technology. Vol. 10. Wiley, New York 1969, p. 710–764
18. DE-OS 1 570 576 (1965) Bayer AG
19. *Schnell, H.:* Poly. Rev. 9 (1964) p. 1
20. *Fox, O. W.*, in: Encyclopedia of Chemical Technology 3. Ed., Vol 18. Wiley, New York 1982, p. 479
21. EP 39 452 (1979) Bayer AG
22. JA-PS 63-205 318 (1988) Asahi
23. JA-PS 63-223 035 (1988) Asahi
24. EP 338 085 (1989) Asahi Kasei Kogyo Kabushiki Kaisha
25. *Margotte, D.*, in: Ullmann's Encyclopedia of Industrial Chemistry. Vol. 19, Verlag Chemie. Weinheim 1981, p. 55
26. *Freitag, D.; Grigo, U; Müller, P. R.; Nouvertné, W.;* in: Encyclopedia of Polymer Science and Engineering. Vol. 11 Kap. Polycarbonates. Wiley, New York 1988
27. *Vernaleken, H.; Millich, F.; Carracer Jr., C. E.* (Eds.), in: Interfacial Synthesis. Vol. 11. New York 1977, p. 65–124
28. *Noguchi, A.:* J. Soc. Org. Syn. Chem. Jap. 21 (1963) p. 928
29. *Kolesnikov, G. S.*, et al.: Vysokomol. Soedin. 7 (1965) p. 129
30. US-PS 3 530 094 (1966) Bayer AG
31. US-PS 3 989 672 (1974) General Electric
32. DE-OS 2 410 716 (1974) Bayer AG
33. *Serini, V.*, et al.: Ang. Makromol. Chem. 55 (1976) p. 175
34. DE-OS 2 211 956 (1972) Bayer AG
    DE-OS 2 211 957 (1972) Bayer AG
    DE-OS 2 063 052 (1970) Bayer AG
35. US-PS 3 153 008 (1964) General Electric
36. EP 284 930 (1988) Idemitsu Petrochem. Co.
    JA-PS 49-028 642 (1972) Idemitsu Kosan
    JA-PS 53-137 298 (1977) Mitsubishi Gas Chemical
37. DE-OS 2 758 520 (1977) General Electric
    DE-OS 2 738 437 (1977) General Electric
    DE-OS 2 738 487 (1977) General Electric
    DE-OS 2 738 488 (1977) General Electric
38. *Nemirowski, J.:* Pract. Chem. 289 (1883) 2, p. 439
39. *Einhorn:* Ann. Chem. 300 (1898) S. 135
40. *Carothers*, et al.: J. Am. Chem. Soc. 52 (1930) p. 314
41. *Carother*, et al.: J. Am. Chem. Soc. 54 (1932) p. 761
42. a) *Schnell, H.; Bottenbruch, L.;* Makromol. Chem. 57 (1962) p. 1
    b) DE-PS 1 300 265 a (1962) Bayer AG
43. US-PS 3 155 683 (1963) General Electric
44. US-PS 3 274 214 (1962) General Electric
45. US-PS 3 220 980 (1964) General Electric
46. a) EP 0 162 379 (1985) General Electric
    b) EP 0 197 351 (1986) General Electric
    c) EP 0 198 338 (1986) General Electric
    d) EP 0 205 042 (1986) General Electric
47. EP 0 224 131 (1986) General Electric
48. EP 0 227 109 (1986) General Electric
49. EP 0 228 699 (1986) General Electric
50. EP 0 230 603 (1986) General Electric
51. a) *Horbach, A.*, et al.: Macromol. Chem. 181 (1980) 1, p. 111
    b) DE-PS 2 937 332 (1979) Bayer AG

3.4 Modification of Polycarbonates Through Additives

3.4.1 Introduction

The properties of a polycarbonate can be modified through the use of various additives: processing can be improved using stabilizers, viscosity modifiers, and release agents, and blowing agents can be used to produce foamed polycarbonate.

More important than these processing additives are additives that impart substantial modification to the final properties of polycarbonate. Examples include offering stabilization against various environmental factors (e.g., heat, light, air, moisture), strengthening through fillers and reinforcers achieving, better flame retardance, and coloring with dyes and pigments.

Combinations of polycarbonate with polymeric additives are not covered in this section. Polycarbonate blends are considered in Section 3.5. Sections 3.4.2–3.4.8 present a short review of the preparation of polycarbonate additives, their structure, and their function. A more thorough discussion of structure–property relationships in polycarbonate is given later (see Section 3.7).

3.4.2 Synthesis of Compounds

3.4.2.1 Preparation of Raw Materials

Only in exceptional cases is unmodified polycarbonate injection-molded into parts or extruded into pellets or sheet. As a rule, additives are added either during primary production or in a separate compounding step. Frequently, heat and UV stabilizers and molding aids are already added in the primary production step, whereas pigments and fillers are added in a second melt step. For each particular additive there is an optimum route, which depends on the dosage requirements, the thermal and mechanical stability of the additive, its miscibility, and even the particular compounding machine. In certain cases premixing is appropriate. The addition of small amounts of additives can be facilitated through utilization of a master batch; this is often useful with respect to homogeneity and exactness of the dose. Because of the sensitivity of these materials to water, drying of the polymers and additives is required. This can be accomplished in a drier attached to the compounding machine.

3.4.2.2 Compounding Process

Polycarbonate and additives are compounded in extruders or kneaders which, depending on the task, can be equipped with different screw configurations containing mixing, pressure, and kneading elements. In the extruder, the melting of the polycarbonate (and other polymeric additives) is followed by the dispersion and distribution of the additive, and finally homogenization of the mixture. The requisite energy is supplied mainly through viscous shearing.

Depending on the application, polycarbonate is supplemented with such additives as stabilizers, viscosity modifiers, release agents, flame retardants, glass fibers, fillers, and pigments. Because polycarbonate is relatively polar, these additives are rather easily dispersed. Compounding with polymeric additives produces blends and alloys with a wide property profile. The compounding conditions (temperature profile, screw design, residence time, throughput, etc.) depend on the molecular weight and melt viscosity of the polycarbonate and on the nature of the additive. Thus the twin-screw extruder configuration and screw design for adding glass fibers would differ from the setup for adding stabilizers or pigments.

As with other amorphous, high T_g (150 °C) polymers, bisphenol A polycarbonate possesses a relatively high melt viscosity. The melt temperature in processing equipment can often be as high as 300–320 °C. Although pure polycarbonate does not degrade at such temperatures, combinations with certain additives can (depending on the kind of additive and its purity) cause a decrease in molecular weight. For this reason, the additives chosen for use with polycarbonates must be stable to 400 °C and dry enough to possess a neutral pH [1, 2].

Sections 3.4.3–3.4.8 present an overview of the patent literature on polycarbonate additives.

3.4.2.3 Compounding Machinery

Various processing systems are used for the compounding of polycarbonate with additives. Examples are kneaders like those from Buss [3] and twin-screw extruders (e.g., ZSK machines from Werner & Pfleiderer [4]). Detailed descriptions of compounding processes are not given here.

3.4.3 Stabilization

3.4.3.1 Stabilization Against Thermal and Thermooxidative Degradation

On the basis of its structure (high content of aromatic hydrocarbons, no secondary or tertiary hydrogen), bisphenol A polycarbonate is quite thermally stable. It can be processed without difficulty up to 320 °C. Under favorable conditions (simple flow conditions) higher melt temperatures can be tolerated. Stabilization against degradation during processing and also during molding of parts is advisable. The stabilizers are added in concentrations of less than 0.5% [5]. The patent literature contains claims for certain organic phosphites [5–10], phosphonites [11], or phosphanes [12], often in combination with epoxides [13] or polydialkylsiloxanes [14]. Organosilicon compounds are also alone added as thermal stabilizers [15]. In contrast to phosphites, the organosilicons do not compromise the hydrolytic stability of the polycarbonate.

3.4.3.2 Stabilization Against Hydrolytic Degradation

Water begins to cleave polycarbonate above 60 °C. Longer contact with boiling water leads to significant decreases in molecular weight, which are accelerated in the presence of thermal stabilizers such as organic phosphites. Epoxides or organic silicon compounds are effective against this degradation [5, 16, 17].

3.4.3.3 Stabilization Against Degradation from UV and Visible Light

The operation of UV light, especially in combination with oxygen and moisture, leads gradually to the photooxidative degradation of polycarbonate compounds. As in thermal degradation, this breakdown results in yellowing and molecular weight decrease. Extensive investigations have been carried out on the mechanism of photooxidative degradation in bisphenol A polycarbonate [5].

For outside applications, UV-stabilized polycarbonate should be used. Benzotriazole and benzophenone systems have been established as especially effective UV stabilizers

[5]. Also named in the patent literature are acrylates [12, 18, 19] and maleic acid derivatives [20, 21].

3.4.4 Lubricants and Release Agents

The addition of lubricants and release agents has proved useful, especially for the production of complex parts. Particularly suitable release agents for bisphenol A polycarbonate are fatty acid esters from one or more alcohols [22–24]. Dialkylamides [25], C_{16} to C_{26} alkanes [26], organopolysiloxanes [27], and perfluorosulfonic acids [28] have been named in the patent literature.

3.4.5 Reinforcing Additives and Fillers

As additives to reinforce polycarbonate, glass fibers are compounded in concentrations from 10 to 40 wt %. An increase in the stiffness (*E* modulus) and a corresponding decrease in the coefficient of thermal expansion results [5]. To obtain optimum properties, appropriate glass fibers and specific compounding equipment must be employed. Appropriate glass fibers are, in general, those that have been surface treated to ensure optimum adhesion between the polymer matrix and the glass fiber. In this respect, the patent literature mentions such additional additives as polyphenylene oxide [29] and organosilicones [30], which improve the impact strength of glass fiber reinforced polycarbonates.

As filler for polycarbonate, the patent literature names wollastonite [31], amorphous aluminum silicate [32], special silicon dioxide [33], and novaculite of certain particle sizes [34].

3.4.6 Flame Retardants

The good flame resistance of polycarbonate can be further improved through the incorporation of appropriate flame retardants. Certain alkaline or alkaline earth sulfonates have proven to be particularly useful because they are added in such small amounts that most of the transparency of the polycarbonate is retained [35, 36]. Other groups of salts, such as sulfonamide salts [37], perfluorborates [38], crown ether salts [39], and calcium titanates [40], have been named. A further degree of flame retardance can be attained through the combination of these salts with an antidrip material such as polytetrafluoroethylene (PTFE) [41, 42] and, if necessary, halogenated compounds like tetrachlorophthalimide [43] and tetrabromobisphenol A oligocarbonate [44]. Upon addition of PTFE, the transparency is lost. Halogen-free flame retardants such as polyphenylene sulfide [45] and polyphosphonate [46] have been reported.

3.4.7 Chemical Blowing Agents

Polycarbonate foam can be formed, of course, by blowing with gas (typically nitrogen). In the last few years, there has been increased interest in chemical blowing agents, which are added to the polycarbonate pellets as a master batch. The chemical blowing agents decompose at processing temperatures to give the desired inert gas, which generates a foam structure in the melt and in the final part [5]. Chemical blowing agents for polycarbonate are 5-phenyltriazole [47], phenyldihydroxyoxadiazone [48], oxadiazone derivatives [49], citric acid esters, and acid anhydrides.

3.4.8 Dyes and Pigments

A pigment or dye used to color polycarbonate should be as chemically inert as possible, to retain the molecular weight of the polycarbonate. Since, however, inorganic pigments such as TiO_2 lead to a slight decrease in notched impact strength and a decrease in molecular weight, pigments are added in combination with stabilizers like organic phosphites [50] and hydrolysis stabilizers like epoxides [51] and organosiloxanes [52]. The dispersion of pigments in polycarbonate can be improved by addition of a small amount of styrene–acrylonitrile (SAN) [53].

References for Section 3.4

1. Encyclopedia of Polymer Science and Engineering. Vol. 4. 2. Ed. Wiley, New York 1986, p. 66
2. *Hellerich, W.; Harsch, G.; Haenle, S.;* „Werkstoff-Führer Kunststoffe“, p. 61, Hanser Verlag, Munich/Vienna 1983
3. Kontinuierliche Misch-, Knet- und Compoundier-Technik. Company brochure Buss AG, Basel 11/78
4. Verfahren und Methoden: Technische Kunststoffe. Company brochure Werner & Pfleiderer GmbH, Stuttgart 1987
5. *Freitag, D.; Grigo, U.; Müller, R. P.; Nouvertné, W.*, in: Encyclopedia of Polymer Science and Engineering. 2 Ed., Vol. 11. Wiley, New York 1988, p. 648–718
6. US-PS 4 335 039 (1978) General Electric
7. US-PS 4 312 803 (1979) General Electric
8. US-PS 4 276 233 (1978) General Electric
9. US-PS 4 102 859 (1976) Bayer AG
10. US-PS 4 073 769 (1971) Bayer AG
11. US-PS 4 297 448 (1979) General Electric
12. US-PS 4 367 303 (1984) Bayer AG
13. US-PS 4 066 611 (1976) General Electric
14. US-PS 4 456 717 (1982) Bayer AG
15. DE-OS 2 920 450 (1979) Bayer AG
16. US-PS 4 556 681 (1985) General Electric
17. US-PS 4 393 158 (1981) General Electric
18. US-PS 3 875 123 (1972) Bayer AG
19. US-PS 3 629 193 (1969) Bayer AG
20. US-PS 4 228 067 (1979) Bayer AG
21. US-PS 4 264 680 (1980) General Electric
22. US-PS 4 143 024 (1976) Bayer AG
23. US-PS 3 836 499 (1972) Bayer AG
24. US-PS 3 784 595 (1970) Bayer AG
25. US-PS 4 554 302 (1985) General Electric
26. US-PS 4 415 696 (1981) General Electric
27. US-PS 4 390 651 (1976) General Electric
28. US-PS 4 041 003 (1975) Bayer AG
29. US-PS 4 048 133 (1976) Bayer AG
30. US-PS 4 147 707 (1975) Bayer AG
31. US-PS 4 269 755 (1979) General Electric
32. US-PS 4 243 575 (1979) General Electric
33. US-PS 4 460 73 (1982) Mobay Chem. Corp.
34. US-PS 4 070 330 (1976) Bayer AG
35. US-PS 3 775 367 (1971) Bayer AG
36. US-PS 4 032 506 (1973) General Electric
37. US-PS 4 254 015 (1979) Dow Chem. Corp.
38. US-PS 4 335 038 (1980) Dow Chem. Corp.
39. US-PS 4 439 556 (1982) Dow Chem. Corp.
40. US-PS 4 352 238 (1982) General Electric

41. US-PS 4 649 168 (1984) Mobay Chem. Corp.
42. US-PS 4 391 935 (1984) General Electric
43. US-PS 4 661 543 (1983) Bayer AG
44. US-PS 4 552 911 (1982) Bayer AG
45. US-PS 4 046 836 (1975) Bayer AG
46. US-PS 4 332 921 (1980) Bayer AG
47. US-PS 3 442 829 (1966) Borg Warner Corp.
48. US-PS 4 097 425 (1975) General Electric
49. US-PS 4 500 653 (1983) Mobay Chem. Corp.
50. US-PS 3 525 705 (1959) General Electric
51. US-PS 3 761 440 (1970) Bayer AG
52. US-PS 3 673 146 (1971) General Electric
53. US-PS 4 464 487 (1982) Dow Chem. Corp.

3.5 Polycarbonate Blends

3.5.1 Introduction

Two series of polycarbonate blends have achieved commercial significance. The first, polycarbonate–ABS blends, is based on an interesting combination of mechanical properties, heat distortion temperatures, and rheological properties. This has led to broad application in the areas of electronics, office machines, and automotive products. Because of the inherent variability of these blends, further development on the product and application side is expected.

The second set of commercial products consists of blends of polycarbonate and polyalkylene terephthalates modified with special graft copolymers. Such blends show the dimensional stability of polycarbonates, the solvent resistance of polyesters, and the typical favorable rheological properties of blends. They are, therefore, often employed in automotive parts having large surface areas.

These two blends have specific property prefiles, which are distinct from the component homopolymers, and are not discussed in Section 3.5. Rather, Sections 3.5, 2–3.5, 7 cover polycarbonate blends that essentially exhibit the properties of polycarbonate but with diminution of certain specific property deficits of the material (e.g., toughness at very low temperatures; crazing in contact with solvents, especially hydrocarbons). From the extensive open and patent literature [1–11] we have chosen polycarbonate blends that have achieved commercial significance or are expected to undergo development.

3.5.2 Polycarbonate–Polyolefin Blends

Essentially the goal of polycarbonate–polyolefin blends is to achieve improved cold temperature toughness, toughness as a function of thickness of the part (critical thickness), crazing resistance (especially against gasoline), flowability, flame retardance, and electrical properties. The following compounds have been named as blend partners: polyethylene, ethylene/propylene copolymers, ethylene/methacrylate copolymers, polypropylene, polybutylene, polyisobutylene, and polymethylpentene [4].

Polycarbonate is not miscible with polyolefins and the polyolefin component is, therefore, dispersed in a polycarbonate matrix. These blends are opaque, owing to the large differences in refractive indices and the semicrystalline nature of the polyolefins.

The incompatibility of polycarbonates and polyolefins leads to delamination at amounts exceeding 5 wt % [4, 11]. The compatibility of polycarbonate and polyolefins is improved through use of ethylene/acrylate or ethylene/vinyl acetate copolymers [4].

Polycarbonate–polyethylene blends have obtained commercial significance [4]. They are distinguished from polycarbonate in their higher toughness (greater critical thickness), decreased tendency to craze, and increased tolerance of certain chemicals. Physical properties of a commercial polycarbonate–polyethylene blend are listed in Table 3.3. Polycarbonate–polyethylene blends are processed at temperatures similar to polycarbonate. Injection molding and extrusion are both possible.

Applications for polyethylene-modified polycarbonate include safety helmets, sporting goods, spools for the textile industry, and bumpers for automobiles [12]. (Other polycarbonate blends are also used for these applications.)

Table 3.3. Physical Properties of a Bisphenol A Polycarbonate–Polyethylene Blend (®Makrolon T 7700)

Properties	Units	Value	ASTM Method
General			
Density	g/cm^3	1.19	D 792
Shrinkage	m/m	0.006–0.008	D 955
Water absorption			D 570
24 h	%	0.12	
Extended	%	0.34	
Mechanical			
Yield strength	N/mm^2	59	D 638
Tensile strength	N/mm^2	59	D 638
Elongation at yield	%	8	D 638
Elongation at break	%	115	D 638
Modulus of elasticity	N/mm^2	2300	D 638
Flexural modulus	N/mm^2	2200	D 790
Izod impact strength			D 256
3.2 mm test specimen	J/m	640–800	
6.4 mm test specimen	J/m	480–640	
Rockwell hardness, A scale		115–118	D 785
Thermal			
Vicat softening temperature (VST/A)	°C	150	D 1525
Heat distortion temperature			
(1.82 N/mm^2)	°C	132	D 648
Electrical			
Dielectric strength	kV/mm	16	D 149
Volume resistivity	$\Omega \cdot cm$	10^{16}	D 257

3.5.3 Blends of Polycarbonate, Polystyrene, and Polystyrene Copolymer

Because of the incompatibility of polycarbonate and polystyrene, blends of these polymers have poor toughness. Despite this incompatibility, however, transparent parts are produced under certain processing conditions. This is because the refractive indices are similar. The combination of transparency and optical birefringence makes these blends suitable for use as matrix materials for laser-written CDs. Polystyrene shows the opposite birefringence to polycarbonate; this means that at specific concentrations,

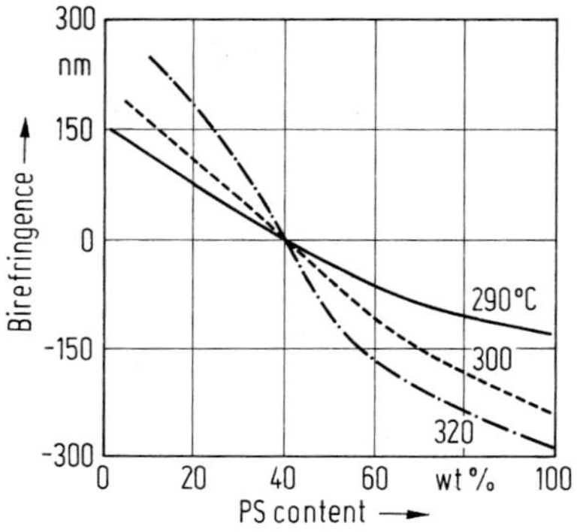

Figure 3.3. Influence of composition and processing temperature on the birefringence of blends of polycarbonate and polystyrene (PS) [13, 14, 17].

birefringence-free blends can be produced [13–16] (see Figure 3.3). Similar effects were observed in blends of tetramethylbisphenol A polycarbonate and SAN [16, 17].

3.5.4 Blends of Polycarbonate and Elastomers

Blends of polycarbonate and elastomers [18–21] are generally based on butadiene and acrylate graft copolymers. Such blends show clearly improved toughness at lower temperatures as the transition from ductile to brittle failure is shifted to lower temperatures. This shift coincides with a shift of the ductile-to-brittle transition to higher wall thickness [22].

The toughness of such blends can be considerably improved through special polycarbonates such as the poly(siloxane-*b*-carbonate), described in Section 3.2.2.2.3 (see Figure 3.4) [23–25]. The heat distortion temperature, processing, molding characteristics, and rheological behavior are not adversely affected. These blends show high energy absorption during accelerated deformation, which makes them suitable for the production of safety helmets or exterior and interior automotive parts (bumpers, dashboards, seat backs, and consoles). In addition, resistance to crazing due to contact with gasoline is considerably improved (Figure 3.5).

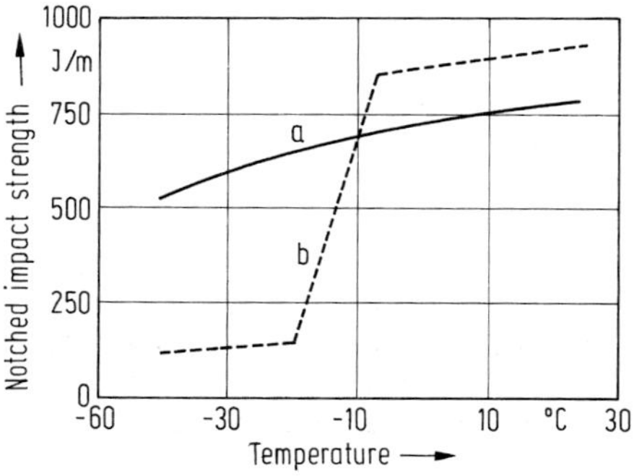

Figure 3.4. Comparison of the notched Izod impact strength of two elastomer-modified polycarbonates: a, Makroblend EC 900; b, standard polycarbonate [25].

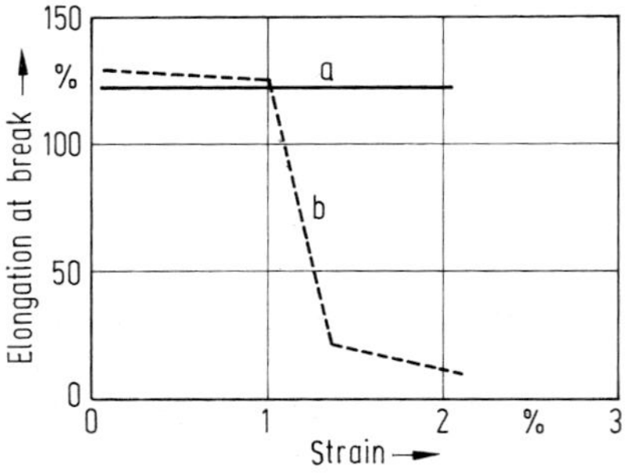

Figure 3.5. Comparison of the environmental stress-cracking resistance of two elastomer-modified polycarbonates: a, Makroblend EC 900; b, standard polycarbonate [25].

An interesting product class consists of polycarbonate blends with special thermoplastic elastomers such as styrene–ethylene–butylene–styrene block copolymers (SEBS) [26]. The SEBS component is obtained through the hydrogenation of styrene–butadiene–styrene block copolymers. In addition to good toughness and processing behavior, the blends with polycarbonate show excellent resistance to oxygen and UV light. In comparison to polycarbonate, they possess clearly superior resistance to gasoline.

Under certain processing conditions, polycarbonate–SEBS blends form interpenetrating networks (IPN) [26]. IPN polymer blends are combinations of different polymers in which each polymer phase is continuous and topologically independent. In situations where polycarbonate is merely dispersed (e.g., as filler in SEBSs), an increase in modulus is barely observed (Figure 3.6). The IPN blend, on the other hand, shows a strong contribution of the continuous phase of polycarbonate. The modulus of the blend is the same as that of the polycarbonate.

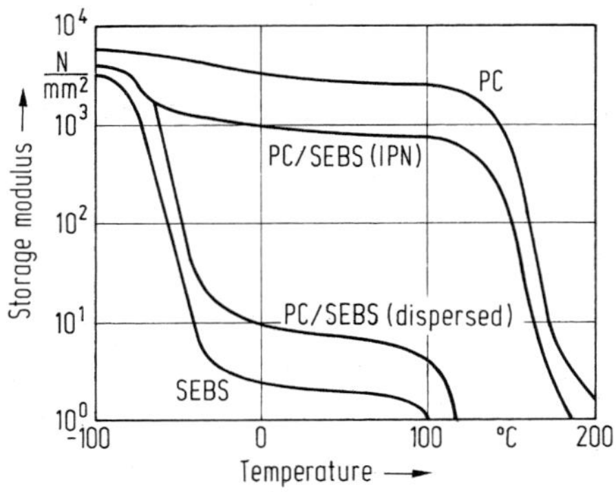

Figure 3.6. Dynamic storage modulus of a blend of SEBS and polycarbonate (volume fraction 70:30) relative to the pure components [26].

3.5.5 Polycarbonate–Polyamide Blends

In the patent literature numerous polyamide–polycarbonate and polyamide–polyester carbonate blends are known. Examples are [27–30], PA-PC (aliphatic) [31], PA (amorphous)–PC [32], PA (aromatic)–PC [33], PA–PC + ionomer [34], PA–PC + copolymers based on conjugated dienes [35], and PA-aromatic polyester carbonates [36–38]. They have not achieved commercial significance.

The goal of this blend development is products that have better solvent resistance than polycarbonate and better heat distortion temperature and notched impact strength than polyamide.

Incompatibility of the blend partners, degradation of the polycarbonate and polyester carbonate (particularly by the amino end groups of the polyamide), and uncontrolled amidation reactions [39] represent serious problems, which must be solved if a commercial blend is to be achieved. Otherwise one obtains products that are not tough and tend to delaminate; notched impact strength is especially bad (even upon addition of typical impact modifiers). Solutions to these problems should be obtainable through the following steps:

- chemical coupling of the polycarbonate, polyester carbonate, and the polyamide, by capping PC end groups with trimaleic anhydride acid chloride [40]
- use of suitable compatibilzers such as polymers with epoxide [41–43] and anhydride end groups [44–47], polyester amides [46, 48, 49], polyether amides [48, 50], polyether

imides [46], polyester carbonates [51], polyurethanes [46], and thermoplastic elastomers [46]
- addition of polycaprolactone to decrease the melt viscosity of the polycarbonate [52]
- use of polyamides with fewer amino end groups [53] or capping of amino end groups with appropriate reagents [54]

The first PA–PC blend has been introduced by a U.S. firm (Dexter) [55].

3.5.6 Polycarbonate–Thermoplastic Polyurethane (TPU) Blends

Especially for special automotive applications, polycarbonate–polyurethane blends enjoy increasing interest because, like elastomer-modified polycarbonate–polyalkylene terephthalate blends, they exhibit good toughness and particularly good resistance to damage from fuel. Additionally the stiffness (E modulus) can be varied widely by changing the composition.

To further improve toughness, particularly at low temperature, rubber can be added to blends of polycarbonate and thermoplastic polyurethane [56]. When the amount of TPU and rubber exceeds 50%, the E modulus lies below 1000 N/mm^2. During compounding and processing of polycarbonate–TPU blends, it should be noted that the limit of the thermal stability of a typical thermoplastic polyurethane lies between 250 and 270 °C. Higher temperatures lead quickly to deterioration of the advantageous properties of these blends. Typical properties of commercial polycarbonate–TPU blends are listed in Table 3.4.

Table 3.4. Properties of Polycarbonate–Polyurethane Blends with Various PC/TPU Contents (Miles Inc., U.S.A.)

Properties	Units	Texin	Texin	Texin	Texin	ASTM Method
General						
Density	g/cm^3	1.21	1.21	1.21	1.21	D 792
Shore hardness		60	65	70	75	D 2240
Mechanical						
Tensile strength	N/mm^2	34.5	34.5	37.9	41.4	D 412
50-% modulus	N/mm^2	17.2	25.5	28.3	31.7	
100-% modulus	N/mm^2	19.3	26.9	31.0	34.5	
200-% modulus	N/mm^2	21.4	31.0	—	—	
Elongation at break	%	325	200	175	150	D 412
Flexural modulus						D 790
−30 °C	N/mm^2	1014	1014	1310	1655	
23 °C	N/mm^2	207	414	690	1034	
70°C	N/mm^2	105	230	483	855	
Flexural strength	N/mm^2	20.7	21.4	31.7	49.0	D 790
Izod impact strength						D 256
23 °C	J/m	n.b.[1]	n.b.	849	886	
−30 °C	J/m	75	120	164	190	
Thermal						
Heat distortion temperature at	°C					D 648
0.46 N/mm^2		55	65	98	108	
1.82 N/mm^2		37	43	59	74	
Vicat softening temperature	°C	146	141	139	140	D 1238
Coefficient of thermal expansion	10^{-5} K^{-1}	12.7	10.5	10.2	9.1	D 696

[1] n.b., not broken.

3.5.7 Other Polycarbonate Blends

In addition to the polycarbonate blends described above, a number of other blends have been reported, particularly in the patent literature. These compounds, which have not achieved commercial significance, include polycarbonate–polyphenylene sulfide blends, polycarbonate–polyphenylene oxide blends, polycarbonate–polysulfone blends, and polycarbonate–LC polyester blends [4]. The first three blends are distinguished relative to polycarbonate by superior flame retardance, toughness, elasticity, and hydrolytic stability. Polycarbonate–polyacetal and polycarbonate–polysulfone blends exhibit decreased crazing tendency. Polycarbonate–LC polyester blends show, with a relatively small amount of LC polyester, a clear increase in stiffness (*E* modulus) [57].

References for Section 3.5

1. *Weirauch, K:* Kunststoffberater 31 (1986) 12, p. 33–39
2. *Schaw, M. T.:* Polym. Eng. Sci. 22 (1982) 2, p. 115
3. *Witt, W.:* Gummi, Asbest, Kunsst. (1987) 4, p. 155
4. *Freitag, D.; Grigo, U.; Müller, P. R.; Nouvertné, W.*, in: Encyclopedia of Polymer Science and Engineeering. 2. Ed. Wiley New York 1987, p. 648
5. *Paul, D. R.; Newmann, S.:* Polymer-Blends. Vol. 2. Academic Pr., New York 1978
6. *Olabisi, O.; Robeson, L. M.; Snow, M. T.:* Polymer Miscibility, Academic Pr., New York 1979
7. *Alfrey Jr. T.; Schenk, W. J.:* Science 208 (1980) p. 813
8. *Heikens, D.:* Kem. Ind. 31 (1982) 4, p. 165–192
9. *Meyer, W. H.:* Chemie i. u. Zeit 21 (1987) p. 59
10. *Olabisi, O.; Kirk-Othmer*, in: Encyclopedia of Chemical Technology. 3. Ed., Vol. 17. Wiley, New York 1983, p. 443–477
11. *Rudin, A.; Braithwaite, N. E.:* Polym. Eng. Sci. 24 (1984) p. 1312
12. Company brochure Mobay Chem. Corp., Pittsburgh, USA 1982
13. *Kämpf, G.:* Polym. 19 (1987) p. 257
14. JA-PS 59-19656 (1984) Mitsubishi Kasei Kogyo Kasashiki Kaisha
15. *Casper, R.; Morbitzer, L.*: Makromol. Chem. 58/59 (1977)
16. EP 0 321 831 (1989) (Bayer AG)
17. *Belaribi, C.*, et al.: Eur. Polym. 22 (1986) p. 487
18. DE-OS 22 59 565 (1974) Bayer AG
19. US-PS 31 30 177 (1961) Borg Warner Corp.
20. EP 00 36 127 (1981) Mobay Chem. Corp.
21. *Neuray, D.; Ott, K.-H.*: Angew. Makromol. Chemie 98 (1981) p. 213
22. *Witman, M. W.; Kamsky, K. L.:* Automotive Plastics Durability Conference Proceedings (1981) p. 91–98
23. *Weirauch, K.:* Kunststoffberater 12 (1986) p. 33
24. *Sternfield, A.:* Mod. Plast. Intern. 17 (1987) 5, p. 8–41
25. Makroblend EC 900. Company brochure Bayer AG Leverkusen 1985
26. *Gergen, W. P.:* Kautsch. Gummi Kunstst. 37 (1984) p. 284
27. US-PS 3 431 24 (1969) General Electric
28. JA-PS 7 328 350 (1970) Unitika Ltd.
29. JA-PS 76 026 936 (1971) Toa Gosei Chem. Ind.
30. JA-PS 59 068 368 (1982) Suwa Seikosha KK
31. DE-OS 3 332 924 (1983) Bayer AG
32. EP A 235 384 (1986) General Electric
33. US-PS P 4 430 484 (1982) General Electric
34. JA-PS 51 151 797 (1975) Asahi Chemical Ind.
35. DE-OS 3 027 957 (1980) Sumitomo Naugatuch Co. Ltd
36. JA-PS 55 050 061 (1978) Mitsubishi Chem. Ind. KK
37. JA-PS 55 133 446 (1979) Sumitomo Chem. KK
38. EP A 235 384 (1986) General Electric
39. *Eguiazábal, J. I.; Nazábal, J.:* Makromol. Chem., Macromol. Symp. 20/21 (1988) p. 255–257

40. US-PS 4 800 218 (1989) General Electric
41. EP A 227 053 (1986) General Electric
42. DE-OS 3 605 573 (1986) BASF AG
43. EP A 304 041 (1988) Nippon Petrochemicals
44. JA-PS 58 008 760 (1981) Dainippon Ink. Chem. KK
45. JA-PS 63 027 556 (1986) Mitsubishi Chem. Ind. KK
46. US P 4 782 114 (1987) Dexter Corp.
47. EP 315 167 (1988) Nippon Oil and Fats Comp. Ltd.
48. JA-PS 62 275 156 (1986) Daicel Chem. Ind. KK
49. US-PS 4 788 249 (1988) General Electric
50. US-PS 4 749 754 (1987) General Electric
51. US-PS 4 788 248 (1988) General Electric
52. *Heggs, R. P.*, et al.: Plast. Eng. (1988) 6, p. 19–32
53. EP A 285 693 (1987) General Electric
54. EP A 285 692 (1987) General Electric
55. Plastics Focus 19 No. (1987) 13, p. 2
56. EP 104 695 (1982) Stamicarbon
57. DE-OS 3 721 239 (1989) Bayer AG

3.6 Polycarbonate Analysis

H. H. Müller and A. Horbach

3.6.1 Spectroscopic Identification of Polycarbonate

Infrared, ^{1}H NMR, and ^{13}C NMR spectroscopic techniques can be employed to identify polycarbonate in an unknown polymeric material.

IR spectra of bisphenol A polycarbonate (Figure 3.7) are characterized by an intense band at approximately 1770 cm^{-1} (C=O stretch), a strong band at approximately 1220 cm^{-1} (asymmetric C=O stretch), a band at approximately 1510 cm^{-1} (aromatic ring modes), and a doublet at 1390 and 1360 cm^{-1} (symmetric CH_3 bending).

The ^{1}H-NMR spectrum of bisphenol A polycarbonate (Figure 3.8) shows the singlet of the isopropylidene group at $\delta = 1.68$ ppm and a symmetric multiple for the aromatic

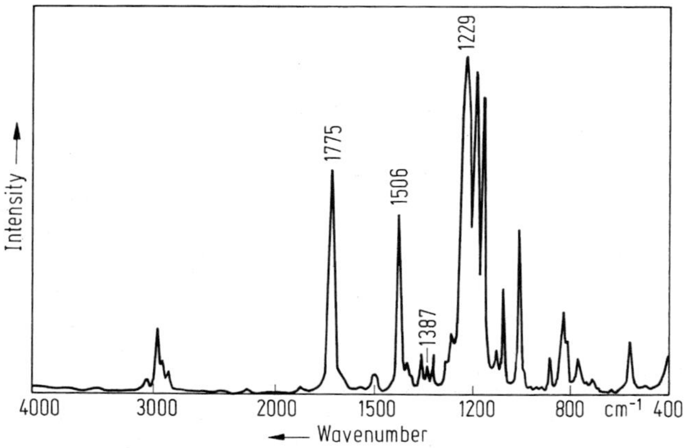

Figure 3.7. IR spectrum of bisphenol A polycarbonate.

signal that forms an AB system with the chemical shifts of δ = 7.25 ppm and δ = 7.13 ppm, and therefore a coupling constant of 9 Hz.

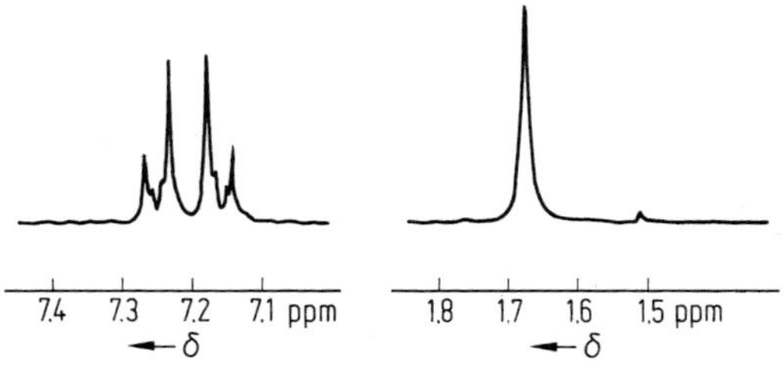

Figure 3.8. ^{1}H-NMR spectrum of bisphenol A polycarbonate (dissolved in $CDCl_3$ with TMS as internal standard, 250 MHz).

In the ^{13}C spectrum of bisphenol A polycarbonate (Figure 3.9), the following resonances have been identified (the aromatic carbon atoms attached to the isopropylidene group are designated C-1): 30.9 ppm (CH_3 group), 42.5 ppm (quaternary C atom of the isopropylidene group), 120.3 ppm (C-3 and C-5), 127.9 ppm (C-2 and C-6), 148.2 and 149.0 ppm (C-1 and C-4), and 152.1 ppm (carbonate C atom).

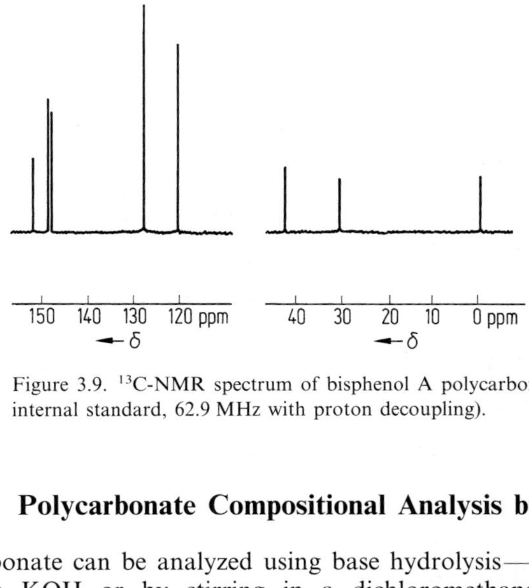

Figure 3.9. ^{13}C-NMR spectrum of bisphenol A polycarbonate (dissolved in $CDCl_3$ with TMS as internal standard, 62.9 MHz with proton decoupling).

3.6.2 Polycarbonate Compositional Analysis by Saponification

Polycarbonate can be analyzed using base hydrolysis—for example, by refluxing with alcoholic KOH or by stirring in a dichloromethane solution containing sodium methanolate. Following saponification, certain amounts of bisphenol A (or the corresponding bisphenol in the polycarbonates) can be analyzed, using high performance liquid chromatography (HPLC) or gas chromatography (GC). For the preparation of samples for GC, the bisphenol can be isolated from the acidified solution by extraction with diethyl ether. For analysis of carbonate content (e.g., in a polyester carbonate), the alkali hydrolysate is acidified and the amount of CO_2 that has been liberated determined using conductivity measurements or gravimetric studies (after absorption in barium carbonate) [1].

In blends based on polycarbonate, such as polycarbonate–SAN blends or impact-modified polycarbonate, the polycarbonate content can be accurately determined by alkaline hydrolysis and CO_2 analysis.

3.6.3 End Group Analysis

In a typical interfacial synthesis of polycarbonate, monofunctional phenols are used to regulate the molecular weight. In principle, this can lead to end groups of the following kinds:

- phenyl and alkylphenyl carbonate end groups, —O—COO—Ar—R (R = H or alkyl)
- phenolic end groups —O—COO—Ar—$C(CH_3)_2$—Ar—OH
- chloroformate end groups —O—COCl

The majority are in the form of phenyl and alkylphenyl carbonate end groups; chloroformate end groups are generally not detected.

The amount of phenyl and alkylphenyl carbonate end groups can be determined using saponification and HPLC or GC as discussed in Section 3.6.2 [1, 2]. Because it is present in negligible amounts, the amount of phenolic OH-end groups cannot be determined using typical titration techniques. A good technique with a relatively low detection limit (<0.001% OH) is the photometric analysis of the complex formed between phenolic OH and $TiCl_4$ [1, 3, 4]. IR spectroscopy (measuring the OH band at 3600 cm^{-1} in a dichloromethane solution) [3] and UV spectroscopy (using the difference in the original polycarbonate and after acetylation with acetanhydride in dichloromethane/pyridine) [5, 6] have been employed to analyze phenolic end groups.

Chloroformate end groups can be analyzed by alkaline hydrolysis (e.g., by using NaOH in ethylene glycol-monomethylether [4] or aniline [7]) followed by determination of the amount of chloride using argentometric titration. If the polycarbonate sample contains traces of inorganic chloride from a two-phase production process, the amount of inorganic chloride must be determined by analysis of an additional blank that does not contain base (to correct the chloride titration) [4].

3.6.4 Analysis of Number-Average and Weight-Average Molecular Weight

3.6.4.1 Weight-Average Molecular Weight

The absolute weight average molecular weight (M_w) of polycarbonate is determined by measuring the dependence of the light-scattering angle. Solvents such as dichloromethane [4], chloroform [8, 9], or tetrahydrofuran [10] are employed. Alternatively, the absolute molecular weight can be determined by ultrafiltration (i.e., using the method of equivalent sedimentation weight) [4, 11].

3.6.4.2 Number-Average Molecular Weight

Normally the absolute number-average molecular weight (M_n) is measured using vapor phase osmometry (VPO), membrane osmometry, or end group analysis. Membrane osmometry of commercial products having M_n less than 20,000 and typical polydispersities can lead to a systematic overestimation of the M_n because of permeation of the membrane by the low molecular weight species. Determination of M_n by end group analysis is achieved by measuring all end groups, followed by calculation of M_n using the following equation:

$$M_n = \frac{10^6 \times n}{\Sigma E_i} \tag{3.14}$$

E_i = end groups: (phenyl or alkylphenyl, phenolic hydroxy, chloroformate) (mmol/kg)
n = number of end groups per molecule.

For the calculation, the number of end groups per molecule must be known (for a linear molecule, n is assumed to be 2). Because the experimental error of the different end group analyses is additive, great accuracy is essential, particularly for higher molecular weight polymers, where the end group content is very small.

3.6.4.3 Solution Viscosity

Of all the methods of molecular weight determination, solution viscosity requires the simplest apparatus. For polycarbonate, the relative viscosity (η_{rel}) in dichloromethane at 25 °C at a concentration of 0.005 g/cm³ is generally reported. For these measurements, Ubbelohde or Ostwald capillary viscometers with a diameter of 0.3 mm and a length of 10 cm are employed [4].

The intrinsic viscosity [η] can be determined from the concentration dependence of the relative viscosity following the Huggins (H) or Schulz–Blaschke (SB) methods. For bisphenol A polycarbonate the concentration dependence constants for dichloromethane at 25 °C are [4]:

$$K_H = 0.43 \quad \text{and} \quad K_{SB} = 0.30$$

The intrinsic viscosity [η] and the molecular weight are related via the Mark–Houwink/Kuhn–Sakurada relationship. For bisphenol A polycarbonates in the range of 15,000–100,000, and moderate polydispersities ($U \leq 3$) the relationship is

$$[\eta] = 2.13 \times 10^{-2}\ M^{0.76} \tag{3.15}$$

for dichloromethane at 25 °C; [η] has units of cm³/g.

For additional solvents (chloroform, dichloromethane, tetrachloroethane, THF, dioxane), the Mark–Houwink constants for bisphenol A polycarbonate are tabulated in the literature [4, 12]. The molecular weight value from solution viscosity (viscosity-average molecular weight) for commercial polycarbonates (in solvents with Mark–Houwink exponents $\alpha \geq 0.65$) corresponds within experimental error to M_w.

3.6.4.4 Size Exclusion Chromatography

Polycarbonate molecular weight is best determined using size exclusion chromatography [SEC; also called gel permeation chromatography (GPC)]; in this respect, polycarbonate resembles many other readily soluble polymers. Separation is carried out using crosslinked polystyrene packing and THF [8, 10, 13–17], dichloromethane [10, 18], or chloroform [19] as the eluent. Differential refractometers of UV photometers serve as detectors.

Although the separation is generally not difficult, the molecular weight calculation can be problematic because polycarbonate standards for calibration are not readily available. In practice, SEC is calibrated using commercially available polystyrene standards and the polycarbonate molecular weight is obtained relative to polystyrene. The result is a value that is higher than the true molecular weight.

If relative values are not sufficient, typical polystyrene standards can converted to a universal calibration curve following the methods of Benoit [8, 10, 13, 16, 18, 19]. With this method, the Mark–Houwink relationship for the polycarbonate must be known or a viscosity detector employed.

When the absolute molecular weight (M_n, M_w, M_z) is required with high accuracy, well-characterized polycarbonate standards are needed. Especially useful are narrow molecular weight distribution polycarbonate standards, which can be obtained by preparative SEC. If there are no polycarbonates having a narrow molecular weight distribution, procedures for calibration with broadly distributed standards [20, 21] can be used (e.g., calibration with a polycarbonate standard whose absolute M_w is known from light scattering and absolute M_n is known from end group analysis).

Another, albeit rather expensive, possibility is utilization of a light-scattering apparatus as a detector [2, 10, 22]. The advantage of this method is that no external standard is needed.

Figure 3.10 shows an SEC chromatogram of commercial polycarbonates of different molecular weights. Polycarbonate A is sold as a relatively low molecular weight polycarbonate for production of compact disks, polycarbonate B is a product for typical extrusion and injection molding applications, and polycarbonate C is a relatively high molecular weight polycarbonate for production of film.

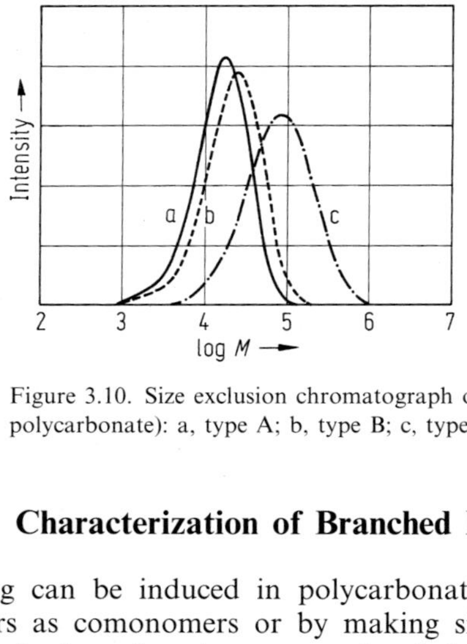

Figure 3.10. Size exclusion chromatograph of polycarbonates (calibrated with bisphenol A polycarbonate): a, type A; b, type B; c, type C (described in text).

3.6.5 Characterization of Branched Polycarbonate

Branching can be induced in polycarbonate either by using tri- or multifunctional monomers as comonomers or by making side reactions on the polycarbonate chain (particularly in polycarbonate produced by melt polycondensaton). The extent of branching can be determined by measuring the solution viscosity of materials in which the M_w or M_n is already known.

Long-chain branches impart to polycarbonate—indeed to all polymers in general—a relatively smaller hydrodynamic volume, and thus a smaller intrinsic viscosity, than obtained with linear polycarbonates of similar molecular weight. By finding the intrinsic viscosity from experiment $[\eta]_{exp}$, the M_w from light scattering, and (from the Mark–Houwink relationship) the calculated intrinsic viscosity $[\eta]_{calc}$ for a linear polycarbonate, one can calculate the ratio of $[\eta]_{exp}$ to $[\eta]_{calc}$, which should be less than 1 in the case of branching. From this ratio, the number of branch points per molecule can be calculated by known methods [23–30]. Specific knowledge about existing branches (e.g., functionality or distribution of branch points) must be possessed or obtained. From the number and kinds of end groups and the absolute M_n, the number of end groups per molecule can be determined (as described in Section 3.6.4.2).

During analysis of branching by solution viscosity, short-chain branches are not considered because they have little effect on the hydrodynamic volume. In contrast, end group analysis gives information on existing short-chain branches.

3.6.6 Characterization of Cyclic Polycarbonate

Under certain conditions of the phase interfacial, cyclic polycarbonate can be formed [7]. The test for cyclic polycarbonate is the same as for branched material: parallel measurement of the intrinsic viscosity and M_w or end group analysis and M_n [7]. Concerning their molecular dimensions, cyclic compounds exhibit behavior similar to that of branched molecules: smaller intrinsic viscosity compared to linear molecules of the same molecular weight, namely, $[\eta]_{cycl} = [\eta]_{lin} \approx 0.66$ in the theta condition [31]. End group analysis of cyclic polycarbonates gives low values of n (<2) because cyclic molecules contain no end groups.

3.6.7 Oligomer Analysis

For the characterization of oligomers in polycarbonate, chromatographic techniques, particularly SEC and HPLC, are quite useful [32, 33]. Using SEC with crosslinked polystyrene as the packing material and dichloromethane as the eluent, polycarbonate oligomers having a degree of polymerization (DP) of less than 10 can be separated [33]. Better separation can be achieved using inverse phase HPLC; oligomers of different kinds can be separated (e.g., oligomers with two hydroxy end groups, with two phenyl end groups, or oligomers with mixed end groups) [33].

3.6.8 Additive Analysis

Depending on the method of production and the expected application, commercial polycarbonates may contain several additives (e.g., processing aids, stabilizers, UV absorbers, flame retardants). Because concentrations often are minimal, the analysis of additives requires special procedures to determine the chemical structure and additive level [34–37]. Chromatographic methods (GC, HPLC, thin-layer chromatography), as well as spectroscopic techniques (UV, IR, NMR, and mass spectroscopy), are especially useful. In many cases it is necessary to prepare the sample to ensure that the additive is separated from the polycarbonate matrix—for example, isolating the additive by extraction or by precipitation of the polymer (in which cases the additive is analyzed in solution).

References for Section 3.6

1. *Müller, H. H.; Bartl, H.; Falbe, J.*, in: Houben-Weyl, Methoden der organischen Chemie, Makromolekulare Stoffe., Vol. E 20, Part III. Thieme, Stuttgart 1987
2. *Horbach, A.*, et al.: Angew. Makromol. Chem. 136 (1985) p. 1
3. *Horbach, A.*, et al.: Makromol. Chem. 88 (1965) p. 215
4. *Horbach, A.; Stuck, W.; Wunderlich, H.*, in: *Snell, F. D.; Ettre, L. S.*: Encyclopedia of Industrial Chemical Analysis. Vol. 17. Wiley, New York 1973
5. *Shchori, E.; McGrath, J. E.*: J. Appl. Polym. Sci., Appl. Polym. Symp. 34 (1978) p. 103
6. *Shchori, E.; McGrath, J. E.*: Polym. Prepr. Amer. Chem. Soc., Div. Polym. Chem. 19 (1978) 1, p. 494
7. *Horbach, A.*, et al.: Makromol. Chem. 181 (1980) p. 111

8. *Bartosiewicz, R. L.*, et al.: Eur. Polym. J. 10 (1974) p. 783
9. *Czlonkowska-Kohutnicka, Z.; Dobkowski, Z.*: Eur. Polym. J. 18 (1982) p. 911
10. *Bailly, C.*, et al.: Polymer 87 (1986) p. 1410
11. *Schulz, G. V.; Horbach, A.*: Makromol. Chem. 39 (1959) p. 93
12. *Bandrup, J.; Immergut, E. H.*: Polymer Handbook. 2. Ed. Wiley, New York 1975
13. *Brzezinski, J.; Dobkowski, Z.*: Eur. Polym. J. 16 (1980) p. 85
14. *Schilling, F. C.*, et al.: Macromolecules 14 (1981) p. 532
15. *Pryde, C. A.*, et al.: Polym. Eng. Sci. 22 (1982) p. 370
16. *Dunn, D. A.; Hagnauer, G. L.*: 43. Proc. Ann. Conf. Reinf. Plast. Compos. Inst. Soc. Plast. Ind. 31 (1985) p. 280
17. *Goetz, H.*, et al.: J. Chromatogr. 349 (1985) p. 357
18. *Abbas, K. B.*: Polymer 22 (1981) p. 836
19. *Dawkins, J. V.*, et al.: Eur. Polym. J. 9 (1973) p. 327
20. *Yau, W. W.; Kirkland, J. J.; Bly, D. D.*: Modern Size-Exclusion Chromatography. Wiley, New York 1979
21. *Dawkins, J. V.*: Chromatogr. Sci. 25 (1984) p. 53
22. *Schorn, H.*, et al.: J. Chromatogr. 353 (1986) p. 273
23. *Elias, H.-G.*: Makromoleküle Hüthig, Basel 1971
24. *Schurz, J.*: Physikalische Chemie der Hochpolymeren. Springer, Berlin 1974
25. *Slade, P. E.*: Polymer Molecular Weights. Part II. Dekker, New York 1975
26. *Billingham, N. C.*: Molar Mass Measurements in Polymer Science. Kogan Page, London 1977
27. *Hoffmann, M.; Krömer, H.; Kuhn, R.*: Polymeranalytik I. Thieme, Stuttgart 1977
28. *Koenig, J. L.*: Chemical Microstructure of Polymer Chains. Wiley, New York 1980
29. *Bohdanecky, M.; Kovar, J.*: Viscosity of Polymer Solutions. Elsevier, Amsterdam 1982
30. *Schröder, E.; Müller, G.; Arndt, F.*: Leitfaden der Polymercharakterisierung. Akademie-Verlag, Berlin 1982
31. *Bloomfield, V.; Zimm, B.*: J. Chem. Phys. 44 (1966) p. 315
32. *Duvdevani, I.*, et al.: J. Polym. Sci., Polym. Lett. Ed. 9 (1971) p. 429
33. *Bailly, C.*, et al.: Polymer 27 (1986) p. 776
34. *Schröder, E.; Franz, J.; Hagen, E.*: Ausgewählte Methoden der Plastanalytik. Akademie-Verlag. Berlin 1976
35. *Crompton, T. R.*: Chemical Analysis of Additives in Plastics. 2. Ed. Pergamon Pr., Oxford 1977
36. *Haslam, J.; Willis, H. A.; Squirrell, D. C. M.*: Identification and Analysis of Plastics. 2. Ed Heyden, London 1980
37. *Crompton, T. R.*: The Analysis of Plastics. Pergamon Pr., Oxford 1984

3.7 Properties of Bisphenol A Polycarbonate

3.7.1 Introduction

Until recently polycarbonates that achieved practical importance as engineering materials were either completely or predominantly composed of bisphenol A. Structural modifications were limited mainly to brominated bisphenol A as a comonomer for flame retardation, trifunctional comonomers targeted for the manufacture of branched polycarbonate, and special end groups for improving flow capability, as well as elastic segment blocks for increasing the impact strength of polycarbonate (see also Section 3.2). Although the mechanical, electrical, and thermal properties of polycarbonates are not significantly altered by the relatively minor contents of brominated comonomer, halogen content causes processing properties to deteriorate. However, in 1989 a new polycarbonate with tailor-made heat distortion temperature based on bisphenol A and bisphenol trimethylcyclohexanone (TMC) was introduced into the market by Bayer AG (see also Section 3.2.3.4).

The major differences between commercial polycarbonates are in molecular weight as well as in the types of additive incorporated. An additive that has a particularly distinct influence on properties is glass fiber. It is customary to differentiate between short and normal (long) glass fiber reinforcement. The average length for short glass fibers is about 100 μm, and for normal glass fibers about 300 μm. Other fillers have only minor importance for polycarbonates. The molecular weight of commercial products for both injection molding and extrusion is between 22,000 and 35,000. Lower molecular weight products with ($M_w < 22{,}000$) show reduced toughness, although flowability is much better. Such high flow products have only recently found commercial application for the manufacture of optical memory disks (compacts disks). Polycarbonates having a molecular weight greater than 35,000 are difficult to extrude. However, it is possible for polycarbonates of high molecular weight (e.g., 100,000) to be cast from solution into sheet or film.

The property discussions in the following sections pertain to average molecular weight injection molding grades of polycarbonate unless otherwise specified. All properties are temperature dependent [1]. However, the corresponding temperature dependencies are explained for selected test parameters. Only virgin pellets were used for manufacturing the test bars on which the properties were measured. In principle, the use of recycled material is possible [2–4]. However, some properties might be altered either through improper processing or as a result of inadvertent contamination of the reground material. During grinding of glass fiber reinforced polycarbonate, additional fiber fracturing causes a reduction in strength [5].

In evaluating material properties, one must always keep in mind that these values have been obtained with test samples of ideal geometry, manufactured under optimal conditions. In production parts, residual stress and orientation conditions are often equivocal. Individual molecules and glass fibers also differ in their orientation. Knit or weld lines further complicate the situation, depending on their location in the parts. Differences in injection molding conditions may also lead to differences in part properties [6]. Properties obtained on test bars cut from finished parts typically differ from those obtained from molded test samples.

The following list summarizes particularly important properties of bisphenol A polycarbonate:

- amorphous structure (and thus transparency)
- high toughness in comparison to other plastics
- relatively high heat distortion temperature (HDT)
- good electrical properties
- ability to maintain its properties, particularly toughness, over a wide temperature range
- high HDT, good electrical properties, higher strength and stiffness, and good toughness of glass reinforced grades [7–13].

It is the combination of these properties that has made polycarbonate a commercially important material [14–16].

The properties of polycarbonate have been independently described in various texts and review articles [17–23]. Section 3.7.2 offers an overview of mechanical, thermal, electrical, and other properties of bisphenol A polycarbonate.

3.7.2 Classification of Polycarbonates According to Rheological Properties

3.7.2.1 Standard Grades

Standard commercial grades of polycarbonates have molecular weights of 22,000–35,000. Depending on the manufacturer, various grades with a more or less tight tolerance are offered in this molecular weight range.

Since molecular weight is known to have a direct influence on rheological and physical properties, a variety of products have been developed, divided according to molecular weight and balancing ease of processing against practical performance. Within the molecular weight range of 24,000–35,000, property differences are barely noticeable. Nevertheless, higher molecular weight polycarbonates are advantageous under certain circumstances—for example, for materials that will be subjected to high stress loads or high impact, and products that may be subjected to stress-cracking solvents or chemicals. The tradeoff is that as the material increases in molecular weight, it becomes more difficult to process because of the accompanying increase in melt viscosity. Depending on actual requirements, a product should be selected that has the best combination of processing ease and end-use performance.

Measurements that reflect the molecular weight and thus represent processibility and mechanical properties are:

- relative solution viscosity (η_{rel})
- flow rate (MFR) and/or volume flow rate (MVR) of the mass
- melt viscosity (η_0)

The relative solvent viscosity is determined according to DIN 51562, Part 3, using an Ubbelohde viscosimeter at 25 °C. A solution of 0.5 g of polycarbonate dissolved in 100 cm^3 of methylene chloride is tested.

The melt flow index is determined according to DIN 53735 (ISO 1133) with a test load of 1.2 kg at 300 °C.

To obtain the melt viscosity, a high pressure capillary viscosimeter with a nozzle length/diameter ratio of 20:1 mm is recommended. The zero viscosity η_0 corresponds to the apparent viscosity η_a in the zero shear range of the flow curve. Apparent and true viscosity of unreinforced polycarbonate grades are practically identical in the previously mentioned molecular weight ranges. The various measuring methods correlate well with one another and permit a differentiation of product types (Figures 3.11 and 3.12).

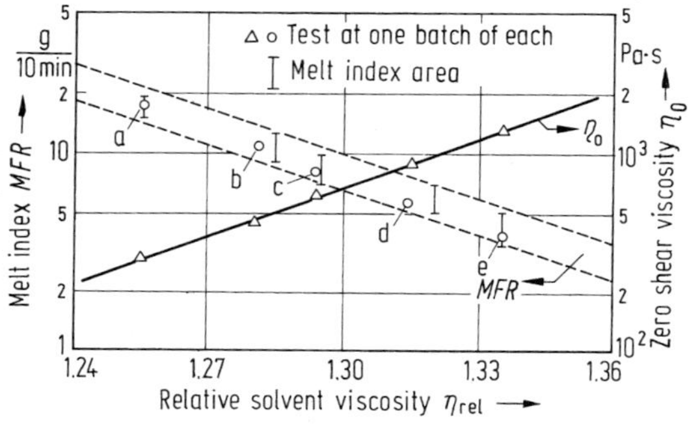

Figure 3.11

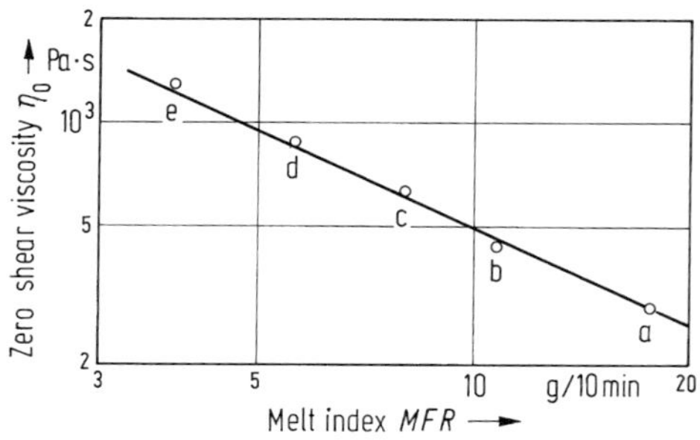

Figure 3.12

Figure 3.11. Correlation between relative solvent viscosity η_{rel} and melt flow index *MFI* 300/11.8 and/or zero shear viscosity η_0 at 300 °C of several Makrolon grades with different molecular weights: a, low molecular weight injection molding grade (2400); b–d, products with increasing molecular weight (2600, 2800, and 3100, respectively); e, extrusion grade with highest molecular weight (3200).

Figure 3.12. Correlation between zero shear viscosity η_0 at 300 °C and melt flow index *MFR* 300/11.8 of several Makrolon grades: a, low molecular weight injection molding grade (2400); b–d, product grades with increasing molecular weight (2600, 2800, and 3100, respectively); e, extrusion grade with increased molecular weight (3200).

Either the relative solvent viscosity and/or the melt flow index can be used successfully for process control in manufacturing. These tests are also recommended for identifying molecular weight loss from processing or other influences. Both the melt flow and solution viscosity values provide information on the potential physical properties of the polycarbonate being measured. The flow curves showing the course of the apparent viscosity as a function of the shear rate appear in Figure 3.13.

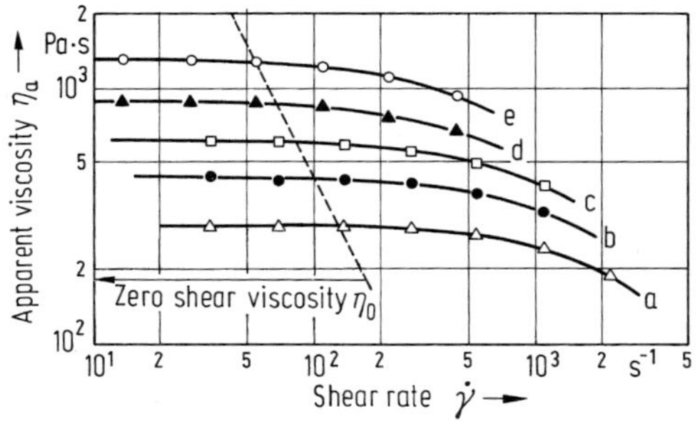

Figure 3.13. Dependence of apparent viscosity η_a at 300 °C on shear rate γ shown for several Makrolon grades: a, low molecular weight injection molding grade (2400); b–d, product grades with increasing molecular weight (2600, 2800, and 3100, respectively); e, extrusion grade with increased molecular weight (3200).

3.7.2.2 Branched Polycarbonate Types

Polycarbonate based on bisphenol A is a linear polymer but can be branched through condensation with multifunctional phenols. The advantage of polycarbonate branching is the creation of a melt that is structurally more stable. Also, while the decrease in the melt viscosity of linear polycarbonate is almost linear with increasing shear rate, branched polycarbonate shows a more precipitous drop in melt viscosity with increasing shear rate (Figure 3.13). The relevance of this behavior is that branched polycarbonate is processible at normal shear rates but has increased melt stability at low shear rates, especially during extrusion. These processing advantages are utilized by manufacturers of twin-wall sheet (see Section 3.8.2.12) as well as for large-volume, extrusion blow-molded containers (see Section 3.8.3) such as 5-gallon water bottles.

As described in Section 3.7.8.2, by branching polycarbonate, it is possible to largely suppress a material's dripping behavior during a fire, which results in a better burn classification. Since viscosity decreases at higher shear rates, parts can be manufactured by the injection molding method without significant processing restrictions.

3.7.2.3 High Flow Rate Polycarbonate Grades

To achieve optimum mechanical and thermal properties as well as to minimize stress-cracking sensitivity and brittleness, polycarbonate with the highest possible molecular weight should be used. Unfortunately, selection in this direction increases melt viscosity, which has a negative influence on the flow behavior necessary for filling molds. The compromise between properties and processibility effectively limits commercial polycarbonates to an upper and lower molecular weight between 22,000 and 35,000.

Recently it has been possible to modify polycarbonate in a unique way that shifts the lower limit of molecular weight downward without sacrificing mechanical properties. Figure 3.14 and Table 3.5 show that these modified polycarbonates, with molecular weights of 18,000–25,000, produce products with good impact strength and improved flow properties [14].

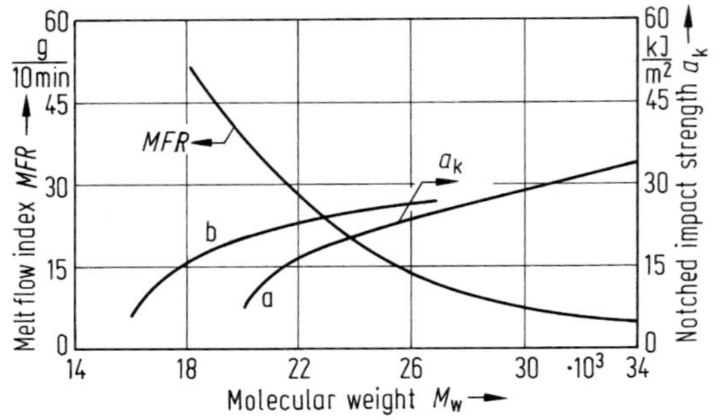

Figure 3.14. Correlation of molecular weight M_w with melt flow index *MFR* (ISO 1133) and notched impact strength (DIN 53453) of conventional (a) and impact-modified (b) polycarbonate.

Table 3.5. Comparison of the Properties of Various Easy-Flowing Polycarbonates [14], Indicating Improvements Achievable Through Targeted Product Optimization

Property	Units	High Flow Grades Newly Developed Products		High Flow Grades Standard Products		Test Method
Melt flow index *MFR* 300/11.8	g/10 min	32–35	17–20	17–20	10–12	DIN 53 735 ISO 1133
Flow spirals at 300 °C	cm	≈54	≈38	≈38	≈33	
Notched impact strength	kJ/m²	>20	>40	>20	>40	DIN 53 453, ISO R 179
Elongation at break after 400 h heat aging at 100 °C	%	10	50	10	30	DIN 53 455, ISO R 527

Flow modifications yield a series of interesting processing advantages in polycarbonate molding applications. Mold filling is simplified in complicated tool geometries, which makes it possible to produce molded parts with thinner walls. Depending on the part, cycle times are shorter, which results in increased productivity and cost savings. In commercial applications, increasing the MFR from 12 g/10 min to 20 g/10 min can reduce the cycle time by about 15%. These polycarbonate grades are gaining an increasing market share as "fast-cycling resins."

Polycarbonates with molecular weight below 22,000 have only limited usefulness as engineering resins. Even with polymer modification, there is a substantial risk of brittle breakage and an increased sensitivity to stress-cracking. Nevertheless, for simple mold geometries, such as laser optical storage disks (compact disks), the mechanical properties are sufficient, even at molecular weights as low as 18,000. Moreover, the improved flow properties can be utilized to mold disks with very low residual stress and low optical birefringence (see also Section 3.10.15) [24].

To overcome the drop in impact properties of high flow polycarbonate, an elastomer modifier or toughener, such as an acrylic rubber, can be added to the resin. However, this type of impact modification influences other polycarbonate properties, altering the typical property profile.

3.7.3 Physical Aspects of Bisphenol A Polycarbonate Polymers [22]

At room temperature, the modulus and strength properties of bisphenol A polycarbonate are comparable to those of other amorphous thermoplastics. However, the toughness of bisphenol A polycarbonate is considerably higher. This is unusual because amorphous thermoplastics are normally stiff and glassy below the glass transition temperature. Most amorphous polymers (e.g., polystyrene, polymethyl methacrylate) show brittle breakage below their T_g. The relationship between macroscopic mechanical behavior and polymer molecular structure is still poorly understood.

3.7.3.1 Dynamic Mechanical Properties

Dynamic mechanical analysis permits conceptual insight into the internal molecular motion of polycarbonate chains. Knowledge of these processes is the basis for understanding the behavior of such materials during large mechanical deformations. Over a wide temperature range, the storage modulus G' of bisphenol A polycarbonate varies little with temperature. The loss modulus G'' reaches a maximum around 150 °C, which characterizes the glass transition temperature T_g (Figure 3.15). Differential thermal analysis leads to a similar value for T_g.

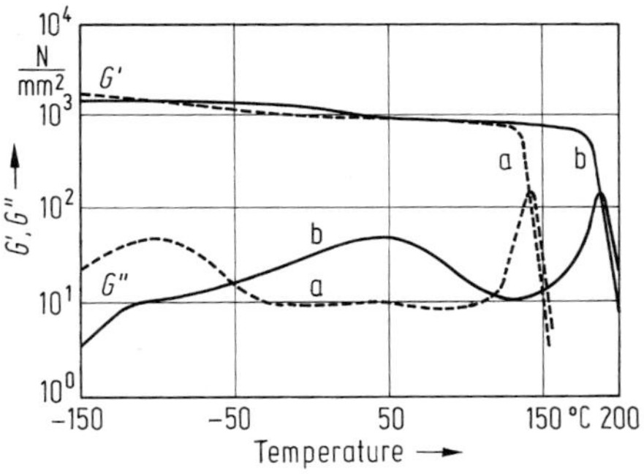

Figure 3.15. Temperature dependence of storage modulus (G') and loss modulus G'' for bisphenol A polycarbonate (a) and tetramethylbisphenol A polycarbonate (b).

In contrast to other thermoplastics such as polystyrene ($T_g = 100$ °C), polybutylene terephthalate ($T_g = 40$ °C), and polyamide ($T_g = 40$ °C), the relatively high glass transition temperature of bisphenol A polycarbonates can be traced back to the hindered rotational processes of phenyl ring containing chain segments. In addition, since the repeat units of such polymer chains are larger and bulkier than those of most other polymers, molecular ordering (e.g., crystallization) is more difficult. A high glass transition temperature leads to excellent dimensional stability, even under load (good creep behavior). The glass transition temperature represents the absolute maximum upper service temperature with respect to mechanical load.

The unusually high toughness of polycarbonate is related to the very wide G'' maximum in the low temperature range (γ relaxation at −100 °C). Preparation of polycarbonate from 3,3′,5,5′-tetramethyl bisphenol A (TMBPA-PC) leads to an increase in the glass transition temperature of about 200 °C and to a shift of the γ relaxation to about 50 °C, as well as a broadening of the relaxation range (Figure 3.15). Both effects are a direct consequence of the additional methyl groups on the benzene rings, which hinder motion, making the rotational process around the carbonate group more difficult.

The upward temperature shift of the γ relaxation is even more dramatic than that of the glass transition. The reduced ductility of TMBPA-PC is a direct result of the increased γ-relaxation temperature.

3.7.3.2 Toughness Behavior

The toughness of polycarbonate can be evaluated by the notched impact test, where temperature plays an important role. Bisphenol A polycarbonate shows a sharp transition from tough to brittle behavior between −10 °C and −20 °C, depending on the grade.

In the notched impact test, toughness is characterized by stable tear propagation of a macroscopic plastic zone during shear deformation. Brittleness is evident when plastic zones can no longer form under shear deformation; relaxation processes, restricted by the lower temperature, are not able to reduce the stress state behind the notch quickly enough. This causes microvoid formation behind the notch (i.e., crazing), which results in macroscopic, unstable tear propagation and brittle failure.

The ductile-to-brittle transition in polycarbonate depends on the following conditions:

- molecular weight
- temperature
- stress rate
- sample thickness (and stress state)
- notch radius (sharpness)
- thermal pretreatment (processing history, annealing)
- additives, external media

The thickness dependency of the ductile-to-brittle transition can be explained by considering the stress state behind the notch. With thin samples, the material behind the notch shows a plane stress condition and can participate in the shear mechanism unhindered, which results in good toughness. Very thick samples show a plane strain behind the notch, which hinders the start of shear deformation and facilitates crazing. This results in brittle behavior. This "critical thickness" transition is just as sharp as the ductile-to-brittle transition observed during the temperature variations. The result is that at increasing sample thickness, the ductile-to-brittle transition temperature shifts toward higher temperatures.

Ductile–brittle transitions as a function of temperature and sample dimensions have been demonstrated with other polymeric and metallic materials. The critical thickness of bisphenol A polycarbonate can be increased through copolycondensation—for example, with bis(4-hydroxyphenyl)sulfide, or with a combination of impact modifiers.

3.7.4 Mechanical Properties of Bisphenol A Polycarbonate

Table 3.6 provides an overview of mechanical properties of unreinforced and glass fiber reinforced polycarbonate.

3.7.4.1 Stress–Strain Behavior in the Tensile Test

Stress–strain diagrams of unreinforced polycarbonate and two kinds of glass fiber reinforced polymers are shown in Figure 3.16 (tensile test DIN 53455 or ASTM D 638). The yield point of an unreinforced polycarbonate is around 60 N/mm^2 and will be achieved at about 6% elongation; in glass fiber reinforced types, it will be at 3–6%

Table 3.6. Typical Values for Mechanical Properties of Natural Polycarbonate

Property	Units	Average Molecular Weight, PC Unreinforced	Reinforced PC 30 wt % Short Glass Fibers	Reinforced PC 20 wt % Long Glass Fibers	Test Method
Density	g/cm³	1.20	1.44	1.35	
Ultimate tensile strength	N/mm²	72	70	100	ISO 527, DIN 53455
Elongation at break	%	120	3.5	3.5	
1% Offset stress	N/mm²	50	65	90	ISO 527, DIN 53455
Strain at 1% offset	%	3.5	2.3	3.0	
Tensile strength at yield	N/mm²	63	75	90	ISO 527, DIN 53455
Elongation at yield	%	6	3	3.5	
Compressive yield stress	N/mm²	75	110	125	ISO 604, DIN 53454
Compressive yield strain	%	7	4.5	5	
Flexural strength	N/mm²	—	135	160	ISO 178, DIN 53452
Flexural yield strain	%	—	6	5	ISO 178, DIN 53452
Flexural stress at 3.5% strain	N/mm²	>70	—	—	DIN 53452
Modulus of elasticity	N/mm²	2400	5800	6000	DIN 53457
Tensile creep strength	N/mm²	>40	—	—	ISO 899, DIN 53444
Creep modulus	N/mm²	2100	—	—	ISO 899, DIN 53444
Impact strength at +23 °C	kJ/m²	n.b.[1]	35	40	ISO 179, DIN 53453
−40 °C	kJ/m²	n.b.	35	40	ISO 179, DIN 53453
Notched impact strength at +23 °C	J/m	900	75	125	ASTM D 256
Ball indentation hardness	N/mm²	110	150	150	ISO 2039, B DIN 53456

[1] n.b., not broken.

elongation and occasionally higher. The type of glass fiber surface treatment (sizing) can influence the strength of reinforced polycarbonate [25].

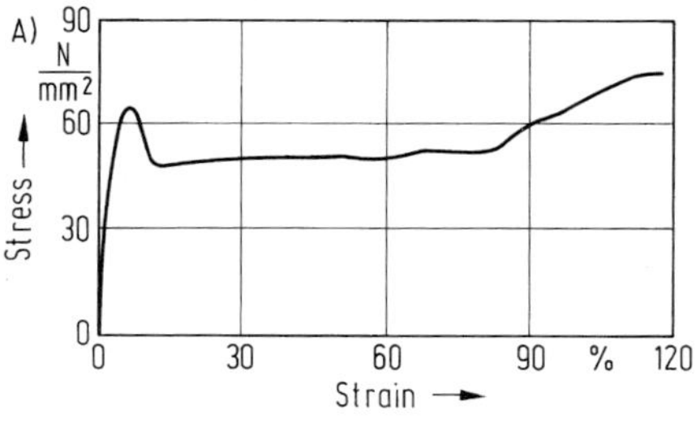

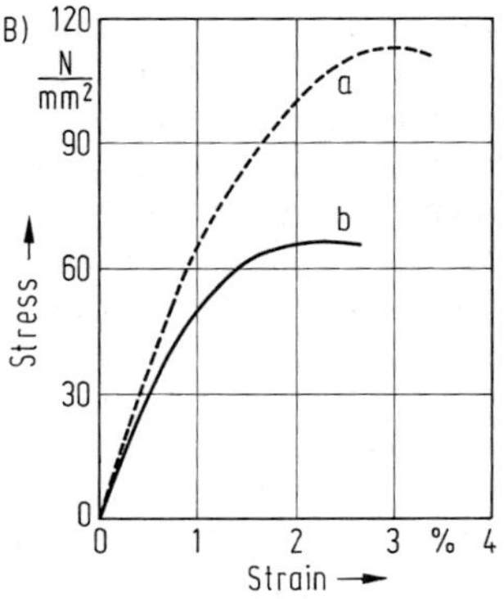

Figure 3.16. Stress–strain diagram of injection-molded bars of unreinforced (A) and reinforced (B) polycarbonate (DIN 53455, 22 °C, 55% relative humidity; tensile bar no. 3, 5 mm/min strain rate): (A) 20 wt % long glass fibers, (B) 30 wt % short glass fibers.

Unreinforced products show distinct yielding and flow deformation at elongations above 6%. The plastic deformation in polycarbonates is related to both low volume contraction [26] and heating [27]. Unaged samples break at greater than 100% elongation. The tensile elongation at break is typically above the tensile yield elongation. Polycarbonate with 10% or more glass fibers (without impact modifiers) usually shows no yielding; the test bar simply breaks when the elongation limit is reached. The high elongation at break of unreinforced polycarbonate decreases after annealing or thermal aging, and simultaneously, the tensile strength at break is reduced. Therefore, the stress–strain curve is a good indicator of the thermal history of test samples. Standard polycarbonates with molecular weights of 20,000–30,000 all show similar stress–strain diagrams.

As with all thermoplastics, the absolute value of the strength is dependent on the test temperature [1] and generally increases with decreasing temperature [28]. Figures 3.17 and 3.18 show the dependence of test temperature on tensile strength at break and on the modulus of elasticity, respectively, of an unreinforced polycarbonate and one with 30% reinforcement with short glass fibers.

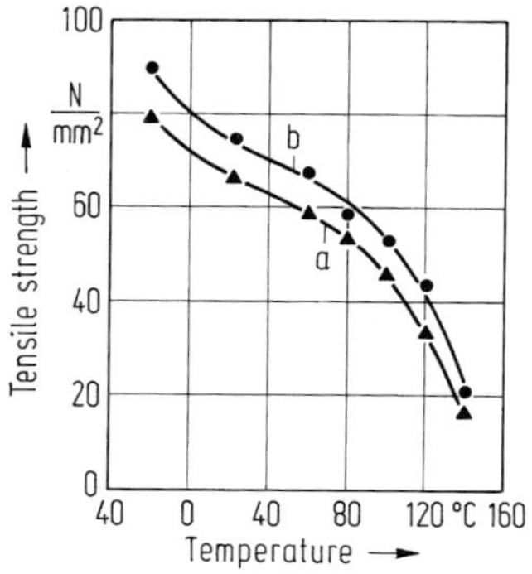

Figure 3.17. Tensile strength at break as a function of test temperature (DIN 53455; tensile bar no. 3): a, unreinforced polycarbonate (feed rate 50 mm/min); b, polycarbonate reinforced with 30 wt % short glass fibers (feed rate 10 mm/min).

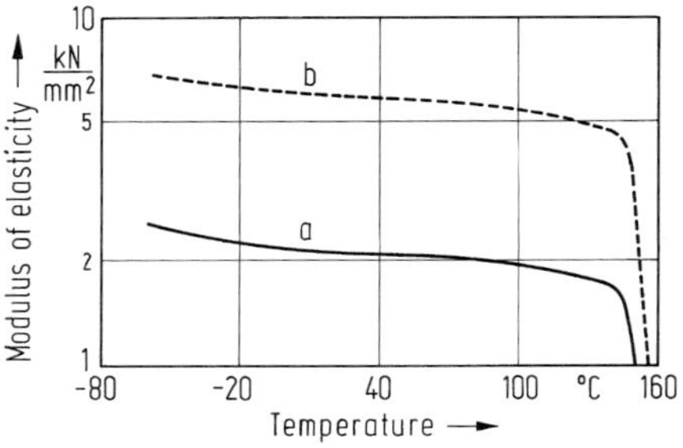

Figure 3.18. Dependence of modulus of elasticity on temperature (DIN 53457; standard test bar no. 3): a, unreinforced polycarbonate (feed rate 50 mm/min); b, polycarbonate reinforced with 30 wt % short glass fibers (feed rate 10 mm/min).

The low dependence of the modulus of elasticity on temperatures up to about 120 °C and the relatively high tensile strength at 100 °C are major reasons for the selection of polycarbonates over other thermoplastics for applications with thermal loads. Carbon fiber reinforcement is seldom used today despite significant improvements in strength and stiffness; a carbon fiber content of 10% achieves almost the same flexural strength as 30% glass fiber reinforcement [29].

Polycarbonate can be oriented by stretching (cold rolling) even at approximately 20 °C with resulting changes in its properties [30, 31]. This makes it possible to introduce orientation by cold-rolling extruded sheet, which significantly increases strength [32]. Such improvements have not been commercially successful because the orientations have an entropy–elastic character and lead to shrinkage as soon as the temperature surpasses a certain value. In cold-rolled samples, the first shrinkage appears well below the glass transition temperature. Stretching above the glass transition temperature (heat stretching) can also lead to orientations and significant property changes.

Since polycarbonate is a highly transparent polymer and maintains this transparency under certain conditions even in the stretched state, optical birefringence measurements may be used to characterize the orientation [33]. However, it should be emphasized that the correlations between birefringence and chain orientation differ in polycarbonate and in polystyrene. Identical birefringence values can reflect very different segment orientations. The reason is that the molecular weight between entanglements (chain length between physical crosslinks in the melt) is considerably smaller in polycarbonate than in polystyrene. Therefore, the same macroscopic elongations lead to higher molecular orientations in polycarbonate than in polystyrene. Orientations are naturally present during the injection molding process but can be minimized by selection of proper process parameters and use of proper polymer melt flow.

Its high chain bulkiness renders polycarbonate practically amorphous during standard injection molding. It can still be brought into a partially crystalline state, however—for example, during film manufacture according to the casting method [23]. Orientation and crystallization can overlap (i.e., oriented crystalline areas can be formed), which leads to additional property improvement in the oriented area. In this way, PC fibers with crystallinity of up to 25% can be achieved through heat stretching (e.g., at 200 °C), which also increases the density in comparison to amorphous polycarbonate [34, 35].

It should be emphasized that in addition to the orientation effects just described, the part receives stresses that are not only entropy–elastic in character but also energy–elastic. The result is molded parts with particular directions, which may show a decline in certain properties. High internal stress states, for example, increase the probability of stress-cracking in air and in certain liquids such as gasoline [36, 37]. Annealing below T_g can help eliminate the energy–elastic stress. The frozen-in entropy–elastic stresses, which are associated with segment orientations, cannot be eliminated until the glass transition temperature has been exceeded. This, of course, leads to distortion of the injection-molded part. Orientation and crystallization also influence the stress–strain behavior as illustrated, in Figure 3.19.

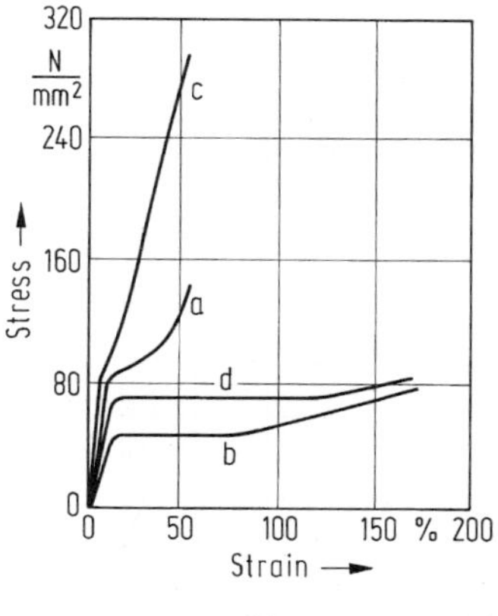

Figure 3.19. Stress–strain behavior of high-molecular weight cast film (DIN 53455, 22 °C, 55% relative humidity): a, longitudinally stretched, amorphous; b, cross-stretched amorphous; c, longitudinally stretched, crystalline; d, cross-stretched, crystalline.

During injection molding of low and medium molecular weight polycarbonates, the flow process causes orientation of the polymer molecules. During proper processing, these orientations are reduced through relaxation. For the manufacture of low birefringence laser optical readable storage disks that are almost orientation-free, high flow polycarbonates are used [24, 38–40].

Unlike polyamides, polycarbonate absorbs very little moisture, and the practical effect on its mechanical properties is slight [41]. A one-year study in a Central European climate found no significant dimensional changes through weathering and/or moisture absorption [42]. Polycarbonate must be thoroughly dried before processing, however, because any moisture absorbed by the material can affect carbonate chemical bonds during processing [43, 44] and can lead to hydrolysis of the polymer chains at standard processing temperatures.

In evaluating glass fiber reinforced polycarbonate, the orientation of the fibers must be considered [45–48]. Test bars are normally molded with material flowing along the length of the mold and thus have a fiber orientation that is favorable for tensile testing. Test bars manufactured from sheet yield tensile test samples with inferior strength and stiffness because of the random fiber orientation.

3.7.4.2 Tensile–Creep Test [48, 49]

Figure 3.20 shows the dependence of elongation on load duration in the tensile–creep test administered at various tensile loads and after removal of all loads. At low loads, there was still no breakage after 1000 hours. The load duration was calculated from the time at which the samples were loaded at full stress.

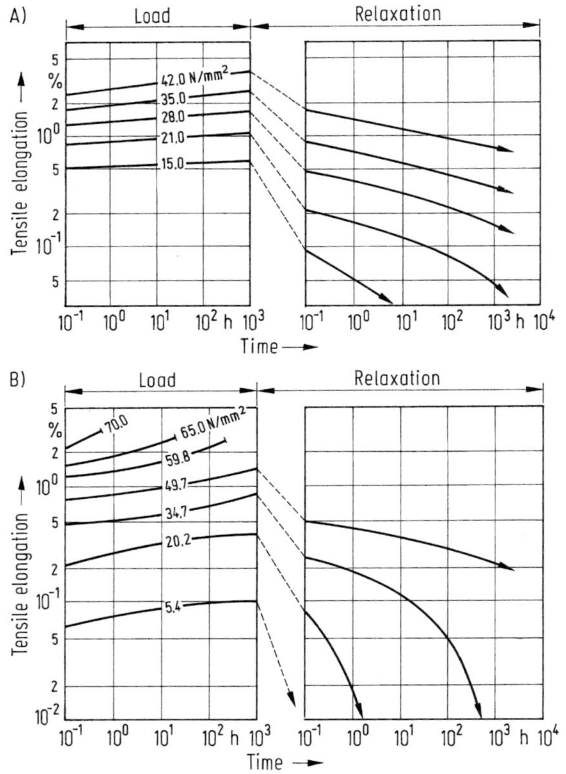

Figure 3.20. Correlation between tensile elongation and load/relaxation times at various tensions in the tensile–creep test (DIN 53455; 55% relative humidity; tensile bar no. 3): (A) unreinforced polycarbonate; (B) polycarbonate reinforced with 30 wt % short glass fibers.

Unreinforced polycarbonate stretches about 4% at a load of 42 N/mm² over 1000 hours; the elongation returns to 0.8% 1000 hours after release of the load. After 1000 hours at a lower load (15 N/mm²), the elongation of 0.6% shrinks to less than 0.3% 10 hours after load release. Figure 3.21 shows how the tensile strength (time–break curves) depends on load duration at various temperatures; Figure 3.22 shows the elongation after 1000 hours in the tensile–creep test as a function of tensile stress at various temperatures (isochronous stress–strain curves for 1000 hours).

Creep tendencies in polycarbonate are very low in comparison to other thermoplastics because polycarbonate deforms very slowly during long-term tensile loading. The elastic component of the deformation is very high. On the other hand, deformation recovery requires an equally long time. At loads below 20 N/mm² over 1000 hours, no residual elongation remains after load release. Thus moderate loads do not lead to

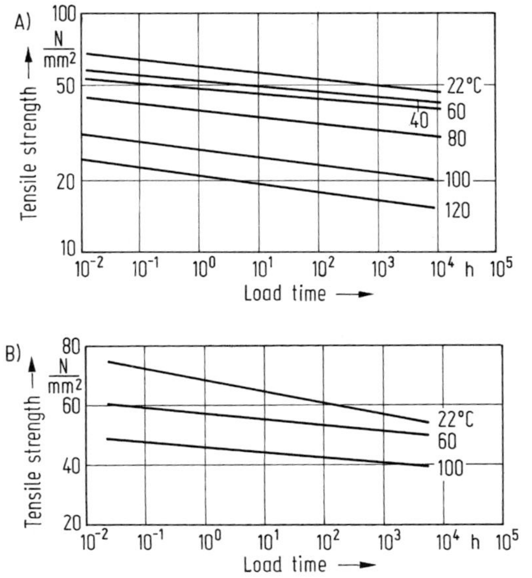

Figure 3.21. Dependence of tensile strength on load time at various temperatures (time fault lines): (A) unreinforced polycarbonate; (B) polycarbonate reinforced with 30 wt % short glass fibers.

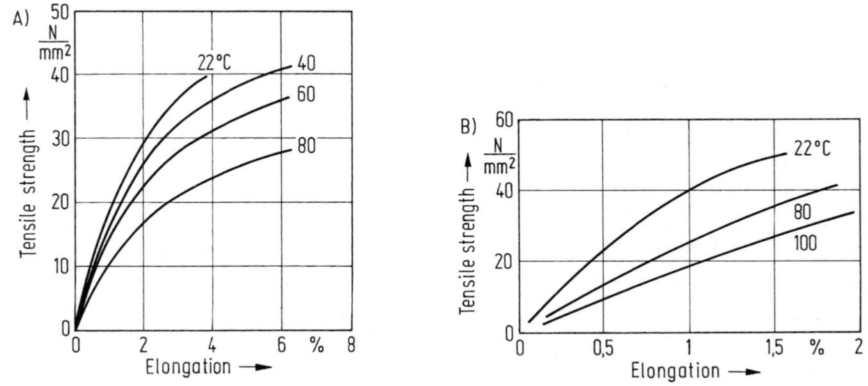

Figure 3.22. Tensile–creep test according to DIN 53444: elongation after tensile load for 1000 hours at various temperatures: (A) unreinforced polycarbonate; (B) polycarbonate reinforced with 30 wt % short glass fibers.

dimensional changes of molded parts. The practical benefits due to this low creep tendency constitute one of the reasons for the multipurpose use of polycarbonate in highly loaded parts.

Low creep tendency can also have a negative effect: stress-loaded molded parts do not reduce internal stresses through creeping. During long-term tensile stress, particularly in the presence of stress-cracking media, the stress may be relieved through crazing or crack formation [48], and such defects usually are associated with part failure.

3.7.4.3 Short-Term Tensile Loading (Hysteresis)

During short-term tensile tests, loading the test bar to just below the yield stress and then applying stress and strain during the rise in stress as well as during the subsequent

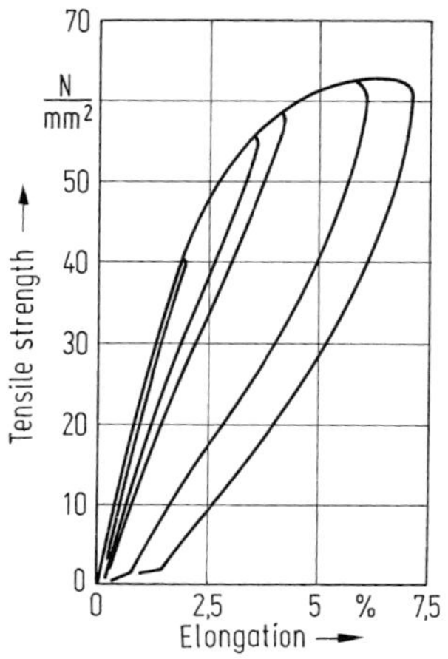

Figure 3.23. Hysteretic tensile test on unreinforced polycarbonate (DIN 53455, 22 °C, tensile bar no. 3; feed rate 25 mm/min).

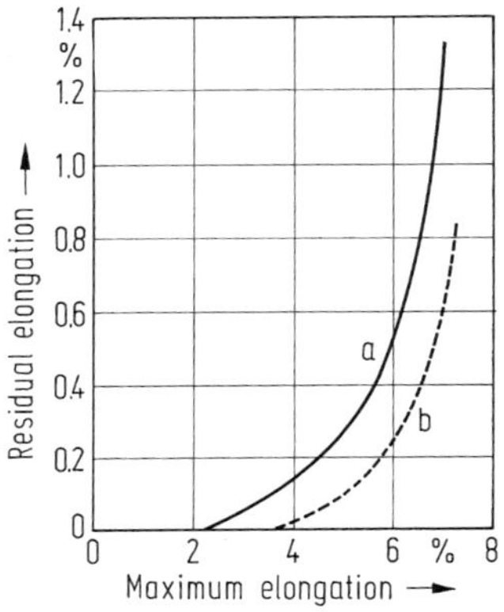

Figure 3.24. Dependence of residual elongation after short-term tensile load in the hysteresis test from initial elongation (tensile bar no. 3, feed rate 25 mm/min): a, immediately after relaxation; b, 1 minute after relaxation.

stress release process results in the stress–strain curves shown in Figure 3.23. The residual elongations measured immediately and 1 minute after load release are shown in Figure 3.24. These results indicate that short-term elongations of 2% after load release return immediately to 0%; elongations of about 3.5% again reach 0 within 1 minute after removal of the load.

3.7.4.4 Compression and Flexural Test

Compression and flexural tests according to DIN 53454 (ASTM D 695) and 53452 (ASTM D 790), respectively, lead to stress–deformation curves very similar to those of the tensile test, although stress values are somewhat higher (Figure 3.25). The yield point at which plastic material flow starts is at about 8% in the compression test. The yield point shows a linear dependence on temperature, yet it is remarkably high even near the softening temperature.

Figure 3.25. Stress–compressive strain behavior in the compression test according to DIN 53454 (a) and in the flexural test according to DIN 53452 (b) of unreinforced and reinforced (30 wt % short glass fibers) grades of polycarbonate, 23 °C.

Flexural strength according the DIN 53452 (ASTM D 790) cannot be defined for unreinforced polycarbonate, since the bars show outer fiber yield before maximum strength is reached. Instead, measurement of flexural stress at 3.5% outer fiber strain is recommended. Glass fiber reinforced polycarbonate breaks typically at outer fiber strains below 5%, so measurement of flexural strength is possible.

Compressive stresses, which appear during compression testing and on one side of a test specimen during flexural testing, do not cause stress-crack formation. As a rule, the flexural modulus is slightly lower than the tensile modulus in glass fiber reinforced polycarbonates. The longitudinal fiber orientation in molded test bars provides additional strength in the tensile test but has much less influence in flexural testing.

3.7.4.5 Impact Flexural Test

3.7.4.5.1 Impact Strength According to DIN 53453 (Unnotched)

In the German standard unnotched impact test, standard test bars (50 mm × 6 mm × 4 mm) of unreinforced polycarbonate show no breakage from room temperature to −40 °C. Even during impact strength tests according to ISO 180/1C on larger universal bars (80 mm × 10 mm × 4 mm), unreinforced polycarbonate does not break.

The unnotched impact strength of reinforced products depends on the type and content of glass fibers. Since the individual glass fibers act as internal notches, the impact strength declines after the addition of even small amounts of fiber. The window from 0 to 5% should be viewed as the critical range. Addition of 10% of normal glass fibers reduces the impact strength from "nonbreak" to about 65 kJ/m^2; at 20% glass, the toughness sinks to about 40 kJ/m^2; higher glass contents show no further significant reduction in toughness. At low glass contents, toughness can be increased through impact modification and by use of specially modified glass fibers. Short glass fibers in amounts of 5–20% show higher toughness than normal glass.

3.7.4.5.2 Notched Impact Strength According to DIN 53453 (Charpy Method) and ISO 180

The notched impact strength (ISO 180/1A) on the universal bar (80 mm × 10 mm × 4 mm) shows lower absolute values in unreinforced polycarbonates than the test according to DIN 53453 on the smaller standard bar. The residual width of the notched universal bar approaches the critical thickness of the ductile/brittle break transition. Some grades do not exhibit notched impact strength according to ISO 180/1A.

The notched impact strength of polycarbonate depends on molecular weight, test temperature, thermal history, notch type, and notch radius [50, 51]. The type of fracture in reinforced products depends on the stress rate. At low rates, the fracture occurs in the polycarbonate matrix; at high rates, the fibers break and/or are pulled out of the matrix [52]. The different failure mechanisms can be recognized through measuring sound emissions during the break [53, 54]. The test according to DIN 53453 uses a standard small bar (50 mm × 6 mm × 4 mm) with a U-notch (radius 0.01–0.2 mm). As shown in Figure 3.26, the notched impact strength depends on the molecular weight and/or the relative solvent viscosity. The toughness values shown in Figure 3.26 cannot be extrapolated to very low molecular weights and/or solvent viscosities, since solvent viscosities below 1.24 cause a disproportionately steep drop in impact strength (compare Section 3.7.2.3). Through addition of suitable additives and/or polymer modifiers, the impact strength and ductile–brittle transition can be improved [55, 56].

Internal stresses in test bars and/or finished parts will influence notched impact strength [57, 58]. Under standard injection molding conditions for polycarbonate, the production process leaves compressive stresses in the outer zone of the test bars and tensile stresses in the core of the molded part. If a sharp notch is applied through the area of compressive stresses, stress-cracking can propagate by forming extremely sharp notches on the base of the notch, particularly at longer storage times; these notches considerably falsify the results of the notched impact strength test. The sensitivity of polycarbonate to very sharp notched makes it necessary to maintain a minimum radius during construction of new parts. Values obtained on finished articles cannot be compared exactly with values obtained on test bars because they are influenced by so many parameters and because of the possibility of stress accumulation [59, 60]. For complicated parts, testing of the finished article under actual service conditions is always recommended.

Notched impact strength is slightly reduced by addition of standard mold release agents; UV absorbers do not influence toughness. High loading of pigments can reduce toughness, but these effects are minimal in higher molecular weight grades. The notched impact strength of unreinforced polycarbonate does not vary significantly between test temperatures of 20 and 100 °C (see Figure 3.27).

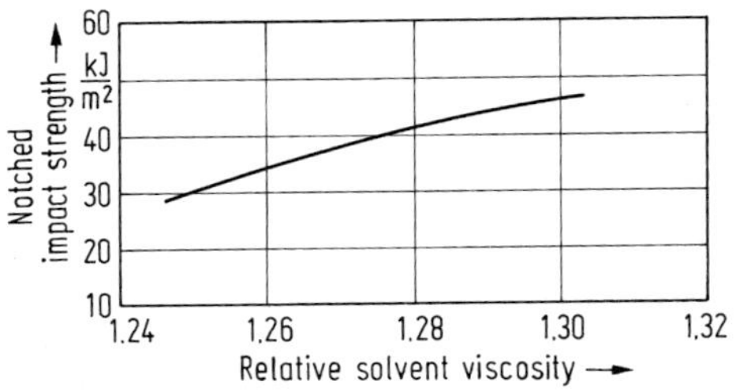

Figure 3.26. Dependence of Charpy notched impact strength (DIN 53453) on relative viscosity (DIN 51562, Part 3).

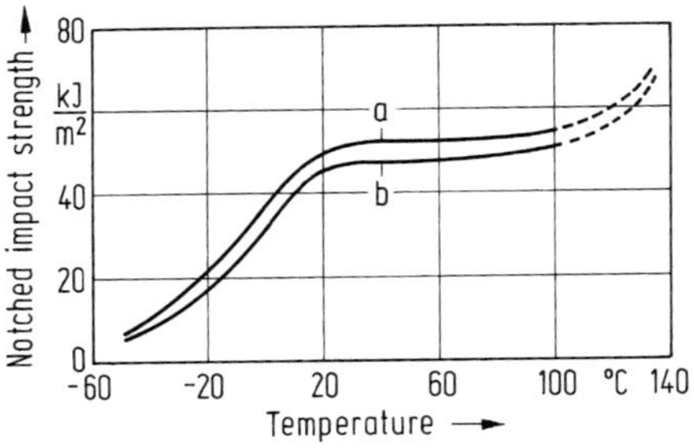

Figure 3.27. Dependence of DIN notched impact strength (notch radius 0.05 mm) on temperature for polycarbonate with two different solvent viscosities (DIN 51562, Part 3): a, $\eta_{rel} = 1.31$; b, $\eta_{rel} = 1.29$.

3.7.4.5.3 Izod Impact Strength and Notched Izod Impact Strength According to ASTM D 256

The Izod method uses test bars sized 63.5 mm × 12.7 mm × 3.2–12.7 mm. Without a notch, unreinforced polycarbonate does not break. The Izod notched impact strength of unreinforced polycarbonate depends to a large degree on the test bar width (notch length). The notched impact behavior of unreinforced polycarbonate shows a sharp dependence on sample thickness. Within these sample widths, extreme differences in impact occur over a wide temperature range. While the widest test bars (12.7 mm) show only brittle breakage, with values of 150 J/m, ductile failure values approximately six times higher are obtained on the thinnest bars at temperatures as low as −30 °C (see Section 3.7.3). Ductile failure results in fissured surface fractures, while brittle breaks are exhibited in smooth surface fractures.

The critical thickness is the test width (notch length) at which the change occurs within a few tenths of millimeter (Figure 3.28A). A similar dependence is shown for the Izod impact strength at different temperatures. The critical temperature is considered to be the temperature at which the failure mode changes from ductile to brittle at a constant test bar thickness (Figure 3.28B).

Within the ductile or brittle failure area, neither the molecular weight nor the presence of various additives has much influence on toughness. However, the transition range between the different failure modes will be affected. The higher the molecular weight, the higher the critical thickness and the lower the critical temperature. Various additives reduce the critical width and shift the ductile–brittle break transition to higher temperatures.

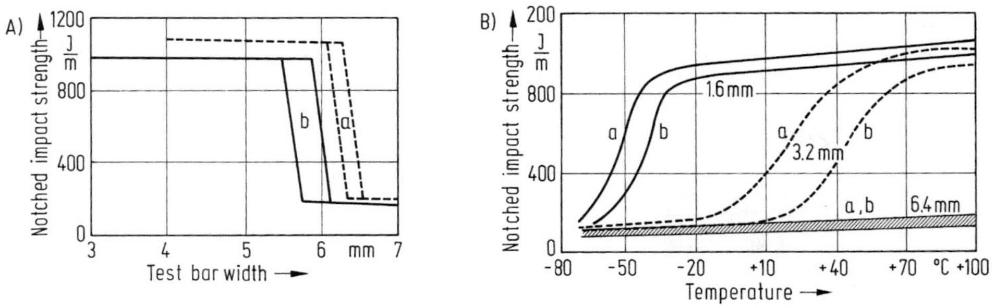

Figure 3.28. Izod notched impact strength of unreinforced polycarbonate at 23 °C and 50% relative humidity as a function of test bar width (A) and the temperature (B): a, $\eta_{rel} = 1.31$; b, $\eta_{rel} = 1.29$.

3.7.4.6 Shot Strength

At a distance of 30 meters, 3 mm plaques, clamped down on both sides, are pierced smoothly, without break or tear formation after being hit with steel-jacketed bullets shot from pistols of various calibers. However, at a distance of 30 cm and an impact angle of 90°, plaques 30 mm thick, clamped down on both sides, will repel a 9 mm steel-jacketed bullet. Since this test is also a type of impact test, the best values are obtained during tests of high molecular grades without additives. A shot strength test for eyeglasses (DIN 58212, Part 4; ANSI Z87) is a more appropriate test of a real application. Since eyeglasses of polycarbonate are also equipped with a hard coat to improve scratch resistance, this test has special significance for such coated polycarbonate plaques. The impact test is conducted using a projectile with a mass of 0.88 g (load classification A3), very similar to the falling-ball test. Polycarbonate samples with molecular weights from 25,000 upward fulfill the requirements of DIN 58212, Part 4, even after coating.

3.7.4.7 Impact Tensile Test

The impact tensile test according to DIN 53448 (ASTM D 1822) must also be included in the methods for determining toughness. Unreinforced polycarbonates show values of 500–800 kJ/m^2; polycarbonates with 20% normal glass fiber show a value of approximately 120 kJ/m^2; polycarbonates with 20% short glass fiber show 60–80 kJ/m^2, significantly lower than that of the corresponding long fiber reinforced product. Within this broad value range, no dependence on molecular weight can be recognized. For all unreinforced polycarbonates, 48-hour annealing at 130 °C leads to values of 600–700 kJ/m^2.

3.7.4.8 Shear Modulus and Mechanical Damping

The method for determining shear modulus and internal damping and/or mechanical loss factor during exposure to low frequencies (between 10^{-2} and 10 Hz) is the measurement of free torsion vibrations with a torsion pendulum, described in DIN 53445. The shear modulus is comparable to the modulus of elasticity. The mechanical loss factor tan δ provides information on energy absorption through molecular processes. The values shown in Figure 3.29 were obtained as a function of temperature on a standard polycarbonate and a glass fiber reinforced type.

The shear modulus changes relatively little up to about 140 °C. This thermostability enables polycarbonate to be used over a wide temperature range, right up to just below the softening temperature. The curve of the logarithmic shear modulus versus temperature is nearly linear over a wide temperature range (40–140 °C). In contrast, partially crystalline thermoplastics, such as polyamide and polybutylene terephthalate, exhibit a sharp decline of the shear modulus at around 40 °C (see also Section 3.7.3)

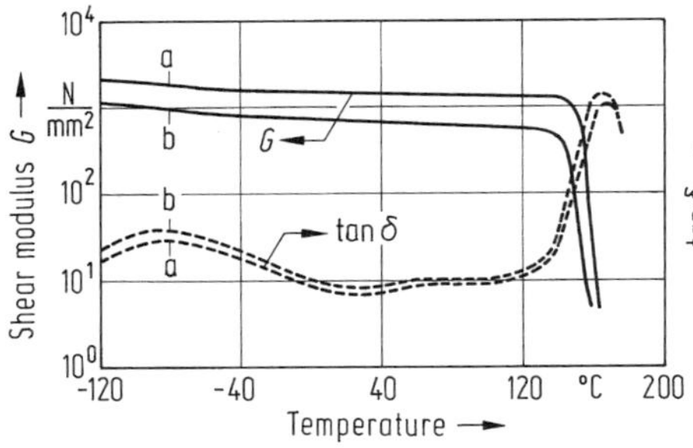

Figure 3.29. Dependence of shear modulus G and tan δ on temperature for polycarbonates: a, unreinforced; b, reinforced with 30 wt % short glass fibers.

3.7.4.9 Mechanical Strength During Dynamic Load [61, 62]

In the fatigue test based on DIN 50100, standard polycarbonate examined under conditions of quasi-constant stress amplitude shows a dependence of alternating stress amplitude on load cycle (Wöhler curve) up to test bar failure (Figure 3.30). The test bar is loaded only in the range of pulsating tensile stress (e.g., cyclic, from a low value near 0 up to a maximum value). The alternating stress amplitude is half the maximum stress with which the sample is loaded during the test.

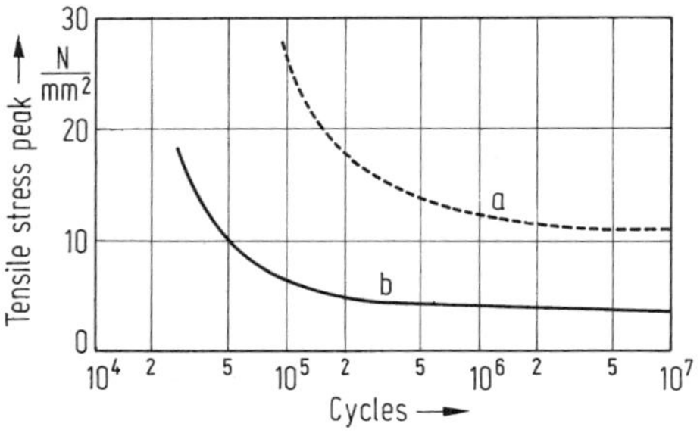

Figure 3.30. Wöhler curves for standard polycarbonate based on DIN 50100 (23 °C, 50% relative humidity, tensile bar no. 3, 7 Hz): a, normal; b, with 3 mm bore.

The tensile fatigue strength after 10^7 load changes at 7 Hz is 20–25 N/mm² for an unreinforced and for 30% short glass fiber reinforced polycarbonate. This corresponds to a reducing coefficient of approximately 3, in contrast to the short-term tensile strength. The fatigue strength is also called prolonged alternating stress strength. For assemblies that are subjected to oscillations, the fatigue strength is valid as the limit for admissible stress.

Fatigue strength depends on temperature, load frequency, load force, existing notches [50], processing conditions [63], and the surrounding media. Test data normally refer to a surrounding medium of air. Stress-cracking liquids or vapors will reduce fatigue strength.

3.7.4.10 Shear Test

The shear test indicates the shear strength of materials, which becomes relevant if, for example, forces in the plane of the cross section react vertical to the axis of the test bar. Shear is differentiated between single and double shear, depending on whether the test bars are simultaneously sheared in one or two cross sections.

According to ASTM D 732, the shear strength of standard polycarbonate up to the yield limit is about 38 N/mm^2, and up to failure about 65 N/mm^2.

3.7.4.11 Sliding and Friction Behavior

Polycarbonate's coefficient of friction against metal is very low, but it rises quickly depending on load and rate. Comparable tests using rotating cylinders (approx. 38 mm diameter) and riders (made of rod stock approx. 3.8 mm diameter) yielded results summarized in Table 3.7. These values are valid for unlubricated polycarbonate. Dry sliding bearings of polycarbonate yield satisfactory results at low loads and relatively low spinning rates. Since the coefficient of friction rises rapidly with increasing temperature, sufficient lubrication should always be used at higher loads and rates. Similar to sliding bearings, toothed gears of polycarbonate have also proven successful at low loads and rates. They are best used in transferring motion and are less satisfactory for transferring forces. Unlubricated gear teeth tend to form powderlike abrasions.

Table 3.7. Coefficient of Friction of Polycarbonate Compared to Polyamide 6 and Steel

Rider	Cylinder	Coefficient of Friction at Circumferential Speed	
		1 cm/s	173–302 cm/s
Polycarbonate	Polycarbonate	0.24	1.93
Polyamide 6	Polyamide 6	0.30	0.29
Polycarbonate	Steel	0.73	0.82
Polyamide 6	Steel	0.49	0.46
Steel	Polycarbonate	0.35	0.45
Steel	Polyamide 6	0.29	0.43

Since polycarbonate gears can be manufactured to high tolerance and are capable of withstanding high initial loads, they are suitable for carrying out rapid switching operations in electronic control mechanisms. At low bearing pressure, the coefficient of friction increases with increasing temperature; at high bearing pressures, however, it is temperature independent [64].

3.7.5 Thermal Properties of Bisphenol A Polycarbonate

3.7.5.1 Introduction

Polycarbonate changes into an elastic material upon warming at approximately 148 °C. At temperatures of 150–155 °C, only negligible plastic flow appears, so that even test bars can be aged for long periods without changes in part geometry. Heat-aging tests are still possible up to 155 °C as long as the test bars are stored on surfaces to which the material does not stick. At 220 °C, heat-softened polycarbonate flows under its own weight. This temperature is defined as the melt point according to the Kofler method.

Since a melt point determined according to the Kofler method represents a characterization of the melt viscosity, comparison to the melt points of other thermoplastics may not be valid.

Polycarbonate can be extruded at approximately 220 °C. Standard extrusion temperatures are between 260 and 270 °C. Injection molding requires higher temperatures: 290–310 °C, more than 150 °C above the softening point.

In the description of mechanical properties, good values even at higher temperatures were referred to several times. For example, from the shear modulus at various temperatures, it is apparent that polycarbonate is a stiff material until just before the softening point. Values for the modulus of elasticity and tensile strength at break at higher temperatures remain high at temperatures above 100 °C. The performance of polycarbonate up to 100–145 °C (depending on the requirement, the equipment used, and the duration of the temperature effect) is a major reason for the multipurpose use of this material for high grade instruments.

An overview of thermal properties of reinforced and unreinforced polycarbonate is shown in Table 3.8.

Table 3.8. Typical Values for Thermal Properties of Natural Polycarbonates

Property	Units	Average Molecular Weight Unreinforced Carbonate	Reinforced Polycarbonates		Test Method
			30 wt % Short Glass Fibers	20 wt % Long Glass Fibers	
Thermal linear expansion coefficient (−50 to 90 °C)	$10^{-6}K^{-1}$	70	30	30	DIN 53752, VDE 0304, Part 1
Heat conductivity	W/(K · m)	0.21	0.22	0.22	close to VDE 0304, Part 1, and DIN 52612
Specific heat capacity	kJ/(kg · K)	1.17	1.09	1.13	
Vicat softening temperature VST/B/120	°C	147	149	149	ISO 306, DIN 53460
Heat distortion temperature					ISO 75, DIN 53461
Method A (1.81 N/mm²)	°C	131	135	138	
Method B (0.45 N/mm²)	°C	138	140	145	

3.7.5.2 Differential Thermal Analysis

The differential thermal analysis provides information on exo- and/or endothermic processes during heating of material samples. Typical thermograms are shown in Figure 3.31. In the range between 145 and 150 °C, standard polycarbonate shows an endothermic jump with an inflection point at approximately 148 °C, which represents a physical definition of the softening point and is also called the glass transition temperature. Above the glass transition temperature, no changes appear up to about 350 °C. At 350–400 °C, decomposition begins. Additives or contaminations can shift the decomposition onset to lower temperatures.

The glass transition will be broadened through thermally induced crystallization or through solvent crystallization, both of which shift it to higher temperatures [65, 66]. In differential thermal analysis, an additional endothermic peak appears at approximately 230 °C, which characterizes the melt temperature of the crystalline components.

Physical aging (e.g., at 130 °C) leads to an entropy relaxation, which becomes noticeable through an endothermic maximum in the differential thermal analysis and is superimposed on the glass transition (glass peak) [67]. This glass peak ceases to appear during repeated heating.

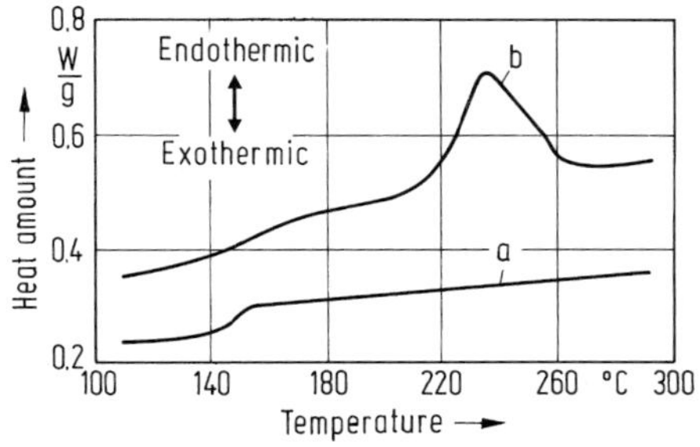

Figure 3.31. Thermogram of standard (a) and of high molecular weight crystalline and stretched (b) polycarbonates (heating rate, 10 K/min).

3.7.5.3 Heat Distortion Temperature

High heat distortion temperature is one of the main factors in the commercial success of polycarbonate. Data on the heat distortion temperature is always connected to various boundary conditions such as duration of thermal load, type of load of the molded part and/or test bar, and failure criteria. During short-term thermal load without mechanical stress, polycarbonate has a heat distortion temperature just below the softening point. In glass fiber reinforced grades, the glass fibers further counteract the plastic flow, causing heat distortion temperature values to be higher by about 5 °C.

3.7.5.3.1 Short-Term Temperature Stability

Freshly molded, unaged standard polycarbonate samples show a Vicat B softening temperature (at a load of 50 N/mm^2) of 146–149 °C. Products with internal mold release agents show slightly lower values. Thus, the Vicat B temperature is in the range of the glass transition temperature. The Vicat A temperature (with a load of 10 N/mm^2) is seldom given for polycarbonate. At 152 °C, it is only slightly above the glass transition temperature. Thermal aging leads to higher Vicat temperature values.

At 1.81 N/mm^2 (method A), the heat distortion temperature of unaged standard polycarbonate according ISO 75, DIN 53461, and ASTM 648 is about 129 °C; for glass fiber reinforced types, it is about 138 °C. According to method B, with a smaller load (0.45 N/mm^2), the values for unfilled and glass fiber reinforced products are measured at 140 and 145 °C, respectively. The heat distortion temperature depends somewhat on molecular weight, the presence of additives, and the thermal history of the test bar.

3.7.5.3.2 Long-Term Temperature Stability According to IEC 216

The short-term temperature stability described above provides information on temperature boundaries for certain short-term compressive and/or flexural loads only. The tests described now assess mainly long-term effects.

Standard test IEC 216 is a suitable method for determining long-term temperature stability. Properties that are relevant for commercial applications are tested at various temperatures until the values for loads of 5000–20,000 hours can be validly extrapolated. Normally, the tensile test is used as a test criterion on sample bars aged for different periods of time under at least three different temperatures (e.g., 130, 140, and 150 °C). For each individual temperature, the aging time is obtained that leads to a 50% reduction of the property value. To eliminate initial heat-aging effects, the property value obtained on a sample bar aged for 48 hours (at a selected aging temperature) is used as the initial value. The measured half-value times are recorded at the temperature chosen (see Figure 3.32) and extrapolated. From the resulting diagram, the temperatures for half-value times of 5000, 10,000, and 20,000 hours can be determined, depending on the requirements [68].

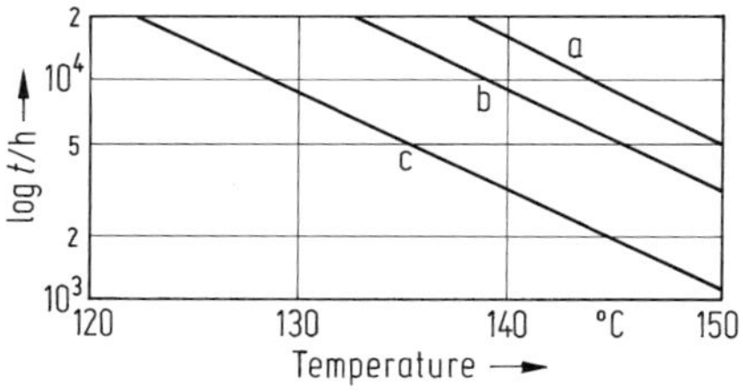

Figure 3.32. Thermal stability of standard polycarbonate evaluated according to IEC 216 (see text for explanation): a, dielectric strength (IEC 243); b, tensile strength (ISO/R 527, DIN 53455); c, tensile impact strength (ASTM D 1822).

Table 3.9 shows typical temperatures at which the dielectric strength, the tensile impact toughness, and the tensile strength drop by 50% after 5000 and 20,000 hours.

Table 3.9. Extrapolated Temperature Boundaries (°C) for Half-Life Value Times of Several Polycarbonate Properties

Property	Temperatures (°C) for Half-Life Value Times			
	Unreinforced PC		PC Reinforced with 20 wt % Glass Fibers	
	5000 h	20,000 h	5000 h	20,000 h
Electrical dielectric strength	150	140	150	140
Tensile impact strength	145	125	150	130
Tensile strength	150	135	150	135

3.7.5.3.3 Long-Term Temperature Stability According to UL 746 B

The temperature indices according to Underwriters' Laboratories (UL 746 B) provide long-term temperature limits for

- applications without impact loads (the tensile test is normally used)
- applications with impact loads (the tensile impact test, e.g., serves as a test criterion)
- electrical applications (the dielectric strength test is typically used)

The temperature indices according to UL 746 B are relative values, which are obtained at comparable measurements of a known material; numerically they agree quite accurately with the temperature limits of the half-peak value times extrapolated for 40,000 hours of aging. In the UL protocol, the temperature indices for electrical properties are never higher than those for mechanical values without impact loads, although in polycarbonate the dielectric strength decreases more slowly than the tensile strength during heat aging. Standard polycarbonates usually have a UL temperature index of

125 °C for applications without impact loads, 115 °C for applications with impact loads, and 125 °C for electrical applications.

UL temperature indices, initially significant only in the United States for electrical/electronics applications, are now recognized worldwide as important characteristics for long-term temperature stability.

3.7.5.4 Thermodynamic Characteristics

3.7.5.4.1 Thermal Conductivity

In the cases of thermal conductivity, specific heat capacity, specific volume, and coefficient of linear thermal expansion temperature dependence shows nonlinear behavior in the glass transition temperature range. For practical applications the room temperature thermal conductivity is of interest; it is 0.21 W/(K · m) for standard polycarbonate. Thermal conductivity rises slightly with increasing glass fiber content [e.g., to approx. 0.25 W/(K · m) at 30% glass fiber reinforcement], since glass fibers conduct heat more readily. Nevertheless, the thermal conductivity of this polymer is low compared to metals.

Thermal conductivity is one of the deciding factors controlling cycle times during the injection molding of polycarbonate, as well as the processing parameters. Since thermal conductivity depends on the temperature [69], the design engineer needs to know the temperature dependence (Figure 3.33A).

3.7.5.4.2 Specific Heat Capacity

Analogous to the thermal conductivity, the specific heat capacity plays an important role in the thermal processing parameters. The plastic melt is more strongly dependent on increasing temperature than the solid material (Figure 3.33B).

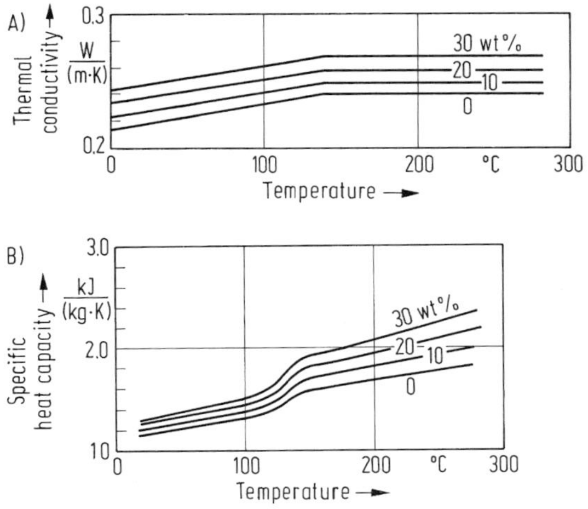

Figure 3.33. Thermal conductivity (A) and specific heat capacity (B) of polycarbonate as a function of temperature and glass fiber content.

3.7.5.4.3 Specific Volume and *PVT* Diagrams

In flow studies, the dependence of the volume on pressure during exposure to various temperatures is on interest. Figure 3.34 shows the dependence of the specific volume on pressure and temperature for standard polycarbonate and a 20% glass fiber reinforced product.

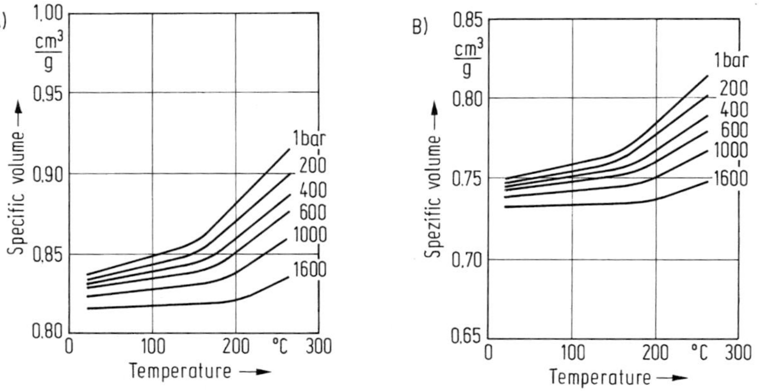

Figure 3.34. Specific volume as a function of temperature and pressure during isobaric cooling: (A) unreinforced polycarbonate; (B) polycarbonate reinforced with 20 wt % glass fibers.

3.7.5.4.4 Linear Thermal Expansion

The coefficient of thermal linear expansion depends slightly on the temperature and lies in the following range:

−70–0 °C	65×10^{-6} K^{-1}
0°–60 °C	70×10^{-6} K^{-1}
60°–140 °C	75×10^{-6} K^{-1}

As with other thermoplastics, the presence of glass fibers reduces the coefficient of thermal expansion for standard polycarbonate (Figure 3.35). With an increasing load of glass fibers, coefficient of expansion of polycarbonate approaches that of metals; this is one reason for the suitability of glass-filled polycarbonate as a substitute for metals. In many applications, metals have already been replaced with glass fiber reinforced polycarbonate.

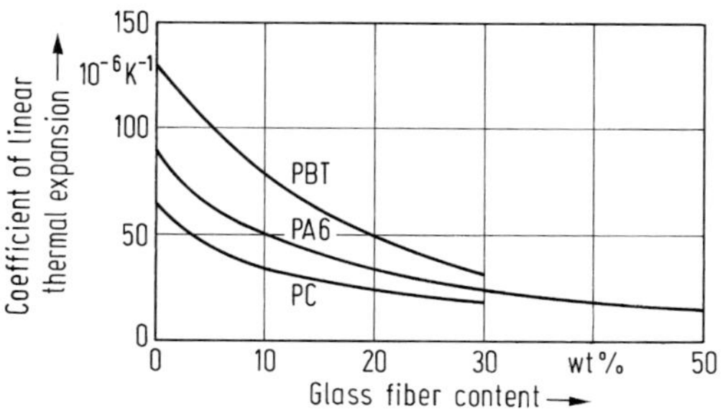

Figure 3.35. Coefficient of linear thermal expansion of polycarbonate, polybutyleneterephthalate, and polyamide 6 as a function of glass fiber content.

3.7.5.5 Shrinkage

Shrinkage, defined as the dimensional difference between an inner mold contour and the outer contour of the molded part, is 0.6–0.8% in unreinforced polycarbonates and depends to a degree on injection molding conditions.

Unlike partially crystalline materials, polycarbonate solidifies to a glasslike material and exhibits no significant morphological changes during cooling.

Glass fiber reinforcement and fiber orientation during processing determine anisotropy and direction-controlled shrinkage. With normal glass fiber reinforcement, the shrinkage is 0.3–0.45% in the flow direction—vertically 0.3–0.5%. Short glass fibers lead to a very low shrinkage anisotropy; shrinkage in all directions is 0.3–0.5%. Its low absolute shrinkage and low anisotropy make short glass fiber reinforced polycarbonate especially suited for high grade precision parts. The dimensional changes to be considered after injection molding are negligible. Over a time span of approximately 2 hours, postmold shrinkage is less than 0.1%. Dimensional changes due to moisture absorption (approx. 0.15% in a Central European climate; even 0.59% during saturation in boiling water) are not significant. Although small dimensional changes are expected during heat aging, no measured values have been published to date on commercial parts.

3.7.5.6 Heat Aging

High processing temperatures or aggressive shearing of high molecular weight polycarbonate leads to initial aging [70] due to lowering of the molecular mass. The presence of residual moisture during processing further contributes to aging. This processing-related aging must be taken into consideration when regrinds are used [71].

Heat aging of polycarbonate at temperatures below the softening point is initially dominated by physical rearrangements alone. Only after an extended period at these temperatures do chemical reactions become noticeable. The process of morphological rearrangements toward more physical ordering during heat aging leads to a structural ordering. Although these structural changes can be detected at the 7° peak in wide-angle x-ray analysis [72], they are not considered to be crystallization. Both the extent and the rate of the rearrangement process increase as the heat-aging temperature approaches the softening temperature [73] (Figure 3.36). This rate reaches its maximum at 130–135 °C. At still higher temperatures, the increasing mobility of the molecules leads to a reversal of the ordering. In differential thermal analysis, this higher order is characterized by an endothermic peak at about 150 °C (the glass peak) (see also Section 3.7.5.2).

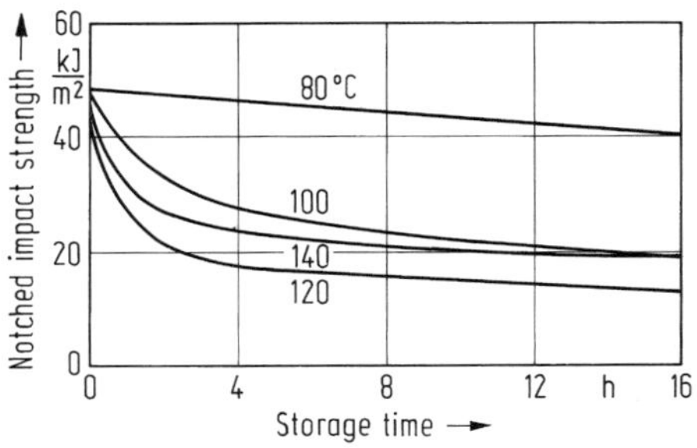

Figure 3.36. Charpy notched impact strength (DIN 53453) of standard polycarbonate using heat aging at various temperatures, measured on standard test bars (60 mm × 6 mm × 4 mm), notched before aging (notched bars after aging have approximately 20% higher values).

Thermally induced structural changes are mirrored in the various property changes that occur [40, 67, 74–83]. The molecular ordering is reversed during short-term heat aging above the glass transition temperature. In some molded parts such reversal can be achieved without any significant dimensional change. Structural changes can also be reversed through cold drawing [84].

Structural changes are coupled with an increase in strength and stiffness as well as a reduction of toughness. The extent of change of the tensile stress at yield and its reversibility during alternating heat treatments are illustrated in Figure 3.37. Since polycarbonate is very oxidation resistant, it shows almost no chemical reactions up to 150 °C.

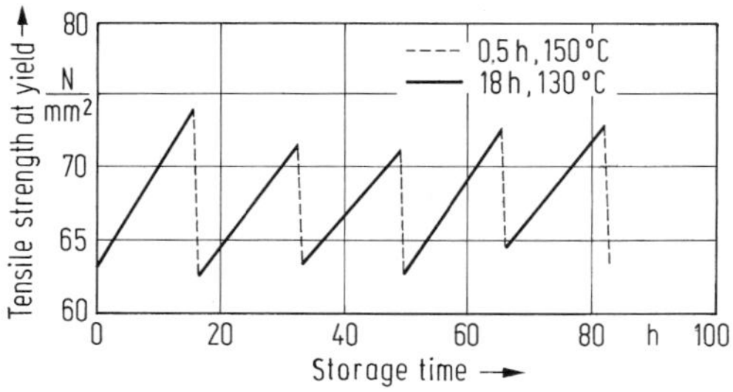

Figure 3.37. Regenerability of heat-aged standard polycarbonate during alternating heat treatments.

The properties described above refer to controlled thermal laboratory tests. Industrial environments often feature a combined attack from light and heat and sometimes from an aggressive surrounding medium, as well. The complexity of these changes can best be obtained using practical tests [85]. The combined influence of UV light and oxygen causes minor photooxidation [86–88], which forms a rough surface [89] (see also Section 3.7.10).

3.7.5.7 Heat Aging in the Presence of Water

Long-term contact of polycarbonate with water at 60–100 °C causes changes through two mechanisms:

- chemical hydrolysis, which results in chain scission with a corresponding drop in toughness [90–92] and a decrease in melt viscosity [93–96].
- Physical aging, as described in Section 3.7.5.6

The purely thermal structural changes occur faster in hot water and/or hot steam, and at lower temperatures, than during dry heat aging. This can be explained by noting the increase in chain mobility caused by the absorbed water (softening or plasticizing effect). Water-saturated polycarbonate samples (e.g., immediately after sterilization by hot steam) undergo the glass transition at lower temperatures. Microcracks may also appear at higher water oversaturation [90, 97, 98]. Because of these aging effects, molded parts that will be exposed to water above 60 °C should be subjected to long-term aging tests.

3.7.6 Electrical Properties of Polycarbonate

Table 3.10 presents an overview of the electrical properties of three forms of polycarbonate.

Table 3.10. Typical Values for Polycarbonate Electrical Properties

Property	Units	Average Molecular Weight Unreinforced PC	Reinforced PC		Test Method
			30 wt% Short Glass Fibers	20 wt% Long Glass Fibers	
Dielectric strength (transformer oil, 50 Hz, 0.5 kV/s, sample thickness 1 mm)	kV/mm	30	30	30	IEC 243, DIN 43 481, VDE 0303, Part 2
Specific surface resistivity	Ω	10^{15}	10^{14}	10^{14}	IEC 93, DIN 53 482, VDE 0303, Part 3
Specific volume resistivity	Ω · cm	10^{16}	10^{16}	10^{16}	IEC 93, DIN 53 482, VDE 0303, Part 3
Dielectric constant (dry) at					
50 Hz		3.0	3.3	3.2	IEC 250,
1 kHz		3.0	3.3	3.2	DIN 53 483,
1 MHz		2.9	3.3	3.2	VDE 0303, Part 4
Dissipation factor (dry) at					
50 Hz		0.001	0.001	0.001	IEC 250,
1 kHz		0.001	0.001	0.001	DIN 53 483
1 mHz		0.010	0.009	0.009	VDE 0303, Part 4
Tracking resistance					IEC 112,
CTI A		100 M	100 M	100 M	DIN 53 480,
CTI B		275	175	175	VDE 0303, Part 1
Electrolytic corrosion effect		A1	A1	A1	IEC 426, DIN 53 489, VDE 0303, Part 6

3.7.6.1 Dielectric Constant and Dielectric Dissipation Factor

The relative dielectric constant ε_r relative permittivity remains almost constant in the temperature range of −20 to +140 °C—that is, the commercially relevant temperature range (cf. Figure 3.38A); it declines negligibly with increasing frequency (Figure 3.38B).

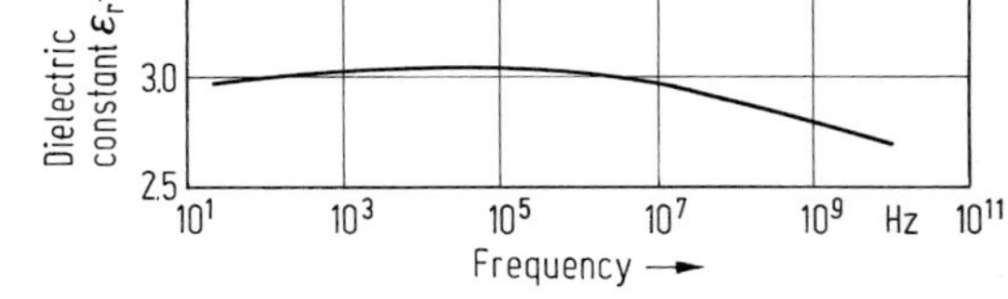

Figure 3.38. Dielectric constant ε_r of standard polycarbonate as a function of temperature (A) and frequency (B).

The dielectric dissipation factor tan δ depends on both temperature and frequency. The temperature dependence at 20 kHz (Figure 3.39) shows two maxima. The low temperature maximum can be traced back to a rotational effect with the carbonyl group, and the

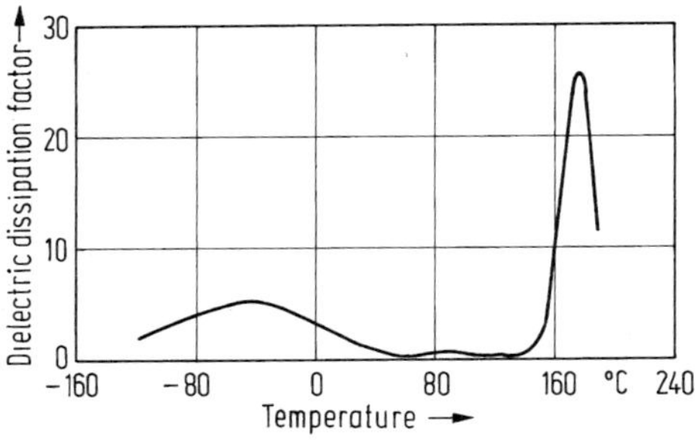

Figure 3.39. Dielectric loss factor tan δ at 20 kHz of standard polycarbonate as a function of temperature.

high temperature maximum to a rotational effect with larger molecular segments. Figure 3.40 illustrates the change of tan δ with frequency [99]. With increasing frequency, the high temperature peak is shifted to higher temperatures. Orientations of high molecular weight polycarbonate shift the high temperature maximum to lower temperatures, and around 80 °C a new, weaker maximum appears. On stretched, crystalline film, the values for the dissipation factor are lower and the main maximum is shifted to higher temperatures by 15–20 K.

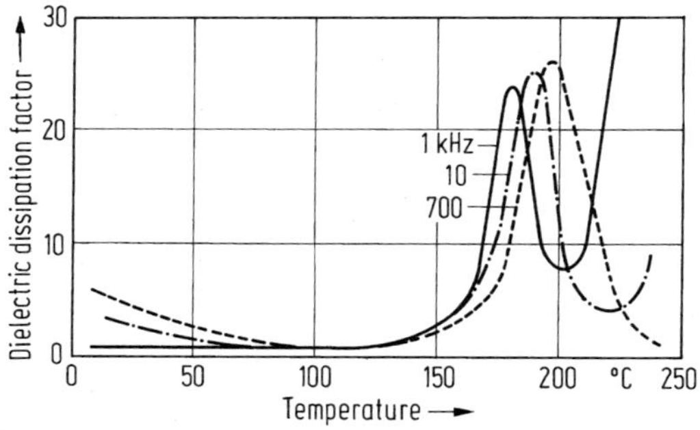

Figure 3.40. Dielectric loss factor tan δ as a function of temperature at various frequencies.

3.7.6.2 Specific Volume Resistivity and Surface Resistivity

The specific volume resistivity according to Part 3 of DIN-VDE 0303 (ASTM D 257) depends on the temperature (see Figure 3.41). In the industrially relevant temperature range, polycarbonate is an excellent insulation material. The specific volume resistivity is increased somewhat by crystallization. The low moisture absorption of polycarbonate does not influence volume resistivity.

The surface resistivity according to DIN-VDE 0303, Part 3 (ASTM D 257), which is 10^{16} Ω/cm, is similar to the volume resistivity and equally independent of moisture absorption and/or aging conditions. Attempts to lower surface resistivity through incorporation of antistatic agents have so far been unsuccessful. However, such lowering can be achieved through posttreatment of the surface of the molded part [100].

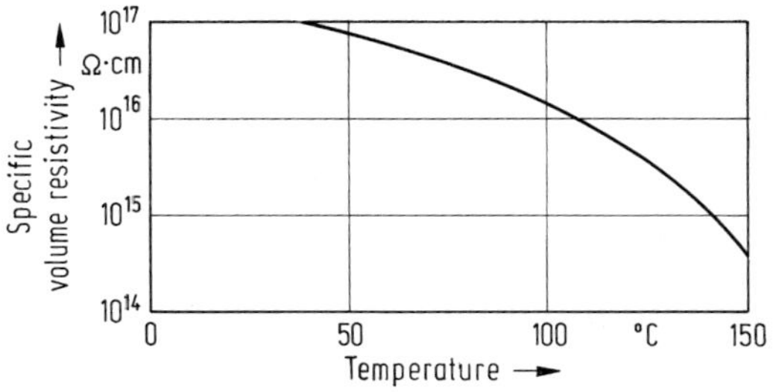

Figure 3.41. Specific volume resistivity as a function of temperature.

3.7.6.3 Dielectric Strength

The dielectric strength according to DIN-VDE 0303, Part 2 (ASTM D 149) (50 Hz, 0.5 kV/s), measured on a 1 mm thick sample bar in transformer oil, is 30–35 kV/mm. In dibutylphthalate the values are considerably higher. However, this test medium is not industrially useful for polycarbonate. Similar to the specific volume resistivity, the dielectric strength depends on temperature. At 140 °C it is approximately two-thirds of the value measured at room temperature. The dielectric strength decreases with increasing thickness of the test bar.

3.7.6.4 Tracking and Electrolytic Corrosive Effect

During determination of tracking according to IEC 112/DIN-VDE 0303, Part 1, standard polycarbonate reaches 100 M (with wetting agent), and 275 (without wetting agent), respectively. During tests with wetting agents, unreinforced polycarbonates with flame retardants achieved comparable values; during tests without wetting agents, the materials reached rates of 200–250. With use of a wetting agent, glass fiber reinforcement showed also no influence on rating. During tests without wetting agent, a rate of 175 is obtained.

Polycarbonate achieves rate A1 in the electrolytic corrosion test according to DIN-VDE 0303, Part 6, and IEC 426. Bromine-containing flame retardants increase the corrosive effect; corresponding types should be rated according to A1 up to A1.2.

3.7.6.5 Electrical Index Values According to Underwriters' Laboratories

Electrical instruments that contain polycarbonate in direct contact with current-carrying and charged areas have proven safe in long-term operation. Individual applications should be tested to determine whether safety requirements are met.

To describe a material's behavior during electrical load, UL instituted the following tests:

Hot wire ignition (HWI) according to ASTM D 3874: simulates the case of a charged part, in direct contact with a sample bar, that is heated through overloading; the time to ignition of the test material is then measured. Polycarbonate achieves times of 30–60 seconds in a sample bar 3.2 mm thick.

High current arc ignition (HAI) according to UL 746 A 43: the sample bar is placed in the immediate vicinity of a high current electric arc. Failure consists of ignition of the sample bar by the electric arc. Polycarbonate ignites only after 15 or 16 light arc ignitions.

High voltage arc-tracking rate (HVTR) according to UL 746 A 25: on the surface of the sample bar, high voltage electric arcs are ignited for 2 minutes, whereupon the electrodes are separated from each other until the arc is extinguished and then brought together again until ignition is once more observed. After 2 minutes the length of the resulting conductive electrical track is measured and divided by 2. Polycarbonate leads to conductive tracks from 8 to 25 mm/min.

Arc resistance according to ASTM D 495 (1973): a sample bar is exposed to a high current electric arc for a maximum of 7 minutes; the time is measured in seconds until the arc is extinguished after a low resistant track has formed on the surface of the test bar. The value for polycarbonate is below 6 seconds.

Comparative tracking index according to ASTM D 3638 (1977): similar to IEC 112, DIN VDE 0303, Part 1 6.94, mentioned in Section 3.7.6.4. Polycarbonate shows values of 275 V in unreinforced standard types and 175 V in glass fiber reinforced types.

3.7.7 Optical Properties

Transparency, combined with good optical properties, gives polycarbonate a multitude of interesting applications in light technology and optics.

The light transmittance of polycarbonate, measured as spectral transmission in the ultraviolet, visible, and infrared ranges, is illustrated in Figure 3.42. In the ultraviolet range, the light is completely absorbed up to wavelengths of about 275 nm. The light transmittance then rapidly rises toward longer wavelengths, reaches a value of 85–90% at the start of the visible range above 400 nm, and then remains almost constant up into the infrared. Light transmittance depends on polycarbonate type and wall thickness (Figure 3.43). Additives (e.g., UV stabilizers) may lower transmission somewhat.

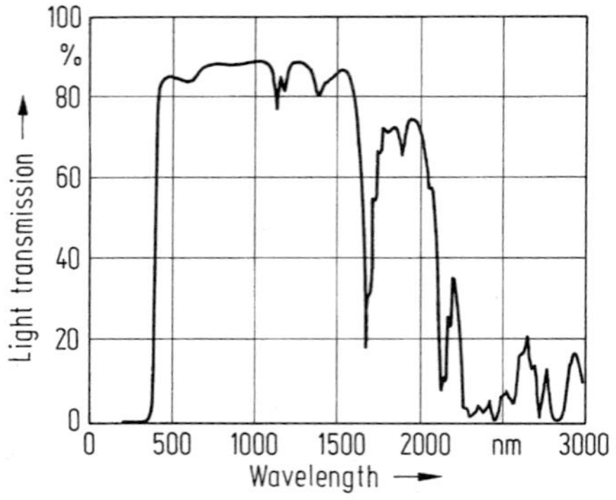

Figure 3.42. Light transmission of polycarbonate as a function of wavelength (4 mm thickness).

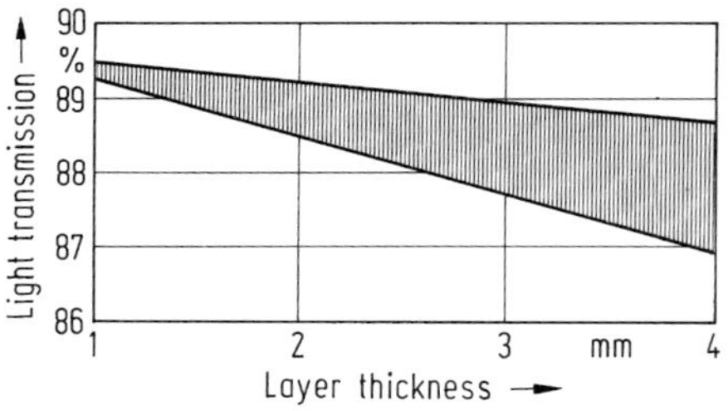

Figure 3.43. Illustration of average light transmission of polycarbonate in the visible region (approx. 400–700 nm) as a function of layer thickness.

The refractive index $n_{D/23}$ (measured at a wavelength of 550 nm) is between 1.582 and 1.586. It is dependent on the viscosity of the polycarbonate. The influence of temperature and molecular weight can be seen in Figure 3.44.

For light-scattering effects, particularly for light globes, which require high light transmission while keeping the light source invisible, polycarbonate manufacturers offer diverse, translucent color shades exactly tailored to specific requirements. The same is true for many colors used in traffic signals (i.e. red, green, amber lights), which cover defined signal ranges according to DIN 6163. Figure 3.45 shows the light transmission of various signal colors based on polycarbonate as a function of wall thickness.

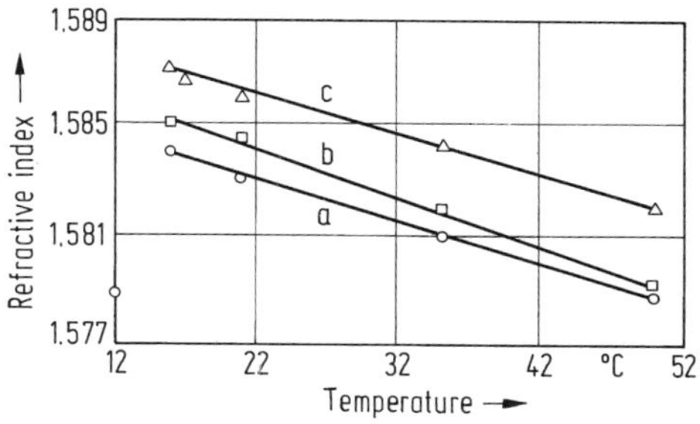

Figure 3.44. Refractive index at a wavelength of 550 nm of polycarbonate as a function of temperature and molecular mass: a, $\bar{M}_w = 20{,}000$; b, $\bar{M}_w = 24{,}000$; c, $\bar{M}_w = 28{,}000$.

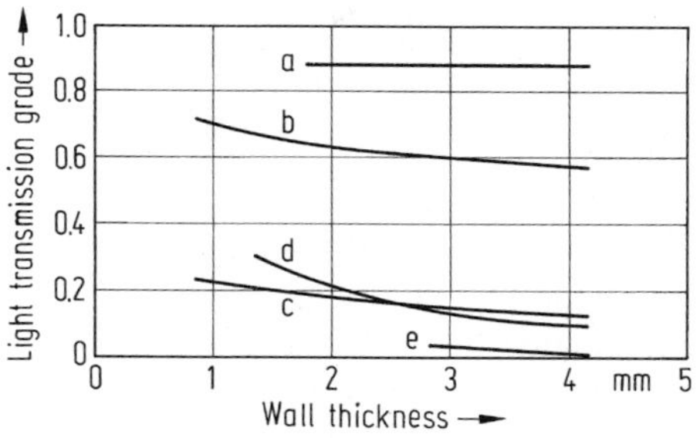

Figure 3.45. Light transmission of polycarbonate signal colors as a function of wall thickness: a, white; b, yellow; c, red; d, green; e, blue.

3.7.8 Combustion Behavior

3.7.8.1 Polycarbonate without Flame Retardant

In comparison to many other thermoplastic materials, bisphenol A polycarbonate shows good resistance to burning even without flame-retardant additives and is already sufficient for many applications (see Table 3.11). According to ASTM D 2863, it has an oxygen index of 26, and it is listed by Underwriters' Laboratories in UL 94 as V-2 (1.6 mm).

Table 3.11. Typical Values for Combustion Properties of Polycarbonate

Property and/or Test	Units	Average Molecular Weight Unreinforced PC	Reinforced PC 30 wt % Short Glass Fibers	Reinforced PC 20 wt % Long Glass Fibers	Test Method
Flammability according to ASTM					
Burning time	s	5	5	5	ASTM D 635-74
Burning length	mm	15	15	15	
Flammability according to UL, 3.2 mm		V2/HB	V1	V1	UL 94
Glowing wire test 2 mm	°C	750	960	960	IEC 695-2-1
3 mm		850	960	960	VDE 0471, Part 2
Red-hot test bar test		IIB	IIb	IIb	VDE 0304, Part 3; DIN 53459
Oxygen index	%	26	32	33	ASTM D 2863-77

3.7.8.2 Polycarbonate with Flame Retardant

Engineering safety aspects—particularly in the field of electrical/electronic technology, where polycarbonate is the preferred material for insulation as well as current-carrying parts—led to a multitude of regulations and standards [101] that extensively reduce the risk of burning. Burn regulations, which over the years have been introduced and/or refined, have made it necessary to develop specially flame-retardant polycarbonate grades. Numerous flame-retardant polycarbonate types now available can be selected for specific requirements in the various industrial areas (Table 3.12) [102].

The first generation of flame-retardant polycarbonate resulted because tests with other plastics had demonstrated flame-retardant properties imparted by modification with halogen carriers, particularly those based on bromine. Initially the polymer chain was modified with tetrabromobisphenol A to achieve a bisphenol A–tetrabromobisphenol A copolymer. These polycarbonate copolymers are still transparent and achieve a UL 94 V0 classification in 3.2 mm wall thickness. In opaque grades, classification V0 is also achieved in 1.6 mm wall thickness. According to the French CSTB standard (epiradiateur test), transparent bromine-containing polycarbonates achieve classification M 3, while transparent standard polycarbonate and transparent polycarbonate types equipped with other flame-retardant systems reach no higher than classification M 4. Only bromine-containing polycarbonates pass the glowing wire test at 960 °C/5 s according to EDF HN 60 E 01 at 2 mm wall thickness.

Despite these advantages, as well as a Vicat temperature of about 3 K higher, bromine-containing polycarbonates have many disadvantages that limit their utility. For example, iron-catalyzed decomposition of the polycarbonate requires the use of special alloy steels in injection molding equipment. In addition, the risk of corrosion that accompanies the splitting off of hydrogen bromide in fire conditions has led to increased interest in the development of bromine-free, flame-retarded products during the last few years.

Bromine-free, flame-retarded polycarbonates can be manufactured by adding special salt compounds; extremely minor amounts show a clear flame-retardant effect. This effect is based on an accelerated reduction of flame-promoting compounds in the lower temperature range of the burn process (400–500 °C) and the formation of a protective crust, which reduces dripping. Suitable salts in the required amounts allow a UL94 V0 classification at 3.2 mm wall thickness, while maintaining transparency as well as other valuable engineering properties of standard polycarbonate.

If a branched (and thus structurally viscous) polycarbonate is used as the base material, burning drips are further reduced and a V0 classification can be achieved even at wall thicknesses of less than 1.6 mm.

Table 3.12. Fire Protection Classifications of Polycarbonate [102]

Flame Protection Classification	Flame Protection System	Achievable UL 94 Classification
1	Bromine compound	V0/3.2 mm transparent
2	Bromine-free, alkaline salt	V0/3.2 mm transparent
3	Bromine-free, alkaline salt (branched PC)	V0/0.8 mm transparent
4	Alkaline salt + antidrip agent	V0/0.8 mm, 5 V/3.2 mm opaque colored
5	Glass fibers	V1/1.6 mm
6	Glass fibers + flame protection system	V0/1.6 mm 5 V/3.2 mm

The effect of alkali metal salts as flame-retardant additives to polycarbonates can be further optimized by adding drip-retardant substances: UL94 classifications of V0 can be achieved up to 0.8 mm wall thickness, and 5V at 3.2 mm. An especially effective antidrip agent has proven to be polytetrafluoroethylene (PTFE). However, such optimal flame-retardant properties come at the expense of transparency.

Opaque polycarbonate types based on a combination of salt and antidrip agent can be used in various applications because they not only pass the important UL classifications mentioned above, but also the M 1 classification of the French epiradiateur test at wall thicknesses of 1–2 mm, and FAR (Federal Aviation Regulation) Test (FAR 25, Paragraph 853a, b), which is relevant for the aviation industry (manufacture of interior parts).

By counteracting the flow of the polycarbonate melt, glass fibers are also capable of retarding burning drips. Since, however, glass fibers do not reduce burn time, a UL94 classification of only V1 can be achieved. By a combination of glass fibers and alkaline metal salts, glass fiber reinforced V0 and 5V products can be manufactured.

The classifications of flame-retardant systems were summarized in Table 3.12 [102]. Based on different fire hazards, the various industries have developed standard simulated burn tests for their respective areas of applications.

3.7.8.2.1 Electrical/Electronics

As carriers of live voltage, overheated electrical structural elements are potentially dangerous ignition sources. Based on such risks, the glowing wire, glowing mandrel, and glowing test bar tests were developed. In addition to a series of national tests, the UL94 test has become increasingly dominant internationally.

3.7.8.2.2 Traffic Sector

The U.S. safety standard MVSS 302 and/or the German FKT Guideline (DIN 75 200), which describe specific tests for the automotive interior, are not critical for polycarbonate. Standard polycarbonate grades pass these tests even without modification with flame retardants. Other regulations (FAR 25, paragraph 853) are valid for applications in the aviation industry. While all flame-retarded polycarbonate types pass the less critical test (paragraph 853b) at 15 seconds' flame exposure, passing of (paragraph 853a) test at 60 seconds' flame exposure can be guaranteed only with a suitable salt/antidrip agent combination or with bromine carriers as flame-retardant systems.

3.7.8.2.3 Construction Sector

In Germany, construction materials are classified according to DIN 4102 into nonburnable (classification A) and burnable (classification B) materials; classification B also has subclassifications. According to DIN 4102 for interior applications, sheets of standard polycarbonate are rated as B1 with the addition of "burning drip." However, large-scale burn tests with polycarbonate sheets after outside weathering have shown a burn behavior corresponding to that of a B1 material, so in principle polycarbonate also fulfills the safety requirements for outside applications. The presence in a formulation of a specific flame retardant does not lead to a better classification in the B1 test.

In the epiradiateur test, required by the French CSTB standard, sheet test samples are enclosed in a heated chamber where the resulting decomposition products are brought to ignition with a pilot flame. According to this test, standard polycarbonate is graded as M4. Transparent polycarbonate can reach grade M3 only when modified with a bromine compound as a flame retardant. The better grades (M1 and M2) can be reached only by opaque product types.

3.7.9 Mold Release Behavior

In complicated molds or when part geometry does not allow sufficient mold taper, mold release problems can appear during the processing of pure polycarbonate. In such cases a mold release agent may be sprayed onto the mold after every few shots, but it represents an additional expenditure.

A better solution is the use of polycarbonate modified with an internal mold release agent compounded into the polymer in amounts less than 1% (see also Section 3.4.4). This approach reduces the energy required for mold opening and part ejection.

To evaluate the effect of a mold release agent, the energy required for mold release must be measured over a large mold temperature range. The reason is that difficult-to-cool mold areas may have locally high temperatures. Figure 3.46 plots mold release energy versus temperature for an easy-release polycarbonate and for a mold-release-free polycarbonate.

When selecting a suitable mold release agent, the additive's effects on the properties of the base polymer (e.g., transparency, notched impact strength, elongation at break, heat distortion temperature, stress-cracking resistance) must be considered. Selecting the wrong mold release agent may considerably reduce these properties. In low voltage electrical engineering applications, the tendency to form contact-damaging accretions must be considered, since these may lead to an increase in the resistance of the switch element.

Such contact-damaging accretions can appear during use of silicone oils, which are normally effective mold release agents for polycarbonate.

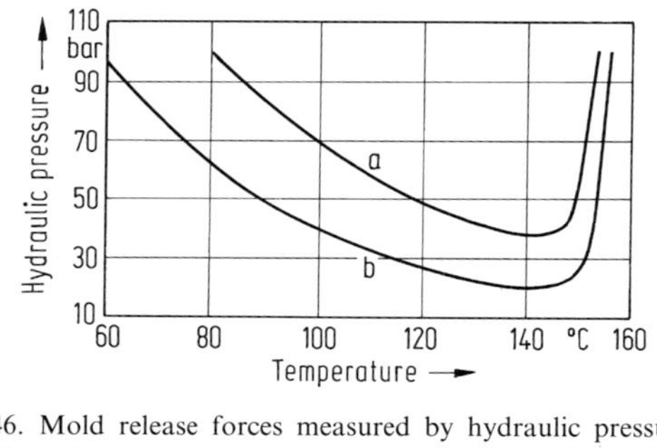

Figure 3.46. Mold release forces measured by hydraulic pressure on ejectors of a light conical mold, as a function of mold temperature at constant cycle times, without (a) and with (b) mold release agent.

3.7.10 Behavior in the Presence of External Chemical and Physical Influences

All tests and properties introduced so far presume pure air as the surrounding medium. In other environments, various properties may show extensive deviations [103].

3.7.10.1 Chemical Resistance

The good toughness of polycarbonate is based on molecular relaxation modes that exist even at low temperatures [104, 105]. Binding of foreign molecules to individual chain segments influences relaxation ability and therefore toughness behavior. In this way, methylene chloride or chloroform binds itself to the carbonate group, which is the reason for the good solubility of polycarbonate in these media [104].

The effects can vary considerably. Crazing and stress-crack formation depend on the surrounding medium as well as the stress condition of the test bar and/or molded part: helium is inert, whereas neon, argon, and oxygen promote stress cracking [106–108].

Knowledge of medium compatibility is very important in the actual use of plastics. Polycarbonate is resistant to mineral acids up to high concentrations, as well as to many organic acids, oxidation agents, and neutral and acidic salt solutions; it also resists many fats, waxes, and oils, saturated aliphatic and cycloaliphatic hydrocarbons, and alcohols (except methyl alcohol) [109–111].

Polycarbonate also is relatively resistant to aqueous solutions of peroxides, persulfates, or acids with oxidizing effects up to high concentrations. Incompatibility with several vinyl plasticizers [112] has particular practical significance. A critical media attack can also be brought about through contact with amines of two-component thermosetting plastics [113]. Polycarbonate is particularly poor in resistance to low molecular weight halogenated hydrocarbons, low molecular weight aromatic compounds (toluene, chlorobenzene), and basic compounds such as amines.

Table 3.13 summarizes the resistance behavior of a selection of compatible and noncompatible media.

Table 3.13. Chemical Resistance of Polycarbonate[1]

Nonresistant Against	Resistant Against
Acetone	Acetic acid, 10%
Alkali solutions	Ammonium sulfate, saturated solution in water
Ammonia	Boric acid
Benzoic acid	Butylalcohol
Benzene	Chromic acid, 20%
Butyl acetate	Citric acid, 10%
Chlorine, wet	Diethylene glycol
Cresol	Ethyl alcohol
Cyclohexanone	Ethylene glycol
Diethyl ether	Gas (aromatic-free)
Dibutyl phthalate	Heptane
Dioctyl phthalate	Hexane
Dimethylformamide	Hydrochloric acid, 20%
Hydrochloric acid, concentrated	Hydrogen peroxide, 30%
Hydrofluoric acid, concentrated	Methane
Methyl alcohol	Nitric acid, 10%
Methyl ethyl ketone	Ozone
Methylene chloride	Perchloric acid, 10%
Perchloroethylene	Phosphoric acid, concentrated
Propionic acid, concentrated	Potassium permangante, 10% in water
Tetrachloroethane	Propane
Tetraline	Sodium carbonate, saturated solution in water
Xylene	

[1] "Resistant" and "nonresistant" are relative terms. The standard test is a 6-day flexural load with outer fiber strain up to 2%; PC will resist a component if the maximum outer fiber strain is not influenced or is influenced only negligibly.

3.7.10.2 Behavior in the Presence of Air, Oxygen, and Ozone

Chemically pure bisphenol A polycarbonate shows high resistance to oxidizing agents, air, pure oxygen, and ozone even near melting temperatures (148 °C), over long periods of time. No significant degradation is observed even after exposure to ozone/air mixtures (2×10^{-6} parts ozone, 1 part air) at room temperature over 6 months, or after aging three months in pure oxygen at 20 bar and 70 °C. During long-term aging of molded parts in air at relatively high temperatures, yellowness slowly increases [114–116] as a function of both temperature and time.

3.7.10.3 Behavior in the Presence of Water and Water Vapor

At room temperature and 60% relative humidity, water absorption of polycarbonate is a maximum of 0.15% (glass fiber reinforced products, 0.13%). The dimensional changes it causes in molded parts are so insignificant that it can be virtually ignored in commercial applications. During water aging the water absorption rate is higher, but still only 0.36% at room temperature (glass fiber reinforced products, 0.29%) and 0.59% in boiling water [115].

The resistance of polycarbonate to water can be classified as good up to approximately 60 °C. At higher temperatures and in long-term contact with water, slow saponification occurs, with consequent degradation of molecular weight as a function of time and temperature. Polycarbonate fares better in repeated, short-term contacts with hot water. Low stress dishware of polycarbonate, for example, can be washed in a dishwasher more than 1000 times without showing detrimental changes. The same is true of polycarbonate parts that are sterilized repeatedly with steam.

3.7.10.4 Weather Resistance

As shown by the light transmittance curve in Figure 3.42, polycarbonate is a good absorber of ultraviolet light. With progressive exposure to UV light, whether during outdoor weathering or indoor exposure to UV-emitting light sources, the surface shows slow yellowing. After further exposure to weathering or radiation, hairline cracks appear on the surface. Depending on product requirements, such cracking can lead to a reduction in strength. In all polycarbonates that must have high UV and/or weathering resistance, UV stabilization is necessary.

The effects of suitable UV stabilizers (e.g., addition of $<1\%$ of a benzophenone or benzotriazole) can be seen in the reduction of light transmission over time, the increase of the yellowness index, and the decline of certain mechanical properties.

Figure 3.47A shows the change in light transmission and yellowness index of standard and UV-stabilized polycarbonate by accelerated weathering in the xenon arc Weather-O-Meter. The reduction of impact strength through accelerated weathering of UV-stabilized or nonstabilized polycarbonate is shown in Figure 3.47B.

The suitability of engineering materials for outdoor application can be determined more accurately by following the changes in properties after several years of direct outdoor weathering. Currently an absolute light transmission loss of 8% (measured on 4 mm thick sheet) can be expected (Figure 3.48A) for UV-stabilized polycarbonate after 10 years of outdoor weathering in a Central European climate. The change in mechanical strength is insignificant. However, the toughness properties are reduced as a result of notch effects due to surface changes. The improved behavior of UV-stabilized grades can be seen in Figure 3.48B.

The good results with UV-stabilized polycarbonate after artificial or outdoor weathering are rarely exceeded by other thermoplastic materials. UV-stabilized polycarbonate has proven itself in many outdoor applications (see Section 3.10). One example is polycarbonate sheet for industrial windowpanes, particularly in areas where impact strength offers an added advantage. Automotive taillights in typical traffic signal colors

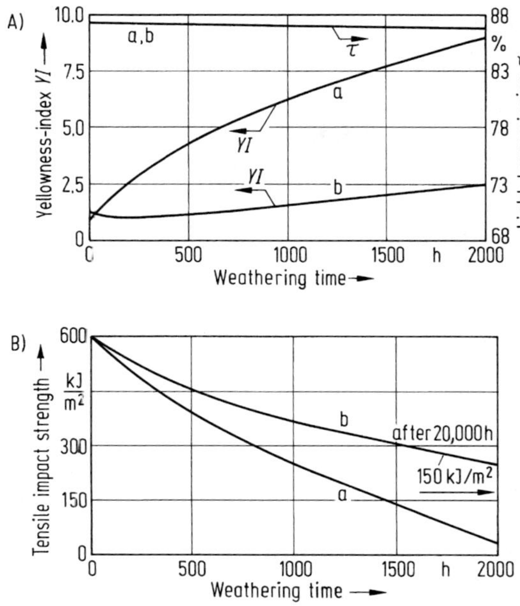

Figure 3.47. Property values of standard (a) and UV-stabilized (b) polycarbonate after artificial weathering (Xenon Weather-O-Meter 17:3, 0.25 W/m²; 50 ± 5 °C blackbody temperature): (A) light transmission and yellowness index (DIN 6167); (B) tensile impact strength (ASTM D 1822, impact rate 3.8 m/s).

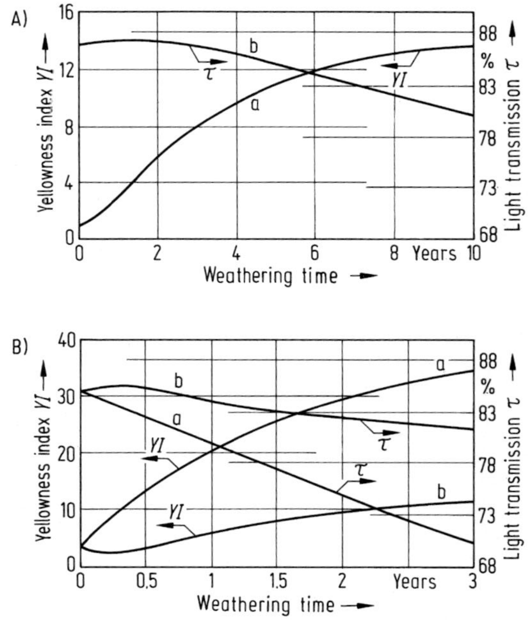

Figure 3.48. Light transmission and yellowness index after outdoor weathering of standard polycarbonate (a) and UV-stabilized polycarbonate without top coating (b) in Central Europe (A) and in Florida (B).

can also be manufactured of polycarbonate. Good stability to weathering also prevents undesirable changes in optical properties (determined by SAE Standard J 576) even after 3 years of exposure to Florida weathering.

Despite the good performance of UV-stabilized polycarbonate, additional enhancements are required in critical application areas such as thin-wall sheets. Although UV-absorbing products promote the formation of a certain inherent protective layer in the outer (approx. 150–200 μm thick) layer, the process introduces undesirable yellowing. For this reason, sheet products receive an additional, high performance UV protection layer, which is capable of counteracting brittleness and almost eliminating yellowing. Such UV protection layers (approx. 20–30 μm thick) are applied either as acrylic coatings, which contain high concentrations of UV absorber, or by coextrusion, in which polycarbonates containing high concentrations of UV absorber are used as the coextrudate. In both cases, continuous coating during the process is possible. The effects achieved with protective UV coating can be verified in accelerated tests (Figure 3.49).

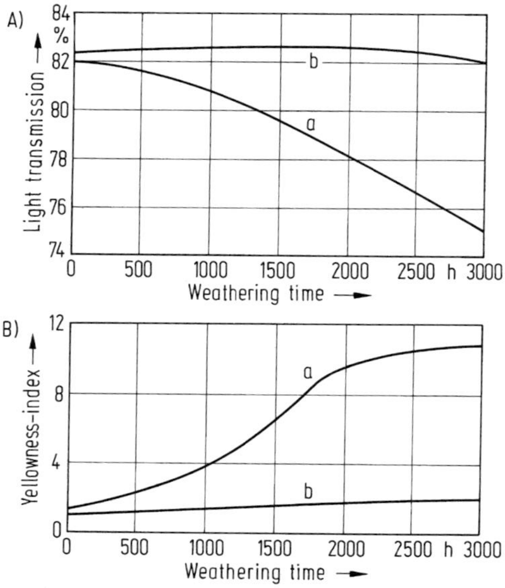

Figure 3.49. Light transmission (A) and yellowness index (B) after artificial weathering (Xenon Weather-O-Meter) of 10 mm thick polycarbonate twin-wall sheet: a, uncoated; b, 25 μm thick coating layer (UV coating 27699a, manufacturer: Diegel).

One particularly critical load for twin-wall sheet is represented by hail impact shock loads between two wall webs. Since twin-wall sheets are used in greenhouses as windowpanes, this type of failure must be minimized. In impact penetration tests, UV-protection-coated twin-wall sheets achieve twice as much energy-absorbing capacity as uncoated sheets after 3000 hours of artificial weathering.

The foregoing techniques of imparting additional UV protection either through coatings or by coextrusion today find increasing applications and open up additional areas for polycarbonate use.

3.7.10.5 Water Vapor and Gas Permeability

The water vapor permeability of bisphenol A polycarbonate depends on the layer thickness (Figure 3.50). Table 3.14 summarizes permeability values of polycarbonate for various gases [117]. The relatively high permeability for oxygen and carbon dioxide limits the use of polycarbonate in food packaging applications.

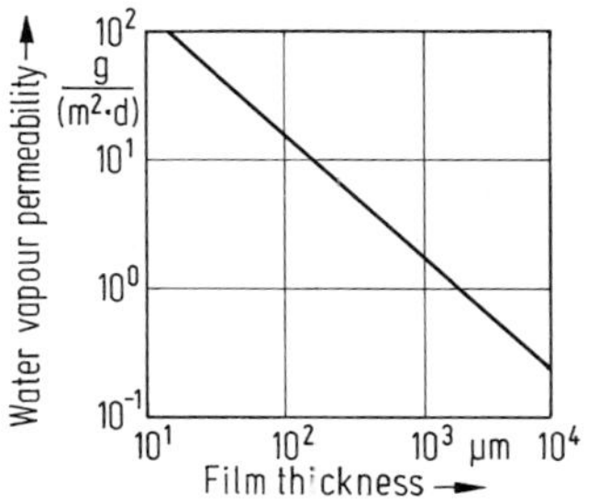

Figure 3.50. Water vapor permeability of polycarbonate film as a function of thickness; measured according to DIN 53122.

Table 3.14. Gas Permeability of Bisphenol A Polycarbonate at Room Temperature, Measured on 100 µm Film [117]

Gas	Permeation $\frac{cm^3\ (NTP)}{m^2 \cdot d \cdot bar}$
Argon	520
Butane	<10
Butylene	<20
Carbon dioxide	6700
Ethane	<30
Ethylene	60
Ethylene oxide	5000
Helium	7200
Hydrogen	7800
Hydrogen sulfide	120
Methane	190
Nitrogen	190
Oxygen	1100
Propane	<10
Sulfur dioxide	6200

3.7.10.6 Behavior in the Presence of High Energy Radiation

In comparison to many other thermoplastics, bisphenol A polycarbonate shows high stability to the effects of high energy radiation. The magnitude of such effects depends on the surrounding conditions as well as the applied radiation dose [114, 118–121].

At approximately 3 Mrad, β- and γ-rays form radicals even at room temperature. The radicals, in turn, cause yellowing of the polymer, and the intensity of this yellowing increases with increasing radiation doses. Radicals scavenged with oxygen can be seen with IR spectroscopy even after low radiation doses [122]. γ-Radiation increases electrical conductivity and dielectric constants [123]. At room temperature the radical concentrations are stable over several weeks; at temperatures above 60 °C, the yellow discoloration largely declines within a short time. UV lighting also accelerates the decline of discoloration. Special additives enable extensive color stabilization [124].

Within a radiation range up to approximately 10 Mrad, polycarbonate shows almost no reduction of mechanical properties [125]. Neutral colored polycarbonate becomes clearly yellow after absorbing 2.5 Mrad, a standard radiation dose for sterilization.

Newer developments with the goal of reducing a shift in color during γ-radiation sterilization (2.5–3.5 Mrad radiation absorption) led to optimized polycarbonate types (e.g., Makrolon RX 2530 from Bayer Corp., USA), where suitable radical-scavening additives are mixed into the polycarbonate. Although these types also show initial discoloration, after several hours the material returns close to the equilibrium color. The total color change is minimal and is operationally tolerable in most cases.

With increasing radiation dose, the molecular weight of the polycarbonate declines [126]. Up to a radiation dose of appoximately 10 Mrad, the resulting effect is so small that no significant property alternation occurs. However, higher doses cause degradation and brittleness of the material, and at radiation doses of approximately 50 Mrad, quality clearly begins to suffer.

3.7.10.7 Resistance to Termite Attack, Microorganisms, and Humic Acid

Polycarbonate is termite-resistant. Only a film that is thin can be bitten through or, if accessible from the edge, pulverized by termites.

Polycarbonate molded parts are resistant to mold fungus [117]. Even after long-term aging of molded parts in garden soil, microorganisms and humic acids show no effect on properties [19].

3.7.10.8 Cleaning, Disinfection, and Sterilization of Polycarbonate

3.7.10.8.1 General

Molded parts made of polycarbonate can be cleaned, disinfected, and sterilized by most commercially known methods. The smooth surface of the molded parts proves to be very advantageous. Under certain conditions, however, contact with cleaning, disinfection, and sterilization agents can cause damage, which is often evidenced by the appearance of stress cracks (see also Section 3.7.10.1), which lead to a general decline of mechanical strength. Still other invisible effects such as microscopic swelling or chemical degradation of the polymer can also affect mechanical strength.

Aside from the type and concentration of the contact medium, the appearance of stress-cracking corrosion depends on the internal stresses experienced by the molded part, as well as mechanical load and temperature during contact.

3.7.10.8.2 Cleaning

In many cases it is sufficient to clean polycarbonate parts with warm or hot water fortified with weak acid, neutral, or weak alkaline cleaning agents.

During cleaning in dishwashers or washing to remove food or food particles, the manufacturer's recommended concentrations should be used (mostly 0.1–0.5% of the normally strong alkaline prerinse mixture), and the temperature should not rise above 60 °C or, better, 55 °C. The subsequent clear-rinse cycle, during which weak acid or neutral substances often remain in the rinse mixture, can be conducted as usual at 80–85 °C because as a rule the contact time is short (<2 min).

Reusable polycarbonate milk bottles can be cleaned in standard glass-washing plants. As with glass, diluted caustic soda (approx. 2%) is suitable as a cleaning agent. During cleaning of animal cages it is important that all the litter material be removed. All traces of the alkaline cleaning agent that is used subsequently should be completely removed as well, to avoid the formation of turbid spots on the polycarbonate surface after the molded parts have been sterilized.

Dust deposits due to electrostatic charges can be reduced to a minimum by applying antistatic agents to part surfaces from time to time. Dipping, spraying, or rubbing with an approximately 1% aqueous solution will suffice. Surfaces soiled with greases such as oil or fat can best be cleaned with a solvent—for example, nonaromatic solvents (petroleum ether, ligroine, test fuel) and alcohols (ethyl, *n*-propyl, isopropyl, butyl, and isobutyl alcohol, but not methyl alcohol), and aqueous emulsions and/or solutions of these compounds are also suitable.

3.7.10.8.3 Disinfection

Depending on their composition, disinfectants can damage polycarbonate materials. A certain amount of caution is recommended for several products that contain aldehyde or phenols as active agents. Otherwise, tests have shown that polycarbonate is compatible with most commercial disinfectants.

3.7.10.8.4 Sterilization

Molded parts that are to be sterilized must first be cleaned. Otherwise the presence of contaminants may lead to the formation of turbidity or cracks during sterilization.

Sterilization with Hot Steam (Saturated Vapor)

The sterilization temperature should not exceed 121 °C; otherwise the molded parts may deform. It should also be determined whether the polycarbonate part will be damaged by additives that may be present in the hot water (e.g., alkaline anticorrosion agents), and proper storage procedures should be used, to prevent the formation of condensation within the part.

As a rule, polycarbonate parts can be sterilized often before slow, chemical degradation causes their mechanical strength to decline to a level insufficient for various applications. Sterilization tests on test bars showed that even after 100 half-hour cycles at approximately 121 °C, impact strength was still comparatively good. This is generally still the case even if, after many sterilizations, random hairline fractures appear and the molded part acquires a slightly milky appearance. Multiple hot steam sterilizations of stressed parts can lead to turbidity (microcracks), which negatively influences the optical quality.

Sterilization with Ethylene Oxide (ETO)

Ethylene oxide may be used either undiluted or in mixtures with carbon dioxide or inert gas. Sterilization temperature should not exceed 65 °C.

Tests showed that with increasing number of sterilization cycles, light craze formation coupled with tear formation can appear. At 55 °C, with pure ETO-treated test bars, after 50 six-hour cycles, the impact strength remained unchanged from its initial level despite slight tear formation.

ETO sterilization is slowly being replaced by hot steam and radiation sterilization.

Sterilization with Energy-Rich Radiation (γ-Radiation)

As explained in more detail in Section 3.7.10.6, radiation doses above 10 Mrad cause changes in polycarbonate properties. Assuming that 2.8 Mrad absorbed energy imparts an acceptable degree of sterilization, polycarbonate can be sterilized approximately 5–20 times, depending on the requirements for the finished part, before prohibitive damage to mechanical strength appears. However, the polycarbonate yellows more each time.

Sterilization with Peracetic Acid

Polycarbonate can be sterilized without problems with 2% peracetic acid.

Sterilization with Hot Air

The maximum temperature of hot air used for sterilization purposes is 135 °C, since polycarbonate parts are subject to deform at higher temperatures.

3.7.10.9 Dyeing of Polycarbonate

Polycarbonate is a polar plastic, hence very difficult to stain on the surface. Polycarbonate parts will not be permanently stained by many natural and synthetic dyes contained in foods (e.g., fruit juices, carrots, coffee, tea). The resistance of such surfaces to staining agents is particularly advantageous in view of the frequent use of plastic as a packaging material, in consumer articles, and as housings for various instruments.

3.7.11 Suitability for Food Contact

The results from tests of the biological effect of bisphenol A polycarbonate on rats have shown that polycarbonate can be classified as nontoxic [127]. In cytological studies of rats, its behavior with respect to cellular cultures has been completely neutral.

There are no reservations against the use of polycarbonate in the manufacture of commodity articles within the context of the German Food and Commodities Law (Paragraph 2, No. 1) [128]. The commodity articles must be suitable for the intended purpose and must fulfill the prescribed regulations in regard to raw material and residual content of foreign substances (Recommendation XI of the Plastics Commission of the National Health Department). The liquid or food that is contained in a polycarbonate receptacle must not take on odor or taste as a result of contact with the plastic.

Bisphenol A polycarbonate can be manufactured in the United States for articles that will come into contact as long as the recipe conforms to the pertinent FDA requirements in the Code of Federal Regulation (Paragraph 21 CFR 177.1580) [129]. A multitude of commercial polycarbonates are available which conform to these U.S. legal requirements for food contact. It should be noted that additional regulations apply to colored products.

Under proper processing conditions, polycarbonate articles have no inherent odor or taste. During contact of polycarbonate with milk, fat, coffee, tea, alcoholic beverages, lemonades, or fruit juices, no traceable influence on odor or taste has been detected.

References for Section 3.7

1. *Oberbach, K.:* Kunststoffberater 3 (1979) p. 87–91
2. *Ryborz, H.; Roll, M.:* Plastverarbeiter 32 (1981) p. 1609–1615
3. *Shea, J. W.*, et al.: Annual Technical Conference 33 (1975) p. 614–617
4. *Ryborz, H.; Ehrengruber, D.:* Plastverarbeiter 34 (1983) 2, p. 119–121
5. *Abbas, K. B.:* Polymer Engineering and Science 20 (1980) p. 376–382
6. *Mosle, H. G.; Criens, R. M.:* Kunststoffe 72 (1982) 4, p. 222–227
7. *Joisten, S.:* Kunststoffe 60 (1970) 12, p. 931–936
8. *Streib, H.:* Kunststoffe 58 (1968) 2, p. 123–130; 4, p. 267–273
9. *Schauf, D.:* Ind.-Anz. 25 (21. 3. 1969)
10. *Streib, H.; Oberbach, K.:* Kunststoffe 56 (1966) 1, p. 15–23; 2, p. 100–104
11. *Streib, H.; Oberbach, K.:* Kunststoffe 55 (1965) 5, p. 309–314
12. *Szigeti, P. R.:* Kunststoffe 70 (1980) 7, p. 420–422
13. *Kircher, K.:* Reinf. Plast. 25 (1981) p. 382–384 and p. 420–422
14. *Müller, P. R.:* Kunststoffe 74 (1984) 10, p. 569–572
15. *Müller, P. R.:* Kunststoffe 70 (1980) 10, p. 636–641
16. *Kircher, K.:* Kunststoffe 77 (1987) 10, p. 993–996
17. *Freitag, D.; Grigo, U.; Müller, P. R.; Nouvertné, W.*, in: Encyclopedia of Polymer Science and Engineering. 2. Ed. Vol. 11, Wiley, New York 1988, p. 648–718
18. *Hechelhammer, W.; Peilstöcker, G.:* Kunststoffe 49 (1959) 1, p. 3–8; 2, p. 93–98
19. *Schnell, H.:* Chemistry and Physics of Polycarbonates. Interscience Publisher, New York 1964

20. Plastverarbeiter 25 (1974) 12, p. 723–732
21. VDI/VDE Richtlinie 2475. VDI Verlag Düsseldorf 1967
22. *Morbitzer, L.; Grigo, U.*: Die Angew, Makromol. Chem. 162 (1988) p. 87–107
23. *Margotte, D.*, in: *Bartholome, E.; Bieckert, E.* (Eds): Ullmanns Encyclopaedie der technischen Chemie. 4. Ed. Vol. 19. Verlag Chemie Weinheim 1980, p. 55–59
24. *Riess, R.; Löwer, H.*: SPE Annual Technical Conference 43 (1985) p. 470–475
25. *Dekkers, M. E.; Heikens, D.*: J. Appl. Polym. Sci. 30 (1985) S. 2389–2400
26. *Spritzig, W. A.; Richmond, O.*: Polym. Eng. Sci. 19 (1979) p. 1129–1139
27. *Maher, J. W.*, et al.: J. Polym. Sci. Polym. Phys. Ed. 18 (1980) p. 2169–2179
28. *Keuerleber, R. H.*: Kunststoffe 70 (1980) 3, p. 167–169
29. *Kompalik, D.; Schmid, B.*: Kunststoffe 78 (1988) 4, p. 308–311
30. *Kato, T.; Kambe, H.*: J. Appl. Polym. Sci. 22 (1978) p. 1767–1774
31. *DeRudder, J. L.; Filisko, F. E.*: J. Appl. Phys. 48 (1977) 10, p. 4026–4031
32. *Murray, A. D.; Rusch, K. C.*: SPE J. 27 (1971) p. 42–46
33. *Wu, M. S.*: J. Appl. Polym. Sci. 32 (1986) p. 3263–3275
34. *von Falkai, B.; Hinrichs, G.*: J. Polym. Sci., Polym. Symp. 58 (1977) p. 225–235
35. *von Falkai, B.*, et al.: Angew, Makromol. Chem. 108 (1982) p. 41–60
36. *Struik, L. C. E.*: Poly. Eng. Sci. 18 (1978) p. 799–911
37. *Dettenmaier, M.*: Polym. Sci. 52/53 (1983) p. 57–104
38. *Kämpf, G.*, et al.: Polym. Eng. Sci. 27 (1987) p. 1421–1435
39. *Anders, S.; Hardt, B.*: Kunststoffe 77 (1987) 1, p. 21–26
40. *Siebourg, W.*: Kunststoffe 76 (1986) 10, p. 917–919
41. *Ito, E.; Kobayshi, Y.*: J. Appl. Polym. Sci. 22 (1978) p. 1143–1149
42. *Kießling, D.; Schaaf, W.*: Plaste Kautsch, 29 (1982) 6, p. 357–361
43. *Knoll, M.*: Plastverarbeiter 29 (1978) 2, p. 75–79
44. *Long, T.; Sokol, R. J.*: Polym. Eng. Sci. 14 (1974) 12, p. 817–822
45. *Menges, G.; Woelfel, U.*: Kunstst. Plast. 32 (1985) 5, p. 31–34
46. *Knutsson, B. A.*, et al.: J. Appl. Polym. Sci. 26 (1981) p. 2347–2362
47. *Woelfel, U.*: Verarbeitung faserverstärkter Formmassen im Spritzgießprozeß. Diss. RWTH-Aachen, 1987
48. *Riess, R.*: Haus der Technik-Vortragsveröffentlichungen 316, Essen 1973 p. 4–16
49. *Paffrath, H. W.; Oberbach, K.*: Kunststofftechnik (1970) 9, p. 316–322
50. *Gotham, K. V.*: Plast. Polym. 41 (1973) p. 273–280
51. *Ryan, J. T.*: SPE Annual Technical Conference 34 (1976) p. 205–207
52. *Jan, B. Z.; Lieu, Y. K.*: J. Appl. Polym. Sci. 30 (1985) p. 3925–3942
53. *Biskup, U.*, et al.: Angew. Makromol. Chem. 108 (1982) p. 113–121
54. *Wolters, J.; Menning, G.*: Angew. Makromol. Chem. 125 (1984) p. 103–119
55. *Sternfield, A.*: Mod. Plast. Intern. 17 (1987) p. 38–41
56. *Kunori, T.; Geil, P. H.*: J. Makromol. Sci. Phys. B18 (1980) 1, p. 135–175
57. *Ram, A.*, et al.: Polym. Eng. Sci. 25 (1985) 9, p. 577–581
58. *Hornberger, L. E.; Devries, K. L.*: Polym. Eng. Sci. 27 (1987) 19, p. 1473–1478
59. *Groshans, V. L.; Takemori, M. T.*: Polym. Eng. Sci. 27 (1987) p. 1006–1014
60. *Oberbach, K.*: Plaste Kautsch. 29 (1982) p. 178–182
61. *Oberbach, K.*: Kunststoffe 63 (1973) 1, p. 35–41
62. *Oberbach, K.*: Haus der Technik, Vortragsveröffentlichung 316 Essen, 1976
63. *Ohishi, F.*, et al.: J. Appl. Polym. Sci. 20 (1976) p. 79–92
64. *Chung, C. I*, et al.: Polym. Eng. Sci. 17 (1977) p. 9–20
65. *Wissler, G. E; Crist, B. jr.*: J. Polym. Sci., Polym. Phys. Ed. 18 (1980) p. 1257–1270
66. *Di Filippo, G. V.*, et al.: J. Appl. Polym. Sci. 34 (1987) p. 1959–1966
67. *Adam, G. A.*, et al.: J. Mater. Sci. 10 (1975) p. 1582–1590
68. *Kaufmann, W.; Thieler, W.*: Kunststoffe 72 (1982) 8, p. 471–475
69. *Dietz, W.*: Colloid Polym. Sci. 255 (1977) p. 755–772
70. *Abbas, K. B.*: Polymer 21 (1980) 8, p. 936–940
71. *Ryborz, H.*: Kunststoffberater 25 (1980) 1/2, p. 40–41
72. *Saffel, J. R.; Windle, A. H.*: J. Polym. Sci. Polym. Lett. Ed. 18 (1980) p. 377–380
73. *Yokouchi, M.; Kobayashi, Y.*: J. Appl. Polym. Sci. 26 (1981) p. 431–440
74. *Morgan, R. J.; O'Neal, J. E.*: J. Polym. Sci., Polym. Phys. Ed. 14 (1976) p. 1053–1076
75. *Broutman, L. J.; Krishnakumar, S. M.*: Polym. Eng. Sci. 16 (1976) 2, p. 74–81
76. *Pixa, R.*, et al.: Polym. Bull. 14 (1985) 1, p. 53–60

77. *Bubeck, R. A., et al.*: Polym. Eng. Sci. 24 (1984) p. 1142–1148
78. *Ryan, J. T.*: Polym. Eng. Sci. 18 (1978) p. 264–267
79. *Ott, H.-J.*: Colloid Polym. Sci. 258 (1980) p. 995–1008
80. *Ito, E.*, et al.: Colloid Polym. Sci. 253 (1975) p. 480–484
81. *Müller, J.; Wendorff, J. H.*: J. Polym. Sci., Polym. Lett. Ed. 26 (1988) p. 421–427
82. *Adams, G. W.; Farris, R. J.*: J. Polym. Sci., Polym. Phys. 26 (1988) p. 433–445
83. *Marton, H.*, et al.: Plastic Eng. (1986) p. 49–52
84. *Matsuoka, T.; Bair, H. E.*: J. Appl. Phys. 48 (1977) 10, p. 4058–4062
85. *Ram, A.*, et al.: Polym. Eng. Sci. 25 (1985) 9, p. 535–540
86. *Abbas, K. B.*: J. Appl. Polym. Sci., Polym. Symp. 35 (1979) p. 345–360
87. *Sherman, E.*, et al.: Polym. Eng. Sci. 22 (1982) p. 457–465
88. *Gupta, A.*, et al.: Macromolecules 11 (1978) p. 1285–1288
89. *Blaga, A.; Yamasaki, R.*: J. Mat. Sci. 11 (1976) p. 1513–1520
90. *Narkis, M.; Bell, J. P.*: J. Appl. Polym. Sci. 27 (1982) p. 2809–2814
91. *Joseph, E. A.*, et al.: J. Appl. Polym. Sci. 27 (1982) p. 4807–4819
92. *Gardner, R. J.; Martin, J. R.*: J. Appl. Polym. Sci. 24 (1979) p. 1269–1280
93. *Kellecher, P. G.*, et al.: Polym. Eng. Sci. 23 (1983) p. 537–542
94. *Martin, J. R.; Gardner, R. J.*: Polym. Eng. Sci. 21 (1981) p. 557–565
95. *Kelleher, P. G.*, et al.: SPE Annual Technical Conference 40 (1982) p. 111–113
96. *Crowley, L.*: SPE Annual Technical Conference 43 (1985) p. 1278–1280
97. *Narkis, M.*, et al.: Polym. Eng. Sci. 24 (1984) p. 211–217
98. *Robeson, L. M.; Crisafulli, T.*: J. Appl. Polym. Sci. 28 (1983) p. 2925–2936
99. *Azis, A. W.; Ab-El-Nour, K. N.*: J. Appl. Polym. Sci. 31 (1986) p. 2267–2273
100. *Heynemann, C.; Gonzales A.*: Kunststoffe 69 (1979) 2, p. 131–132
101. *Troitzsch, J.*: Brandverhalten von Kunststoffen, Hanser, Munich 1982
102. *Kircher, K.*: Kunststoffe 74 (1984) 12, p. 727–731
103. *Ries, R.*: VDI-Z. 12 (1975) p. 568–574
104. *Varadarajan, K.; Boyer, F.*: J. Polym. Sci. Polym. Phys. Ed. 20 (1982) p. 141–154
105. *Kato, T.; Kambe, H.*: J. Appl. Polym. Sci. 22 (1978) p. 1767–1774
106. *Sung, N. H.*, et al.: Polym. Eng. Sci. 23 (1983) p. 328–336
107. *Brown, N.; Imaai, Y.*: J. Appl. Phys. 46 (1975) p. 4130–4135
108. *Wright, D. C.; Gotham, K. V.*: Polym. Eng. Sci. 23 (1983) p. 135–139
109. *Gnauck, B.*: Kunststoffe 71 (1981) p. 51–57
110. *Khokhlov, A. A.*, et al.: Int. Polym. Sci. Technol. 4 (1977) p. 854–860
111. *Wyzgoski, M. G.; Jacques, C. H. M.*: Polym. Eng. Sci. 17 (1977) p. 854–860
112. *Haag, E.*: SPE Annual Technical Conference 40 (1982) p. 134–136
113. *Loos, W.*: Werkstofftech. 6 (1975) p. 314–324
114. *Christopher, W. E.; Fox, D. W.*: Polycarbonates. Reinhold New York 1982
115. *Hechelhammer, W.; Peilstöcker, G.*: Kunststoffe 49 (1959) 1, p. 3–8; 2, p. 93–98
116. *Peilstöcker, G.*: Kunststoffe 51 (1961) 9, p. 509–512
117. *Vieweg, R.; Goerden, L.* (Eds): Kunststoff-Handbuch. Vol. 8. Hanser, Munich 1973, p. 45–46
118. *Barker, R. E., Jr.; Moulton, W. G.*: J. Polym. Sci. 47 (1960) p. 175–191
119. *Giberson, R. C.*: Mod. Plast. 39 (1962) p. 143, 146, 228
120. DB-PS 1 085 668 (1960) Bayer AG
121. *Klein, A. J.*: Plast. Design Forum 12 (1987) 11/12, p. 35–58
122. *Gordienko, V. P.*, et al.: Int. Polym. Sci. Technol. 8 (1981) p. 36–38
123. *Amin, M.*, et al.: Angew, Makromol. Chem. 150 (1987) p. 81–87
124. *Lundy, C.*, et al.: SPE Annual Technical Conference 47 (1988) p. 1348–1351
125. *Maslyar, K. D.; Thomas, J. R.*: ANTEC 1980, p. 619–621
126. a) *Acierno, D.*, et al.: Radiat. Phys. Chem. 17, (1981) p. 31–34
    b) *Golden, J. H.; Hazell, E. A.*: J. Polym. Sci., Polym. Chem. Ed. 1 (1963) p. 1671–1686
    c) *Golden, J. H.*, et al.: J. Polym. Sci., Polym. Chem. Ed. 2 (1964) p. 4787–4794
127. *Bornmann, G.; Löser, A.*: Arzneimittelforschung 9 (1959) p. 9–13
128. Bundesgesundheitsblatt Nr. 16 (1973) vom 15. 6. 73, 96. und 126. Mitteilung. Empfehlung XI sowie die Ergänzungen in der 96., 126. und 130. Mitteilung
129. Food and Drug Chemical Horizons 2 (1963) 2, p. 91

3.8 Processing of Polycarbonate

3.8.1 Injection Molding

S. Anders, A. Kaminski, and R. Kappenstein

3.8.1.1 General Comments

A characteristic of polycarbonate is its relatively high melt viscosity. To allow thermoplastic processing, the temperature of the melt must exceed the softening temperature by 100 K for extrusion and 150 K for injection molding. Even at these elevated temperatures, the melt is very viscous, and machines with a high driving power and torque are required. Wide flow channels and short flow paths are design choices that favor melt flow in the mold.

The mold temperature is relatively high (approx. 90 °C) and has a significant effect on the cycle time. The susceptibility of polycarbonate to sharp edges has to be considered during the mold design stage. The creep modulus of polycarbonate is very high. The relaxation of applied stress or molded-in stress occurs very slowly, however—a property related to environmental stress-cracking in certain environments (e.g., halogenated solvents).

The tendency to experience environmental stress-cracking can be minimized by polycarbonate-specific part and mold design as well as by polycarbonate-specific processing conditions (e.g., mold overpacking).

The following discussion is a general overview of injection molding and extrusion processes. More specific detailed information can be found in the referenced literature. Sections 3.8.1.2–3.8.1.4 illustrate specific polycarbonate-related processing details.

Polycarbonates can be used in all processing machines that are suitable for thermoplastic materials. They belong to the class of thermoplastic resins that must be dried prior to processing to prevent surface streaks due to hydrolysis. Because of their broad processing windows, polycarbonates can be processed via injection molding and extrusion as well as by special-purpose processes (e.g., the production of compact disks or injection blow molding).

3.8.1.2 Processing Steps

3.8.1.2.1 Storage and Drying [1]

Generally polycarbonates are shipped and stored in plastic containers that are not moisture-tight. Water absorbed may damage the polycarbonate during processing. Such damage impairs the mechanical properties of the finished parts and may cause bubbles and surface streaks. Any excess moisture must, therefore, be removed before processing, or at least during melting (devolatilization).

In the case of pellets that are not kept in moisture-tight packaging, storage in a dry place reduces moisture absorption, hence drying times. To avoid the condensation of atmospheric moisture onto the surface of the granules, containers stored at very low temperatures should be allowed to reach room temperature in the workroom prior to opening. This applies particularly to material packed in moisture-proof containers. Pellets are dried continuously or batchwise in hot air dryers. To achieve the permissible moisture levels, the drying step can be carried out with fresh air (i.e., in an open system) as well as by means of recirculation and feeding in of a certain portion of fresh air or, in cases of excessive humidity (summer, tropical climate), by dehumidified fresh air. The drying temperature is 120 ± 5 °C. The required drying times are, depending on the efficiency of the dryers, usually 2–4 hours. The permissible moisture content is 0.02% for injection molding and 0.01% for extrusion. This requires a drying time of 4–12 hours in air circulation dryers or 2–5 hours in "quick" dryers.

Reuptake of moisture after the drying step can be avoided by using hopper dryers or hoppers with a small volume for pellets, which can then be processed very fast. Inadequate predrying of the material manifests itself in distension of the melt strand and a certain degree of bubble formation, depending on the moisture content. It is, therefore, particularly easy to detect this condition in transparent material.

There are several laboratory methods of determining the moisture content of molding materials: methods that involve elaborate apparatus, and other, simpler approaches that can be employed in the production plant (e.g., the Karl-Fischer test, the TVI Test) [2].

One such simple method requiring little apparatus or time yet providing an indication of dryness sufficient for processing purposes, described in the technical literature as the TVI (Tomasetti's volatile indicator) and method, is suitable for checking the dryness of nonreinforced polycarbonate. The following simple apparatus is required: one hot plate, which can be set at a surface temperature of 270 °C $\pm$ 5 °C for polycarbonate; one pair of tweezers; one wooden ruler approximately 20 cm long; and a number of glass slides (size 35×75; thickness ~ 1 mm).

In the TVI test, two slides are heated on the hot plate at 270 °C. After 1–2 minutes of heating-up time, tweezers are used to place three or four granules of the thermoplastic to be tested on one of the glass slides; the second slide is then placed on top and the two slides pressed together using the edge of the ruler until the specimen formed by the granules has a diameter of approximately 10–12 mm. After a further 60 seconds on the hot plate without application of pressure, the specimen is cooled in air. Any bubbles visible in the test specimen indicate that the material tested contained too much moisture.

One possible way of avoiding the step of predrying the granules is to remove the moisture during melting with the aid of a venting unit. This has been common practice for some time now in injection molding. Raw material suppliers usually recommend processing temperature ranges for their product grades and at least a 30% usage of the unit's maximum shot capacity. This helps to avoid degradation and melt overflow at the vent part. In the case of extrusion processing, the devolatilization step is supported by vacuum.

3.8.1.2.2 Material Changeover, Break in Production, Cleaning

For changeovers from other resins to polycarbonate, purging (mostly PMMA resin in the first step, followed by a mixture with 50% polycarbonate) is recommended. In the case of gas-nitrided steel for screws and cylinders, it is advisable to maintain a temperature of 160–180 °C during breaks in production, since solidifying polycarbonate can tear steel particles from the brittle nitride layer. These steel types favor the formation of a thin black-brown film containing a degraded polycarbonate. When the melt solidifies, this film separates from the cylinder wall and manifests itself in the moldings in the form of tiny spots or specks. With modern abrasion-resistant plasticating units, this type of defect can be prevented.

A thorough cleaning of strongly contaminated plasticating cylinder is possible with the above-mentioned purging pellets or powders. Although cleaning is possible in the assembled state [3], heavily contaminated systems require the disassembly of the unit.

Prior to the disassembly the unit is purged with high viscosity polyethylene or polypropylene. The individual parts can then be cleaned with wire brushes while still hot, and polished with cleaning pastes. The cleaning of individual parts can also be carried out in aluminum oxide fluid beds [4] with ultrasonic sound [5] or in oil [6]. The use of highly abrasive cleaning powders and welding flames should be avoided. Contaminants can be burned off in an oven at temperatures above 400 °C, but this is not recommended for all steel types.

3.8.1.3 Injection Molding

3.8.1.3.1 Machine Components [7]

General Remarks

Polycarbonates can be processed on all commercially available injection molding machines. In comparison with other, low viscosity thermoplastic resins, polycarbonates require a screw drive with high torque. In calculating the necessary clamp force for proper mold design, the pressure in the mold can be assumed to be approximately 500 bar. The size of the plasticating unit should be considered, to allow a short residence time for the melt in the system during a given cycle time. When heat-sensitive resin grades are being processed, lower processing temperatures, shorter cycles times, and lower shear rates should be considered.

Screw Geometry and Nonreturn Valves

Special screw geometries are not required for processing polycarbonates. Screw types, such as progressive screws and short compression screws, are used but are not recommended.

Throughput and the quality demands made on the melt are constantly increasing. In the processing of polycarbonate, three-zone screws with an L/D ratio of about 18:1 to 22:1 and a flight depth ratio between 2:1 and 2.5:1 have successfully been used for homogeneous plastication of the resin [8]. Zone distribution and flight depth ratio correspond to the concept of a universal standard screw (see Figure 3.51). All commonly

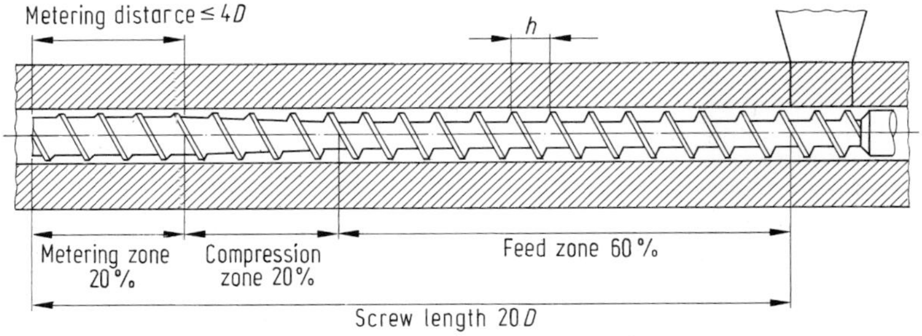

Figure 3.51. Division of standard screw in zones (b-screw according to EURO MAP).

used nonreturn valve designs (e.g., plates, balls, or ring systems) can be used; ring valves are very successful. The antibackflow valve must be shaped so as not to disturb the flow; otherwise small amounts of melts may become trapped in "dead spots." Furthermore, it is desirable to have a large flow cross section.

Abrasion Protection [9]

Partially or completely gas-nitrided screws and cylinders fail to meet today's requirements. Compared with other thermoplastic resins, polycarbonates give rise to few abrasion problems; however, abrasion-resistant plasticating units offer qualitative and quantitative advantages. As a rule of thumb, general protection (universal protection) against abrasion and corrosion is recommended. The protection measures listed in Table 3.15 show good results.

Bimetallic cylinder tubes are coated on the inside with a 2 mm armored layer, which guarantees sufficient corrosion and abrasion resistance. Occasionally, when needed in cases of repairs, bimetallic tubes are shrunk into a carrier cylinder.

Different materials can be chosen for screws. For combined abrasion and corrosion protection, hardened steel highly alloyed with chromium is used. These surfaces tend not

Table 3.15. Selection of Materials to be Used for Wear-Resistant Injection Molding Machine Components

Cylinders	Centrifugal deposition of a suitable wear-resistant coating (usually based on Ni-Co-Cr-W-B; nearly iron-free), alloyed and unalloyed carbon steels for the supporting Insertion of centrifugally coated bushes; supporting tube, nitrated steels, e.g., 34 Cr Al Ni 7 (1.8550) 31 Cr Mo V9 (1.8519) Boride diffusion coatings; small diameters
Screws	High Cr-alloyed depth-hardened steels (up to about 60 mm diameter and a length of 1500 mm); partly additionally ion nitrated, e.g.: X 155 Cr V Mo 12 1 (1.2379); X 165 Cr Mo V 12 (1.2601); X 210 Cr 12 (1.2080); X 220 Cr Mo 12 2 (1.2378); X 210 Cr W 12 (1.2436) Stelite screw flight hard-facing with ion nitrated Cr steels, e.g.: X 35 Cr Mo 17 (1.4122); hardened; small and medium diameters X 22 Cr Ni 17 (1.4057); hardened; small and medium diameters Stellite harde facing of flights and chromium-plated root surface and flanks; e.g., 31 Cr Mo V 9 (1.8519) Boride diffusion coatings; small diameters
Cylinder head	High alloy Cr steels; ion nitrated under the heading "Screws" Standard nitrated steels, hard chromed, e.g., 31 Cr Mo V 9 (1.8519)
Nonreturn valve	Tip and thrust ring – flights of wing tip always hard-faced with Cr Ni B alloy with added carbide (a) High alloy Cr steels; ion nitrated if necessary under the heading "Screws") (b) High Cr alloyed depth-hardened steels under the heading "Screws") Locking ring (a) High Cr-alloyed steels with good toughness depth hardened or tempered, e.g., ion nitrated. X 155 Cr V Mo 12 1 (1.2379) X 35 Cr Mo 17 (1.4122) X 40 Cr Mo V 5 I (1.2344) All structural elements made of – hard materials – boride coated – CVD/PVD coated (chemical vapor deposition/physical vapor deposition of TiN)

to be subject to "tear-offs" caused by adhering polymers when the polymer melt solidifies. (It is not necessary to keep the machine at 180 °C during breaks of production time when polycarbonate is processed.)

Another advantage of abrasion-resistant surfaces is the absence of a tendency to form layers of degraded polycarbonate (discoloration and particles in the finished part). Chromium-alloyed, heat-treated steel with a Stellite armor plating on the sensitive screw flight tips and additional ion nitration of the complete surface is particularly useful. Stellite-plated flight tips with a chromium-plated screw core are sensitive to rough handling and to abrasion—that is, the chromium layer may chip off.

Flight or total surface armoring with Stellite is used in case of repairs for standard plasticating units. The reliability of such measures is not clear, however. Cracks, chip-offs, and warpage are frequently reported, with the result that it may be economical to replace screws rather than repair them. The criteria for the materials chosen for the nozzle and the antibackflow valve are the same as those for the screw.

Longer and trouble-free production, as well as higher machine efficiency, improves the economics of the process by a factor of 3–6.

Vented Plasticating Units [10]

Trouble-free functioning of vented plasticating units for moisture removal depends not only on operating conditions such as temperature profile, cylinder capacity usage, and back pressure; it also depends significantly on the design factors. An L/D ratio of 20:1 is employed with a shear element in front of the volatilization zone. To avoid melt overflow at the vent port, a feed pocket should be integrated in a plug that can be disassembled easily. It is very useful to have a declining temperature gradient from the feed zone to the venting zone. The temperature profile after the venting zone is determined by part and mold design. Processing polycarbonate with a vented cylinder poses fewer problems than for injection molding than for extrusion.

Nozzle

The relatively high melt viscosity of polycarbonate facilitates the use of so-called open nozzles. Disadvantages sometimes accompanying the use of needle or slide shutoff nozzles include additional sheer zones or unfavorable flow corners such as in shutoff nozzles with a perpendicular rotary or slide valve. In general, a screw decompression stroke after the metering step can prevent any melt from dripping out.

Temperature Control [11]

One important parameter governing the quality of injection-molded parts is the temperature of mold cavity. Particular properties such as molded-in stress, warpage, mold part tolerance, and surface properties are controlled by the temperature of the mold cavity. No amount of elaborate process control on the injection-molding machine can compensate for errors resulting from inappropriate temperature control (wrong design, temperature fluctuations).

3.8.1.3.2 Processing Parameters

Melt Temperature

The melt temperature is the average temperature of the molten polymer in the pool in front of the screw after the plasticating process. Since a temperature gradient exists in both axial and radial directions [12, 13], exact measurement of the temperature is extremely difficult. Indeed, it is practically impossible to measure the temperature during processing. It is therefore necessary to rely on approximate temperature values. Usually, the temperature is measured with an immersion thermocouple in the melt that has been purged out of the nozzle. More reliable (and not affected by subjective errors) is the use of a special thermocouple in the nozzle [14].

Actual local temperature peaks in the shear and mixing zones cannot be directly measured in the nozzle. Temperature peaks, even when they are apparent for only a short time, can damage temperature-sensitive polymers significantly. Such products must be processed at low temperature in the case of injection molding machines with high shear generating screw elements. Depending on the flowability of the polycarbonate grades and the property requirements of the molded part, the temperature range for injection molding is between 280 and 330 °C. Specialty grades may have to be processed at lower or higher temperatures.

The relatively broad melt temperature range has resulted in a variety of polycarbonate grades for injection molding. These grades vary in melt viscosity and in other respects; there are unfilled grades, reinforced grades, filled grades, and specially modified grades. Easy-flow grades are generally processed at lower temperatures. Higher processing temperatures are recommended for high viscosity products as well as for glass fiber reinforced polycarbonate grades. Longer residence time in the plasticating screw, due to long cycle times or very low capacity, requires a lower melt processing temperature if thermal degradation is to be avoided. The design of the molded part—in particular, the ratio of the flow length to the wall thickness—is critical. For molded parts with a high ratio of flow length to wall thickness, and when the injection mold cavity needs to be replicated with great accuracy (e.g., in the injection molding of compact disks: see Section 3.8.1.3.3), a higher melt temperature must be used.

The temperature of the melt in front of the screw is the result of dissipation, thermal conductivity processes in the screw channel, and heat transfer from the cylinder and the nozzle to the resin. The general temperature profile setting is shown in Figure 3.52. The profile can vary up or down (arrows) because the melt temperature also depends on the melt viscosity of the polycarbonate, the screw design, the screw speed, and the back pressure. When polycarbonate is injection-molded with a vented cylinder, it is very important to maintain the correct melt temperature (approx. 280 °C) in the venting zone, to prevent the melt from rising into the vent hole.

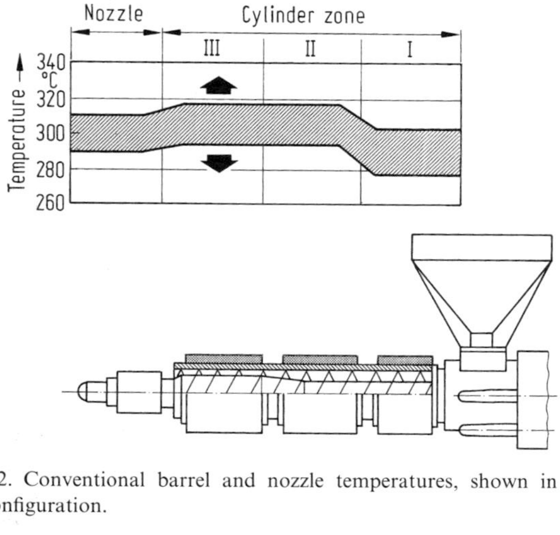

Figure 3.52. Conventional barrel and nozzle temperatures, shown in conjunction with a typical injection molding configuration.

Mold Temperature

Temperature distribution in the injection mold is not naturally homogeneous because the temperature fluctuates both locally and periodically. The average mold temperature can be determined from the periodic thermal fluctuation measured with a thermocouple closely positioned beneath the mold cavity surface. The mold temperature is a parameter having a decisive influence on cycle time and on the quality of the molded parts. Low mold temperatures result in rapid cooling and higher output rates; the quality of the molded parts, however, is adversely affected. Specifically, there are increases in degree of orientation, internal stress, and postmold shrinkage, whereas the surface quality is reduced. A compromise must, therefore, be found between quality and output. For polycarbonate, an average mold temperature of 80–130 °C is recommended. To achieve a high surface quality, the mold temperature should be 100 °C or more. This is recommended for glass fiber reinforced grades or when high replica accuracy of mold surface structures is required.

A prerequisite for a good quality molding is, however, uniform cooling. This requires a good temperature control system to ensure homogeneity of temperature distribution along the mold cavity wall.

Injection Pressure, Speed, and Holding Pressure

The important parameters for the injection and holding pressure phases must be set prior to processing, following the same general rules as for other thermoplastic resins.

To avoid "jetting," the injection is often started with low injection speeds. The ratio of flow path to wall thickness determines the injection speed: higher ratios require higher speed to fill the mold cavity. Limitations are set by the available injection pressure and the dissipation energy (shear) generated in narrow flow channels. To adapt for changing cross sections along the flow path, the injection speed profile can be set as a function of stroke. Good results are obtained with a constant melt front speed. Toward the end of the mold-filling phase, a reduction of injection speed is recommended to help avoid pressure peaks in the holding pressure phase and overfilling of the mold. It is also important to select the right point (depending on path, time, or pressure) to switch from the injection phase to the holding pressure phase (Figure 3.53). Several computer programs that simulate the mold-filling process are available [15, 16].

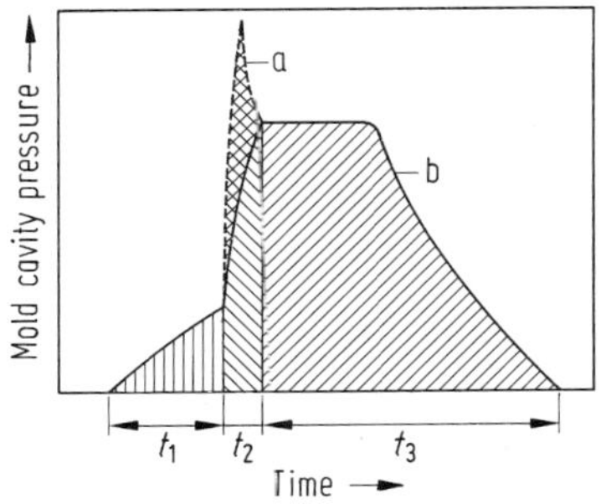

Figure 3.53. Courses of the inner mold pressure: a, unfavorable course; b, favorable course, t_1, injection phase; t_2, compression phase; t_3, back-pressure phase.

The holding pressure as a function of time should be selected such that it compensates for shrinkage during the cooling period until the demolding temperature is reached. Overloading of the mold due to high holding pressure or sink marks due to low holding pressure or short holding time should be avoided. Computer simulations for modeling the holding pressure phase are being developed [17]. Frequently an overloaded mold is the reason for molded-in stress, which can cause stress-cracking in various environments. The right choices of pressure and speed via process optimization lead to a significant improvement in the quality of the molded parts.

Screw Speed, Back Pressure

The rotational screw speed should be selected from the screw circumferential speed in a way that this speed v_u falls between 0.05 and 0.2 m/s. However, it should not exceed 0.3 m/s (Figure 3.54). The back pressure, which supports the uniformity of plastication, is generally about 100 ± 50 bar (hydraulic pressure usually 5–15 bar).

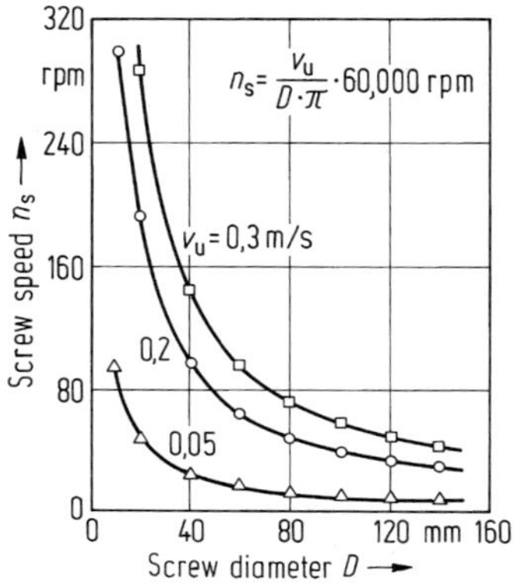

Figure 3.54. Screw speed v_u as a function of screw diameter and screw peripherial speed.

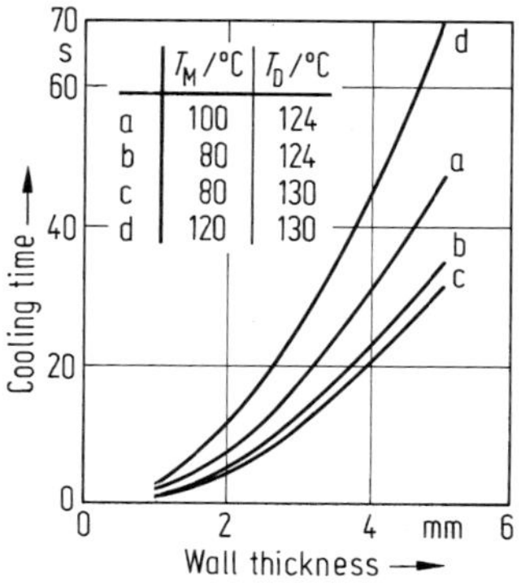

Figure 3.55. Cooling time as a function of wall thickness for unfilled standard polycarbonate (melt temperature T_m = 300 °C, mold temperature, T_M, T_D = average molding temperature).

Cooling Time

The cooling time has a major impact on the cycle time; thus it also has a significant bearing on productivity. Figure 3.55 shows the effect of wall thickness and mold temperature on cooling time.

The thermal conductivity for polycarbonate is 0.1 mm²/s. A polycarbonate demolding temperature guideline is 130 °C, but depending on the resin grade, this value can vary. With increasing glass fiber content, the thermal conductivity increases. Because higher melt and mold temperatures also enter the picture, however, cooling time is not necessarily shortened.

Residence Time

If a small amount of intense dye is added into the hopper while transparent polycarbonate is being processed, the dye will be eluted over a period of time. The melt shows a residence time distribution for the machine [18, 19]. The maximum intensity of the dye in the resin represents the average residence time. A theoretical estimation of the residence time is possible. The higher the cycle time and the lower the utilization degree of the available shot volume of the injection-molding machine, the higher the residence time.

Knowledge of the residence time distribution of the resin in the injection molding unit is important in determining bottlenecks and weak spots. Polycarbonate that remains for excessively long periods in dead spots of the plastication unit can be degraded. If this happens, color changes, black specks, or brown or silver streaks may result.

3.8.1.3.3 Special Processes

Injection Molding of Compact Disks

Compact disks are optical data storage devices that are made of polycarbonate; the information is retrieved by means of a laser. Although the most popular applications of this technology are in audio CDs and computer data storage (CD-ROM), devices for digital video storage are also being developed.

A CD is a round disk with a diameter of 120 mm and a thickness of 1.2 mm that is injection-molded onto a nickel stamper: that is, the "bumps" on the stamper surface are replicated as pits on the disk surface. The disk is rendered playable after metallizing and the application of a protective overcoat. Processing conditions and mold parameters are determined by the overall requirement of low birefringence in the disk (originating from stress and orientation components) as well as accuracy of replication of the pits on the stamper surface, which are about 0.1 μm in depth (Figure 3.56). Audio CDs require a birefringence of 50 nm or less.

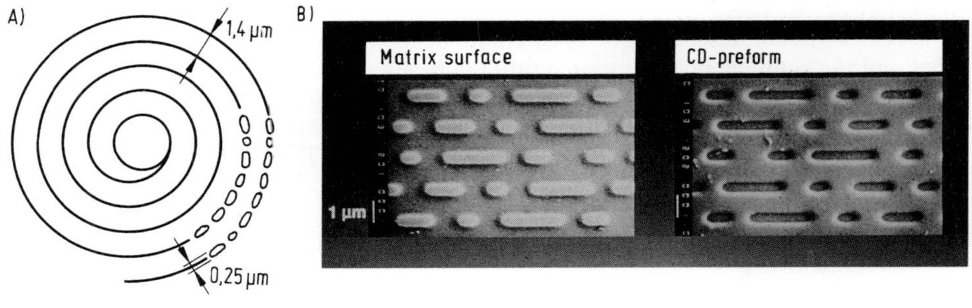

Figure 3.56. Pit structure of a compact disk: (A) schematic drawing and (B) scanning electron micrographs (left: stamper surface; right: disk surface).

High precision injection molding machines that are capable of good process control and reproducibility are required for the manufacture of CDs. To this end, the ability to measure and control various process parameters is essential. A clamping force of approximately 600 kN is necessary.

The mold is in general fitted with a cold sprue together with a long machine nozzle. The sprue is removed by an internal punch in the mold, which punches out the center hole (15 mm diameter). Normally, the mold is allowed to ventilate during the injection stage. Experience has shown that the amount of ventilation allowed should not be influenced by the stamper thickness. Evacuation of the cavity before injection is the best way to avoid a product-damaging excess of ventilation.

The processing parameters and drying conditions used for CDs are related to those used for standard polycarbonate. A melt temperature of 320–350 °C is used for this easy-flowing PC grade. To achieve the required low birefringence, the specific process parameters may be varied within a relatively large range; normally, they depend on the overall injection system–mold combination [20]. By optimizing process conditions, disks relatively free of orientation- and stress-induced birefringence may be molded with cycle times under 7 seconds. Figure 3.57 shows such a disk under polarized light and a plot of the resulting birefringence.

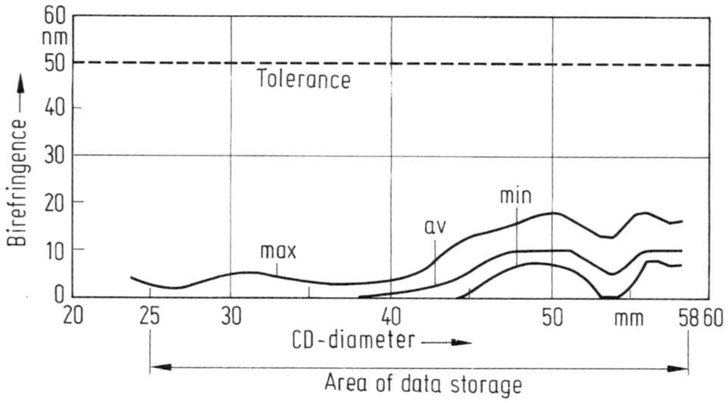

Figure 3.57. Compact disk with low birefringence seen in polarized light and the measured pitch in the area of information storage: max, maximal; min, minimal; av, average pitch.

Two-Component Injection Molding [21, 22]

Two-component injection molding is used for parts that require the performance characteristics of two different plastics. In this process, a core material is encapsulated by a skin material. Viscosities of both components are adjusted to be similar to each other and the components are injected into the cavity in the same cycle; the mold layout ensures that they are distributed evenly (Figure 3.58). It is important that the processing temperatures and adhesion properties of the two components be very similar. Polycarbonate has exhibited good adhesion to ABS, ASA, PBT, TPU, PVC-P, SAN, PET, and

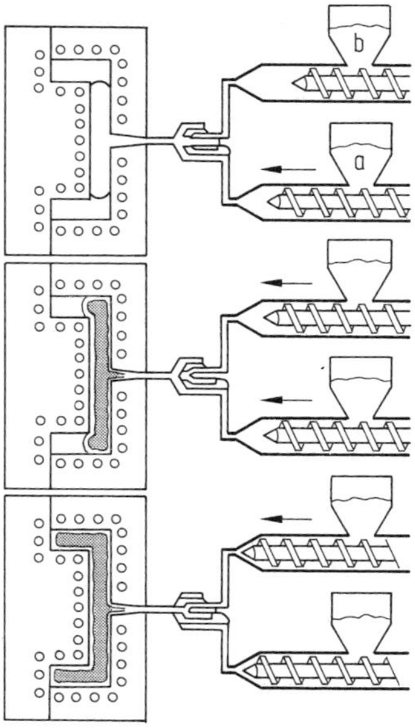

Figure 3.58. Two-component injection molding process flow with simultaneous phase for shell components (a) and core components (b).

PPSU, as well as with polycarbonate blends with ABS and PBT. Poor adhesion was seen in two-component parts made with polycarbonate and polystyrene and polyolefins.

Originally, this technique was used to mold thick-walled parts with foamed cores and smooth skins. Today, the process is also used for thin-walled parts with compact, unfoamed cores (e.g., glass containing polycarbonate as a stiffening core material). An interesting variation is the use of soft thermoplastic urethanes as the outer layer around a hard polycarbonate core. Earlier used primarily in the furniture and business machine sector, this technique is finding applications today in other sectors such as electrical engineering and transportation.

Thermoplastic Foam Injection [23, 24]

Special foam grades of polycarbonate are used for numerous parts, especially in the lighting sector, for business machine housings, and for electrical housings. Materials of these types usually contain a small percentage of glass that not only stiffens the part but also acts as a nucleating agent. Foam molding is usually a low pressure process. However, this process usually does not always give a smooth, fissure-free surface without relatively much loss of void volume (i.e., greater weight). The raw material supplier provides the foaming agent as a concentrate containing 2–10% of the foaming chemical. Properties of the foamed part depend on void volume, cell structure, and surface quality, which in turn are influenced by the quality of the foaming agent and by the processing parameters. A void volume (cell volume) of 30% is targeted.

Clamping force is adjusted to withstand a pressure of 80–100 bar inside the mold. Higher injection speeds favor a uniform density distribution but require a large capacity for mold ventilation.

Foam injection has the advantages of using low pressure machines (Figure 3.59) with preplastication, affording faster injection rates (through the use of pistons driven by a hydraulic accumulator), and allowing the use of larger platens for the low tonnage required. Foam injection may also be carried out on more traditional molding machines. The use of a shutoff nozzle is imperative to minimize premature foaming of the melt in the cylinder and leakage of melt during and after plastication. Because the foaming agent may be corrosive, hence able to cause cracking in the plastication unit and mold, special preventive measures are recommended. (Refer also to Section 3.8.1.3.1.)

Injection Blow Molding

Hollow articles from 100 ml capacity and spherical light covers up to 500 mm in diameter are among the items that may be made from polycarbonate using the injection blow molding process. As the name implies, the process is a combination of injection and blow molding. The process takes place in three steps (Figure 3.60):

- injection molding of the preform (parison) in the melt temperature region
- blow molding of the preform in the thermoelastic temperature region
- removal of the hollow article and cooling with air

The process is normally carried out on special injection–blow molding machines in which these three steps are executed sequentially. The machines may consist of two or three stages, with two 180° or three 120° changes of direction for the mold core.

Injection blow molding offers various advantages:

- highly accurate dimensional reproducibility, with respect not only to the opening of the hollow article but also to its circumference and length (i.e., injection molding quality)

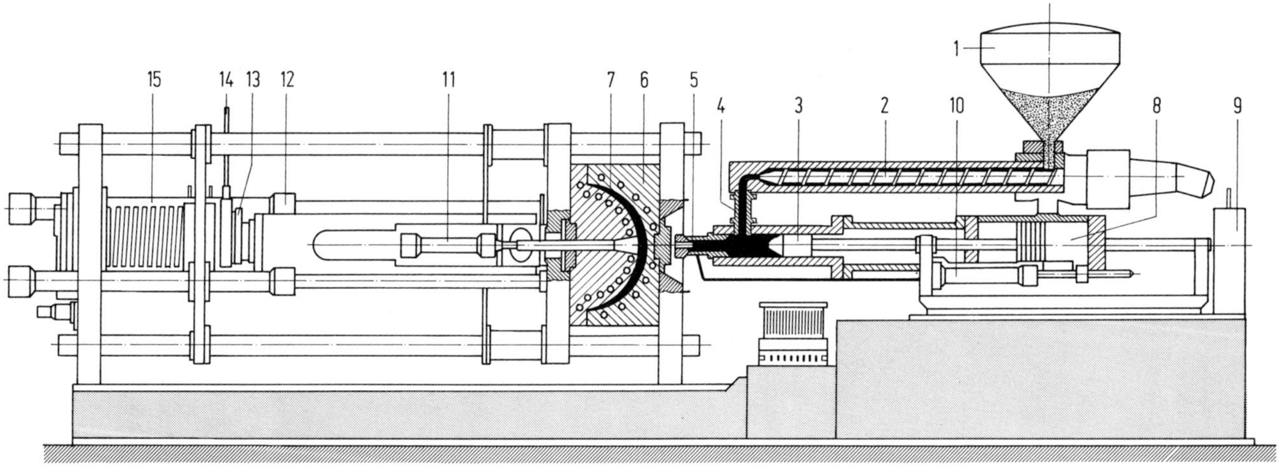

Figure 3.59. Low pressure injection-molding machine with melt accumulator (preplastification): 1, hopper; 2, screw; 3, transfer cylinder with plunger; 4, nonreturn valve; 5, hydraulic die; 6, force; 7, article; 8, injection cylinder; 9, plunger reservoir; 10, traveling barrel injection unit; 11, ejector; 12, traveling barrel closing plate; 13, membrane; 14, traveling barrel membrane; 15, injection molding cavity height adjustment.

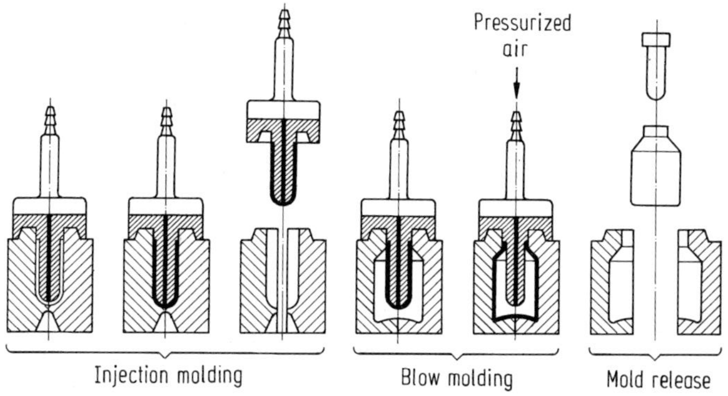

Figure 3.60. Processing technology of injection blow molding.

- uniform wall thickness distribution (saving of material); no distorted edges or thick areas (hence uniform stress distribution)
- minimum weight and volume tolerances
- good reproducibility (since no seams or other markings in the moldings)
- optimal transparency with amorphous plastics
- excellent surface quality
- no rejects due to necking or material that is pinched off at the top and the bottom, as seen with extrusion blow molding
- polycarbonate injection molding grades can be used (since high melt strength is not required)

However, certain stringent conditions must be met for successful injection blow molding:

- Compared to extrusion blow molding, the raw materials must exhibit improved flow and thermal stability.
- Containers with asymmetric openings, numerous openings, or molded-in handles cannot be made.

These restrictions are necessary given the relationship between the width and height of the hollow article. For example, these dimensions for a rectangular cross section should be in a ratio of 1:2.5 or less. Requirements of extrusion blow molding are also applicable for injection blow molding (e.g., minimize sharp edges and corners to obtain better final part properties). The regions between the neck and the shoulder, between the slope of the shoulder and the outer flat surface and between the outer flat surface and the bottom should all have the largest possible radii. Every area with a smaller radius increases the difficulty of forming the article in a uniform manner.

Final wall thickness is directly related to parison. Wall thickness decreases with increasing blow ratio. Normally, wall thicknesses between 0.4 and 0.8 mm for small articles made of polycarbonate can be achieved without difficulty. The minimum attainable wall thickness is limited by the flow properties of the parison (which in turn depend on the raw material) and by the blow ratio required to achieve the core-to-article diameter. This means that parisons having thinner walls can be molded only with a special length. These correlations are influenced by the mold and parison temperatures.

So far the upper limit of the blow ratio for polycarbonate, defined as the ratio of the blown article to that of the parison, is 3:1. Blowing the parison beyond the point at which the surface is set up causes local stress points in the blown article. Stresses may be minimized by decreasing the blow ratios or by blowing at higher temperatures. For good results (i.e., uniform wall thickness) with injection blow molding, one requires:

- exact processing of the core (to minimize wall thickness variations)
- uniform temperature distribution in parison and mold

Core temperature plays an important role in the blow molding of polycarbonate. The core heats up automatically as the heat is applied, usually spiral or fountain cooling must then be applied. The temperature control apparatus should provide for high throughput and should be capable of maintaining a constant temperature. Effective cooling channels—in additon to the blowing air channel—can be provided only for core diameters larger than 15 mm. If such conditioning is impossible, a uniform temperature distribution must be achieved by means of heat exchange. Sometimes it may be necessary to air-cool the tip of the core from the outside. The mold surface must be highly polished and preferably hard chrome plated. Figure 3.61 shows a heating/cooling channel arrangement that promotes a uniform initial parison temperature.

An attempt has been made to classify the blown article into discrete zones, assumed to be independent, which undergo thermoplastic deformation independently during blowing. Such a classification would permit comparison of the different temperatures called for in the longitudinal and cross-sectional directions. The temperatures given in

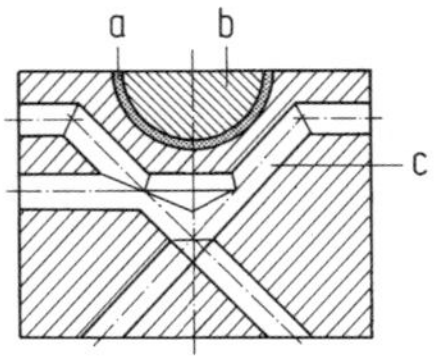

Figure 3.61. Example of a favorable tempering channel in an injection blow mold: a, preform; b, core; c, tempering channel.

Table 3.16 are applicable for the injection blow molding of a medium viscosity polycarbonate.

Parisons are normally provided with tapers from 2 to 2.5 mm in diameter. Closed nozzles or melt decompression must be used to minimize stringing. In special cases (e.g., large hollow containers such as housings for lighting fixtures), the parison is molded separately in another molding machine, then brought up to blow molding temperature in an intermediate heating stage and introduced into the blow molding machine.

Table 3.16. Typical Startup Data for an Injection Blow Molding Unit (blowing pressure 7–10 bar)

Parts	Temperature °C
Plastification cylinder	
Feeding zone	265–280
Compression zone	265–280
Metering zone	265–290
Die	265–280
Hot runner	255–275
Hot runner die	250–265
Injection molding cavity	
Core	130–150
Bottom area	40–60
Chase and shoulder area	100–140
Thread area	80–100
Blow molding cavity	60–80

3.8.1.3.4 Injection Molding Faults

Polycarbonate is used predominantly for value-added parts. Hence, defects encountered in the molding of polycarbonate parts should be addressed in a timely manner; it is obviously better to minimize their occurrence in the first place. The most commonly encountered defects, such as loss of molecular weight, surface disturbances, color changes, internal stress, and sink and burn marks, are discussed below.

Faults are possible during drying of the pellets (i.e., before the thermoplastic melt is processed). Injection molding at high temperatures without proper drying leads to a reduction in molecular weight (i.e., a decrease in viscosity). This can cause a loss of physical properties of the molded part and mositure-related streak marks (cosmetic defects) on the surface (Figure 3.62).

Molecular weight also may be reduced if processing is done at very high temperatures and also if the melt is subjected to very high shear rates. Such treatment causes burn marks and cloudlike cosmetic defects along with microcracks on the surface. Air trapped in a mold that is not properly vented can also give rise to burn marks and

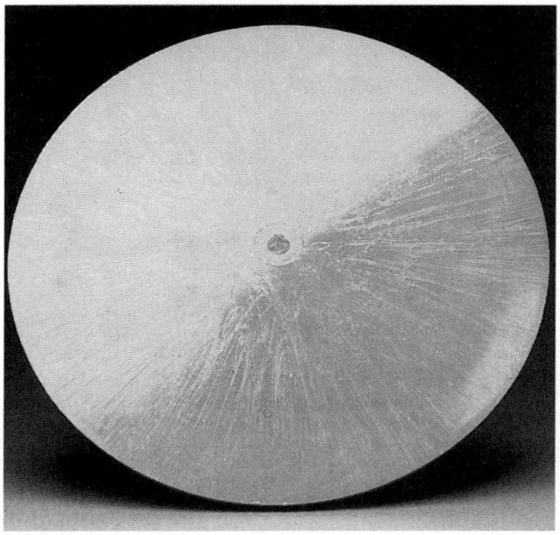

Figure 3.62. Moisture streaks.

degradation. Such defects may be minimized by decreasing the injection speed. Most of the time, a change of the flow pattern of the melt or proper venting is required.

"Dull spots" can result from superficial melt layers that are cooled at the mold cavity surface and pushed along by the continuing melt. These particles, unlike the melt itself, do not adhere to the mold wall well and end up at the surface of the molded part as a "dull spot." If the melt stagnates and cools for a short while in the flow channel, "record groove" marks are seen on the surface of the molded article (Figure 3.63).

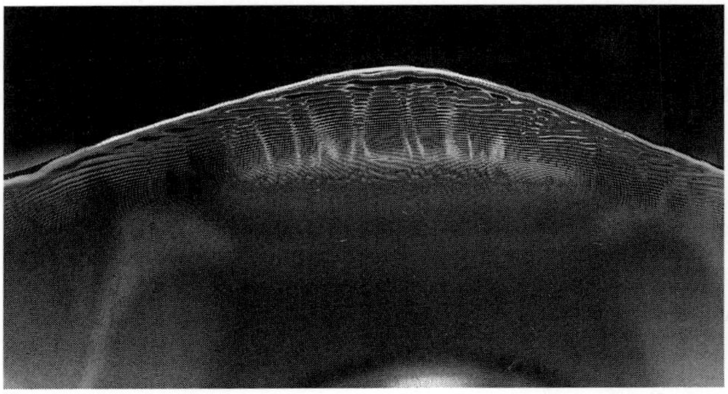

Figure 3.63. Record grooves.

These defects are due mainly to too slow an injection speed or too low a melt temperature. Dull spots may also be caused by disturbances of the melt flow in the mold (e.g., at changes in cross section or at sharp edges that lie perpendicular to the flow direction). If the injection speed is too high, if the gating system selected is inappropriate for the mold, or if the melt, upon entry into the mold, is not guided or "broken" by a wall or "post," "jetting" may develop (Figure 3.64).

Sink marks and voids may be the result of local thick sections or general greater wall thickness. To minimize such defects, melt flow to the mold must be maintained; that is, the holding pressure must be maintained until the melt has solidified in the mold. This

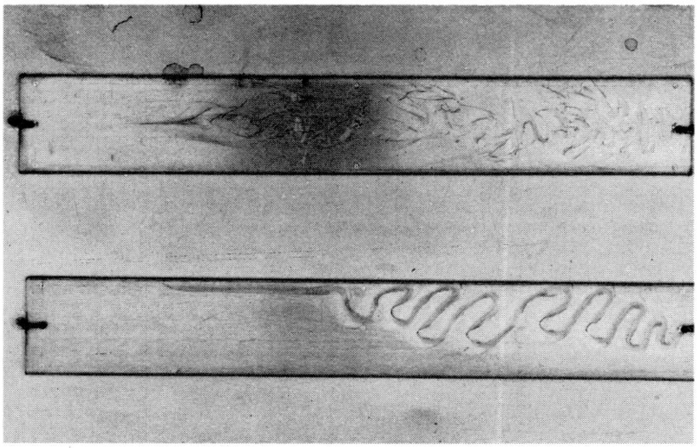

Figure 3.64. Jetting.

requires an adequate level of holding pressure. In particular, it must be ensured that the gate will not freeze off prematurely.

A low mold temperature is frequently the cause of surface disturbances and cosmetic defects. Nonuniform mold wall temperature, on the other hand, causes uneven shrinkage, hence nonuniform stresses and warpage in the molded part.

3.8.1.4 Part and Tool Design: Construction Details

3.8.1.4.1 Part and Mold Design

Determination of Thicknesses

In designing a new part, the layout of the thickness is important. In most applications, 2–3 mm is used. The lowest realized wall thicknesses are 1–1.4 mm (CDs) and the highest are 6–12 mm (filter bowls); both are used in actual parts. For reasons of economy, it is important to keep the wall thickness as low as possible, to reduce weight and costs.

To achieve the lowest possible thickness, it is necessary to consult a chart showing flow lengths and wall thicknesses (Figure 3.65), or melt flow data from the raw material supplier. The chart can also be used to define the number of necessary gates.

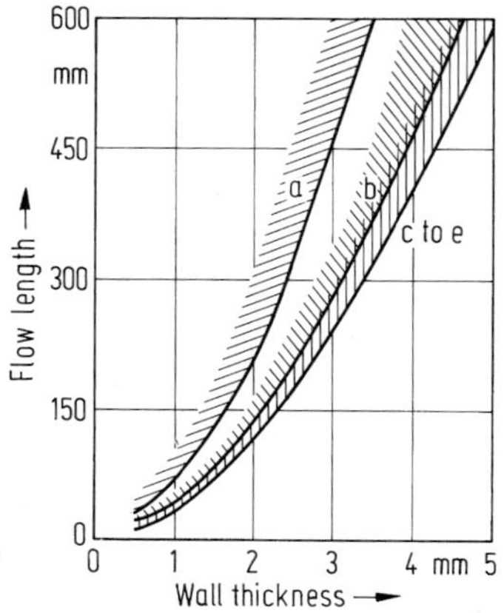

Figure 3.65. Plot of flow length versus wall thickness for polycarbonate of various molecular weights: a, easy flowing injection molding grade; b, standard PC; c, slightly higher molecular weight; d, PC with 10% glass fibers; e, PC with 30% short glass fibers.

The Relation of Ribs to Wall Thickness

For optical reasons, it is important to keep the surface of the parts free of sink marks (light refraction) and other imperfections. An important point here is the relation between ribs and wall thickness, which is 1.2–2.5 for polycarbonate. If thick ribs cannot be omitted, the risk of shrink marks can be reduced by using the following corrective actions:

- higher afterpressure during molding
- no ribs placed in flat surface areas
- surfaces designed to be rough or textured
- light colors
- maximum radius of 0.25 mm at the transition of rib to wall

In most cases, visible shrink marks can be reduced only with constructive corrective actions. Another method for omitting shrink marks is a slight foaming of the melt, which is feasible only for glass fiber reinforced products. The previously mentioned rib-to-wall thickness relation is also the ratio for sidewall connections and bosses, which are used for self-cutting or self-tapping screws. Bottom connections are defined by the type of screw and the necessary pullout forces. Normally the boss diameter is 2–2.5 times higher than the screw diameter *D*. For optical requirements, this relation can be reduced to $1.6 \times D$ by reduced strength. In case of higher optical requirements, the bottom area of such parts has to be ejected with side ejection devices.

Radii and Demolding Draft

Radii at ribs and transition areas should be big enough to prevent notch effects, but not too large to cause sinks by using the larger radius. Figure 3.66 shows the influence of different notches on the notched impact resistance according to ASTM D 256, method D. Clearly notch designs b and c have a big effect on impact resistance. Thus to ensure the required impact properties, radii of 0.25–0.5 mm should be specified in the design. To avoid shrink marks at rib transitions, the radius should not be too large.

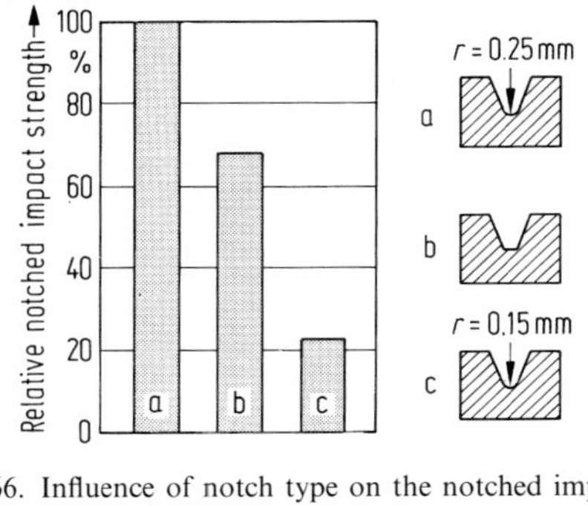

Figure 3.66. Influence of notch type on the notched impact properties of PC.

In general, demolding drafts of 1 °C should be used to ensure proper mold ejection. Difficulties can be expected for heavily ribbed parts and for bosses; such designs call for tooling that incorporates additional ejector pins and stripper sleeves. In most cases, an angle of 0.5° can be used for unreinforced polycarbonate (0.75° for reinforced). It is even possible to work without demolding drafts when the wall thickness (stiffness) and working surface are large enough to accommodate the placement of sufficient ejector pins (e.g., cylindrical parts that do not allow any draft to reach the inner diameter).

Surfaces

In addition to the high surface gloss given to molding parts by polishing the tool surface, a structured surface is required. This can improve resistance to macroscopic scratches and optical appearance, as well. For parts with textured surfaces that will be laid in the direction of demolding, a larger draft angle is necessary. For the case of electric erosion, the mold angles should be chosen from among those in Table 3.17. Scarred surfaces in general result in low gloss effects. For polycarbonate, a silk gloss cannot be avoided.

Table 3.17. Minimum Draft Angle at Spark Eroded Surfaces, Depending on the Surface Roughness at $S = 2$ mm (choose one step higher for glass-reinforced products)

Charm No.[1]	Average Roughness Value R_a (μm)	Average Surface Roughness R_z (μm)	Draft Angle PA Grade	PC Grade	ABS Grade
12	0.40	1.5	0.5	1.0	0.5
15	0.56	2.5	0.5	1.0	0.5
18	0.80	3.3	0.5	1.0	0.5
21	1.12	4.7	0.5	1.0	0.5
24	1.60	6.5	0.5	1.5	1.0
27	2.24	10.5	1.0	2.0	1.5
30	3.15	12.5	1.5	2.0	2.0
33	4.50	17.5	2.0	3.0	2.5
36	6.30	24.0	2.5	4.0	3.0
39	9.00	34.0	3.0	5.0	4.0
42	12.50	48.0	4.0	6.0	5.0
45	18.00	69.0	5.0	7.0	6.0

[1] Roughness classification per VDI 3400.

With embossed letters and designs, which require a deeper contour in the tool, the ratio of letter depth to width should be of 1:2. The depth should be as flat as possible, to prevent the formation of air bubbles behind the letter. Letters with sloping or curved contour edges are advantageous for good demolding; cores and slides should be polished in the direction of demolding. In general, chrome-plating of the tool is acceptable as corrosion protection, but this measure will not result in surface improvement.

Mechanical Connections

Snap connections are a cheap and easy method of joining parts. Plastics and polycarbonates in particular are suited well for this type of connection because of the high flexibility of these materials. Many basic calculations for arriving at snap shapes are available in the literature as well as from suppliers of raw materials. The basic maximum elongation for a single-use snap is 4% for unreinforced polycarbonate and 2% for reinforced polycarbonate; for more frequent snapping activity about 60% is recommended. In all snap connections only very low tensions should remain after the connection has been made. When using screws to join polycarbonate parts to steel, the higher thermal expansion of the plastic and a greater decrease in torque must be taken into consideration. Figure 3.67 shows the decrease of the pretension force (of torque) for polycarbonate. Self-cutting screws as joining elements provide relatively high tear strength. After several joining procedures, however, the threads are destroyed, hence no longer usable.

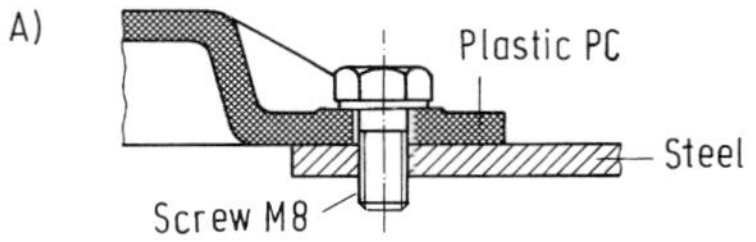

Figure 3.67. (A) Typical bolted joint (B) Decrease of pretension strength of screws, measured in polycarbonate (head of screws is shown in (A)).

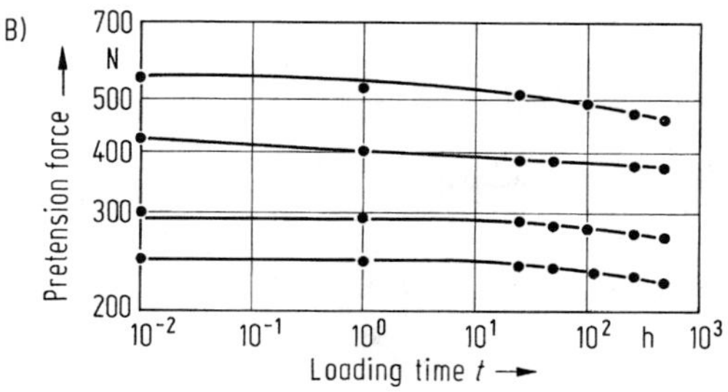

When the VDE requirement (unit can be assembled 10 times without thread damage) must be met, self-forming screws may be used. For polycarbonate and its blends, only plastic screws having a small pitch angle ($<45°$) and small core diameters ($<0.6 \times$ screw diameter D) should be used. Screws with pitch angles of 60° or more (e.g., the steel screw DIN 7970) are not recommended. The core hole diameter should be $1.8–2 \times D$ for polycarbonate. For glass-reinforced polycarbonate and its blends, the diameter is $0.9 \times D$. The thread length should exceed $2.3 \times D$.

Molding with Metal Inserts

Most plastics are not suited for the insert molding of metal inserts because their coefficient of thermal expansion is significantly higher than that of metal. This value for unreinforced polycarbonate is $60 \times 10^{-6}\ K^{-1}$ and therefore six times higher than that for steel, three times higher than that for brass, and twice as high as for aluminum and lead. This means that plastic shrinks more than metal. The insert molded part prevents the plastic from shrinking. There are stresses formed in the plastic. If, on the contrary, the part temperature increases during use, the metal, which heats up more rapidly, experiences thermal expansion before the plastic does.

The following guidelines are helpful in reducing stresses in polycarbonate parts that are molded with metal inserts.

1. If the metal inserts are heavier than 1.5 g or have a diameter of more than 6 mm, they must be preheated up to about 200 °C.
2. Such preheated metal parts can be made from any metal. Brass and aluminum blends are preferred because of their higher coefficients of thermal expansion.
3. Avoid sharp edges and knurls, to reduce the risk of notch formation in the plastic. Inserts that absorb some of the shrinkage stresses (slitted inserts) are particularly advantageous.
4. The surrounding plastic diameter must be sufficient: for cylindrical inserts, the polycarbonate thickness should be 0.5–0.7 d ($d =$ highest diameter of the insert).
5. The mold temperature should be at least 85 °C.
6. All metal parts should be free from oils, grease, and dirt, to avoid formation of stress-cracking; cutting oil in particular is very difficult to handle.

Even smooth, cylindrical inserts are fixed well in polycarbonate because of the considerable shrinkage stress during the cooldown period. This shrinkage stress is almost permanent. An additional securing device to avoid shifting is normally not necessary. Notch effects created by knurls can be avoided by slightly roughening the surface, by attaching a rib (small, plane surface), or by choosing hexagonal inserts.

Better conditions can be obtained using glass-reinforced polycarbonate with the coefficient of thermal expansion of $25–30 \times 10^{-6}\ K^{-1}$. Here the difference in comparison to brass is small. Nevertheless, the foregoing rules should be observed, in particular points 3 and 4. If possible, the thermoplast should be allowed to shrink freely, as explained in connection with the metal sleeve in Figure 3.68. In general, metal parts that are assembled later offer optionally better solutions.

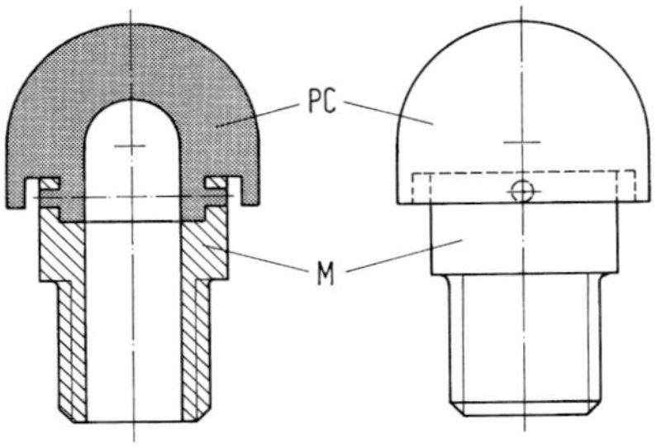

Figure 3.68. One way of remolding a metal shell (M) to avoid shrinkage stress.

Film Hinges

Classic film joint hinges used with polypropylene and polyamide are seldom usable with polycarbonate because of the lower flexural fatigue strength of this material. One or two bending processes are acceptable in well-designed polycarbonate joints however (e.g., when assembling parts in which a film hinge is used for just one process).

Thread Design

Normal thread designs, as well known with metals, can be used in only a few polycarbonate parts because of the formation of stress-cracking due to notch effects. For heavy-duty plastic parts made out of polycarbonate (e.g., filter belts, water faucets), curved threads (Figure 3.69) are preferred.

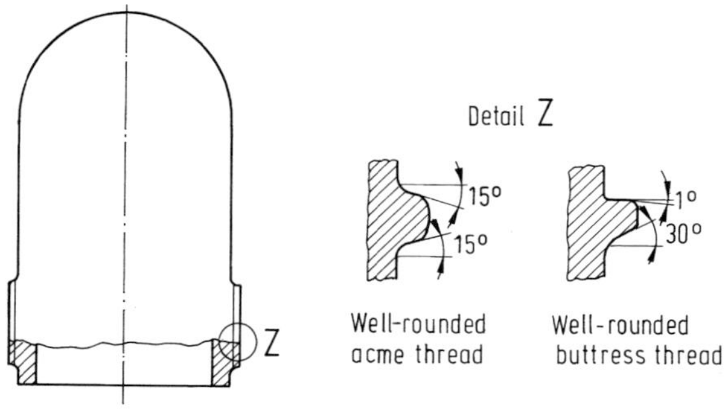

Figure 3.69. Thread formation for heavy-duty plastic parts.

Breakout Windows

Breakout windows—preformed, thin-wall areas, which can be removed easily after molding—are sometimes necessary. Here, part design, can present a problem because the breakout walls should be very thin but also thick enough for proper filling of the windows. For unreinforced and reinforced polycarbonate, wall thicknesses of 0.6–0.7 mm have proven to be satisfactory. The middle area of the window should be at least two-thirds of the basic wall thickness (Figure 3.70).

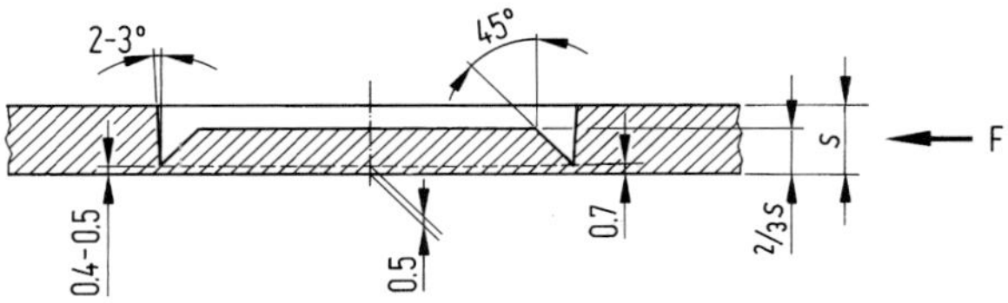

Figure 3.70. Formation of breakout windows from reinforced and unreinforced polycarbonate: *F*, feeding direction; *S*, overall wall thickness.

Shrinkage

Being an amorphous thermoplastic, polycarbonate shrinks little and uniformly. Unreinforced polycarbonate shrinks 0.6–0.8%, as displayed in Figure 3.71A. Injection-molded parts shrinking on a core show around 0.6% shrinkage; in use at elevated temperatures, the small difference in shrinkage (compared to unhindered shrinkage) is offset. The shrinkage of moldings from thick-walled parts may run higher than 1%. The boundaries of shrinkage are most heavily influenced by injection pressure, hold pressure, and hold time. The placement of the gate does not influence shrinkage, which is approximately the same in the flow direction and in the perpendicular direction. Postmold shrinkage at room temperature is relatively small (<0.1%).

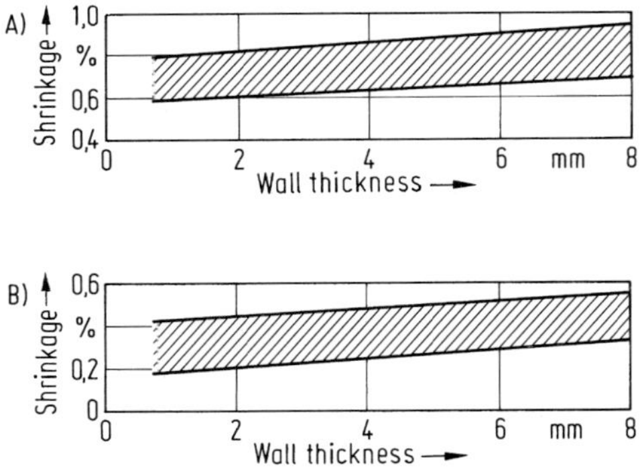

Figure 3.71. Shrinkage of standard polycarbonate (A) and of polycarbonate reinforced with 20–30 wt % glass fibers (B).

With glass-reinforced polycarbonates, as with other glass-reinforced thermoplastics, shrinkage is influenced by the direction of flow: it is lower in the flow direction than in perpendicular direction, and the overall average is about 0.35%, with the lower and upper boundaries of approximately 0.3–0.45% (Figure 3.71B). The further the measuring point of shrinkage is removed from the injection point, the greater the increase in shrinkage perpendicular to the flow direction. Thus in large, thick-walled parts, a value of greater than 0.5% may be detected. The hindered shrinkage is about 0.1% less.

With injection-molded parts that call for particularly high dimensional stability, a production mold normally is used and provision is made for a correction in the final mold. Should this not be feasible, a trial should be made on a special mold to ensure proper tolerance of important dimensions.

Tolerances

Basically manufacturing tolerances set forth in DIN 6901 can be used for polycarbonate. It is well known that for injection-molded parts under optimized conditions, the tolerance is ± 0.1; for 30% glass fiber reinforced polycarbonate, tolerance should be within 0.05%.

Warpage

Like amorphous thermoplastics, unreinforced polycarbonate shrinks in a relatively uniform manner and there is hardly any tendency to warp. Warpage, if it occurs, is usually due to varying wall thicknesses in the molded parts, and varying amounts of shrinkage throughout the part result. Only for glass fiber reinforced polycarbonates with a glass content over 10% can a small shrinkage difference be observed (see Section 3.7.5.5). The behavior needs to be taken into consideration when one is laying out the injection molds (Figure 3.72).

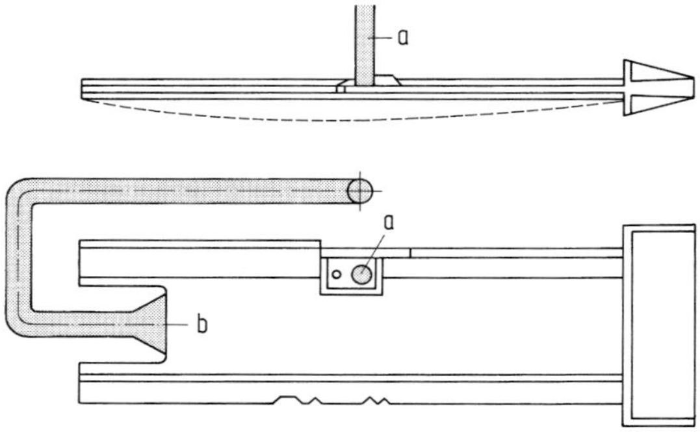

Figure 3.72. Avoiding distortion by varying the gating point: gating point at "a" causes distortion of the part by diagonal orientation of the glass fibers; gating point at "b" (main orientation in longitudinal direction) causes no distortion.

3.8.1.4.2 Mold Design

In the design and manufacture of molds, one can refer to the general conditions in construction of molds for thermoplastics. The details mentioned in the paragraphs that follow should be considered in the construction of molds to be used for polycarbonate, fiber glass reinforced polycarbonate, and polycarbonate blends.

Mold Materials

In principle, the same kinds of tool steel commonly used in the construction of thermoplastic molds can be used for polycarbonate (Table 3.18). The recommendations of steel producers should be ascertained.

Polished mold cavity surfaces having sufficient hardness do not show abrasive wear from unreinforced polycarbonate. With surface hardnesses greater than 55 HRC (Rockwell C) more than a million injection cycles can be expected; generally a hardness of 52–55 HRC is sufficient. Textured surfaces are by nature more susceptible to wear. Velvet-textured surfaces especially show wear at higher pigmentation and glass fiber levels, as indicated by a loss of luster in the parts. For textured molds, therefore, a surface hardness of 55–58 HRC is recommended.

Table 3.18. Steel Choices for Injection Molding Cavities

Abbrevation	DIN-Material Number	Material Description
C45 W	1.1730	Unalloyed tool steel for tool stands
21 Mn Cr 5	1.2162	Hardened steel for tools produced by machining
X 19 Ni Cr Mo 4	1.2764	Hardened steel for tools produced by machining
X 6 Cr Mo 4	1.2341	Hardened steel for tools for swaging (cold sinking)
40 Cr Mn Mo 7	1.2311	Heat-treated steel for etching engraving
40 Cr Mn Mo S 86	1.2312	Heat-treated steel with improved machining properties
X 40 Cr Mo V S 51	1.2347	Heat-treated steel with improved machining properties and improved service durability
X 45 Ni Cr Mo 4	1.2767	Through-curing steel with high toughness
X 38 Cr Mo V 51	1.2343	Through-curing steel with good tempering properties to nitration
X 155 Cr V Mo 121	1.2379	Lebedúritic abrasion-resistant tool steel SAE/AISI D2
X 210 Cr 12	1.2080	Lebedúritic abrasion-resistant tool steel SAE/AISI D3
X 42 Cr 13	1.2083	Corrosion-proof tool steel for normal corrosion strain applications
X 36 Cr Mo 17	1.2316	Corrosion-proof steel for high corrosion strain applications
X 3 Ni Co Mo Ti 1895	1.2709	Curable special steel, distortion-free

Often the existence, due to insufficient cooling, of extreme high temperatures at small mold cores can create small pits at mold surfaces, resulting in the sticking of the parts to the mold core. To prevent this, materials that are better heat conductors (e.g., beryllium, copper, heat-conducting steel) or coated surfaces should be used. For the above-mentioned steel surface coatings, simple coatings from copper or relatively similar coatings have been successful. Titanium nitride (TiN) seems to impart additional improvement for bonding, demolding, hardness, and surface temperature.

For items molded of polycarbonate with halogen flame retardants, corrosion-resistant steels are recommended for regions that come in contact with the molded part.

Part Gating and Gate Location

Gate location is generally determined by the flow of the material, the surface requirements, of the orientation (in the case of glass-reinforced products). Most gate designs (direct gate, pinpoint gate sprue, ring gate, tab gate, tunnel gate, submarine gate, etc.) can be used for the injection molding of polycarbonate. Large designs such as sprues and film gates seem to work especially well with polycarbonate. Another reason for choosing relatively large gate diameters (e.g., in comparison to pinpoint and submarine gate designs) is that filling problems can occur with the smaller diameters, possibly resulting in surface imperfections. Figure 3.73 gives some general guidelines.

The injected material entering the mold cavity must meet a reflecting wall to avoid jetting. Therefore injection above the comprehensive submarine gate is efficient.

For higher optical requirements a visual gate point is generally not acceptable. Instead, a direct sprue to the back side of the part on an auxiliary core is often recommended. Unconventional injection types such as the side or bent gates and curved tunnel gates are used for glass fiber reinforced and polycarbonate blends.

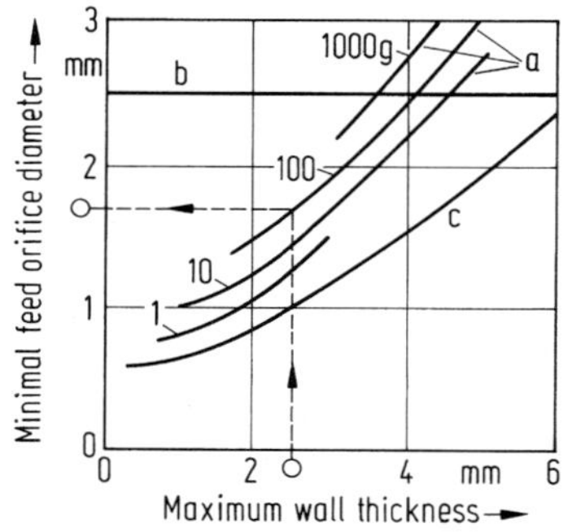

Figure 3.73. Guidelines for the choice of feed orifice diameter: a, molding part weight; b, upper limit; c, lower limit.

Hot Runners

The application of hot runner elements in injection molds is possible for polycarbonate. Hot runner nozzles are recommended, and hot runner blocks should have rheologically uncomplicated runners—that is, no torpedoes, internally heated runner bridges, or dead corners should be allowed to inhibit the flow of the molten material. Furthermore, all elements must have precision thermal control. Thermocouple elements are necessary to control the maximum temperature close to the hot runner, and regular monitoring of a closed-loop system for heat control is essential. Large contact points to the tool parts should be avoided because of the associated heat loss. The use of heat-insulated plates is recommended.

High alloy tempered steel is the preferred material for the construction of hot runner elements, which must serve at high temperatures without losing dimensional tolerances. For bromine-based flame retardants the steel should have high chrome content, as well, to impart corrosion protection. Copper alloys are not recommended for this application, since in practice decomposition of the polycarbonate may result.

Venting

For consistent filling of the mold cavity, there must be provision for the immediate venting of the mold cavity containing the molten polymer material. Typically, vents are placed on the surface at the mold parting line.

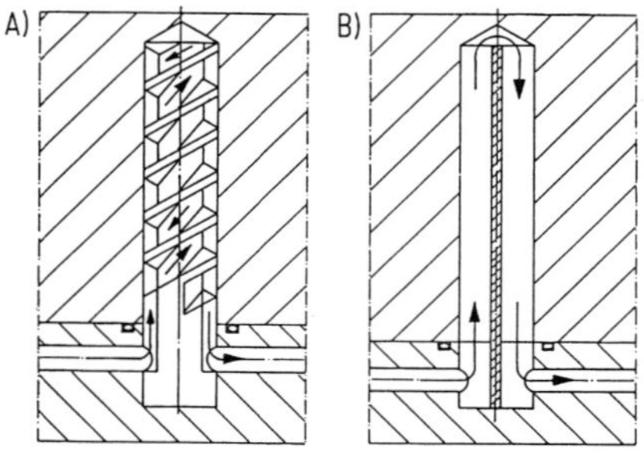

Figure 3.74. Temperature regulation possibilities for deep-seated tool areas: A, finger cooler with inserted spindle; B, finger cooler with baffle plate.

Mold Cooling, Temperature Regulation

Since mold temperatures of 80 °C or more are used in producing polycarbonate items, it is important that all mold areas be uniformly regulated. If a tool is not cooled properly by normal control devices, so-called finger coolers bring good results (Figure 3.74).

At relatively high molding temperatures (approx. 300 °C), some tool steels may begin to pit, causing parts to stick. In such cases it is necessary to use heat conducting materials (see Section 3.8.1.4.2).

Demolding/Release

The incompletely cooled thermoplastic adheres strongly to the mold core, usually requiring high ejection forces, especially for very hot parts. Ejector pins should have large surface areas; if possible, stripper plates should be considered.

3.8.2 Extrusion Procedure for Polycarbonate

F. J. Brinkschröder

3.8.2.1 Introduction

Modern extrusion processes and sophisticated applications require special polycarbonate types, to accommodate not only the desired properties of the molded part but also the type of process chosen.

The extrusion process calls for a polycarbonate of high purity, low "fish-eye" content, and constant viscosity. The optimum viscosity is dependent on the processing and application. Branched-type products are well suited. Figure 3.75 plots melt viscosity

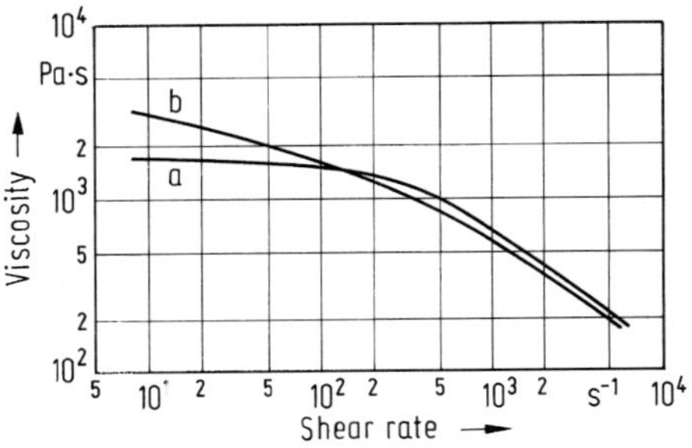

Figure 3.75. Melt viscosity of polycarbonate extrusion grade as a function of shear (melt temperature 280 °C) for linear polycarbonate (a) and branched (non-Newtonian viscosity) polycarbonate (b).

with respect to shear rate for linear and branched polycarbonate types. Whereas linear polycarbonate has a constant melt viscosity over a large shear rate range, the melt viscosity of the branched material depends heavily on shear rate. The latter polymer has the following advantages during extrusion:

- *High melt strength at low shear rates*: this property has advantages for the extrusion behavior of the melt between the end of the slot die and calibrator or calender rolls as well as in the slot die itself.

- *Low viscosity in areas of higher shear speed in the extrusion screw*: this property allows a much broader processing range and improves the feeding and homogenizing process in the extruder.
- *In the case of coextrusion:* there is a better adjustment of the polymer viscosity to the viscosity of the second product.

3.8.2.2 Preparation of the Material

Polycarbonates can be processed in normal extrusion devices and in vented-barrel-type extrusion equipment. The pellets must be dried before molding. A very low amount of moisture can cause the following problems:
- brown, dripping residues at the end of the nozzle; visible blisters or microscopic zipperlike blister formations
- formation of stripes
- strong adherence of the melt in the channel containing the adapters, the screen changing device, and the die, which can cause discoloration of the extruded material if buring is allowed to occur
- nonuniform feeding in the extruder

Permissible pellet moisture content is limited to amounts not exceeding 0.01%. Therefore the requirements for the extrusion of polycarbonate are higher than for injection molding. A sufficiently low moisture content can be reached through drying the pellets at 120 °C in an air-circulating oven for 4–12 hours or in an accelerated drying oven for 2–5 hours. Temperatures below 120 °C are not sufficient for drying. Dry storage of the nonairtight bags in which pellets are supplied facilitates the drying process. Cold-stored pellets should be placed in the working environment and allowed to warm to room temperature before opening to avoid condensation of humidity on the pellet surface. Various methods for testing moisture content were described in Section 3.8.1.2.

3.8.2.3 Scrap

Trimming, start-up, and purging generate significant amounts of polycarbonate scrap, and reprocessing is desirable. It must be realized, however, that the previous application and the repelletizing process itself can have negative thermal degradation effects and can introduce other impurities that may influence the optical quality of the recycled material. A mole-mass change also changes the flow properties, with resulting impairment of the blending of the recycled material, the feeding properties of the extruder, and finally the extrudate quality.

3.8.2.4 Extruder Requirement

3.8.2.4.1 Feeding Zone

Polycarbonate can also be processed on extruders featuring feed grooves and tempered feeding zones (Figure 3.76). The advantages of this extruder concept—namely, consistency of the feeding properties at viscosity deviations and high throughput at low setting temperatures—can be achieved for polycarbonate only if the following provisions have been considered.
- The barrel must be provided with half-circle feeding grooves ($r = 1$ mm) over a length of 3.5D. Lunate, rectangular, or sawtooth feed grooves are less suited because they can cause the pellets to be fed at too high an intensity.

- The grooved bush must be subjected to intense heating from outside (with oil or ceramic heater bands).
- The hopper must be insulated against the high cylinder temperatures.

Only the combined action of the grooves together with the outside heating process results in sufficiently stable feeding and high, consistent throughput. It must be borne in mind that in extruders with an effective feeding zone, the melt is exposed to higher shear strength than in conventional equipments. Particularly in the case of very special polycarbonate types whose composition makes them temperature sensitive, shear can cause property changes.

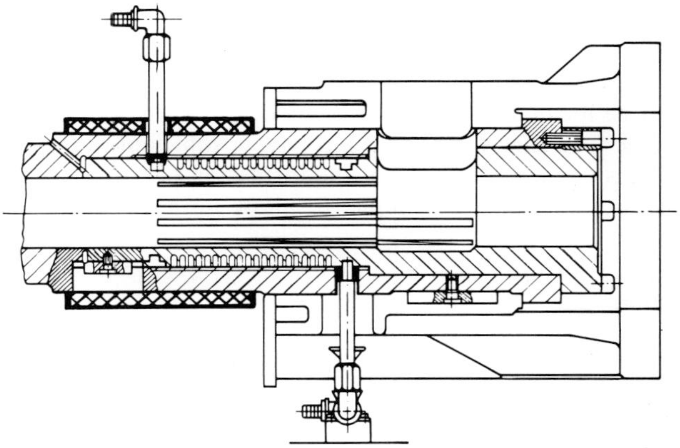

Figure 3.76. Barrel with grooved bush for efficient material feed.

3.8.2.4.2 Temperature Control

Air- or water-cooled equipment can be used, and an exact temperature profile of all temperature control equipment is required for good product quality. To ensure that such a temperature profile is obtained, the extrusion cylinder must be fitted with four to six independently controllable heating zones.

3.8.2.4.3 Driving Power

As extruder drive, a direct current, shunt-wound electric motor with thyristor feed is recommended. This type of drive fulfills all processing requirements, such as consistent rotational speed, infinitely variable control for varying speed in a wide range (maximum 1:20), and good adjustment to the power requirements of the screw. An automatic power-limiting device can be used to avoid excessive torsion of the screw due to high torque.

3.8.2.4.4 Power Requirement

The power that is needed depends on the material viscosity and the processing temperature that will be used. The specific power requirement for polycarbonate types is 0.2–0.35 kW · h/kg. The power requirement for the extruder depends on the screw diameter as follows:

Screw diameter	mm	30	45	60	90	120
Driving power	kW	7	10–20	30–40	60–70	120–130

The use of an efficient feeding zone should make it possible to take advantage of the driving power indicated. The standard value for a single-screw extruder used in polycarbonate processing is between 5 and about 110 rpm. Profile extrusion equipment require a maximum of 60 rpm.

3.8.2.4.5 Materials for Barrels and Screws

A bimetallic centrifugally lined material is used for the cylinder; either hardened high chromium steels or chrome-plated steels are recommended for the screws. The materials used in the machine components for injection molding described in Section 3.8.1.3.1 are also used for extrusion.

3.8.2.4.6 Venting Hole on Vented Extrusion Cylinder

The venting hole should be located to one side on the cylinder; for practical reasons, it should be easily accessible. The suction tube should have a sufficient diameter and must be heatable, since heating is necessary to prevent the formation of condensate in the vent tube. Assembly and disassembly must be simple, to allow purging to be accomplished quickly.

3.8.2.5 Screw Selection

The geometry of the screw has an important influence on a molding material's feeding and melting properties. It influences evenness of melt flow, output, melt temperature, and material homogeneity. For the processing of polycarbonate, both normal three-zone screws and venting screws are used successfully.

3.8.2.5.1 Three-Zone Screws

In the case of smaller screw diameters and frequently changing production and working conditions, three-zone screws are recommended. Figure 3.77 shows the zone diagram of

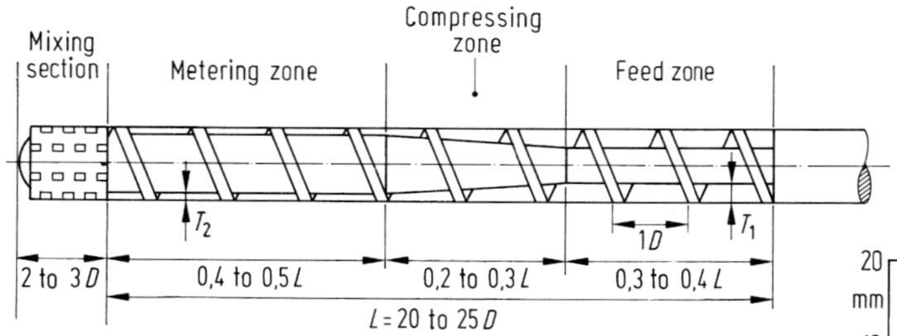

Figure 3.77. Zone distribution of three-zone screws for processing polycarbonate.

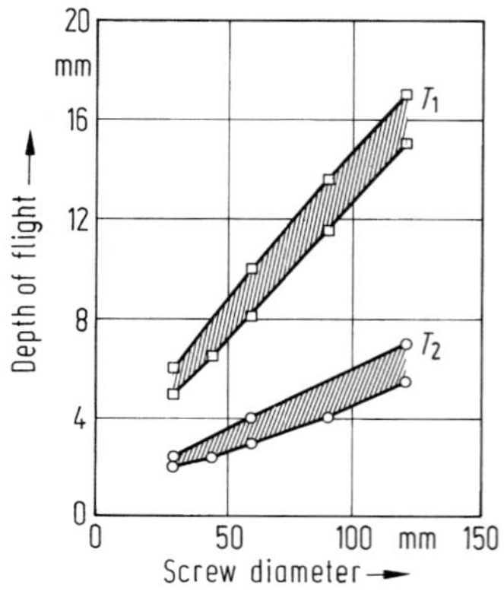

Figure 3.78. Screw flight depths for the processing of polycarbonate: depth of flight feed zone (T_1) and depth of flight metering zone (T_2).

such a screw. Mixing elements at the end of the metering zone enable considerable improvement in homogeneity, resulting in an even flow in the die, which is an especially favorable feature during the production of profiles such as twin-wall sheet. Mixing parts are available in different geometries. Screw flight ratios of 1:2.2 to 1:2.8 are recommended. Figure 3.78 shows the screw flight depths of the feeding and metering zones of a three-zone screw.

3.8.2.5.2 Vented Screws

Vented (degassing) screws are recommended for extruders having a screw diameter that exceeds 60 mm. They enable continuous production at constant processing parameters, unaffected by variations in air humidity.

The production of solid sheet and twin-wall sheet entails extreme conditions for constant processing parameters over longer production periods. Drying of resin in separate dryers is the most common cause of inconsistency, since the high capacity required is not possible under these conditions. Therefore, vented extruders are recommended to extrude solid and twin-wall sheets. Separate predrying of the resin is not necessary.

The minimum length for vented screws is 30*D*, although 33*D* is better. The vent hole is located 20*D* away from the feed opening.

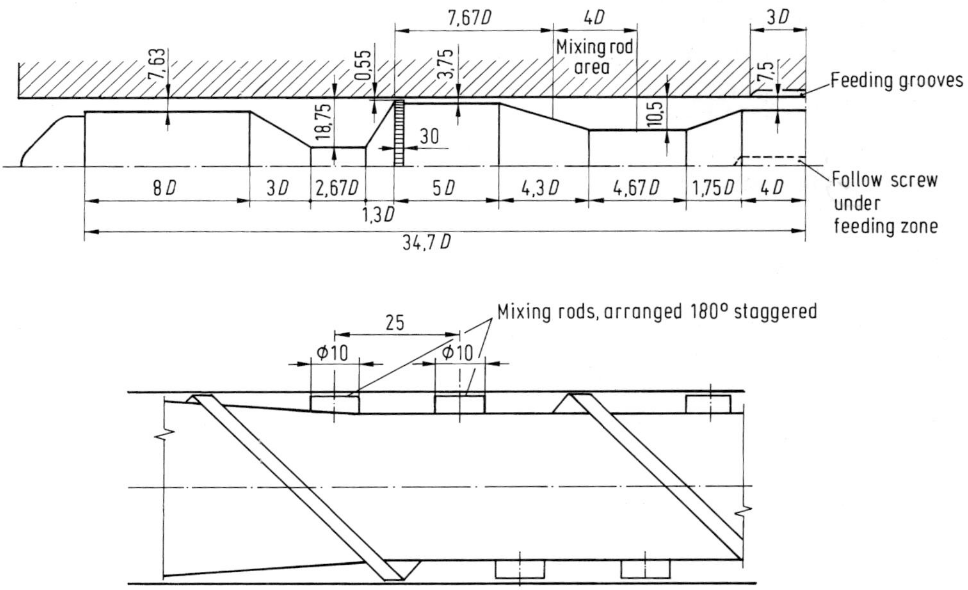

Figure 3.79. Degassing extruder concept with outer screw diameter of 75 mm (lead of screw flight, unit diameter).

Figure 3.79 shows a degassing (vented) extruder design used successfully with screw diameters of 75 mm and larger. The screw produces a stable throughput of 1.6–1.8 kg/h per revolution. This screw can process resins of both low and high viscosity.

3.8.2.6 Screen Changer, Melt Pump, Static Mixer

If high optical quality of the extruded product is required, the use of a continuously operating screen changer with the following screens is recommended:

- support screen with 200–300 mesh/cm^2
- fine screen package with three screens at 2500 mesh/cm^2
- support screen with 200–300 mesh/cm^2

Short, chromium-plated melt channels without sharp corners are recommended. Melt pumps will improve throughput uniformity and facilitate the combination of regrind with virgin resin. It generally is practical to integrate a melt pump into an extrusion line for both solid and twin-wall sheet production units.

3.8.2.7 Cleaning of Extruder and Tools

Processing of polycarbonate always requires a clean work environment. If a fast, safe, and inexpensive means of cleaning of the extruders and tools is not found, material and production losses may exceed cleaning costs.

The following procedure for cleaning screws and parts covered with polycarbonate melt has been proven useful:

- Purge extruder and die with high density polyethylene at the polycarbonate processing temperature.
- Disassemble screen changer, static mixer, adapter system, melt pump, and die, and clean these units by hand. If manual cleaning is not possible, the parts can be cleaned in a heated oil bath or a fluidized aluminum oxide bath.
- Purge the extruder screw with chopped polymethyl methacrylate, which may be mixed with chrome polish or CORATEX (a cleaning compound from Feldmühle AG, Wesseling, Germany). Make sure to vent the fumes properly.
- Remove the extruder screw and clean it thoroughly with a rag.

3.8.2.8 Temperature Profiles

When polycarbonates are processed on extrusion lines, it is preferable to have the temperature profile diminish from the hopper toward the die (Figure 3.80). Such a temperature profile mainly reduces the driving torque (energy) required for the screw. The high temperatures in the feed zone are not likely to cause feed problems with polycarbonate. The temperature of the melt exiting the die depends on the item to be produced. For pipes, tubes, profiles, hollow parts, and so on, the melt temperature should be between 250 and 260 °C. One way to increase the melt viscosity for complex designed profiles is to decrease the melt temperature. The melt temperature desired should be set at the end of the metering zone and in the adapter. If a high melt viscosity is not required or if the die openings are narrow, of course the melt temperature can be increased to 280–300 °C. The result is a low flow resistance in the die, and melt fracture is avoided.

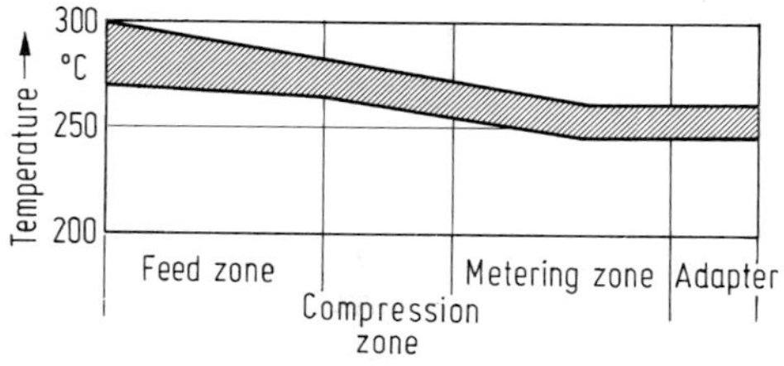

Figure 3.80. Temperature profile of an extruder for processing polycarbonate.

The surface of the emerging melt should be smooth and shiny. A rough surface almost always indicates too low a die temperature close to the die exit; only rarely is it a sign of too low a melt temperature.

When polycarbonate is processed in vented extruders, the temperature of the first metering zone (M_1) is frequently set higher than that of the M_2 (see Figure 3.81), to ensure efficient venting of the material.

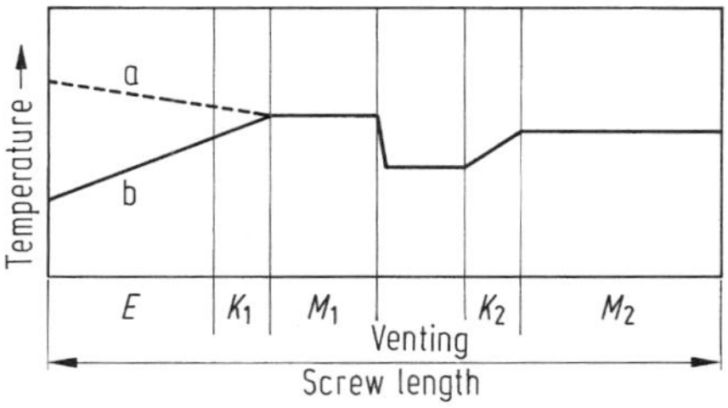

Figure 3.81. Temperature profiles in a vented extruder for polycarbonate (a) and other resins (b): E, feeding zone: K_1, K_2, compression zones; M_1, M_2: metering zones.

3.8.2.9 Slot Dies for Sheet and Film Extrusion

Slot dies with a manifold designed in the form of a coathanger are used for sheet and film extrusion. Compatible shear rates and viscosities are used in the manifold and in the throttle area to assure uniform flow and relaxation of the melt throughout the entire die width. In practice it is quite common for one die to be used for processing various formulations and materials. For this reason such a single die is provided with a choke bar, which enables the pressure to be equalized to a certain extent, ensuring that constant flow rates can be maintained.

The die lips can be fine-tuned to meet the narrowest tolerances. The most precise setting is obtained with flexible lips. Flexible lips with adjustment screws of the push–pull type are preferred.

A tapered die, which can be positioned quite close to the roll nip of the calender stack, must be used in the processing of polycarbonate. With this design, the distance between the die orifice and the entrance to the roller nip can be kept as small as possible, to prevent the melt from sagging and accumulating on the roll. Figure 3.82 shows a cross section through a tapered slot die with flexible lips.

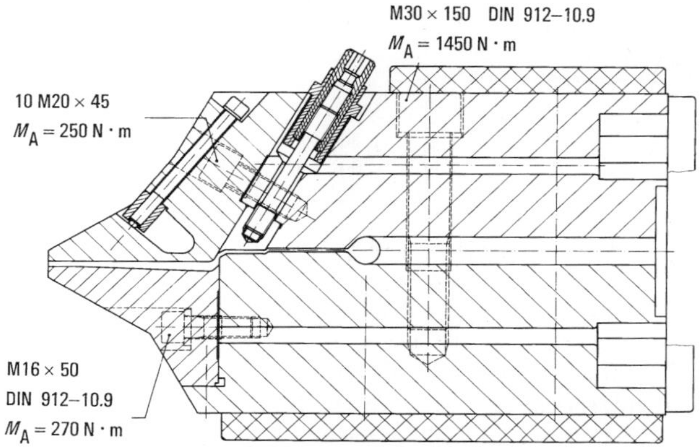

Figure 3.82. Diagram of a slot die for processing polycarbonate.

Apart from effectively spreading the melt, the slot die must guarantee an even distribution of temperature over the width of the unit. Dies with a high volume of metal provide the best conditions for stable temperature and unimpeded production.

The foregoing considerations apply as well to slot dies for flat film production. Dies without a restrictor bar are used in film extrusion. The back pressure from the narrow die gap makes it possible to adjust the film thickness with the flexible lips alone.

Accumulations of material, which form at the edge of the die lips and then proceed to grow back inside the die, causing the film to develop streaks, continue to present a problem.

By using a narrow die gap with a high flow rate and slightly tapered lips (3–5 °, only over the last 5 mm), it is possible to induce a self-cleaning effect that will retard the forming of accumulations and increase trouble-free production time. The lip land length should be between 30 and 50 mm for tapered models. The lips usually have a radius of about 0.5 mm at the front edge.

Design Information

- The cross section of the zone in which the melt enters the die should be rectangular. This is a very important premise for coextrusion.
- The lip heater should be used only to preheat the die. Continuous temperature control applied to the lips could cause thickness variation of the extruded part.
- The assembly bolts should be tightened only with a torque wrench.
- Silicone rubber between the die body and the restrictor bar will make adjustments of the restrictor bar easier. Filled lubricants, which could seep out of the gaps and contaminate the melt, should not be used.
- The steel for the die should have a high chrome content and should take a good polish.
- The distribution channels need not be chrome plated, but the die lips must have a highly polished surface. With newer polishing methods a surface finish of 0.1 μm or less peak-to-valley height is possible.

3.8.2.10 Manufacturing of Solid Sheet

Setup guidelines for the production of polycarbonate sheeting by extrusion are found in Table 3.19.

Table 3.19. Guidelines for Sheeting Liner Layout

Screw Diameter (mm)	Output (kg/h)	Die Width (mm)	Mass Temperature (°C)
200	750–900	3500	280–310
150	500–700	2500	
120	330–400	2000	
90	200–250	1500	

3.8.2.10.1 Calender

Positions of calender rolls used for sheet extrusion are displayed in Figure 3.83. Calender designs e and f are used in the production of high quality optical sheeting. The adjustable position of the third roll makes a major contribution to sheet quality. Calender designs a and b of Figure 3.83 lead to lower optical quality sheeting. A complete film and sheet extrusion line (film thickness > 100 mm) is shown in Figure 3.84.

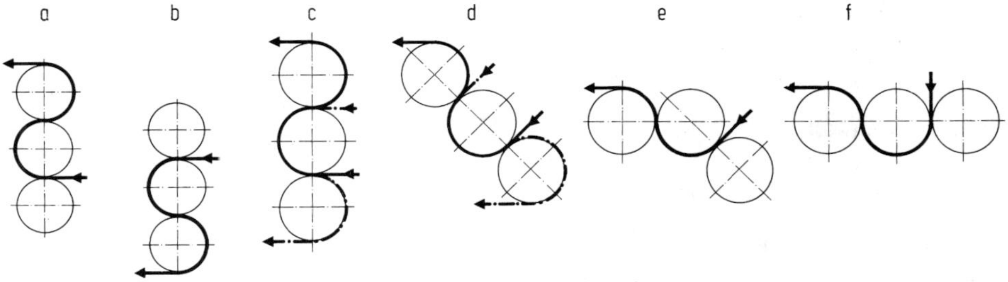

Figure 3.83. Common roll arrangements for calender stacks (arrows indicate melt input and output): a, vertical roll arrangement with horizontal melt input at the bottom cylinder; b, vertical roll arrangement with horizontal melt input at the middle cylinder; c, vertical roll arrangement with melt input at either the middle or the bottom cylinder; d, diagonal roll arrangment with variable melt input at the middle and bottom cylinders; e, angular roll arrangement with melt input 45°; f, horizontal roll arrangement with vertical melt input.

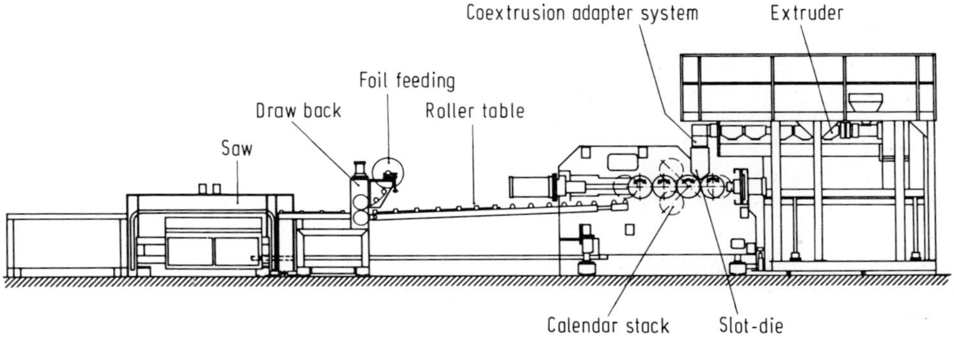

Figure 3.84. Schematic presentation of an extrusion plant (Breyer system).

The calender has four horizontally positioned rolls. When producing optical quality sheet, the melt is fed through the first nip. Special attention is needed with respect to the point at which the melt first touches the roll and at the rolling bank. Sheet quality depends significantly on the way the rolling bank is formed. It is first necessary to eliminate defects such as gels, bubbles, and die lines, which show up directly out of the die. Sheet quality also depends on melt temperature, melt viscosity, sheet thickness, roll temperature, ratio of roll to die opening, distance of roll to die lips, and other variables. These correlations must be determined empirically. Indispensable requirements for cost-effective production of glass clear sheet include a safe extrusion line and well-trained operators.

Special attention should be given to the following aspects of the sheet production by extrusion:

- movement of the sheet through the calender
- residence time of the sheet in the calender
- quality of the roll surface
- synchronization of the roll drives
- tight tolerances in the roll bearings
- roll temperature controls (e.g., temperature distribution during the extrusion process)
- adjustability of residence time in the calender for throughput and sheet thickness changes
- roll position adjustments
- safety of operation
- electrical controls

3.8.2.10.2 Roll Temperature Settings

Two basic profiles can be used to empirically determine the suitable roll temperatures (see Table 3.20). A roll temperature profile in accordance with situation a, which will deliver the optically best and flattest sheet surfaces, can be achieved only if the residence time of the sheet through the calender can assure a flat sheet as well as an average sheet temperature of no less than 145 °C at the departing point of the last calender roll. The temperature must be constant across the entire cross section of the sheet.

The roll temperature profile of situation b of Table 3.20 is used only with a vertical roll stack. There is a slight reduction in optical sheet quality due to sticking of the sheet against the last calender roll.

Table 3.20. Roll Temperature Profiles for Determining the Suitability of Optical Sheeting

Situation a	Cooling Effect
Roll 1: 110–120 °C	Line contact between the melt and roll 1
Roll 2: 125–135 °C	First residence time on roll
Roll 3: maximum 145 °C	Second residence time on roll
Roll 4: maximum 145 °C	Third residence time on roll
Situation b	Cooling Effect
Roll 1: 110–120 °C	Line contact between the melt and roll 1
Roll 2: 130–140 °C	First residence time on roll
Roll 3: 140–150 °C	Second residence time on roll
Roll 4: 170–185 °C	Third residence time on roll

Roll temperatures above 140 °C are possible only when oil is used as the heat transfer medium. Thickness tolerances in production are usually between ±3 and ±5%. To minimize dust, it is advisable to separate the dust-free extruder and the calender area from the cutting and packaging area. Dust-free plasticizing and calendering are essential for good sheet quality.

3.8.2.11 Polycarbonate Film Production

A chill-roll takeoff is used to produce film less than 100 μm thick. Figure 3.85 shows a chill-roll takeoff using a tangential melt feed onto the chill roll. A thickness measuring device, a film preconditioning unit, an edge-trimming station, and a suction device for the trimmed edge are among the components of a chill-roll production line. Horizontal and vertical adjustments are required for optional positioning of the point at which the melt hits the chill roll. An additional temperature-controlled roll or cooling roll is an advantage.

The film can be cooled gradually as it passes from the chill rolls to the last temperature-controlled or cooling rolls. This improves the flatness of the film, and defects are reduced. It is also possible to equip the chill-roll takeoff with a third (or smoothing) roll, for use with film thicknesses exceeding 100 mm.

Figure 3.86 illustrates a compact, flat film production. If relatively thick films (100–800 μm) are to be produced, the use of a heavy-duty takeoff unit is recommended. A special surface finish can be achieved by using a pressure roll with the desired surface structure.

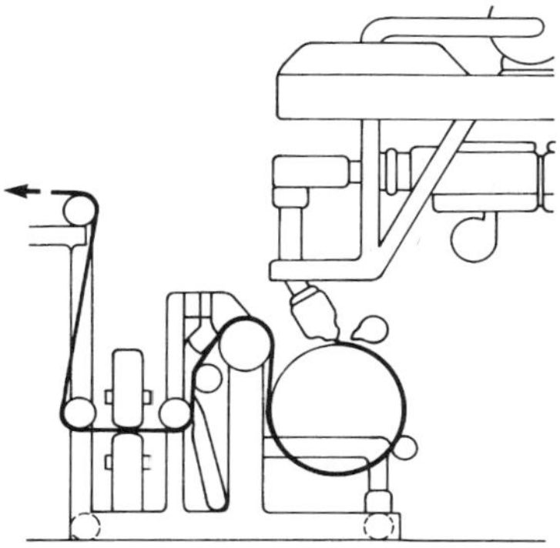

Figure 3.85. Chill-roll drawback with tangential feeding.

When producing single-layer as well as coextruded films (< 100 μm), the distance between slot die and the chill roll should not be much greater than 10 mm. This is possible only if the extrudate is directed onto the chill roll tangentially at an angle of less than 30–45° to the horizontal. Care must be taken to ensure that the emerging material is not drawn off over the edges of the die lip.

Air jets are used to give stability to the lateral edges, and the edges of the film are said to be "nailed down" onto the chill roll. Sometimes the film, is electrostatically charged to ensure an even contact with the chill roll.

The cooling roll temperature must be determined by trial for each individual size of machine.

The diameter of the first roll, henceforth referred to as the chill roll, is of critical importance because it may account for 90–95% of the total cooling capacity. The chill roll must be large enough to ensure that any irregularities in the melt temperature (the emerging melt is frequently hotter in the middle than at the edges) can be equalized on the cooling roller. Thus the diameter of the roll must be as large as possible, because a relatively lengthy dwell time on this roll produces an even temperature in the film as it leaves the roll.

To assure an almost even temperature across the whole width of the film at the die exit, the melt temperature in the last metering zone of the extruder should be equal to the temperature of the material which has just left the extruder.

A chill roll for line speeds of greater than 100 m/min should have a diameter of at least 600 mm. The diameter of the second roll is usually between 250 and 450 mm.

Relative Speeds of the Rolls

The basic speed of the machine is determined by the temperature-controlled roll. The chill roll, on the other hand, runs more slowly, to ensure that the film is lying smoothly.

A peripheral presetting is made if a film that shrinks during cooling is to be drawn from the chill roll in the absence of tension.

All the deflector and contraction rolls have direct drive and must be geared to the speed of the squeezer roll takeoff.

The chill roll has a separate direct drive, which must be adjustable within ±5% relative to the line speed. The chill roll should principally be equipped with a temperature control system that can handle temperatures up to 190 °C because the polycarbonate film extrusion might require those temperatures in isolated cases.

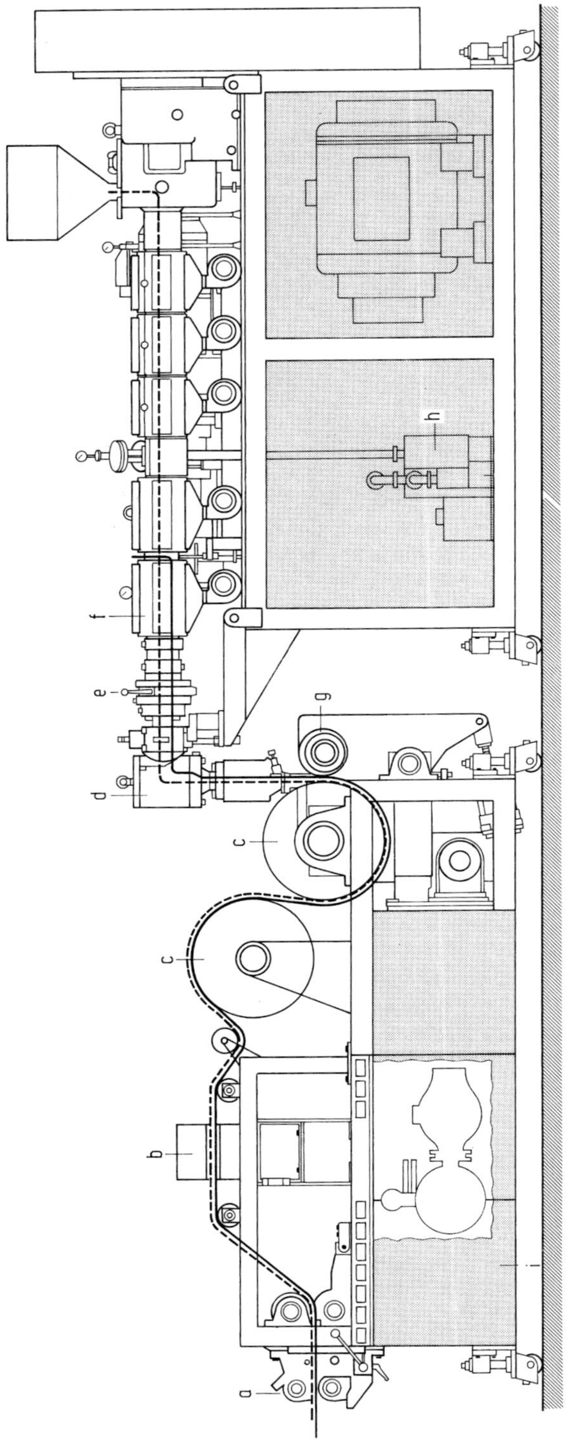

Figure 3.86. Sheet for thermoforming compact plant with three-roll calender stack (Reifenhäuser system): a, drawback and cutting device; b, thickness gauge, c, cooling roll; d, slot die; e, adapter; f, extruder; g, pressure roll; h, vacuum plant.

Polycarbonate film is best wound with direct-driven center surface winders. The winder motor should be a thermally controlled dc motor.

It is not usually necessary to oscillate the extruder or the winder, since edge trimmings can amount to about 25%.

Various firms supply units that recycle the edge trimmings from films and feed them directly to the hopper. Separate reworking for injection molding is a simple alternative.

3.8.2.12 Production of Twin-Wall Sheets

3.8.2.12.1 Geometries

The production of twin-wall sheets from polycarbonate has such great economic importance today that the process is described in detail. Figure 3.87 shows cross sections of sheets and profiles in this configuration.

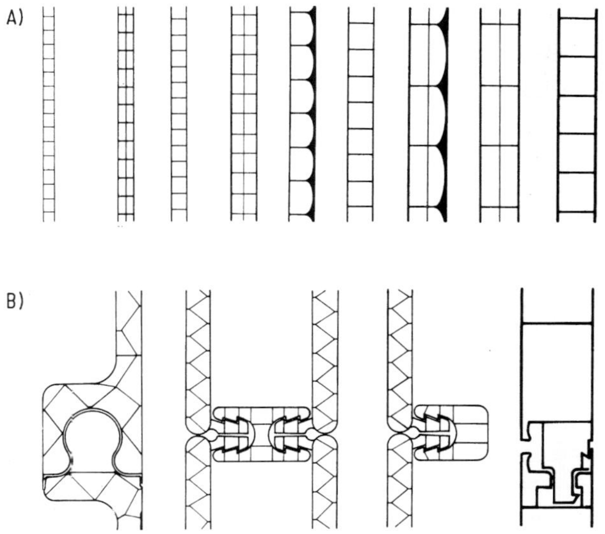

Figure 3.87. Different cross-sections of twin-wall sheets (A) and twin-wall profiles (B).

A significant improvement in physical properties (e.g., weathering behavior) was achieved with tunnel geometry and coextrusion of a UV-resistant coating. Figure 3.88 is a basic diagram of this optimized two-wall sheet with UV coating. In contrast to the

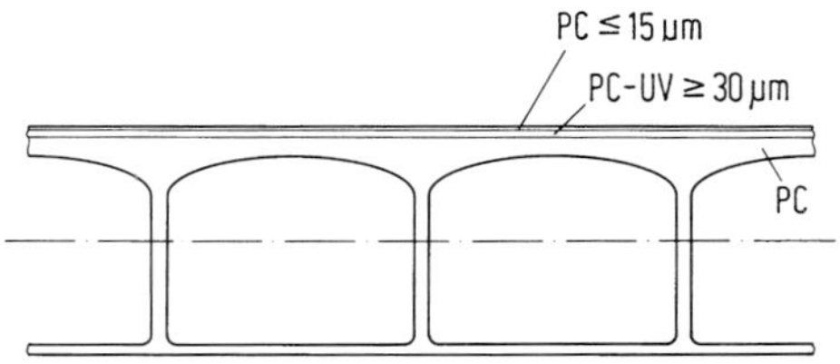

Figure 3.88. Cross-sectional view of a twin-wall sheet with tunnel geometry and PC-UV coating (patented).

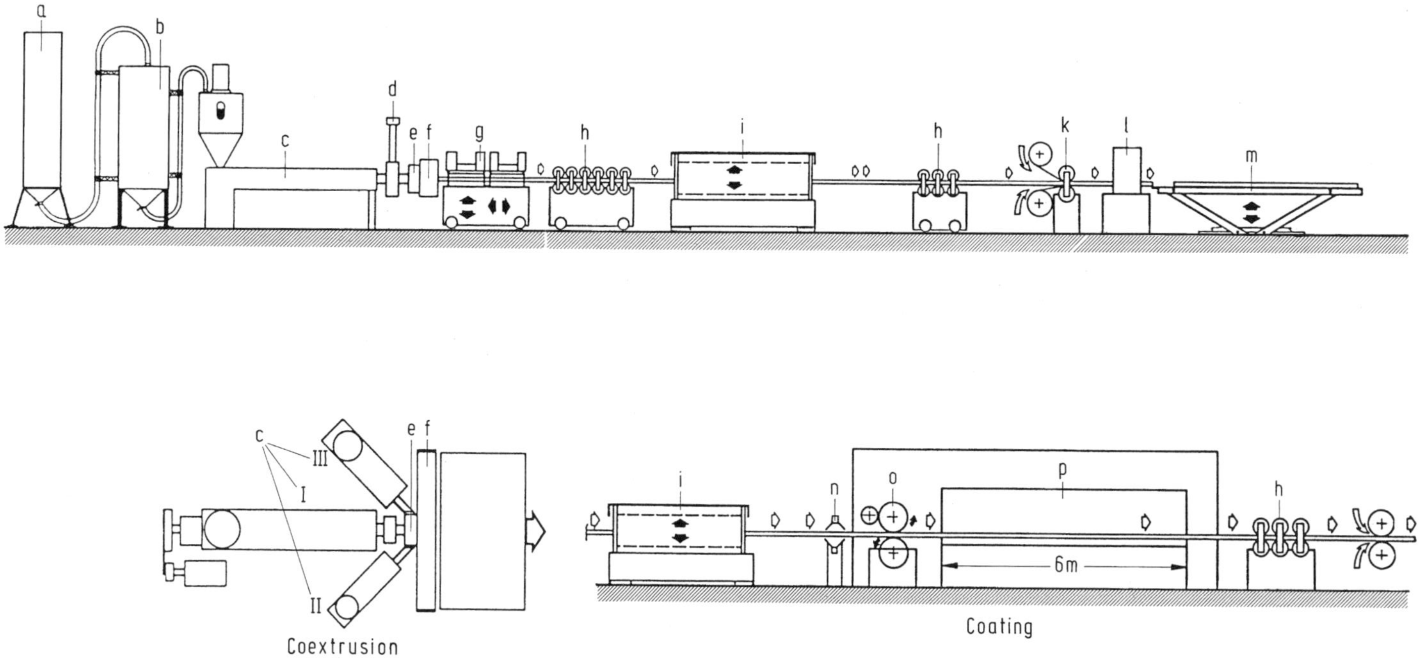

Figure 3.89. Twin-wall sheet plant: a, silo; b, predrying tank; c, extruder; d, screen changer; e, coextrusion adapter; f, die; g, calibrator; h, drawback; i, heating unit; k, foil input; l, saw; m, sheet delivery table; n, ionization; o, application roll; p, air oven.

conventional box-type profile, the optimized vault-type profile significantly reduces the critical tensile stress at the corners between the belt and web under stress. The protective coating of pure polycarbonate can be omitted when special UV absorbers are used.

Table 3.21 gives some technical information about extrusion lines, which are used in the economical production of the majority of twin-wall sheets. Table 3.22 presents the linear takeoff speeds of various sheet profiles and sheet weights. Figure 3.89 shows schematically how twin-wall sheeting is produced.

Table 3.21. Technical Data for the Production of Profile Sheets

Kind of Profile	Profile Height (mm)	Screw Diameter (mm)	Throughput (kg/h)	Maximum Die Width (mm)
Twin-wall profile	Variable	60	50–70	400–500
Twin-wall sheet	10	90	150	2100
Triple-wall sheet	16	90	150	1400
Triple-wall sheet	16	120	250	2100
Triple-wall sheet	16	150	400	2500

Table 3.22. Drawback Speeds for Different Profile Crosscuts and Weights

Kind of Profile	Profile Height (mm)	Area Weight (kg/m^2)	Drawback Speed (m/min)
Triple-wall sheet	16	2.7–3.0	1.0–1.1
Twin-wall sheet	10	1.7–2.0	1.6–1.8
Twin-wall sheet	8	1.5–1.7	2.0–2.2
Twin-wall sheet	6	1.3–1.5	2.5–2.7
Twin-wall sheet	4	1.0–1.3	3.0–3.5

3.8.2.12.2 Design of Plants for the Production of Twin-Wall Sheets

Extruder

The design of extruders for polycarbonate processing was described in Section 3.8.2.4. Use of vent-type extruders is recommended for the production of twin-wall sheets. The devolatilization provides good process consistency over longer production times with respect to the large material quantities to be dried. The cleanest possible work environment will be achieved if the extruders are supplied from silos.

Nozzle

Figure 3.90 shows the cross section of a single-plate die for the production of twin-wall sheet. The melt, which is distributed through two distribution channels located one above the other, is controlled by a choke bar. The dies are equipped with a system of channels that ensures the uniform distribution of the support air required for calibration of the melt. Also important here is the rectangular arrangement of the melt channels at the nozzle inlet throat.

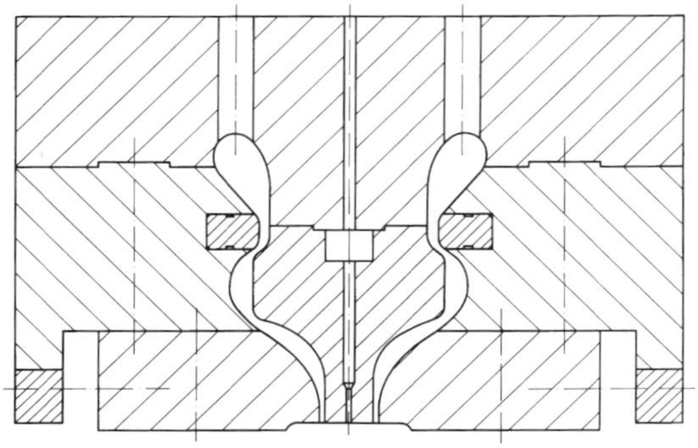

Figure 3.90. Cross-section of a twin-wall sheet die (Omipa system).

Calibrator

Vacuum calibrators are used for the calibration of twin-wall sheets. The following points are essential for proper calibration:

- The vacuum must be directed through the cross-sectional slots to the profiles.
- The first vacuum zone in the inlet feed zone must be controlled separately.
- Uniform temperature control of the sheets is critical.
- A uniform low pressure distribution is necessary to ensure an even sheet surface.
- The cleaning of vacuum and ventilation channels must be easily and quickly accomplished.
- The channel connecting block must be maintained absolutely airtight.
- It must be impossible for cooling media to enter the vacuum channels.
- Depending on the takeoff speed and the outer wall thickness of the twin-wall sheet, two or three sequentially arranged calibrator sheets (individual lengths approx. 600 mm) must be provided.
- The surface of the calibrator should be lightly sandblasted and given a matte chromium plating.

Annealing

To reduce frozen-in stresses as much as possible, the twin-wall sheet must be annealed after it has passed through the calibrator. The annealing is done in ovens with a length of 6–8 m. The sheets are cooled down to 50 and 60 °C, and slowly heated to 135–140 °C in a continuous oven, to reduce warpage due to heating too quickly and unevenly.

3.8.2.13 Production of Profiles and Tubes

Different profiles are shown in Figure 3.91. Tools and calibrators must be specifically designed for the profiles to be produced. The design specifications can be summarized as follows:

Tubing and Tubelike Profile Tools

- Lip length for profile thickness < 1.5 mm, $15 \times$ wall thickness; for profile thicknesses > 1.5 mm, $20 \times$ wall thickness.
- Die diameter, 20% larger for a tube diameter < 30 mm; 15% for tube diameters > 30 mm.

- Die wall thickness, 20–25% greater than the part thickness.
- Outside calibration dimension, approximately 1–1.2% larger than finished part size.
- Support air holes (for start-up) through the mandrel must be provided.

Twin-Wall Profile Dies

- Lip length as for tubing, above.
- Lip gap width corresponds to the profile thickness.
- Profile width and height in the calibration units, approximately 0–3% larger than the finished profile size.
- Support air ventilation through all cores is necessary for start-up.
- The supply of support air must be extremely uniform over the entire width.
- A number of air holes are located in the mandrel support ribs for uniform supply of the support air.
- Additional control of the melt flow is provided via die temperature control for wide dies.

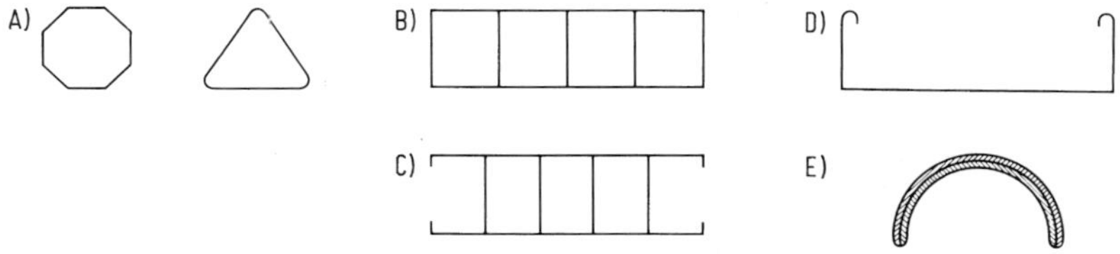

Figure 3.91. Kinds of profile: (A) tubular profiles; (B) hollow profiles with compartments; (C) hollow profiles with compartments and lugs; (D) solid profiles; (E) composite profiles.

The support air is used only during start-up of the extrusion process. The open channels allow pressure equalization after the support air has been shut off. In the absence of support air, the profile would collapse during start-up, and calibration would be impossible. To obtain a uniform melt distribution in large-surface, wide-mouthed sheet dies, extrusion against a narrow feldspar damper is recommended. A choke bar is advantageous for extrusion of wide, lightweight panels.

Dies for the production of solid profiles correspond to those for the production of hollow wall profiles regarding lip gap and lip length.

Calibration

Selection of the appropriate calibration process is decisive for the quality of the extruded profile. The calibration process determines size, surface, and production speed. These variables, in turn, must be adjusted to the specific material properties as already mentioned (e.g., low melt resistance, melt solidification, and adhesion of the melt to steel).

The length of the calibrator is important, as well as its surface and the cooling method. With indirect cooling and small wall thickness (<1.5 mm), calibration is complete after approximately 150–300 mm of calibrator contrast, by which point the material is so stiff that no further influence on the profile shape is possible. The takeoff speed has a great influence and must be adjusted to each wall thickness.

High takeoff speeds (>2 m/min) are advantageous for profile extrusion, especially for transparent profiles. Transverse waves or other surface imperfections that occur at certain frequencies can be stretched and evened out by increasing the takeoff speed.

Vacuum connections must be monitored with sensitive vacuum gauges, since good pressure consistency is required. The measurement points must be selected to ensure precise instrument readings at low pressure (200–450 mbar).

Vacuum cooling tank calibration is used with geometrically simple profile tubes. Intense cooling is achieved through direct contact between the plastic and water. At the same time, a water film forms around the profile to be calibrated. The adhesiveness of the melt to steel is greatly reduced, and the melt glides easily through the calibrator. At the same time, the surface is protected against scratching during passage through the calibrator. A profile shrinkage of 1–1.2% is assumed in the establishment of calibrator diameters.

Vacuum cooling tank calibration is not suitable for the production of geometrically complicated hollow chambers and hollow profiles. Here, vacuum calibration with indirect cooling is appropriate (Figure 3.92). Chromium-plated or polished surfaces are preferred for profile control. The vacuum must be advanced to the profile via the cross-sectional channels. To ensure uniform ventilation, vacuum channels with close spacings (minimum distance 20 mm) must be installed in the inlet zone of the calibrator.

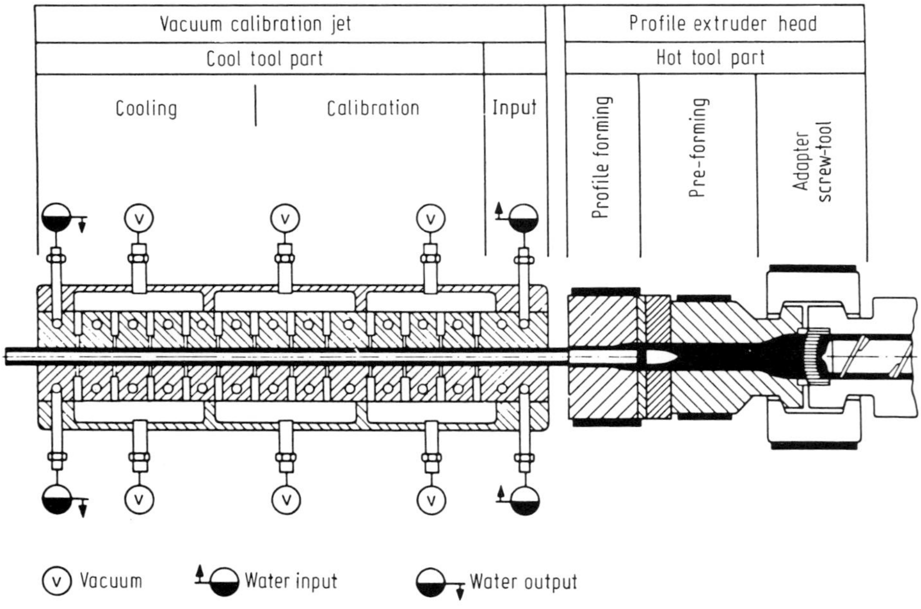

Figure 3.92. Vacuum calibration for profiles.

Calibrator lengths of 250–300 mm are sufficient. Takeoff speeds between 2 and 5 m/min can be achieved with profile wall thicknesses of 0.3–0.8 mm (twin-wall sheets). The takeoff speed is approximately 1–1.5 m/min for wall thickness profiles between 1 and 1.5 mm. The profile form has some influence on the takeoff speed.

Most U-shaped solid profiles from polycarbonate are produced in extrusion plants that make polymethyl methacrylate or polystyrene profiles. The extruder, in this case, requires higher drive power for the same profile form, and the temperature control and the embossing temperature must be set differently. With U-type profiles the walls are usually extruded downward, in a hanging position. The calibration can be performed by temperature-controlled rolling calibrators or through drawing plates.

With drawing plate calibration, the part size is based mainly on the preformed profile in the die. Tube-forming processes have been successfully used with polycarbonate for geometrically simple symmetrical U-type profiles. With this method, sheet metal profiles from which the desired profile is cut out are attached in sequence to a cooling device. The extrudate is led through the openings in the sheet metal profiles, which are normally separately mounted on movable rods or parallel guides. The profiles are air-cooled during forming.

Smaller solid profiles are calibrated with rotating disks, which are submerged in water. Eventually, the tank can be put under vacuum. This process is especially suited for thermoplastics with low melt strength and those that tend to stick to metal. Nearly mark-free profiles can be produced with this process at high speed. The cooling medium forms a lubricating film between the profile surface and the calibrator walls, which greatly reduces adhesiveness. A small gap between the die and the calibration device is an advantage for low melt strength materials. Because for plastic material does not stick to the calibrator, a small gap between calibrator and nozzle is possible. The profile passage is cut approximately 0.2 mm larger than the desired wall thickness of the profiles. The calibrator length is between 100 and 200 mm.

3.8.2.14 Coextrusion

Coextruded products, both sheet and film, have become increasingly important in the market. The objectives of coextrusion are variable. For example, better weather resistance of the sheets is achieved through appropriate coextrusion when twin-wall or solid sheets are being produced. Figure 3.93 shows cross sections of polycarbonate sheets to which protective UV-absorbing layers have been applied through coextrusion.

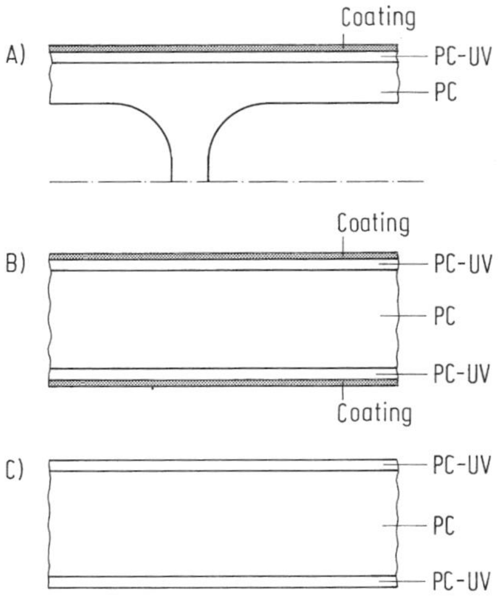

Figure 3.93. Cross-sectional views of UV-protected twin-wall and solid sheets: (A) twin-wall sheet with two coatings; (B) solid sheet with two coatings; (C) solid sheet with one coating.

The coextrusion of an additional layer on top of the UV-protective layer is required when the UV stabilizer vaporizes prior to the calibration process. The coextruded layer contains a high percentage of UV absorber, to provide very good UV protection for the coextruded goods with a minimum coating layer of 30 μm. Since during processing the UV absorber tends to diffuse into the supporting layer, as well as the top layer, the top coating receive sufficient UV protection. The top layer can be no thicker than 15 μm.

It is essential for coextrusion that the coating materials be matched in their flow behavior to the support material, to minimize wall-pushing effects or uneven thickness distribution, which can cause surface defects. New UV-protective coating materials suitable for one-layer coextrusion make the additional top layer obsolete. Application for coextruded film sheets with polycarbonate can be found, for example, in the packaging industry. The presence of polycarbonate in polypropylene/coupling agent/barrier materia/coupling agent sandwiches gives bonds high temperature resistance and transparency. The same or similar systems are also used for hollow profile blow molding applications.

Coextruded bonds with polycarbonate also can be produced using the multimanifold slot die and adaptor processes, as described in Sections 3.8.2.14.1 and 3.8.2.14.2.

3.8.2.14.1 Multimanifolds Sheet Die Process

The layers to be coextruded may be supplied from separate flow channels and a multimanifold sheet die used to combine them. This design described for the two-layered die is shown in Figure 3.94A. The supply channels feed each layer through coathanger-shaped distribution channels. After the throttle zone, a choke bar is arranged in a slanted position, which can be adjusted. Dead spots in the melt flow are eliminated through the slanted arrangement in the supply and distribution zone of the choke bar.

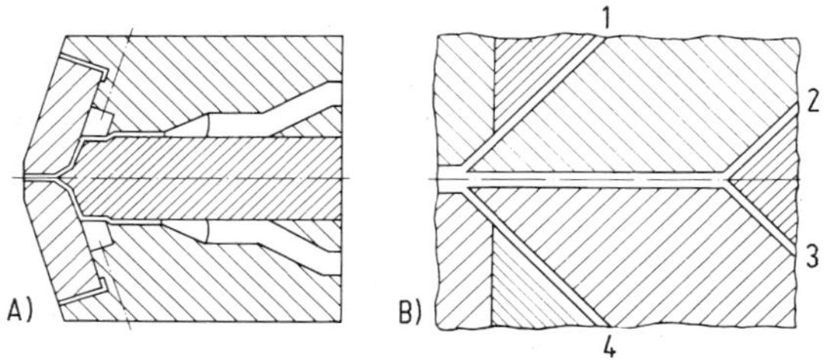

Figure 3.94. Multilayered slot dies: (A) two-layer die with single canals and die restriction (Reifenhäuser system) and (B) schematic presentation of a four-layer die.

Both melt streams are combined prior to the common lip gap. Both lower and upper lips can be adjusted to permit modification of the flow velocity of the individual layers where the stream is combined.

Figure 3.94B shows a quadruple layer combination. The lips of the discharge gap must be adjusted to the desired layer thickness for control of the total sheet thickness. The bottom side of the tool is slanted to ensure that the gap between the die and the chill roll is as small as possible.

3.8.2.14.2 Adapter Process

In contrast to the multimanifold sheet die process, in the adapter process the coextruded melts are combined in an adapter (black box). Figure 3.95 shows lengthwise and cross-sectional views of an adapter for the production of twin-wall sheets from polycarbonate. The adapter process has advantages, especially for the coextrusion of more than four layers and thin layers, either as the top layer or as the middle layer (barrier layer).

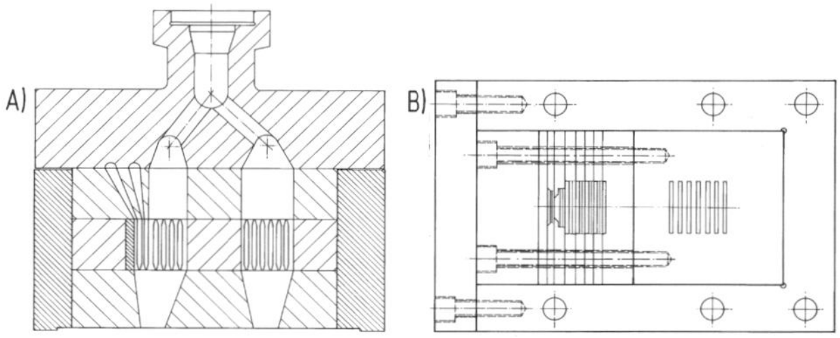

Figure 3.95. Adapter for twin-wall sheet extrusion: (A) longitudinal view; (B) cross-sectional view.

Laminar adapters, which permit adjustment of the number and location of the layers, are advantageous in certain cases. In addition, the module packages can be exchanged, as can individual components. In this way, the flow characteristics can be matched to the rheological behavior and the operating parameters. Figure 3.96 shows a twin-wall coextrusion sheet plant.

Figure 3.96. Coextrusion plant for the production of twin-wall sheets. (Photo: Bayer, Leverkusen).

Based on the experience with twin-wall sheet extrusion, an adapter system was developed for the coextrusion of solid sheets from polycarbonate with UV protection. The illustration in Figure 3.97 makes clear that various combinations can be manufactured with simple changes: for example, three layers (ABS/regrind ABS/polycarbonate) for ski boxes. Other systems for the coextrusion of polycarbonate layers include commercial adapters from Cloeren and Reifenhauser, whose sytems make it possible to adjust the flow parameters of the several melt streams to the rheological characteristics of the materials and the process parameters.

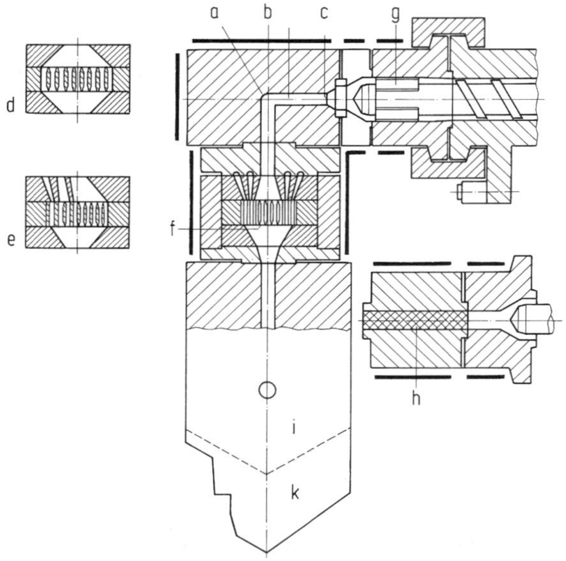

Figure 3.97. Coextrusion–adapter system for solid sheets: a, deflector; b, pressure gauge; c, temperature control; d, module block monoextrusion; e, module block three-layer extrusion; f, module block five-layer extrusion; g, planetary mixer; h, static mixer; i, Keller 600; k, IKV 500 jet system.

3.8.3 Extrusion Blow Molding of Hollow Profiles

F. J. Brinkschröder

3.8.3.1 General Characteristics

Polycarbonate has the major advantages of glass, such as transparency and freedom from odor and taste, without the fragility. Figure 3.98 shows typical extrusion blow-molded containers from polycarbonate. In combination with other polymer materials, such as polyethylene terephthalate, the barrier characteristics can be balanced.

3.8.3.2 Polycarbonate Types for Extrusion Blow Molding

Extrusion blow molding requires high melt strength. Therefore, highly viscous types of polycarbonate are primarily used. Linear, highly viscous materials can be used up to a bottle size of 1.5 liters. Better suited, however, are specially developed high melt strength polycarbonates (see also Section 3.8.2.1), which excel in the lower shearing velocity range ($<100\ s^{-1}$) with a significantly higher viscosity than the linear types. Properties that favor extrusion (see Section 3.8.2) are also favorable for extrusion blow molding.

3.8.3.3 Extruder Layout

The design guidelines specified in Section 3.8.2 are also applicable to extruder and screw design for extrusion blow molding.

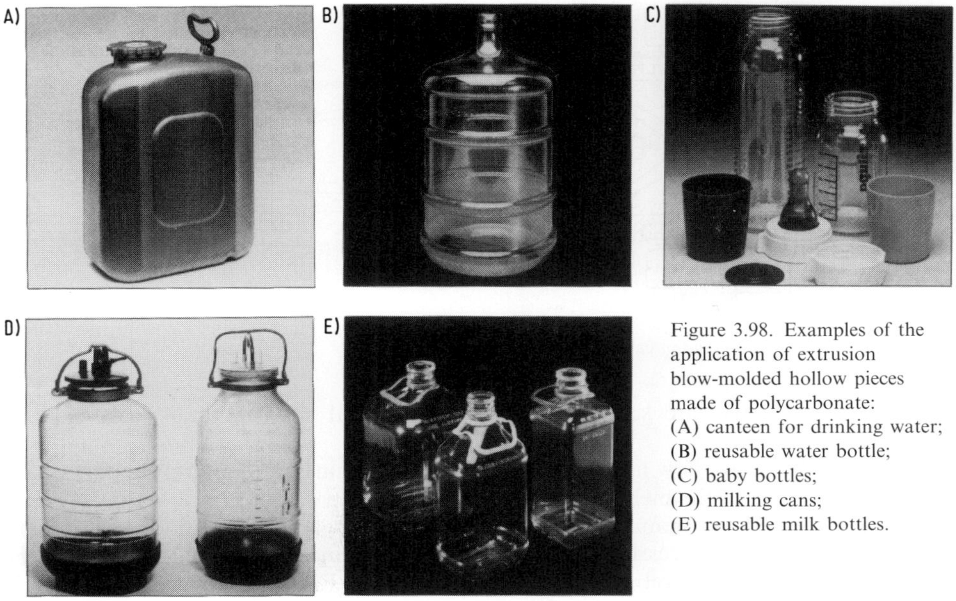

Figure 3.98. Examples of the application of extrusion blow-molded hollow pieces made of polycarbonate:
(A) canteen for drinking water;
(B) reusable water bottle;
(C) baby bottles;
(D) milking cans;
(E) reusable milk bottles.

Polycarbonate requires a machine design different from that used for polypropylene and polyethylene. Extrusion blow molding manufacturers offer standard extruders with slotted and cooled feed zones for the processing of these thermoplastics. Tests and production with polycarbonate must often be carried out on such machines. In these cases, the feed zone must be modified to withstand temperatures up to 300 °C. Otherwise, the danger of overloading the extruder drive exists. See Section 3.8.1.3.1 regarding the material selection for the screw and cylinder.

3.8.3.4 Design

3.8.3.4.1 Molds for Continuous Extrusion Blow Molding

Containers of up to 5 liters can be produced by continuous extrusion blow molding. Blow molding dies that distribute the melt through a heart-shaped curve into the die opening are suitable for this. With other designs, the movable mandrel is fastened to a support device. The melt is diverted by 90° in a central adapter tube prior to the mandrel. A conventional mandrel support tool is illustrated in Figure 3.99. Improved designs include spiral distribution devices. All developments have the goal of producing nearly streak-free products.

3.8.3.4.2 Molds for Discontinuous Extrusion Blow Molding

Containers with volumes exceeding 5 liters are best produced discontinuously (i.e., using storage or accumulator dies). The melt coming from the extruder is accumulated and, at the appropriate moment, quickly discharged with the help of a hydraulic piston. Depending on the die type, the mandrel support separates the melt into two or more spatial streams, which must be combined again to form the desired hollow extrudate.

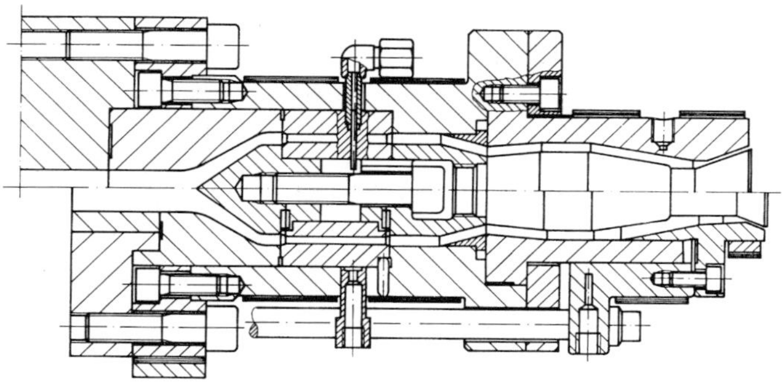

Figure 3.99. Mandrel carrier (Kautex system).

These partial streams are sometimes visible as vertical streaks, or at least exist as structural seams.

Figure 3.100 shows flow lines in a 20-liter water container. A mandrel support device without separation of the melt stream is not possible. The design principle for dies must therefore be based on minimizing the vertical streams and material orientation, or blurring them. Other successfully used systems are a centrally supplied annular piston storage with a wall mandrel support device (Battenfeld-Fischer), and a melt supply in the control storage wing die (Bekum).

Figure 3.100. Flow works in a 20-liter container made of polycarbonate.

3.8.3.4.3 Blow Molding Dies

Die lips that are conical toward the outside are best suited for extrusion blow molding done in conjunction with polycarbonate processing. Uniform heating of the mandrel and die body is necessary to prevent the emerging melt from curling up on the outside or inside. Therefore, when larger die sizes are used, heat should be supplied to the mandrel by means of a heating cartridge installed inside the mandrel. The mandrel temperature is kept constant by the flowing melt during production.

3.8.3.5 Wall Thickness Programming

Wall thickess programming is required for the production of larger blow-molded polycarbonate parts. Differences in wall thickness become larger with increasing length of the blow-molded part because the weight of the preform causes it to stretch after being discharged from the nozzle lip. In addition, the nonuniform cooling of the preform affects the distribution of wall thickness. Thin spots in the upper end of the part (neck area) must be prevented by wall thickness programming.

Wall thickness controls are based on die lip adjustment. The gap width between the conical die body and the mandrel is changed during the extrusion of the preform by movement of the mandrel. Dies with a conical angle of up to 90° are found to offer the advantages of a quick gap width change with a small adjustment path, and therefore, smaller displacement effects inside the tube head. Figure 3.101 shows the wall thickness variation on a 20-liter container with and without wall thickness control.

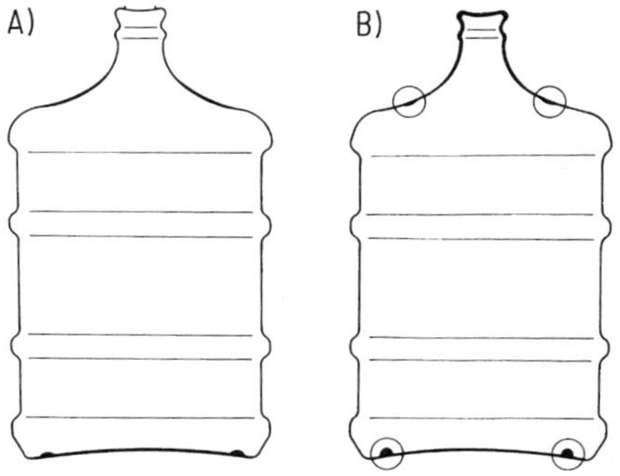

Figure 3.101. Wall thickness pattern with (A) and without (B) wall thickness programming. Without wall thickness distribution, mass accumulates at the circled areas in particular.

The bottom area of a blow-molded part is controlled differently. The preform is flat and compressed on the seam. Large wall thicknesses at the end of the seam are the result of a mold bottom that is round because of uneven expansion conditions. These large wall thickness differences often create stress problems, which can lead to premature fracture of polycarbonate blow-molded parts. The problems can be minimized through proper part design as well as by achieving an optimum flow of the melt stream in the mandrel–nozzle zone.

3.8.3.6 Blow Molding Tools

The following properties of polycarbonate impose specific requirements on the design of blow molding equipment:

- comparatively high heat conductivity [(0.21 W/(m · K)] and therefore a comparatively high solidification rate
- a small forming temperature range
- stress-cracking sensitivity
- high reproducibility in the melt of mold surface defects

Otherwise, the generally known design guidelines are valid for blow molding equipment.

3.8.3.6.1 Pinch-off Seam

The formulation of mold parting lines and pinched-off edges represents a compromise between separation and connection. From the point of view of economy, a clean separation is necessary. Imprecise parting lines require expensive follow-up cleaning. The pinch-off edge geometry as shown in Figure 3.102 proves to be optimal for polycarbonate. Following the crimp pocket, the tool is milled freely. The excess or sealed-off material should be compressed only on the edge toward the mold. This action partially cools the sealed-off material, which can contact the mold during demolding without sticking.

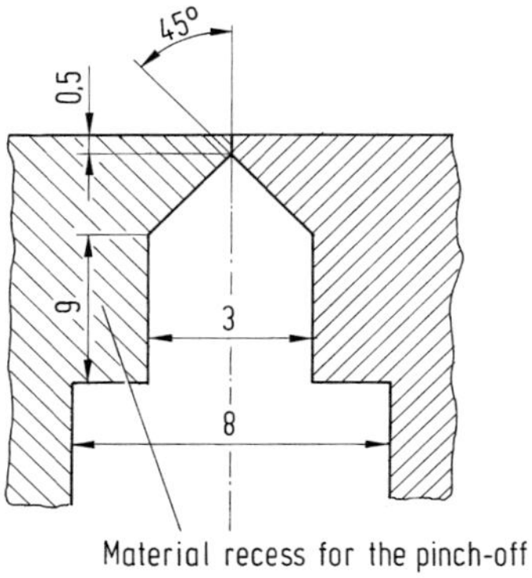

Figure 3.102. Pinch-off geometry for polycarbonate blow molding cavity.

3.8.3.6.2 Mold Surface and Ventilation

Polycarbonate conforms very precisely to the shape of the mold, as well as to any mold surface defects (scratches, depressions). To obtain good surface qualities (e.g., transparency), proper mold surfaces are necessary. The mold surface must not be distorted as a result of this work (e.g., polishing), however, because even very large surface areas or small depressions can influence the visual appearance.

The air trapped between parison and mold wall must be displaced during blowing. The parting planes must be equipped with venting slots. In addition, the circulating reinforcement bars must be equipped with additional channels (approx. 0.1–0.2 mm diameter).

It is recommended that the surface of larger mold (e.g., for 20-liter water bottles) be roughened somewhat to completely force out the air. This measure results in a small diminution in transparency. Roughening is done by microsandblasting with fine grains at 1–1.5 bar blasting pressure.

3.8.3.6.3 Temperature Control

The use of three separately controlled temperature zones is recommended for larger vessels. The mold for the production of a 20-liter water bottle, for example, would be controlled separately in the neck and bottom areas. The mold temperature should be between 50 and 70 °C.

3.8.3.7 Processing Instructions

Some typical temperatures for an extruder with a blow molding die are as follows:

Control adjustment T_1, 280 °C; T_2, 270 °C; T_3, 260 °C

Metering zone T_4
Adapter
Diversion
Blowing die
Nozzle
} 250 °C (start-up temperature, 260 °C)

Table 3.23 lists cycle times as a function of bottle volume or weight.

Table 3.23. Cycles Times Depending on Bottle Size and Weight

Bottle Content (liters)	Weight (g)[1]		Cycles Time (s)[1]	
1	75	(79)	13	(10)
2	150	(150)	15	(12)
3	190	(180)	15	(15)
4	210	(210)	22	(18)
5	230	(230)	22	(20)
10	420	(420)		(36)
20	800	(800)	60	(45)
40	1000	(1500)		(60)

[1] Numbers in parentheses are values for the Krupp-Kautex machine.

3.8.3.8 Annealing of Extrusion Blow-Molded Reusable Containers

Stresses are built into blow-molded parts from three processes: extrusion, molding, and subsequent cooling. In some plastics these inner stresses diminish over time without substantial change in the outer shape of the item. With polycarbonate, however, the inner stresses remain mostly intact. After each cycle of use, reusable polycarbonate bottles are put through a cleaning process (see also Section 3.7.10.8.2) that creates inner stresses, which in turn can lead to cracks, especially in the pinch-off area. Therefore, polycarbonate bottles intended for reuse should be annealed. The following values serve as a guide for tempering times and temperatures:

Container size	liters	1–5	6–20
Temperature	°C	$\approx 130 \pm 3$	$\approx 130 \pm 3$
Time at temperature	min	≈ 30	≈ 50

3.8.3.9 Coextrusion Blow Molding of Hollow Profiles

The viscosity and transparency properties of polycarbonate are even better utilized for hollow profiles in combination with other thermoplastics. Such material combinations as polypropylene with adhesion promoters or barrier materials have opened up the packaging sector as an application area.

Coextrusion–blow molding molds of different manufacturers are used. These systems use heart-shaped melt diverters or spiral distributors. The spiral distribution system is illustrated in Figure 3.103. Combinations with polycarbonate are manufactured today with all systems. Accumulator systems are not yet used.

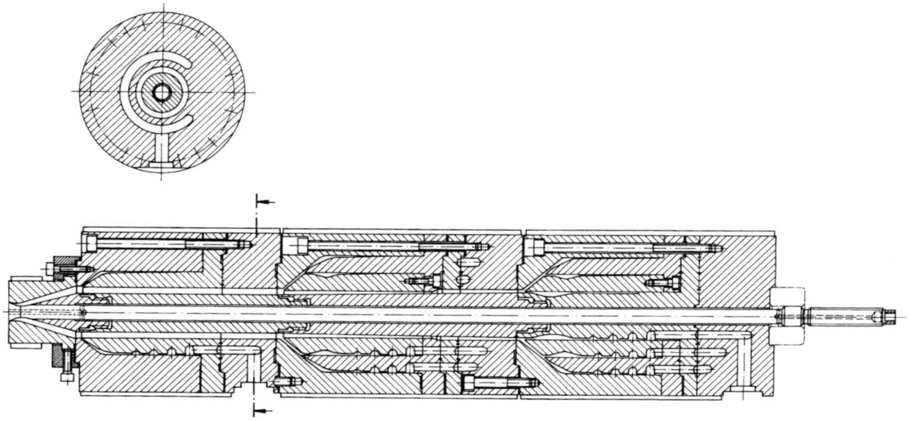

Figure 3.103. Coextrusion blowhead with helix distributor (Battenfeld system)

3.8.4 Regrinding and Recycling

The processing of polycarbonate is accompanied by a small amount of decomposition, which leads to wider molecular weight distribution. Regrind can contribute additionally to decomposition. The tensile and bending properties, as well as the thermal properties, are changed only to a small extent and thus are of no significance. The regrind of standard polycarbonate is generally reusable despite a slight decrease in viscosity.

Polycarbonate properties are often utilized under extreme service conditions. In such cases, especially when toughness is important, or when the products are safety components, the use of a regrind is not advisable. Consideration must also be given to the potential of adding impurities. For slightly overdesigned parts, 20–30% clean, dry regrind material can be used. As a rule, 100% regrind is used only for less demanding parts.

Because of the good stability of polycarbonate, its molecular characteristics are changed only slightly during part use. Transparent polycarbonate undergoes chemical change only in a very thin outer layer upon high UV-ray exposure, or after long service outdoors, but the extent of the changes through the entire molded thickness is insignificant. Recycling of worn-out molded parts is therefore conceivable. Incompatible components must not be introduced into the polycarbonate during the second processing, however.

Molecular weight decomposition is more pronounced during processing fiber glass-reinforced polycarbonate than in work with standard types. The processing is accompanied by measurable fiber glass breakage, and regrinding contributes to the decomposition of polymers and fiber glass alike.

For product types with fire-retardant additives, the effectiveness of the fire retardant may be reduced.

References for Section 3.8

1. Die Trocknung von Kunststoffen. Application leaflet Bayer AG, Leverkusen Nr. 185/78
2. *Backofen, W.*, et al.: Plastverarbeiter 20 (1969) 8, p. 554–562

3. Company brochures released by
   - Alfred Engelmann, Wiedemark, about Rozylit-Ganulat
   - Ramisch-Kleinewefers, Krefeld, about Zimmerpaste
   - Engineering Chemicals B. V., Steenbergen, Niederlande, about Polypurge-Pulver
   - Feldmühle AG, Werk Wesseling, Wesseling, about CORATEX-Emulsion
4. Wirbelbett-Reinigungsbad. Company brochure Schwing-Verfahrenstechnik GmbH, Neukirchen-Vluyn 1986
5. Ultraschall-Reinigungsanlagen. Company brochure Karl Roll GmbH & Co., Mühlacker-Enzberg 1986
6. Thermoplast-Reinigungs-System. Company brochure C. Christ, Munich 1986
7. *Johannaber, F.; Stoeckhert, K.:* Der Kunststoffmaschinenführer. 2nd Ed. Hanser, Munich 1984
8. Schneckengeometrie der Bayer-Thermoplaste. Application leaflet Bayer AG, Leverkusen Nr. 173/86
9. Verschleißschutz beim Spritzgießen. Application leaflet Bayer AG, Leverkusen Nr. 458/84
10. *van Hooren, R.; Kaminski, A.:* Plastverarbeiter 31 (1980) 8, p. 441–446
11. *Schwittay, D.; Sajben, J.:* Plastverarbeiter 33 (1982) 11, p. 1366–1370; 12, p. 1477–1480
12. *Fingerle, P.:* Kunststoffe 63 (1973) 7, p. 418–422
13. *Kohler, R.:* Plastverarbeiter 28 (1977) 6, p. 314–318
14. *Hardt, B.:* Plastverarbeiter 34 (1983) 3, p. 219–221
15. *Wübken, G.; Bangert, H.:* Kunststoffe 74 (1981) 1, p. 46–51
16. *Bernhardt, E. C.* (Ed.): CAE Computer Aided Engineering for Injection Molding. Hanser, Munich 1983
17. *Klenz, R.:* Graduation Thesis at Institute für Kunststoffverarbeitung an der TH Aachen 1985
18. *Tadmor, Z.; Klein, I.:* Principles of Plasticating Extrusion. Reinhold, New York 1969
19. *Pinto, G.; Tadmor, Z.:* Polym. Eng. Sci. 10 (1970) 5, p. 279–288
20. *Anders, S.; Schmid, H.; Sommer, K.:* Kunststoffe 79 (1989) 1, p. 55–59
21. *Eckardt, H.:* Plastverarbeiter 34 (1983) 4, p. 316–319
22. *Eckardt, H.:* Kunststoffberater 28 (1983) 1/2, p. 26–32
23. *Cohnen, W.; Meyer, W.:* Kunststoffrundschau 20 (1973) 10, p. 425–434
24. *Eckardt, H.:* Schweizer Maschinenmarkt 73 (1973) 14. 3. 1973

3.9 Processing

3.9.1 General Characteristics

Polycarbonate enjoys the same advantages offered by other thermoplastically processible materials in that parts can be manufactured directly into usable products and/or finished parts ready for final assembly without the need for secondary operations. Many methods require merely the removal of the gate through stamping or cutting, which is then completed by milling or polishing if parts of high optical quality are to be produced. Whereas conventional materials such as metal and wood are routinely subjected to machining, this process step plays a comparatively minor role in polycarbonate; even in comparison to thermosets, machining is rare.

Machining operations for polycarbonate (in addition to gate removal) are required in the following areas:

- refining through surface treatment and tempering
- joining
- reshaping of semifinished products, particularly sheet
- sawing, drilling, cutting, and grinding during machining of sheet
- prototyping and short-run manufacturing from semifinished products

Polycarbonate is easily machined by many different methods. The following ground rules must be observed:

1. Prior to the thermal cutting or reshaping (between 160 and 200 °C), polycarbonate must be dried exactly the same as prior to injection molding (see Section 3.8.1.2.1). Depending on the storage conditions, polycarbonate contains 0.1–0.25% water, which can lead to undesirable bubble formation at temperatures above 150 °C. Since the diffusion path of water is greater in semifinished products than in pellets, semifinished products may require longer drying times (depending on their thickness). Drying times at 120 °C are as follows: 4 hours at a layer thickness of 2 mm, 13 hours at 4 mm layer thickness, and 30 hours for 10 mm.
2. Reshaping below the glass temperature (about 148 °C) often leads to high internal stresses, which have a negative influence on the toughness and load behavior. It should be noted that only minimal residual stresses remain in the finished part from machining operations such as stamping, screwing, thread cutting, drilling, and joining.

3.9.2 Annealing

The annealing of molded polycarbonate parts is a thermal processing step that in some applications increases the quality of finished parts. Annealing promotes relaxation of processing-induced stresses and reduces the tendency of cracks to form during contact with aggressive media.

It is possible to reduce stress at 80–95 °C and a residence time of at least 12 hours. At 130–135 °C annealing time is shorter. The area farthest away from the outside wall is brought to 130 °C for a few minutes. The annealing process is useful or necessary for several application cases.

While annealing at 80–95 °C hardly influences toughness and elongation at break, at 130–135 °C, depending on the duration of the process, annealing reduces both notched impact strength and elongation at break. The reduction in toughness does not randomly increase with increasing annealing time; rather, it approaches a limiting value. Annealing of polycarbonate parts is useful only when the advantage of tension relaxation outweighs the disadvantage of toughness reduction. One such example is provided by the reusable milk bottle, manufactured by extrusion blow molding, where annealing clearly improves durability. Material accumulations in the bottle mouth area and nonuniform temperature distribution in the bottom of the bottle lead to considerable frozen-in stresses. During repeated washing with caustic washing solutions and during contact with clear-rinse aids, these stresses may lead to crack formation and premature failure of the molded part. When these reusable bottles are annealed at approximately 130 °C (e.g., with infrared radiation), cracks do not form in the bottom area.

3.9.3 Surface Treatment

The texture of a molded part can be influenced with the aid of several methods for changing the surface. Table 3.24 lists the various surface treatment methods and their practical applications.

3.9.3.1 Cleaning

A flawless surface coating is possible only on a surface from which all dust, dirt, oil, and grease have been removed. For this purpose, suitable cleaning and rinsing agents are those that do not attack the plastic or reduce its utility. Aromatic hydrocarbons

Table 3.24. Surface Treatment Methods for Polycarbonate

Method	Area of Application
Coating	To change the visual appearance of the molded part To increase resistance to UV light To improve resistance to aggressive aspects of the environment To improve scratch resistance
Metallizing	To increase reflection capability; to create a conductive layer with electromagnetic shielding effect
Cleaning	As a step that precedes coating
Printing, and polishing, and roughening	To identify or label; to change the visual appearance of a molded part
Grinding	To obtain targeted surface effects

(e.g., benzene, toluene, xylene) or chlorinated hydrocarbons (e.g., chloroform, trichloroethylene, carbon tetrachloride) either dissolve polycarbonate or may cause undesirable crack formation. The cleaning of dusty parts is best done with clean water. In stubborn cases, weak concentrations of acidic, neutral, or weakly alkaline rinsing or washing agents may be added. Hard-to-clean greasy or oily surfaces should be washed with aromatic-free gasoline fractions, such as petroleum ether or ligroin. Alcohols such as ethyl alcohol, isobutyl alcohol, or isopropyl alcohol can also be used, but not methyl alcohol.

3.9.3.2 **Coatings** [1]

Polycarbonate molded parts can be coated with all standard coating methods. To maintain good polycarbonate properties and avoid stress cracking, the coatings must be tailored to the property profile of the polycarbonate. For this purpose, the following area of sensitivity must be noted.

- Polycarbonate is sensitive to sharp notches. During overelongation of a part coated with a brittle coating system, microcracks form on the surface layer. If rigid bonding between the coating and polycarbonate should cause such microcracks to propagate into the substrate material, very sharp notches will develop in the polycarbonate, reducing the toughness of a previously unnotched molded part. Therefore, elastic coating systems are preferred for polycarbonate. If the desired effect (e.g., scratch resistance) can be achieved only with brittle hard coats, a coating with an elastic primer layer, which prevents the propagation of microcracks from the surface layer, is recommended.
- Polycarbonate is sensitive to components that promote stress-cracking. It dissolves and/or swells with exposure to several standard solvents used in coatings chemistry. On the other hand, some swelling of the polycarbonte surface is desired and promotes good bonding between polycarbonate and the coating; on the other hand, most such components tend to form stress cracks in polycarbonate. Good coating results are obtained if the polycarbonate is low in stress and the aggressive solvents, instead of causing stress-cracking, evaporate quickly and completely after application of the coating. Physical drying of one-component polyurethane coatings and chemical hardening of two-component polyurethane coatings are approaches especially well suited for polycarbonate parts. Coatings so treated show excellent adhesion to polycarbonate parts.

Two-component polyurethane coatings show such a high reactivity at relatively low temperatures that complete drying can be achieved with a heat treatment lasting around 45 minutes at only 80 °C, a temperature well below the heat distortion temperature of polycarbonate.

Two-component polyurethanes coatings protect the polycarbonate substrate against many influences, including mechanical (scratching), chemical (dissolving, stress-cracking corrosion), and meterological (ice, hail). By selecting individual system components, the elasticity of the two-component polyurethane coating can be adjusted to the polycarbonate. This makes it possible to eliminate, even at low temperatures, drawbacks of finished-part properties stemming from brittleness.

Since polycarbonates can be dyed to almost any shade, the application of internal dyes is always the most cost-effective method for coloring them. Coating is required for polycarbonate structural foam with higher surface requirements since the natural surface is rough and marbled. Solvent-resistant surface coatings, such as polyurethane coatings, protect the polycarbonate part from an aggressive environment. Such a layer of coating blends the good chemical resistance of the coating with the other good properties of the polycarbonate.

3.9.3.2.1 UV-Protective Coating

A similar, synergistic combination of coating and polycarbonate is achieved through UV-protective coatings. The single purpose of such coating layers is to enrich the polycarbonate surface with light-protective agents.

In most applications, polycarbonate is sufficiently protected against UV light by incorporation of UV absorbers distributed in the material. In applications characterized by extreme UV light exposure (e.g., high altitudes, lights and UV emission) and in those calling for color stability during outdoor weathering (e.g., brilliant-white or translucent-white colors), additional UV protection is required. Additional UV protection is also useful in the very thin surfaces of twin-wall sheet. Polycarbonate flat sheet is also available with a UV-protective coating for high performance applications.

In flat sheet extrusion, the UV absorber enrichment on the polycarbonate surface can be achieved through coextrusion (see Section 3.8.3.14). In complicated geometries, this enrichment can be achieved only through coating. The preferred and best-suited coatings are those based on acrylics with high concentrations of UV absorber. Through copolymerization, the hardened coatings can be made sufficiently elastic. Following are some examples of suitable UV-protective coatings:

- LS-123 and L-446 (Morton International Co., Illinois, U.S.A.)
- Diegel coating 38 391 (E. Diegel, Alsfeld, Germany)
- Light-protective coating RE 2830 (Dr. Renger GmbH, Strullendorf, Germany)

3.9.3.2.2 Scratch-Resistant Coating

For many applications, unmodified polycarbonate is sufficiently resistant to scratching. For some applications the scratch resistance must be further improved. The use of a scratch-resistant coating of polycarbonate, particularly in conjunction with a UV-protective coating, is gaining increasing significance in applications such as automotive lenses, helmet visors, eyeglass lenses, and industrial window glazing.

In principle, scratch resistance can be improved by the addition of a hard acrylic coating, which also serves to improve resistance to UV light. However, this type of scratch resistance is not sufficient for most applications. Good results can be achieved with silicone coatings. However, commercial polysiloxane top coatings alone do not give the polycarbonate sufficient UV protection. Since they absorb little UV light, UV rays penetrate almost unhindered into the interface layer between polycarbonate and the coating, leading to a slow degradation of the polycarbonate in the interface. The result is a slow loosening of the surface layer from the substrate and subsequent diminution of optical quality.

Silicone coatings are best applied after the polycarbonate has been primed with a UV-stabilized, elastic acrylic coating. Such a system imparts to polycarbonate a high scratch resistance along with excellent UV protection. The elasticity of the primer effectively guards against crack propagation from the surface layer into the polycarbonate. Visors and eyeglasses, which are equipped with standard scratch resistance systems, are more or less tough depending on the elasticity of the primer coat. Good primer-coating systems guarantee almost the same toughness as uncoated polycarbonate.

Some examples of commercial coating systems for polycarbonate are listed below:

- SHP 300/SHC 3000, SHP 401/AS 4000 (GE Silicones, General Electric, United States and Europe)
- Lensgard I and II (Morton International Co., United States, Morton Int./Dr. Renger & Co., Germany)
- Acryking K 101, PH 220, PH 328 and PH 700 (Mitsubishi Rayon, Japan)
- SRC-UV'T 146 and 200 (Red Spot, United States; Sonneborn & Rieck, Ltd. Great Britain)
- UV-A + B (Fujikura-Kasei, Japan)
- S28 and S30 (Exxene Corp., United States)
- Irdalon 300-1 (Irdal, Usine de Daillères, France)
- KASI PCN and KASI PC-Flex (KRD Sicherheitstechnik GmbH, Germany)

3.9.3.3 Printing

Clean surfaces of polycarbonate parts can be easily imprinted using low pressure, offset or screen printing methods. Printing colors have been developed for bisphenol A polycarbonate which, in addition to excellent adhesion and abrasion resistance, shows sufficient stability to light and heat.

3.9.3.4 Coloring

While it is relatively simple to dye polycarbonate by adding suitable dyestuffs and pigments to the melt, subsequent surface coloring of molded parts from solution dipping is possible only with special organic pigments, and special methods must be used. Acceptable, uniform coloring of the clean surface of molded parts is possible with among other agents, solvent-soluble dyestuffs. This method is used in special cases (e.g., coloring of eyeglass lenses) but has not found widespread application.

3.9.3.5 Metallization

Metal coats can be applied to polycarbonate surfaces using various techniques.

High gloss metal coats with uniform layer thickness can be achieved by high vacuum metallization (cathode sputtering). The adhesion of these layers is often very good and fulfills expected requirements. This method is also used for metallized film, light reflectors, parabolic mirrors, and laser-optical readable data carriers (e.g., compact disks). A surface coating should be applied to protect the thin metal layer from mechanical failure. In compact disks, for example, a surface coating protects the aluminum reflecting layer.

Although polycarbonate is an excellent material for electronic housings, it is a poor shield against electromagnetic waves. The rapid increase in the use of polycarbonates for electrical instrument housings requires a cost-effective metallization technology that can be applied to complicated part geometries. In principle, shielding for polycarbonate housings can also be achieved by incorporating conductive components such as carbon black, aluminum flakes, or nickel-plated graphite fibers. A more cost-effective method is electroless surface metallization of a nickel and/or copper–nickel coating. Polycarbonate metallized according to this method offers noise suppression of 65–120 dB.

Other methods for metallizing polycarbonate surfaces are the application of a conductive coating and pot galvanization. These methods are seldom used.

3.9.3.6 Sanding

Sanding belts have proven to be excellent for grinding polycarbonates. The circumferential speed of the tape should be 20–30 m/s. Contact pressure should be low. Cooling is not necessary, although alkali-free lubricants can be advantageous. Sanding tapes with 400–500 grit are suitable for dry grinding and waterproof grinding tapes for wet grinding. Naturally, hand grinding is also possible. For this purpose fine granulated sandpaper and soft, ceramic-bonded silicone carbide disks, or lamellar disks, are suitable.

3.9.3.7 Polishing

Polycarbonate sheet can be polished under certain conditions. Best results are achieved using medium density polishing disks with an approximate circumferential speed of 30 m/s. An alkali-free paste (e.g., chromoxide green) is used for prepolishing. For final polishing, only a clean disk without additives should be used.

3.9.4 Joining

Polycarbonate can be joined by form locking (rivets, snap fits), material bonding (adhesives and welding), and frictional connection methods (screws). All joining methods must take into consideration the inherent material properties of polycarbonate.

It is important that no aggressive components, such as solvents, amines, and aromatic softeners from sealers or adhesives, come in contact with any joining areas of the polycarbonate. It must be also noted that some joining methods (e.g., friction welding) may lead to distinct temperature gradients between the joined surface and the part interior, possibly causing increased inner stress. The joining methods must be conducted in such a way that the inner stresses that arise are kept as low as possible. Internal stresses should be minimized by using sensible joining methods that cause little or no inner stress formation.

3.9.4.1 Mechanical Connections

Section 3.8.1.4.1 describes mechanical joining methods and the polycarbonate-specific properties that must be considered.

Mechanical connections often require elastic connecting links, either for uniform distribution of compressive stress (e.g., as a damping element) or for gas- and/or moisture-proof encapsulation. Polycarbonate sheets are joined in part by profile bands.

Suitable sealing materials are silicone rubber, elastic polyurethane, and polyolefins, as well as other polymers. Sealants must be free of solvents or migrating softeners, stabilizers, and other additives that may damage polycarbonate. During peroxide crosslinking of elastomers, peroxides should be selected whose fission products do not cause stress-cracking. Similarly, no amines (decomposition products of the accelerator) should remain after sulfur vulcanization.

Benzene may migrate out as a decomposition product from nonannealed silicone rubber sealants that were vulcanized with benzoyl peroxide. Benzene can cause stress-cracking in the polycarbonate. Residual content of diethylamine in rubber vulcanized with thiuramdisulfide and sulfur may damage polycarbonate by aminolysis. Aminolysis is particularly likely to occur when the amine is not removed through evaporation or outgassing. Under certain conditions, such as increased temperature and higher humidity, locally high amine concentrations may lead to corrosion in the molded polycarbonate part.

3.9.4.2 Bonding

Polycarbonate parts can be joined by several adhesion methods. The quality of the joint is excellent in most cases. In comparison to joining via welding, the adhesion method is seldom used. The main reasons for the neglect of this approach are the high cost of the materials and the undesirable requirement for application of solvent. Adhesives for polycarbonate are preferred in the following cases:

- joining methods for larger, bulky parts (e.g., joining of sheet)
- small production runs
- nonuniform joining surfaces, where an adhesive simultaneously serves as a filler or putty
- applications in which adhesives represent a type of flexible damping element

Before adhesion can be initiated, the surface must be thoroughly cleaned of dirt and foreign materials. This can be accomplished by washing with warm water, to which may be added weakly acidic, neutral, or weakly alkaline rinsing agents (pH 4.5–7.5). Greasy or oily surfaces are cleaned with ethyl, isopropyl, isobutyl, or other aromatic-free alcohols.

3.9.4.2.1 Bonding with Adhesive Solvent and Adhesive Coating

Polycarbonate can be bonded with the aid of methylene chloride (boiling point 41 °C) or ethylene dichloride (boiling point 84 °C). For this adhesion process, one of the areas to be joined is immersed in the adhesion solvent for 5–10 seconds, then pressed with a contact pressure of 0.5–1.0 N/mm^2 to the second joining area. After a brief period of maintaining the contact pressure, the solvent agent is evaporated at 80–90 °C for approximately 6 hours. If the adhesions are to be exposed for a short time to high temperatures or boiling water, annealing for at least 5 hours at 120 °C is required.

As an adhesive, methylene chloride is more difficult to handle, but after the joining process it can be removed more quickly and more completely. In the interest of easier handling, a mixture of ethylene and methylene chlorides is often used, to prevent the adhesive solvent from evaporating as quickly as it does with pure methylene chloride. In all applications requiring that the solvent evaporate very slowly out of the adhesive solvent, a solution of approximately 5–8% polycarbonate in methylene chloride is used as an adhesive "dope." It is important that the adhesive dopant be thinly applied. Here, too, the solvent must be removed through annealing. Bonding with adhesive solvent or adhesive dopes yields high strength bonds after complete removal of the solvent. Joining polycarbonate with other materials is possible with this method only if the second material can also be dissolved by the solvent.

3.9.4.2.2 Bonding with Reactive Adhesives

Various companies have developed specific reactive adhesives, which can also be used successfully with polycarbonate. Reactive adhesives make it possible to bond polycarbonate to many other materials.

The application of reactive adhesives is simple and fast in comparison to adhesive solvents and dopes. The requirements with regard to accurate fitting of the joint areas are not as high. Reactive adhesives with elastic properties after curing are used in the automotive industry (e.g., for gluing lenses of transparent polycarbonate to metal-reflecting, opaque polycarbonate).

Reactive adhesives for polycarbonate based on epoxy resin must be free of low molecular weight amines. Polymeric amino amides can be used as hardeners. Possible reaction of residual amino groups with polycarbonate must be avoided by ensuring that the amino group reacts completely with the epoxy group.

Two-component and one-component polyurethane adhesives have also proven successful in joining of polycarbonate, but they must be free of solvents and amines. Silicone adhesives are particularly well suited as joint–gap–filler systems (e.g., for glazing of industrial and greenhouse windows). Cyanoacrylate adhesives, which are very suitable for quick attachment of polycarbonate, should be used only to glue stress-free parts that will not be subjected to hydrolytic loads during use.

Typically, bonding of polycarbonate does not introduce additional stresses into parts to be joined. This is an advantage in comparison to various welding methods because it permits the use of bonding in molded parts that can tolerate only minimal inherent stresses.

3.9.4.3 Welding [2]

Polycarbonate is very suitable for welding and can achieve weld-line strengths comparable to the strength of the material. Several methods are available for welding of polycarbonate semifinished products, molded parts, and film.

Additive such as dyestuff and fillers as well as special grades may reduce the suitability of welding, but the presence of standard mold release agents does not affect this process. Different polycarbonate grades (e.g., unreinforced and glass fiber reinforced polycarbonate) can easily be welded to each other. Polycarbonate can also be welded with its blends or even with plastics of other types, such as polymethyl methacrylate.

In general, local thermal loads in polycarbonate parts due to the welding process lead to frozen-in stresses in the weld-line area. Therefore, for low stress requirements, it is useful to anneal the welded parts (compare Section 3.9.2) or, if possible, to weld molded parts when they are warm.

In the thermoelastic and/or thermoplastic state, the welding process is conducted at temperatures above 150 °C. Therefore, to avoid bubble formation, parts to be welded must have low moisture content in the joining zone. This sometimes requires predrying of the molded part.

Hot-gas welding and hot plate welding are mainly used in welding semifinished polycarbonate products. In large-scale production runs, polycarbonate molded parts are predominantly welded using the heating element and ultrasonic methods. Although vibration welding is not used very often, this technique may gain in significance. Depending on the application, polycarbonate film can be joined mainly with special types of hot plate welding or by ultrasonic welding. The use of high frequency welding is currently being discussed for film, but to date it has no industrial significance.

3.9.4.3.1 Hot Plate Welding (Mirror Welding) [2–5]

Hot plate welding is used for welding semifinished products as well as for the joining of polycarbonate parts manufactured in mass production volumes.

The processing parameters depend considerably on part size, width of weld line, and specific joining requirements. The temperature of the heating element is 250–270 °C. Since polycarbonate melt shows high adhesion strength to metal, the heating element is coated with a film of polytetrafluoroethylene. Higher temperatures cannot be used with metal elements coated with polytetrafluoroethylene. After warm-up times of 5–30 seconds, depending on the joining requirements, the molded parts are joined with compressive pressure of 1–2 N/mm^2. The hot plate welding method yields high grade weld lines with relatively low frozen-in stresses in the weld-line area. Small parts welded by this method require unfavorably long cycle times.

3.9.4.3.2 Ultrasonic Welding [2, 3, 6–10]

Ultrasonic welding is very good for joining polycarbonate production parts. However, the method is limited to special, precise geometries of the areas to be joined. For polycarbonate, the required heat energy is introduced into the welding zone at a frequency of 20 kHz with amplitudes of 25–40 μm. In small parts, cycle times of less than a second are possible.

Ultrasonic welding is suitable for special molding (e.g., spot welding) but not for joining semifinished polycarbonate parts. The maximum weld-line length per oscillation unit is about 300 mm.

Achievable strengths are comparable to material strength. Ultrasonic weld lines often contain a sharp notch, however, and frozen-in stresses may be greater. Because of the appearance of sharp notches, safety factors must be made part of strength calculations. Frozen-in stresses can be reduced through annealing (see Section 3.9.2).

3.9.4.3.3 Friction Welding [2, 3, 11–14]

By maintaining suitable joint–seam geometries, polycarbonate is easily welded by rotation–friction welding, vibration welding, and angular welding. Being able to use high welding energy makes it possible to achieve cycle times of several seconds even with large molded parts. Vibration and angular welding of polycarbonate is done at a frequency of 240 Hz and contact pressure of 1–3 N/mm^2. During rotation welding of polycarbonate, circumference speeds of 10–15 m/min and contact pressure of 1–3 N/mm^2 are used.

3.9.4.3.4 Hot Gas and/or Hot Air Welding [2, 3, 15, 16]

The use of welding methods based on hot gas or hot air is limited to semifinished product processing. For polycarbonate, the temperature of the airstream 5 mm in front of nozzle is 450–500 °C. As a rule, this welding method does not achieve inherent material strength in the joint zone, and even when conducted most conscientiously, especially in glass fiber reinforced polycarbonate, it produces welds that are low in strength. It should be noted that the weld area must be thoroughly dried beforehand.

3.9.4.3.5 Filament and/or Resistance Welding [2, 3]

During most welding methods, the escape of minute polycarbonate particles from the weld line is unavoidable despite covered weld-line design (particularly during friction and ultrasound welding). If the finished part requires high cleanliness (e.g., parts for use in the medical market, large construction parts with angular welding surfaces), polycarbonate can be welded successfully by the resistance welding method.

3.9.4.3.6 Welding of Polycarbonate Film

Because value for the dielectric constant and the dissipation factor (tan δ) are unacceptably low, high frequency welding at standard frequencies cannot be used for polycarbonate film. In the packaging industry, thin polycarbonate film is bonded by hot sealing at approximately 240 °C. Welding a film 0.25 mm thick requires about 2 seconds.

During heat welding, the polycarbonate is pulled over a wedge-shaped heating element and a subsequent roller unit uses a defined joining pressure to maintain contact between the two pieces.

Polycarbonate film can also be welded by ultrasound if the weld-line geometry is suitable.

3.9.5 Shaping without Cutting [17–19]

3.9.5.1 General

At temperatures between approximately 150 and 220 °C, polycarbonate shows a wide thermoelastic range that makes it possible to thermoform molded parts by bending, vacuum or compressive forming, extrusion, or stamping. To avoid undesirable bubble formation during thermoforming, the molded parts must be dried to a moisture content of less than 0.02% (compare Section 3.8.1.2.1).

3.9.5.2 Heat Forming of Film and Sheet

Polycarbonate sheet and film can be readily formed in vacuum deep-drawing, stretch-drawing, or other similar methods because of the wide thermoelastic range of these materials at temperatures roughly 30–60 K above the melt temperature. Film and sheet can be quickly heated with an infrared radiator. Because of polycarbonate's insensitivity to oxygen, the surface does not become discolored, even at high temperatures. At low deformation degrees, one can also work at 145–160 °C. A relaxation time of 0.5–1.0 hour should be allowed at 145–160 °C to reduce stress. Minor residual forces should be considered during mold dimensioning.

In deep-drawing molds for polycarbonate, surfaces that lie in the direction of the deformation should have a draft angle of about 3–5° to avoid release problems. Back-tapering must also be avoided. The mold surface must be perfect, since the material can form exactly to the surface texture.

3.9.5.3 Bending

If polycarbonate stock (e.g., film, sheet, tubing) is heated to temperatures above melting, preferably 160–180 °C, parts can be quickly formed. Although bending processes on cold semifinished products are possible, they lead to reduced toughness.

The bending of film or sheet to curved bodies is best accomplished by placing the heated, softened molded body under light pressure over a countermold or mandrel of the desired radius. To achieve a shiny surface, polished metal molds are used. To avoid sticking, molds with textile surfaces are also used. Molds of other materials, (e.g., wood, epoxy resin), are also suitable for prototyping or short production runs.

3.9.5.4 Stamping, Pressing

Letters or designs can be stamped into the surface of molded parts of bisphenol A polycarbonate in various ways. In general, however, letters are stamped into sheet or other molded parts with a stamp heated to 120–140 °C. To avoid unnecessary stresses, this method should not be used for stamping depths exceeding 0.1–0.2 mm. By using commercial stamping film, decorative letters can be imprinted. Film or sheet can be stamped continuously if the sheet is run through a stamping calender immediately after manufacture.

3.9.6 Cutting and Machining [17–19]

3.9.6.1 General

Polycarbonate can be easily machined with commonly used steel machining tools. Thanks to its high toughness and relatively high softening temperature, the tendencies toward running or adhesion are low.

As is customary with other plastics, polycarbonate must be machined with very sharp cutting tools made of hard alloys, using cutting angles such as those used for machining alloy steel. Properly set cutting conditions create smooth, even, continuous chips and smooth cutting surfaces that show no macroscopically visible grooves. Experience has shown that this can be achieved with tools of varying cutting angles by varying the cutting speed and feed rate.

General process recommendations are as follows: high cutting rates, slow feed, and relatively great chip depth. Cooling is done with compressed air or pure water. Polycarbonate is not resistant to the oil emulsions and cutting oils generally used during metal machining. Thus to avoid the formation of cracks on the part, these oils should not be used for cooling. It is very important to achieve excellent surface quality for tools.

Stock material such as round bars may contain inherent production stresses. Therefore, semifinished products are stress-relieved before machining by annealing at approximately 130–135 °C. The recommended annealing time is about 30 minutes per initial 5 mm of wall thickness. The cooling, and possibly also the warming up, should be conducted slowly.

3.9.6.2 Turning, Milling

Polycarbonate is easily machined on a turning lathe. For this purpose, cutting steels customary for metal machining are also used for polycarbonate. The tip of the tool should be well rounded and should have a radius of approximately 0.5 mm. Smooth surfaces are achieved with a very sharp cutting steel at cutting rates between 100 and 300 m/min, and feed of approximately 0.3–0.5 mm/revolution.

Polycarbonate can easily be milled with a milling machine. Most advantageous are cutting rates between 100 and 500 m/min at a feed of 0.1–0.5 mm/revolution. The miller should have a side slope of up to 10° as well as a front clearance angle of 5–10°. Cutter heads have also proven successful in the machining of plastics. The most preferred are plain milling cutters with sloping blades.

3.9.6.3 Drilling

Semifinished polycarbonate products (e.g., sheet) can be machined with commercial metal drills. At small wall thicknesses, no cooling is necessary; larger part thicknesses require air cooling. Drilling and cutting oils normally used during metal machining should not be used. Spiral drills should have a front clearance angle of 5–8°, a side slope of 3–5°, an acute angle of 60–90°, and a twisting angle of 12–16°. For larger holes, a commercial disk cutter with a positive side slope is suitable. The cutting rate is 20–60 m/min at a feed of 0.1–0.5 mm/revolution.

3.9.6.4 Sawing

Polycarbonate parts can be machined with bandsaws, circular saws, and hacksaws. During hand sawing, wood- or metal-cutting saw blades can be used. Experience has shown that the best circular saws have slightly crossed or back-sloped blades 1.0–2.5 mm thick, with the thickness depending on the diameter of the blade. Corresponding to the thickness of the part, the best saw blades are those with a tooth distribution of 4–8 per millimeter. The side slope should be 5–15° and the front clearance angle 20–40°. The cutting rate is generally between 2000 and 4000 rpm.

Band saws yield good results at cutting rates between 600 and 1000 rpm at sheet thicknesses of 0.7–2.5 mm, and belt widths between 10 and 20 mm at a tooth distribution of 2.5–3.5 per millimeter.

3.9.6.5 Screw Cutting

For screw cutting of polycarbonate, threaded drills of the type usually used for steel are recommended. These drills must be very sharp and must not have been used for metal machining.

Predrilling should be approximately 0.1 mm larger than normally is done with steel. Lubricant cutting oil must be avoided. Since the plastic encounters the drill with considerable resistance, the drill will become dull relatively quickly; therefore, molybdenum disulfide or paraffin oils may be used for lubrication. When using special self-tapping screws, the surrounding wall must be sufficiently thick; no aggressive agents are permitted in the work environment.

3.9.6.6 Stamping and Cutting

Polycarbonate parts can be stamped up to a thickness of approximately 3 mm with standard metal stamping tools. The stamp should be heated to 140–175 °C and should have a horizontal slope of about 5–6°. At 3–4 mm sheet thickness, the cutting slit should be 0.01 mm.

Using hammer shears, sheet with a wall thickness of 1–2.5 mm can be cleanly cut at a lip angle of 40° and play of 0.01–0.2 mm between the blades.

References for Section 3.9

1. *Schönfelder, M.:* Deutsche Farbenzeitschrift (1979) 2, p. 59–64
2. DVS-Taschenbuch, DVS-Richtlinien und Merkblätter. DVS-Verlag, Düsseldorf 1987
3. DIN 1 910, Part III. Beuth-Verlag, Berlin (1977)
4. *Potente, H.:* Kunststoffe 67 (1977) 2, p. 98–102
5. *Potente, H.; Kreiter, J.; Mohrmann, A.:* Kunststoffe 78 (1988) 1, p. 84–87
6. *El Barbari, N.:* Diss., IKV Aachen 1988
7. *Menges, G.; Michel, J.; El Barbari, N.:* Ermittlung von Materialkennwerten und Arbeitsdiagrammen für den Einsatz von Ultraschall bei der Verbindung von Kunststoffen. Abschlußbericht zum AIF-Vorhaben Nr. 5385, IKV-Aachen, 1985
8. Fügen von Formteilen und Halbzeugen aus thermoplastischen Kunststoffen mit Ultraschall. Zentralverband der Elektrotechnischen Industrie e. V., Frankfurt 1983

9. *Land, W.; Erlenkämper, E.:* Kunststoffe 75 (1985) 6, p. 315–320
10. *Land, W.:* DVS-Bericht, Vol. 84. Deutscher Verlag für Schweißtechnik, Düsseldorf 1983
11. *Zasada, H. U.:* Konstruktion 29 (1977) p. 397–403
12. *Potente, H.; Reinke, M.:* Verbindungstechnik 12 (1980) p. 65–67
13. *Potente, H.; Michel, P.; Ruthmann, B.:* Kunststoffe 77 (1987) 7, p. 711–716
14. *Stokes, V. K.:* Polym. Eng. Sci. 28 (1988), Part 1, p. 718–727; Part 2, p. 728–739; Part 3, p. 989–997; Part 4, p. 998–1008
15. *Menges, G.; Alf, E.:* Kunststoffberater 18 (1973), Part 1: 3, p. 149–156; Part 2: 4, p. 241–245; Part 3: 5, p. 345–349; Part 4: 6, p. 427–431
16. *Vohwinkel, H.:* Kunststoffe 71 (1981) 10, p. 753–757
17. Makrolon-Platten – Bearbeiten. Company brochure Röhm GmbH, Darmstadt 1980
18. Makrolon-Platten – Umformen. Company brochure Röhm GmbH, Darmstadt 1980
19. *Szigeti. P. R.:* Kunststoffe 70 (1980) 7, p. 420–422

3.10 Application Areas for Polycarbonate

Because of the above-described good properties as well as the capability of tailoring polycarbonate grades toward specific requirements, this material has found multiple application possibilities in almost all industrial areas.

When bisphenol A polycarbonate was introduced in 1959 (Bayer AG), its high heat distortion temperature and its suitability as an electrical insulator were utilized in the replacement of metal housing and ceramic insulators in electrical–electronics technology. The additional transparency as well as high strength and stiffness of glass fiber reinforced grades and their relatively easy thermoplastic processibility enable a broad substitution for conventional materials. Although the replacement of metal and ceramic is already virtually completed in Europe, the substitution of thermoset plastics still holds promise.

The balanced property profile in connection with an almost limitless design freedom has led to substitution processes that were not obvious at first. In numerous applications today, the less expensive resin is replaced by the relatively more expensive polycarbonate because polycarbonate enables the manufacture of functionally better molded parts that can be transported more easily and at less expense. Substitution of glass by the much tougher polycarbonate is also occurring. The low weight of polycarbonate glazing is often the deciding factor.

3.10.1 Electrotechnology

The special advantages that have destined polycarbonate for use in the electronics market are its electrical insulation values, which are independent of surrounding conditions, and its burn behavior, high heat distortion temperature, impact strength, transparency, and dimensional stability.

Optimal insulation resistance is important in parts carrying live current for protection against electric shock and separation of electrical elements from each other. Polycarbonate shows excellent properties with regard to surface resistance, dielectric strength, and specific volume resistivity. The values are usually independent of temperature and humidity.

Fire protection is the basic idea behind the constantly increasing demands for product made of fire-resistant grades of polycarbonate. Basic grades achieve UL 94 V2 rating starting at 0.8 mm. Specific fire-resistant product types achieve V0 at 0.8 and/or 1.6 mm (see also Section 3.7.8).

Electrical instruments used in industry may reach temperatures exceeding 100 °C; this service must not negatively affect functional fitness and dimensional stability. Polycarbonate can be exposed to low mechanical loads up to 135 °C; in glass fiber reinforced form, loads may be imposed at temperatures as high as 145 °C. The ball indentation test according to VDE 0470 yields values of 136–138 °C.

In many cases electrical instruments must absorb impact and shock stress without safety risk. Polycarbonate has excellent impact strength over a wide temperature range and is rarely surpassed by other materials in this regard.

Transparency is significant where visual control of installed instruments is required. For this reason, transparent, colorless polycarbonate is often used for housings and/or instrument covers. It is the combination of requirements and achievable properties that makes possible the multitude of applications that can be realized with polycarbonate. Some examples are housings for relay equipment, lamp holders, fluorescent strip sockets and holders, connector and plug strips, relays, safety covers, high voltage plugs, coil formers, electrical meters, distributor box covers, and housings (Figure 3.104), street light lamp covers (Figure 3.105), and various electrical household appliances (Figure 3.106).

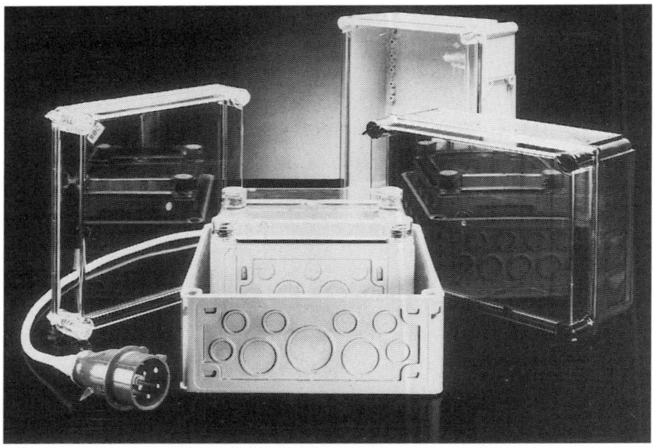

Figure 3.104. Distributor box cover (transparent polycarbonate) and housing of 20% glass fiber polycarbonate.

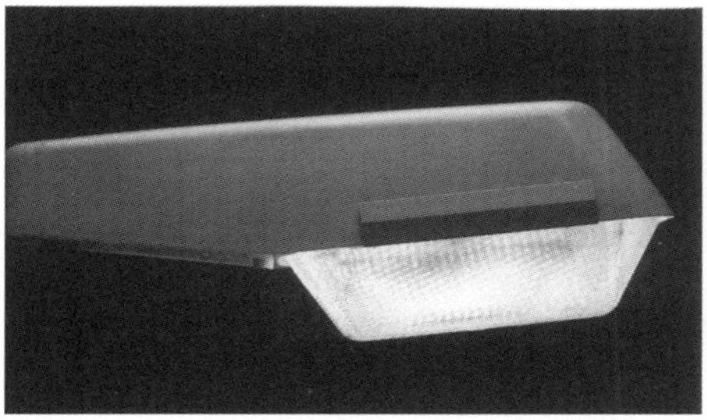

Figure 3.105. Sheetlight covers of UV-stabilized polycarbonate.

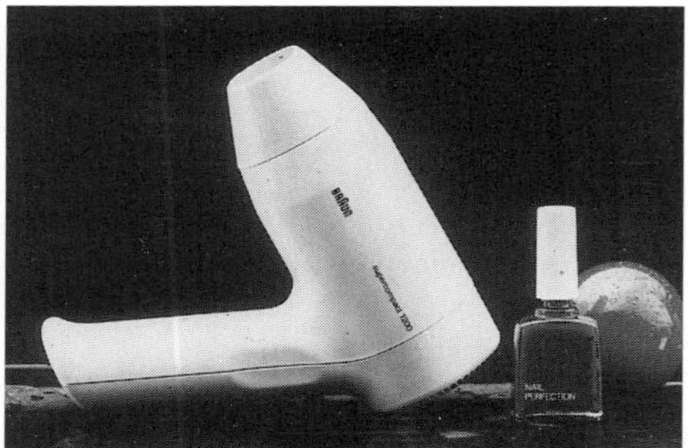

Figure 3.106. Hair dryer of flame-retardant, unreinforced polycarbonate.

Insulation-encapsulated distributor equipment with bottom boxes of glass fiber reinforced polycarbonate, and covers made of transparent polycarbonate instead of metal, molded plastic, and ceramics were the first (1959) applications for polycarbonate.

In addition to cost-effective processibility, the material requirements were as follows: insulation capability, freedom from contact corrosion, sufficient flame retardance, good dimensional stability, high service temperature, and a favorable balance of strength and toughness. Component parts such as switches and safety fuses can be mounted directly on a chassis plate. Even after 30 years, no better material than polycarbonate has been found for this application.

Good insulating properties alone are not a sufficient argument for using polycarbonate in electronic technology. Only the combination of at least one other positive property leads to the selection of polycarbonate.

For connectors, the first function of the embedding material is insulation and fixation. Stringent requirements for dimensional stability as well as flame retardance point to polycarbonate as a high performance material. For luminescent diodes, polycarbonate serves as an encasement for optical-electronic construction components in the form of a transparent and/or translucent insulator with a heat distortion temperature that enables the soldering of joints [1]. In this case, polycarbonate can be more economically processed than epoxy resins.

For film and slide projectors, polycarbonate has replaced metal castings and sheet iron. Current-carrying parts can be mounted on baseplates of reinforced polycarbonate, as well. In addition to good electrical insulation values, baseplates and housings must show high heat distortion temperature and flame retardance. Good insulation properties and flame retardance in connection with abrasion resistance and high dimensional stability of 10% glass fiber reinforced polycarbonate were arguments for the selection of this material as identity and credit card readers.

Although polyamide is superior to polycarbonate during fatigue tests, the latter has better dimensional stability and meets the stringent requirements for drill housings.

In addition to insulation requirements, industrial and municipal lamp holders must exhibit dimensional stability at high temperatures. Light-tube holders of polycarbonate, which fulfill these requirements, have been used in Europe for many years, replacing melamine resins for electrical outlets and wall switches in industrial and household uses. The broad design freedom offered by polycarbonate in comparison to other molded plastics makes it possible to meet the increasing call for optically demanding switch elements. For picture elements in the mosaic picture technology, polycarbonate shows high dimensional accuracy and stability as well as flame retardation properties.

The vast number of electrical and electronic instruments leads to a multitude of required property profiles. Because electrical kitchen equipment comes into contact with food, and depending on the function of the part, such products also must comply with different national regulations and/or standards. Coil formers that serve at high temperatures require stiff, thin-walled molded parts capable of handling long-term stress; glass fiber reinforced polycarbonate with the highest possible molecular weight has proven successful. A very special property profile is known from satellite technology, which requires a low temperature insulation material from which almost no components outgas into the vacuum. Here too, polycarbonate has proven successful.

New requirements that have surfaced recently include flame-retardant materials characterized by low corrosion and toxicity of flue gas. The very minor amounts of flame-retardant additives of many commercial polycarbonate products do not increase the corrosivity of flue gases during fires; according to DIN 57472, these materials can be classified as halogen-free for cable encapsulations. The switch to such polycarbonate grades reduces secondary damage from fires, such as damage to sensitive electronic instruments or corrosion or fastening irons reinforced concrete. In nuclear power plants, flame-retardant polycarbonate cable channels have replaced cables made of polyvinyl chloride.

Approximately half the polycarbonate produced worldwide is used in the electronics area. The substitution of molded plastic for conventional materials such as glass, metal, and ceramics is not yet complete. Further substitution of these products, additional substitutions for polyvinyl chloride, as well as continued innovations and growth of the electronic sector, will lead to a continued increase in the use of polycarbonate.

3.10.2 Polycarbonate Sheet, Glazing [2]

The combination of high transparency with high heat stability at a competitive price makes polycarbonate an ideal material for transparent sheet. Although glass and polymethyl methacrylate are superior to polycarbonate with regard to scratch resistance, polycarbonate has no equal when fracture resistance is also required.

Drawbacks to standard polycarbonate include tendency to turn yellow, loss of light transmission, and reduction in toughness after long-term outdoor exposure. These drawbacks are largely eliminated by applying a UV-protective covering (coating, coextrusion). For applications requiring increased scratch resistance, polycarbonate sheet can be equipped with special scratch-resistant coatings, preferably based on silicone.

Polycarbonate is processed via extrusion into flat sheet, twin-wall sheets, and/or triple-wall sheet. Material consumption is approximately the same for flat and walled types of sheet. Applications characterized by greater surface area per unit weight favor twin-wall sheet. Flat sheet is used whenever other window glazing cannot handle the necessary weight or safety requirements. An excellent example of the use of sheet to meet requirements of optical quality and toughness is the perimeter surrounding the field of play in ice hockey stadiums. Comparable alternatives are unthinkable. Flat sheets of polycarbonate are used as glazing of very-large-area roof constructions. During remodeling and replacement of outdated construction materials, the low weight of polycarbonate sheet is an advantage, especially if it eliminates the need to reinforce sheet mountings weakened through age. Within the play areas of schools, kindergartens, and similar buildings, the use of polycarbonate sheet contributes to increased safety. Its toughness, for example, prevents the kinds of injury due to glass splinters. Polycarbonate sheet is preferred as glazing in areas subject to vandalism, as well. The only strength limits of such window glazing materials are encountered if the unreinforced polycarbonate sheet can be pushed from its mounting by bending.

Polycarbonate twin-wall sheet is preferred for use in greenhouses and industrial window glazings. Its excellent thermal insulation, ease of installation, and low surface weight at high part stiffness form a property combination that has resulted in a considerable and continuously increasing market during the last several years. By transforming the initial box profile into an arch profile with reduced tensile stress in the transition areas between belt and web (see above: Figure 3.88), considerably tougher sheet can be manufactured. UV coextruded or surface coating simultaneously counteracts degradation and notching due to microcracks in the critical surface. This optimized sheet fulfills the requirements for long-term weathering use and resistance to puncture-like impact loads (e.g., hail).

Twin-wall sheet, which is manufactured with a multitude of chamber and connection profile geometries (see Figure 3.87), is used mainly for glazing with thermal insulation functions. The heat transfer coefficient of 3.9 $W/(m^2 \cdot K)$ for a 4 mm twin-wall sheet [in comparison to 5.7 $W/(m^2 \cdot K)$ for a 4 mm glass sheet and 5.3 $W/(m^2 \cdot K)$ for a 4 mm thick polycarbonate flat sheet] shows considerable energy savings during periods requiring indoor heating. Volumewise, the most significant application is greenhouse glazing. With only slightly reduced total light transmission from approximately 86% of a 4 mm flat sheet to approximately 80% for a 4 mm twin-wall sheet, plant growth is virtually unaffected. Greenhouses with multiple-wall sheet can be found everywhere, from polar regions (for protection against low temperatures) to desert regions (for protection against dehydration).

Another significant share of multiple-wall sheet is found in numerous private hobby applications and in noncommercial greenhouses. As an economical and functional construction element, the twin-wall sheet has multiple uses in many industrial areas. Because of its special geometry, it fulfills many flammability classifications for the construction sector. Additional applications are winter gardens and temporary roofs. The sheet market, which accounts for one-fourth of worldwide polycarbonate production, is a very significant application. The replacement of glass by polycarbonate for glazing purposes is not yet complete. It is expected that the market significance of polycarbonate sheet will increase during the coming years, with a resulting increase in polycarbonate production of more than 25%.

3.10.3 Transportation

Applications in the transportation sector make use of the impact strength, heat distortion temperature, and, for outdoor applications, weatherability of polycarbonate. This versatility is demonstrated in a series of applications such as reflectors, taillight covers (Figure 3.107), motorcycle windshield visors (Figure 3.108), vent and cooling grids, interior lights, covers for electrical powerhouses, windshield wiper brackets, car fuses, equipment and coverings used inside aircraft, park clocks, and traffic signals with inside lights.

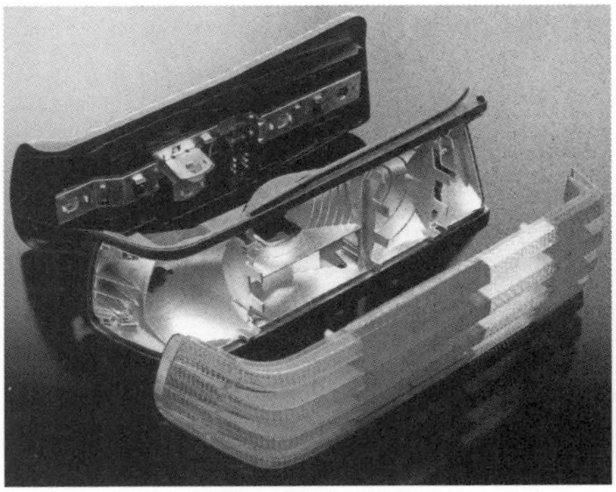

Figure 3.107. Taillight covers of UV-stabilized, easy-flowing polycarbonate.

Figure 3.108. Motorcycle windshield visor of UV-stabilized polycarbonate with medium viscosity.

Many transportation applications utilize the transparency of polycarbonate, particularly in the signal area and as headlight lenses with scratch-resistant coatings. Because it is only poorly resistant to contact with fuels and solvents, polycarbonate is predominantly used in modified form as well as in the form of blends for exterior parts. Even though large volumes of polycarbonate are used for bumpers, these items consist exclusively of blends. With headlight lenses (Figure 3.109), the required scratch-resistant coating of the polycarbonate affords protection against aggressive chemical attack. The great design freedom available with polycarbonates promises increased future sales. The increasing integration of headlights into the automotive chassis is expected to result in a substantial increase in volume for polycarbonate use.

Figure 3.109. Headlight lenses of easy-flowing, UV-stabilized polycarbonate.

The multitude of currently possible polycarbonate applications in the transportation market is approximately 11% of overall polycarbonate production (without blends). In addition, polycarbonate in the form of blends with ABS or polyester, preferably polybutylene terephthalate, has greater significance in the automotive area.

3.10.4 Film

Applications for film are divided into the following areas:

- extremely high molecular weight film, which can be manufactured only via a solution as cast film
- high molecular weight extrusion film
- coextruded composite film

The thermoplastic processing of polycarbonates with a molecular mass exceeding 35,000 is no longer practical. However, these materials can be cast from a solution directly into film. The process of achieving optically flawless film is difficult. By using proper cast and stretching techniques, high molecular weight cast film can be manufactured in the longitudinal as well as the crystalline, longitudinally stretched form. The difference is in the tensile strength. Isotropic cast film shows, according to DIN 53455, a tensile strength greater than 80 N/mm^2; longitudinal shows a tensile strength more than 100 N/mm^2. A crystalline as well as longitudinally stretched film shows a tensile strength more than 220 N/mm^2. Crystallization also improves the heat distortion temperature.

Major applications for cast film are in the areas of electrical insulation film for line transformers and aluminum-metallized film for condensers.

Other application areas are diaphragms for loudspeakers and telephone receivers, identity and credit cards and coating film for ski. The solution technology of cast film manufacture provides the interesting possibility of incorporating additives and aggregates that cannot be mixed into such highly viscous systems by compounding in the conventional sense. One practical example is the incorporation of conductive carbon black in an amount that permits a specific volume resistivity of $0.8\,\Omega \cdot cm$ and thus yields film that can be described as electrically conductive.

Such conductive film can find application as a high grade antistatic medium as well as in the manufacture of so-called typing tapes. The typing tapes are designed to consist of a three-layered composite: a carbon-black-filled cast film with a vacuum-metallized, thin aluminum layer, onto which a dyestuff layer is applied as the third component.

Over a typing head with electrically conductive elements, a current flow is created between the typing head and the aluminum layer; this current heats the metal layer and, by melting the dyestuff, makes the dyestuff layer available for a printing process (electrothermal ribbon).

Measured in tonnage, cast film has relatively minor significance in the overall polycarbonate film market. However, since cast film is used at very low thicknesses (preferably 15–100 μm), a large proportion of the film surface is used.

Extruded polycarbonate films with molecular mass in the 35,000 range are used in thicknesses of 125–500 μm. They are available as glass-clear products with various surface structures and as film with high inner light scattering due to special filler materials; they also come equipped with flame-retardant and antistatic systems. These films are used as technical high grade and decorative information carriers in many industrial areas. Examples include check-cashing and credit cards, decorative and functional instrument screens, advertising displays, commercial and industrial signage, surface covers for film keyboards, instrument dials and scales, and control and warning signals for automotive instruments. In the nonoptical area extruded polycarbonate film can be found as telephone and loudspeaker diaphragms, cover plates and packaging, and film switches, as well as insulation and various types of vacuum thermoformed parts.

New, transparent film for vacuum thermoforming with permanent antistatic equipment makes safety packaging possible for high grade electronic instruments. The specific optical, mechanical, and thermal properties of polycarbonate film enable the highly accurate reproduction of larger formats and color drying at higher temperatures, high color brilliance, and during reverse-side printing with the protection of the printed area from scratches and abrasion. Adhesion is also possible, as well as the use of punched film parts as self-adhesive labels.

With a good stretchability, extruded polycarbonate films can be thermally deformed without problems. Vacuum formed films show little backspring as well as low relaxation and have, therefore, high outline sharpness and dimensional stability.

A third application of polycarbonate films, which today is less significant economically, involves coextruded blow-molded and flat films. Various coextrusion composite films can be realized, such as:

- polypropylene/adhesive/polyvinylidene chloride/adhesive/polycarbonate
- polypropylene/adhesive/polycarbonate
- polyethylene/adhesive/polycarbonate
- ethylene–vinyl acetate/polycarbonate
- polyethylene terephthalate/polycarbonate.

In these coextrudates polycarbonate takes over the function as the deep-drawable, heat distortion temperature carrier and permits high form stability during hot filling and steam sterilization of foodstuffs. The second component, as a barrier material, counters the disadvantage of polycarbonate's high oxygen and carbon dioxide permeability. Depending on the composite construction, it is possible to obtain transparent, glossy films that accept printing well and have excellent stiffness and heat distortion temperature.

To adjust the viscosity of the polycarbonate melt to the melts of the adhesion and barrier materials, low molecular weight types of polycarbonate are used ($M_w = 22{,}000$). Such low molecular weight polycarbonates alone can be extruded only under certain preconditions, but coextrusion is readily possibe in composites, with the barrier material heated up to approximately 230 °C. Polyethylene can even be coextruded at coating temperatures of approximately 310 °C without problems.

Coextrusion with polyethylene is also possible on standard extruders for polyethylene processing. Starve feeding of the screw during start-up should be watched for when grooved intake bushings are used.

The developement of polycarbonate as a coextrusion partner in packaging film is anticipated. Because of the higher stiffness of polycarbonate in contrast to polyethylene, such composites permit wall thickness reductions and thus lower packaging weight, in addition to the thermal advantages already noted. The question of packaging disposal will open more opportunities for polycarbonate as polymers containing halogen atoms become less favorable.

Paper–polycarbonate composites are being tested with success. The coating of paper–aluminum composites can be achieved, introducing the possibility of additional applications in the packaging market.

3.10.5 Structural Foam [3]

Polycarbonate can be foamed via direct gassing of the melt with nitrogen or with the aid of chemical blowing agents during injection molding (see Section 3.8.1.3.3). Molded foams show an integral density distribution with almost pore-free surface skin and closed-celled light foam in the core. During normal injection molding of foamable thermoplastics (FTP), the surface is rough and marbled and for most applications requires a decorative surface coating. Two-component polyurethane coatings have proven to be especially effective. The gas–counterpressure method yields almost smooth surfaces, but volumetric weights (mostly > 1.0 g/cm^3) are relatively high. FTP processing yields foam densities of 0.9 g/cm^3 (approx. 30% density reduction) and sometimes even less.

Sink marks, which may appear during the manufacture of various molded parts with rib/wall thickness ratios greater than 2.5:1, can be avoided by light foaming and a density reduction of less than 5%. This application is a specialized form of standard injection molding. With polycarbonate molded foams in the general sense, the primary purpose of the density reduction of 20–30% is not to promote savings in cost and weight; it is used much more as a technical aid for the manufacture of large-area or large-volume parts. The scope of polycarbonate molded forms includes the manufacture of molded parts that either cannot be made at all by standard injection molding or would, if produced, have lower quality.

The following can be used as decomposition foaming agents: 5-phenyltetrazole, 3-phenyloxadehydrodiazinon-(6), isatoacid anhydride, citric acid esters. The preferred method of application is not pure blowing agents, but rather by easy-to-handle concentrates, which can be purchased from polycarbonate manufacturers. The concentrates are added to the polycarbonate at 2–10%, depending on the type of concentrate and the foaming task.

The best foamable polycarbonates are average molecular weight types with 5% glass fibers as both reinforcement and nucleation material. Even product types with 10–30% glass fibers are offered for foaming. The latter show higher stiffness and strength, but lower toughness, and they allow limited foaming only to densities below 1.0 g/cm^3. However, eliminating glass fibers does not improve toughness, since the bubble structure causes internal notching, which is accompanied by a loss in toughness.

Important industrial properties of a typical polycarbonate foam are summarized in Table 3.25. In comparison to other thermoplastic foams with a density of 0.9–

Table 3.25. Typical Values for Important Industrial Properties of Polycarbonate Structural Foam with 5 wt% Glass Fibers, Using the Example of Makrolon SF 800 Z

Properties	Units	Value	Test Method
Mechanical Properties			
Tensile-strength at break	N/mm	37	ISO/R 527, DIN 53455
Elongation at break	%	6	
5% Flexural stress			
23 °C	N/mm^2	70	ISO 178, DIN 43452
60 °C		60	
Modulus of elasticity	N/mm^2	1900	DIN 53457
Flexural modulus			
23 °C	N/mm^2	2200	DIN 53457-B3
60 °C			
Charpy impact strength			
23 °C	kJ/m^2	70	ISO 179, DIN 53453
−20 °C	kJ/m^2	70	
Izod impact strength			
23 °C	J/m	500	ASTM D 256,
−20 °C	J/m	500	Test bar 6.4 mm
Total penetration energy	N.m	50	DIN 53443 Part 2
Ball indentation hardness H 30	N/mm^2	50	ISO 2039/B,
Thermal Properties			
Coefficient of linear thermal expansion	10^{-4} K^{-1}	0.6	DIN VDE 0304, DIN 52752-B
Thermal conductivity	W/(m · K)	0.17	DIN 52612
Vicat softening temperature VST/B 120	°C	135	DIN ISO 306, DIN 53460
Heat distortion temperature			ISO 75, DIN 53461
Method A (1.80 N/mm^2)	°C	126	
Method B (0.45 N/mm^2)	°C	131	
Flammability according to UL 94 (wall thickness: 6.4 mm)		V-O	UL 94, IEC 707
Electrical Properties			
Dielectric strength E, (transformer oil) (test bar: 6.4 mm)	kV/mm	>8	IEC 243, DIN VDE 0303, Part 2
Specific surface resistance	Ω	10^{16}	IEC 93, DIN VDE 0303, Part 3
Specific volume resistivity	Ω · cm	10^{16}	IEC 93, DIN VDE 0303 Part 3

1.0 g/cm^3, polycarbonate represents a tough, stiff, very temperature-resistant material with good flame-retardant properties and a balanced price/performance ratio. The application possibilities of this material class are surely not yet fully utilized. One major current application for polycarbonate structural foam consists of cable distributor sockets as well as cable distributor boxes (Figure 3.110). Individual parts for this application weight more than 15 kg. The suitability of polycarbonate structural foam for this area of application was evaluated by the German Federal Postal Service in local tests under various environmental conditions. During industrial use the material has proven successful during outdoor weathering for many years. The thermoplastic design of these molded parts achieved considerable weight reduction in comparison to concrete and metal construction used formerly; in addition, the polycarbonate equipment is much easier to transport and install.

Another important area of application is housing parts for electronic instruments, particularly office machines. The high toughness of polycarbonate structural foam

Figure 3.110. Cable distributor box. Housing with base-molded support of polycarbonate structural foam; density approx. 0.95 g/cm^3. (Plant photo: Stewing, Dorsten).

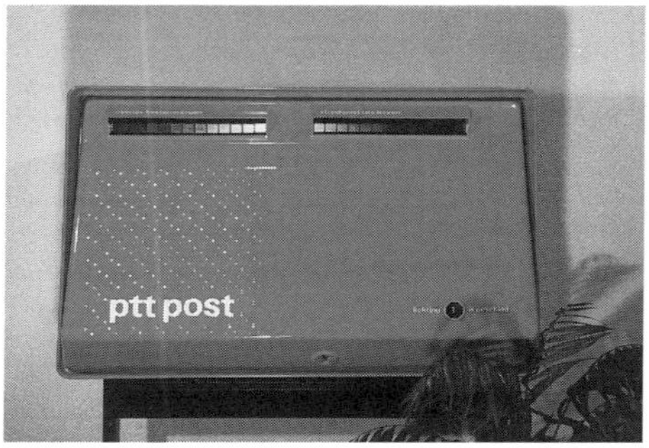

Figure 3.111. Mailbox of polycarbonate structural foam (Plant photo: Hogro BV, Waalwijk/Holland).

Figure 3.112. Streetlight housing of polycarbonate structural foam. (Plant photo: Hogro BV, Wallwijk/Holland).

provides the high grade instruments with the necessary protection with regard to flame retardation as well as a heat distortion temperature high enough to ensure safety.

As with standard polycarbonate parts, good electrical insulation values are utilized with structural foam to connect current-carrying parts directly with a foamed polycarbonate housing and/or chassis. Polycarbonate structural foam is used for mailboxes (Figure 3.111), large-area parts from traffic signs, streetlight parts (Figure 3.112), light poles, and bases for television sets. Car roofs of this material have already been tested. Commercial production of telephone booths begain in 1989.

Polycarbonate structural foam can also be manufactured by extrusion; densities below 0.6 g/cm^3 can be realized at very good surface quality. Good extrudability in combination with good thermal and mechanical properties, even at this density, opens for this material class a completely new range of application possibilities.

3.10.6 Medical Technology

Polycarbonate is often used for manufacturing medical instruments. Applications are in surgery and organ support, in nursing and wound care, as medication dispensers, and in laboratory apparatus. Polycarbonate brings to these tasks transparency, toughness, and sufficient temperature stability for hot air and/or hot steam sterilization.

Instruments such as blood oxygenators, dialysis machines, and connector elements (Figure 3.113) for blood and infusion solutions must allow unhampered visual monitoring to ensure freedom from bubbles. Such unrestricted visibility is possible only with a clear, transparent material such as polycarbonate. Many instruments for the medical

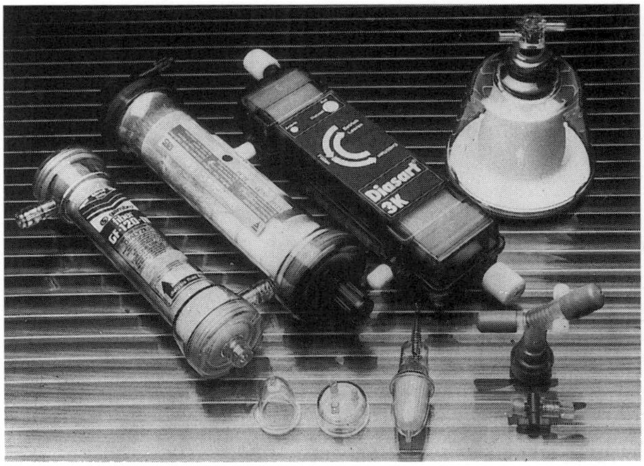

Figure 3.113. Medical instruments, blood centrifuge, dialysis instruments, and hose connector of medium viscosity polycarbonate with mold release agent.

market consist of safety parts that require high strength against failure. They must also withstand disinfection by standard sterilization methods (e.g., hot steam, hot air, radiation, and ethylene oxide sterilization). Polycarbonate fulfills these requirements; specific product types fulfill the requirements of the Class VI test for plastics of the U.S. Pharmacopeia.

Culture flasks, animal drinking bottles, urine bottles, sterilization bottles, and animal cages in particular require ability to undergo repeated sterilization, as well as transparency, toughness, and economical processibility. Here again polycarbonate's high purity and its property of neither absorbing nor emitting substances are of significance. Animal cages are subjected to an especially demanding environment: without reduction in utility, they must resist the gnawing of laboratory animals and withstand mechanical loads during use, as well as the cleaning process and numerous hot steam sterilization cycles.

Oxygen masks, baby bottles, breast pumps, and medicine flasks are all applications in which the material comes in contact with mucous membranes, in addition, the device serves as a receptacle for food or medicine. The high toughness of polycarbonate guarantees much greater finished-part stability without the danger of injuries from glass breakage, and indeed, the polymer has almost completely replaced glass in these applications. In housings for blood oxygenators as well as filter housings and end caps for dialysis canisters, polycarbonate is in direct contact with blood. Polycarbonate blood pumps are also exposed to additional mechanical loads. No material excels polycarbonate in offering the required property combination of compatibility with blood, transparency, sterilizability, low weight, and high toughness and/or strength. Additional applications that entail contact with blood and with infusion solutions are hose connections and/or distributing elements, connectors, injection parts, and shutoff nozzles, as well as plasma bottles and drainage pumps. Now that sterilization media other than ethylene oxide are being used, it has become possible to substitute polycarbonate for materials with low heat distortion temperature (polycarbonate can undergo hot steam sterilization at 121 °C several times). In this regard, additional applications are opening up for polycarbonate.

No application should be undertaken without a consideration of the sensitivity of polycarbonate to amines, several solvents, and strong bases, as well as to components capable of causing stress-cracking. For example, animal cages must be thoroughly cleaned before hot steam sterilization. Primary or secondary amines adhering to the surface can lead to a type of localized corrosion of the polycarbonate; the same is true for locally high concentrations of strong caustic solutions. Some of the commercially available disinfectants may cause stress-cracking in polycarbonate.

It must be further considered that dioctylphthalate in polyvinyl chloride causes stress-cracking in polycarbonate. Direct contact of these materials must be avoided if the polycarbonate surface contains tensile stresses at the contact point.

3.10.7 Packaging

Because polycarbonate fulfills virtually all food contact regulations, certain grades are suitable as packaging materials for a variety of foods. In addition to physiological suitability, many applications also require transparency, toughness, ease of cleaning, and a hygienic appearance.

The polycarbonate baby bottle, replacing glass, has been a large-volume application for many years worldwide. The high toughness of polycarbonate, its high safety and low weight were deciding factors in this substitution. A glass bottle is better than a bisphenol A polycarbonate based bottle only when many sterilizations at high temperatures (e.g., >121 °C) are required.

Polycarbonate is used for beverage packaging in the form of reusable bottles (Figure 3.114)—units that are cleaned in standard glass-bottle cleaning plants and refilled. The high breakage resistance of stress-relieved bottles permits reuse of 100 times and even more. Reusable polycarbonate bottles are used exclusively for noncarbonated beverages destined for immediate consumption [4].

Figure 3.114. Reusable water bottle of structure-viscous polycarbonate.

Figure 3.115. Large-size kitchenware of medium viscosity polycarbonate.

The advantages of polycarbonate for bottles are retention of dimensional stability and failure to crystallize or become turbid in hot (approx. 80 °C) washing solutions. The disadvantage is high permeability to oxygen and carbon dioxide. Today it is possible to use polycarbonate in the manufacture of packaging for carbon dioxide containing soft drinks and fruit juices by combining barrier plastics [e.g., PE, PP, PET, E/VAL (ethylene vinylalcohol) or PVDC (coating)] in the form of multilayered compounds. These layered compounds can be manufactured by lamination coating, extrusion coating, coextrusion, or coextrusion blow molding.

Applications for single-layer bottles have been on the market since 1975 in the form of reusable milk bottles [5], and for more than 15 years as reusable drinking water containers with a volume capacity up to 20 liters for areas with water potability problems such as parts of California, Thailand, and South America. The manufacture of transparent polycarbonate milking buckets is one of the oldest applications of this material for transport containers. Other applications include bottles that hold the concentrate in fruit juice vending machines, water bottles, juice cans, and food containers of varying sizes (Figure 3.115), as well as cans for automatic dispensing of milk and juices, thermal cans and storage containers for food in private households, and food storage units for use aboard aircraft (Figure 3.116).

Figure 3.116. Food storage unit of polycarbonate; specialty types are used that are suitable for food contact and also pass the FAR 25 Paragraph 853b (USA) burn test.

3.10.8 Household and Consumer Articles

In addition to polycarbonate properties such as good impact strength and high heat distortion temperature, the household and consumer article market may also require transparency, and special polycarbonate grades also may be subject to food contact laws of various countries. Polycarbonate manufacturers offer product types which, in their property composition, conform to the recommendations of the German Federal Health Department and to the U.S. FDA standards, and are on the approved list of the French government.

There is a variety of polycarbonate applications in this area. A modern household is equipped with many electrical appliances in which polycarbonate plays a special role (see Section 3.10.1). Examples are housings for pasta machines, the food-contacting containers in food processors, egg boilers, tanks for flat irons, coffee makers, coffee filters, attachments for electrical grills and microwave ovens, hair rollers, hair-dryer and razor housings (Figure 3.117), various video and sound system parts, as well as lamp parts.

Because it is suitable for food contact as well as sufficiently resistant to hot water, polycarbonate is also used in the manufacture of plastic dishes and silverware of all types, cans, bowls, and even colorful, decorative kitchen sinks (Figure 3.118). It is always important to make sure that polycarbonate parts are stress-free: for example, stressed parts can lead to stress corrosion during many rinsing cycles. In comparison to metal, porcelain, and glass, one must also consider polycarbonate's high susceptibility to scratching and its nonresistance to such solvents as nail-polish remover and various cleaning products and disinfectants. Dishes made of stress-free plastics can be cleaned in household dishwashers almost indefinitely, provided solution concentration and temperature are not too high, and clear-rinse aids incompatible with polycarbonate are not used.

Figure 3.117. Razor housing of unreinforced polycarbonate with back-molded polyurethane.

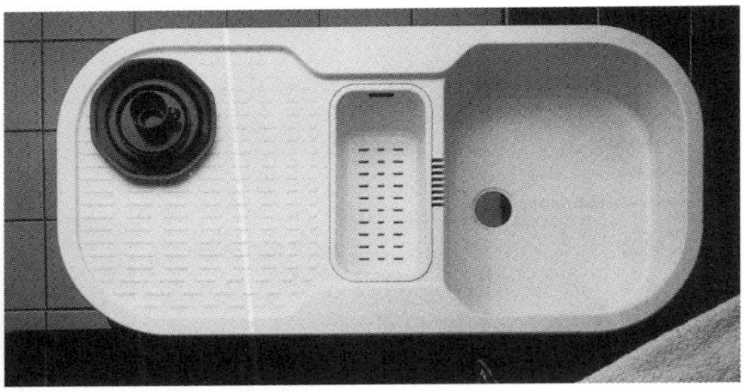

Figure 3.118. Kitchen sink of average viscosity, unreinforced polycarbonate.

Figure 3.119. Camera housings of polycarbonate with 30% short glass fibers.

A novel application is microwave cookware made of polycarbonate. Important requirements here are low weight, low staining from cooked food, low dielectric heating, and high toughness up to a maximum service temperature of approximately 140 °C. Pure fats should not be heated in such vessels, however, since temperatures may rise above 140 °C. The relatively low heat conductivity of the polycarbonate reduces heat losses; the cooler surface makes it possible to handle freshly microwaved containers with bare hands.

3.10.9 Film, Photo, and Optical Instruments

Glass fiber reinforced polycarbonate is an ideal substitute for metal in optical instruments. It can be cost-effectively processed into complicated housing parts with numerous functional elements and ribbing to provide high toughness combined with low tendency toward dimensional changes due to thermal or mechanical loads. Requirements for the use of these materials in lens mountings and camera housings are low creep tendency and dimensional independence from the ambient moisture content. Short glass fiber reinforced polycarbonate is almost unmatched by other thermoplatics in its precise dimensions and dimensional stability. Therefore, it is used to manufacture lens mountings, still and moving cameras (Figure 3.119), binoculars, microscope parts and flashlight instruments. Slide and video projectors made of polycarbonate also utilize the electrical insulation properties and high heat distortion temperatures of this material.

3.10.10 Sport and Leisure

The many different demands on materials used in the sport and leisure market often can be met by polycarbonate. High toughness is the most important argument for polycarbonate in this market, but other important features such as strength, transparency, and weather stability are also deciding factors.

Typical applications for winter sports are skid carriers for skates, the transparent frames of ice hockey arenas, ice hockey helmets with transparent visors, ski helmets, ski goggles and glacier glasses, carrying cases for skis, and snowmobile covers. Typical summer sports applications are running lights for boats, compass housings, ship's lanterns (Figure 3.120), surfboard parts (Figure 3.121), underwater lights, housings for bicycle-mounted computers, and safety helmets (helmet and visor).

Figure 3.120. Ship's lamp of UV-stabilized, average viscosity polycarbonate.

Figure 3.121. Injection-molded surfboard fins of polycarbonate.

Figure 3.122. Motorcycle safety helmets of polycarbonate with higher viscosity; often these polymers are further modified to impart additional impact strength.

3.10.11 Safety

High strength, toughness, and temperature stability, as well as transparency in some applications, are arguments for the use of polycarbonate in the safety area. Typical applications are helmets (Figure 3.122) for various purposes.

Safety factors make polycarbonate the preferred material in the most diverse areas (e.g., as a replacement for glass to avoid injuries). The high flame-retardation capability of special polycarbonate types is another obvious advantage.

One application in which the reliable toughness of polycarbonate becomes very noticeable is in the safety shields used by police departments.

3.10.12 Glasses, Headlight Lenses

As a tough and clear–transparent organic glass, polycarbonate is an ideal substitute for silicate glass in the manufacture of lenses for safety glasses and sunglasses, as well as for lenses for corrective glasses. Although silicate glass has the advantage of somewhat higher transmission, its disadvantages with respect to polycarbonate are its tendency to break more easily and its high specific weight.

Polycarbonates used for eyeglass lenses must be especially pure, with a minimum transmission of better than 87% at 4 mm thickness; they must be free of specks and just offer minimum light scattering. Other specialty grades must limit the transmission of wavelengths above 400 nm, to protect the human eye by absorbing critical UV irradiation. An additional specialty grade makes it possible to manufacture welding safety glasses with a transparent absorber for radiation in the IR range.

Polycarbonate is processed in considerable volumes into corrective lenses, particularly in the United States. The desired dioptic properties are achieved partly by polishing of blanks and partly by precision injection molding. A larger share of polycarbonate is currently processed into safety glasses and sunglasses. To impart improved scratch resistance, most such items must be coated.

The determination of new and/or refined standard requirements on the strength of eyeglass lenses with regard to impact loads will lead to an increase in use of polycarbonate for this application. These stiffer requirements in many areas will not be matched in the future either by glass or by any other alternative material. Polycarbonate with at least medium molecular weight is required for applications requiring scratch-resistant lenses. To avoid undue reduction in toughness of the total composite, the scratch-resistant coating either must possess considerable elasticity itself or must be suitable for application to the polycarbonate surface after elastic priming.

3.10.13 Building Sector

Aside from electrical insulation parts made of polycarbonate, the building sector is mostly limited to window glazing (see Section 3.10.2). Additional applications are in the areas of sanitation (water spiggots, shower door glazing) and lighting. Up to now, the low price of polyvinyl chloride has prevented the use of polycarbonate in other door and window applications (e.g., window frames).

Only in certain conditions characterized by extensive use of halogen-free flue gases is flame-protected polycarbonate in use. Cable channels for nuclear power plants provide one example.

3.10.14 Office Technology

Housing parts (Figure 3.123) for data entry and data transmission instruments of various types pose a great application potential for thermoplastic materials and are still in the early developmental phase of a growth market. The major requirements for housing materials are as follows:
- flame retardance
- light stability (which enables economical use without additional coating)
- impact strength (for high grade instruments)
- heat distortion temperature engineered to the heat dissipation requirements accompanying constriction of the device.

Figure 3.123. Office machine housings of flame-protected, unreinforced polycarbonate.

Suitable polycarbonate grades that fulfill all requirements have been developed. Cost reasons dictate the use mainly of ABS, polystyrene, and a polycarbonate–ABS blend, as well as modified polyphenylene oxide. Polycarbonate is mainly used in high performance applications and as structural foam in very large-area coverings and chassis.

In addition to housing parts, polycarbonate is used for keyboards, jackets for pipe-post, writing instruments, and drawing templates or rulers.

3.10.15 Laser-Optical Data Storage Systems [6, 7]

The first step in the realization of a data storage system whose data are read by a laser beam, and for which polycarbonate is the best-suited material, consisted of the develop-

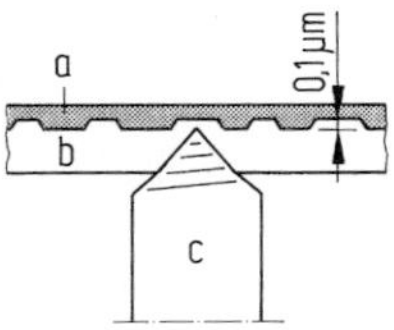

Figure 3.125. Schematic representation of a polycarbonate CD: a, protective layer; b, transparent carrier material; c, laser beam.

Figure 3.124. Audio (compact) disk of easy-flowing polycarbonate.

ment of the digital audio compact disk (Figure 3.124) by Philips and Sony. After the shellac disk and PVC long-playing record, the audio compact disk represents the third generation of recording disks that enable static-free music replication.

A schematic representation of a compact disk is shown in Figure 3.125. The digital information is stored on the surface of the transparent polycarbonate in the form of microscopically small pits (see above: Figure 3.56B). These pits are imprinted with a stamper in the mold during the injection stamping and/or molding process.

Laser technology poses the following extremely high requirements on the carrier material:

- high transparency
- low water absorption (to avoid warpage)
- mechanical stability and good heat distortion temperature
- extremely good flow behavior (to achieve stress-free disks, prerequisite for low birefringence)
- extremely high cleanliness (to guarantee error-free reading)

For compact disks, special easy-flowing polycarbonate was developed. Given an appropriate polycarbonate grade as well as good processing, the optical path difference, which is directly related to the birefringence, is approximately 15 nm—far below the maximum admissible limit of 50 nm (Figure 3.126).

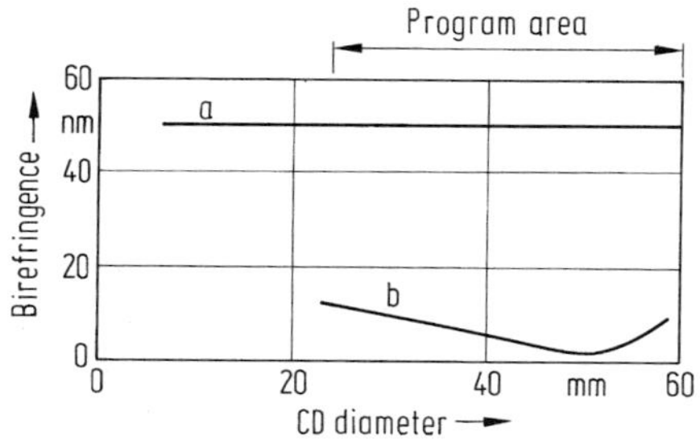

Figure 3.126. Birefringence of a polycarbonate, injection-stamped disk: a, maximum admissible limit; b, measured birefringence.

An expansion of the laser technology for data storage is in full swing. The following systems are under development:

- ROM (optical read-only memory); same technology as audio CD
- DRAW (direct read after write); here the user "writes" with a pulsating laser beam along stamped tracks; the information consists of "holes," which are burned into a metal oxide layer
- E-DRAW (erasable direct read after write); this system allows the user to write, read, and erase data at will; various techniques are used (e.g., reversible phase changes of metallic or metal-oxide layers) in connection with absorption differences or magneto-optical methods

Polycarbonate—possibly in modified form—has a good chance of being used in these applications. Laser-optical data storage disks represent a significant future growth market for polycarbonate.

3.10.16 Application of Glass Fiber Reinforced Polycarbonate [8]

One very important application for polycarbonate is metal substitution in the sense of a true substitute for former or current metal applications or in applications once served by metals.

In many cases, such substitutions require high stiffness of the molded part. When the natural stiffness of polycarbonate is not sufficient, specific construction designs featuring bracing ribs or box profiles in the molded parts can increase stiffness. Another possibility is increasing material stiffness with glass fiber or carbon fiber reinforcement. While carbon fiber reinforced polycarbonates have very limited use because of the cost/property ratio, glass fiber reinforced polycarbonates can be found in large quantities in almost every area except applications calling for transparency. The change of the modulus of elasticity and impact strength with glass fiber content is plotted in Figure 3.127. Toughness and impact strength can be increased by adding modifiers up to a glass fiber content of approximately 10%, which, however, reduces stiffness somewhat. The impact strength decreases with increasing glass fiber content.

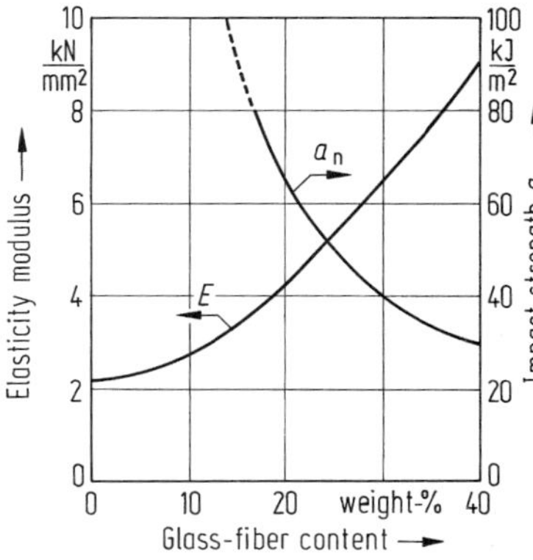

Figure 3.127. Dependence on glass fiber content of the impact strength (according to DIN 53453) and elasticity modulus (according to DIN 53457) of reinforced polycarbonate.

Short glass fibers with fiber lengths of approximately 100 μm, and long fibers or normal glass fibers of approximately 300 μm, are used in the granulate. Longer glass fibers increase stiffness considerably. It is not economical to incorporate more than 40% of normal glass fibers into a highly viscous polycarbonate melt. Such a glass fiber content, which leads to a modulus of elasticity of approximately 11,000 N/mm^2, still does not achieve the modulus of elasticity of metals. If the case requires, the molded part can be stiffened even more by suitable ribbing.

Another positive effect of glass fiber reinforcement is the decresae in coefficient of thermal expansion that occurs with increasing fiber content (see above: Figure 3.35). Polycarbonate with high fiber content shows coefficients of thermal expansion similar to those of several metals. This permits the direct molding of prewarmed metal inserts without the appearance of great stresses in the polycarbonate during cooling of the melt. Glass fiber reinforcement increases hardness and further improves dimensional stability and accuracy with respect to size. Glass fiber reinforced polycarbonate is used wherever higher stiffness and strength as well as improved dimensional stability is required in comparison to unreinforced polycarbonate, and where reduced impact strength can be tolerated. Current areas of applications were pointed out in earlier sections.

Characteristic examples of applications already cited are switch boxes, spools, loudspeaker enclosures, baseplates for film and slide projectors, drill housings (all in Section 3.10.1), lens mountings, and camera housings (Section 3.10.9).

References for Section 3.10

1. *Schmuck, G. W.; Schnitzler, A.:* Kunststoffe 68 (1978) 10, p. 657–660
2. *Nising, W.:* Kunststoffe 76 (1986) 11, p. 1095–1097
3. *Kircher, K.; Piontek, J.:* Kunststoffe 78 (1988) 12, p. 1181–1186
4. *Heidenreich, K.:* Kunststoff 73 (1983) 4, p. 22–26
5. *Heidenreich, K.:* Deutsche Molkerei-Zeitung 34 (1978) p. 1184–1186
6. *Kämpf, K.:* Techn. Rundsch. 78 (1986) 10, p. 72–86
7. *Anders, S.; Schmid, H.; Sommer, K.:* Kunststoffe 79 (1989) 1, p. 55–59
8. *Kircher, K.:* Reinforced Plastics 25 (1981) 11, p. 382–384, 25 (1981) 12, p. 420–422

3.11 Economic Significance

Despite above-average growth rates for polycarbonate worldwide from the early 1950s until today, there is no end to this material's volume development in sight. The reasons for its steady growth within the context of market dominance and expansion include the following important factors:

- *Favorable property combinations of transparency, heat distortion temperature, toughness, flame retardance, and physiological harmlessness*: these balanced technical qualifications are not achieved by any other cost-effective thermoplastic. Materials that are superior to polycarbonate in one or another property do not achieve the level of performance exhibited by polycarbonate in other technically important properties.
- *Broad application spectrum*: there is hardly a market area in which polycarbonate is not represented.
- *Possibilities of customized product modifications with, for example, copolymers, blends, and special additives for new areas of applications*: the last-mentioned factor is of utmost importance for the expansion of polycarbonate. Although the largest area of application is the electrical engineering market, where polycarbonate has a better than 40% share (worldwide), underrepresented areas of applications are beginning to gain market share (Table 3.26). In view of the known applications in these areas as well as ongoing projects, the future is likely to bring a shift in market share in favor of new polycarbonate growth markets at the expense of the established and most heavily represented area, namely, electrical engineering technology.

Table 3.26. Classification of Worldwide Polycarbonate Consumption (Unblended) by area of Application

Application	Share of Consumption (%) 1982	1983	1987
Electric and electronics industry including data carriers	59	55	46
Disks and film	18	20	23
Automotive industry	6	6	11
Packaging and medical	5	5	5
Office machines	2	3	5
Helmets	2	2	2
Other	8	9	8

The following application areas should be pointed out as new polycarbonate growth markets:

- *Twin-wall sheet*: particularly for industrial glazing and greenhouses, as well as flat sheet for construction glazing and site protection. Upon optimization of the weathering behavior (by coating with suitable light-stable coatings or by coextrusion with UV absorber concentrates based on polycarbonate), twin-wall sheet has conquered the market and shows high growth rates. In addition to good light transmission, high stiffness at low weight, and excellent impact strength, the thermal insulation and thus possible energy savings are crucial for the market.
- *Compact disks as audio and data storage media*: by reason of its transparency, toughness, low water absorption, and low birefringence, polycarbonate has established itself in this new area of application, which today uses more than 10,000 tons of polycarbonate worldwide. Data storage disks which, similar to audio CDs, have laser scanning, have in some cases already been realized with polycarbonate.
- *Headlight lenses*: many U.S. car models are already equipped with polycarbonate headlight lenses, in some cases, combined with polycarbonate reflectors. In Europe, polycarbonate headlight lenses will be introduced in the near future. This trend should increase the replacement of glass by polycarbonate for this area. Important for success is the combination of the polycarbonate headlight lens with a scratch-resistant coating. In addition to good weathering stability and scratch resistance, the cost-effectiveness of polycarbonate represents an important selection criterion. A series of good coating systems is already available.

Three-quarters of the worldwide volume is produced by Bayer and General Electric Plastics. Additional manufacturers are in Japan (Mitsubishi Engineering Polymers, Teijin, Idemitsu), in Italy (Enichem), in Brazil (Polycarbonatos do Brasil, since 1985), in the United States (Dow, since 1985), and in Korea (Sam Yang, since 1991).

Polycarbonate consumption in 1987 was approximately 420,000 tons, not including the volumes used as blend partners. Of this amount, approximately 30% fell to Western Europe, roughly 43% to the United States, 15% to Japan, and 12% other countries.

Extrapolation of the volume increase for the years 1986 to 1989 points toward a considerable increase in the market significance of polycarbonate. As a construction material, polycarbonate contributes to determining technical advances. Its exclusion from modern life is unthinkable.

3.12 Commercial Products

Polycarbonate was offered in 1989 by the companies listed in Table 3.27. Table 3.28 provides an overview of grade descriptions of standard commercial polycarbonates; the products are classified according to their viscosity. Measured as MFR 300/1.2, (melt flow rate 300 °C temperature and 1.2 kg load) products with low molecular mass each show melt viscosities greater than or equal to 15 g/10 min (ISO 1133); products with higher molecular weight show MFR values of approximately 5 g/10 min. Although the individual viscosity stages of the different commercial products are similar, they are not as a rule identical.

Bayer manufactures high molecular weight polycarbonates (M_w 90,000–150,000) for use in the production of high molecular weight film. In addition to basic types, every manufacturer offers variations with special attributes; several examples are shown in Table 3.28.

Supplementing the above-mentioned standard grades, various specialty types are offered. The manufacture of compact disks requires such specialty grades as Makrolon CD 2005, Lexan OQ1020L, Jupilon H 4000, and Panlite AD 5503.

Table 3.27. Trade Names of Polycarbonates by Manufacturers (1989)

Trade Name	Manufacturer
Makrolon	Bayer AG, Germany
Makrolon	Bayer USA, Inc./Miles Inc., U.S.A. (now Bayer Corp., USA)
Calibre	Dow Chemical Company, U.S.A.
Sinvet	Enichem, Italy
Lexan	General Electric Company, U.S.A.; General Electric Plastics, Europe
Idemitsu Polycarbonate	Idemitsu Petrochemical Company—Japan
Novarex	Mitsubishi Kasei Corp., Japan
Jupilon	Mitsubishi Gas Chemical, Japan
Panlite	Teijin Chemicals Ltd., Japan
Durolon	Polycarbonatos do Brasil S.A., Brasil[1]

[1] Joint venture between Idemitsu Petrochemical (Japan) and Pronor Petroquimica and Proquigel Group (Brazil).

Table 3.28. Type Description of Commercial Standard Polycarbonate, Divided According to Viscosity Classification (1989)

Trade Name	Molecular Weight of Products		
	Lower	Average	Higher
Makrolon	2400	2600, 2800, 3100	3200
Lexan	121	141, 161, 101	131
Calibre	300-22	300–15, 300–10, 300–6	300–4
Sinvet	201	221, 251, 271	301
Novarex	7022 A	7025 A, 7027 A	7030 A
Jupilon	H 3000	S 3000, S 2000, S 1000	E 2000
Panlite	L 1225	L 1250, K 1285	K 1300
Idemitsu Polycarbonate	A 2200	A 2500, A 2700	A 3000
Durolon	2200	2500, 2700	3000

Table 3.29. Types of Equipment in Viscosity Classes Exemplifying Medium Molecular Product Types (1989)

	With Mold Release Agent		For Contact with Food		
Basic Type	With UV Absorber	Without UV Absorber	With UV Absorber	Without Mold Release Agent	With Mold Release Agent
Makrolon 2800	2803	2805	2807	2808	2858
Lexan 161	163	161 R	163 R	164	164 R
Calibre 300–10	302–10	301–10	303–10	200–10	201–10
Sinvet 251	253	R 251	R 253	254	R 254
Novarex 7027 U	7027 R	7027 R		7025 FD	
Jupilon S 2000	S 2000 U	S 2000 R		S 2001	
Panlite L 1250	L 1250 Z	L 1250 Y			
Idemitsu Polycarbonate A 2500	V 2500	R 2500			
Durolon 2500	V 2500	R 2500			

Structurally viscous grades are offered by General Electric (Lexan 151) and by Bayer (Makrolon 1239). Specialty grades offered in the manufacture of twin-wall sheet include Lexan ML 3324 and Makrolon 1143.

Bromine-containing, flame-inhibited polycarbonate grades belong to the product mix of most manufacturers; some of these are Lexan 2014R, Calibre 700-6, Novarex 7025 N J, Jupilon N-3, Panlite LN 1250, and Toughlon NB 2500. Table 3.30 lists bromine-free, flame-protected types. In addition to the above-mentioned grades, there is available a series of speciality grades.

Table 3.31 shows an overview of glass fiber reinforced commercial products. In addition to polycarbonate producers, a number of independent compounders offer a wide product range, particularly of specialty products. The most important European compounders are LNP Plastics (Holland), Wilson Fiberfill (Belgium), Lati (Italy), and Eurostar (France).

Table 3.30. Bromine-Free, Flame-Protected Polycarbonate

Type Description	Transparent	V0/3.2 mm	V0 ≤ 1.6 mm
Makrolon 6555 and 6557	+	+	–
6870	+	+	+
6465	–	+	+
Lexan 920 A and 940 A	+	+	–
920 and 940	–	+	+
Calibre 700–6 and 700–10	+	+	–
800–6 and 800–10	–	+	+

Table 3.31. Overview of Short or Normal Glass Fiber (GF) Reinforced Polycarbonate (Type Description)

Trade Name	Fiber Reinforced				
	Short Glass[1]		Normal Glass		
	20 wt %	30 wt %	10 wt %	20 wt %	≥30 wt %
Makrolon	8030 8025	8030 8035	9415	8325	8345 (35% GF)
Lexan	ML 3512	ML 3513	500	3412	3414 (40% GF)
Calibre	–	–	551	–	–
Sinvet	–	–	2510	2520	2530 (30% GF)
Novarex	7025 G20	7025 G30	–	–	–
Jupilon	–	–	GS 2010 M	GS 2020M	GS 2030M (30% GF)
Indemitsu Polycarbonate	–	–	G 2510	G 2520	G 2530 (30% GF)
Panlite	–	–	GN-3115	GN-3120	GN-3130 (30% GF)

[1] Several Japanese companies offer also short glass fiber reinforced polycarbonate with 10 wt % glass fibers.

4 Polyacetals

H.-D. Sabel, H. Schlaf, P. Unger, U. Ziegler

4.1 Introduction

H. Schlaf

4.1.1 Definition

In the terminology of organic chemistry acetals are the reaction products of aldehydes and alcohols whose characteristic structural feature is the acetal group

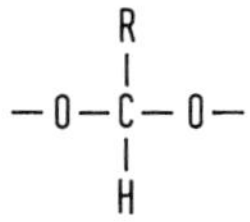

which is formed from the carbonyl group. Since the class of polymer to be discussed here displays the repeating acetal building block

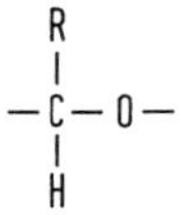

they are referred to as *polyacetals.*

Polyacetals comprise homo- and copolymers of aldehydes and cyclic acetals whose characteristic building blocks –CHR–O– are arranged into essentially linear chains of alternating carbon and oxygen atoms. The polyacetal structure can be interrupted by incorporation of other polymerizable monomers.

In the last decades, only polymers based on the monomers formaldehyde and the cyclic trimer of formaldehyde, trioxane, have achieved technical significance. On the basis of their repeating oxymethylene units, $-CH_2-O-$, the long-chain polymeres of formaldehyde and trioxane were also designated by Staudinger as polyoxymethylenes. The names polyformaldehyde and polytrioxane are less common. For more detailed description, one also speaks of formaldedye or trioxane homo- or copolymers. The established English term for polyacetals is *acetal resins.*

4.1.2 History of Polyacetals

Staudinger and his school, in particular Kern [1], were the first to carry out systematic investigations of the polymerization of very reactive formaldehyde in the 1920s. Even though Walker [2], Carruthers, and Norrish [3] concerned themselves further with polyoxymethylenes and recognized their favorable properties, technological exploitation at first remained closed to this class of polymers because of their thermal instability.

After intensive work carried out from the end of the 1940s, chemists at E.I. Du Pont de Nemours and Co. in the USA achieved a breakthrough in 1956 with the preparation of high molecular weight, thermally stable polyformaldehyde that could be thermally processed [4]. The decisive step forward consisted of blocking the unstable hemiacetal end groups by esterification with acetic anhydride. After announcement of the new product in 1956, small experimental quantities become available in 1957. In 1959 a production facility with an annual capacity of ca. 7000 tons per year came on-stream in the USA. Since that time the product has been on the market under the tradename Delrin®.

Thermally stable copolymers of trioxane that contain small amounts of cyclic ethers or formals as comonomer units, and that exhibit the additional benefit of alkali stability, were developed as the alternative to the end group stabilized homopolymer of formaldehyde by the Celanese Corp. in the USA [5] and Hoechst AG in Europe [6].

Production at the Celanese Corp. in a 7000 t/yr plant came on-stream in 1962 in Bishop, TX. Since that time the product has been sold under the tradename Celcon®. In 1962, Hoechst and Celanese formed a joint venture, Ticona Polymerwerk GmbH, which has produced and sold a trioxane copolymer under the name Hostaform® since 1964. In a similar way, Polyplastics Co. Ltd. emerged in Japan as a joint undertaking of Diacel Ltd. and Celanese Corp. The acetal copolymer produced there is sold under the tradename Duracon®.

Since the introduction of polyacetals as thermoplastic materials, a number of well-known chemical firms worldwide have taken up research an development activities in this area. Only a few have succeeded so far in developing marketable products. Besides the three firms mentioned, Ultraform GmbH, a communal undertaking of BASF AG and Degussa AS has produced and sold a trioxane copolymer under the tradename Ultraform® since 1971. In 1972 the Asahi Chemical Group joined the circle of polyacetal manufacturers as another Japanese producer with the formaldehyde homopolymer Tenac®. The product palette was broadened in 1985 with the copolymer Tenac C. In 1976 the production of trioxane-dioxolane copolymer, Tarnoform®, was begun in Tarnow, Poland. In 1981, a third Japanese firm, Mitsubishi Gas Chemical Corp. (MGC), started up a 10,000 t/yr plant in Yokkaichi for the production of the trioxane copolymer Jupital®.

In 1983 Ube Industries Ltd., Japan, announced successful completion of a gas-phase polymerization of formaldehyde to thermally stable polyoxymethylene [7]. Up to now, however, there has been no further information concerning the market introduction of corresponding products.

References for Section 4.1.2

1. *Kern, W.* In: *H. Staudinger* (Ed.): Die hochmolekularen organischen Verbindungen. Springer, Berlin 1932
2. *Walter, F.:* J. Am. Chem. Soc. 55 (1933) p. 2821
3. *Carruthers, J. E.*, et al.: Trans. Faraday Soc. 32 (1936) p. 195
4. *Schweitzer, C. R.*, et al.: J. Appl. Polym. Sci. 1 (1959) p.158
5. US-PS 3027352 (1960) Celanese Corp.
6. DE-PS 1124703 (1959) Hoechst AG
7. Japan Chemical Week, 9. 3. 1983

4.2 Chemical Structure of Polyacetals

H. Schlaf

As already mentioned in section 4.1.1, the essential structural feature of the polyacetals is a polymer chain of alternating carbon and oxygen atoms. Strictly speaking, this structure can only be obtained via polymerization of formaldehyde or its cyclic oligomers such as trioxane or tetroxane. Nevertheless, the term polyacetals can also be used when the acetal chain is interrupted by other structural elements. Of significance here are homo-, co-, and terpolymers.

4.2.1 Homopolymers

Homopolymers are the product of the polymerization of a single monomer. The chemical structure of the resulting homopolymers is determined by the nature of the monomeric building block. Some examples of homopolymeric acetals are: polyformaldehyde, polytrioxane, polydioxolane, poly-1,3-dioxepane (poly(butanediol formal)), and poly-1,3,6-trioxocane (poly(diethyleneglycol formal). Homopolymers obtained from formaldehyde and trioxane exhibit identical structures. Polyformaldehyde has particular technical significance.

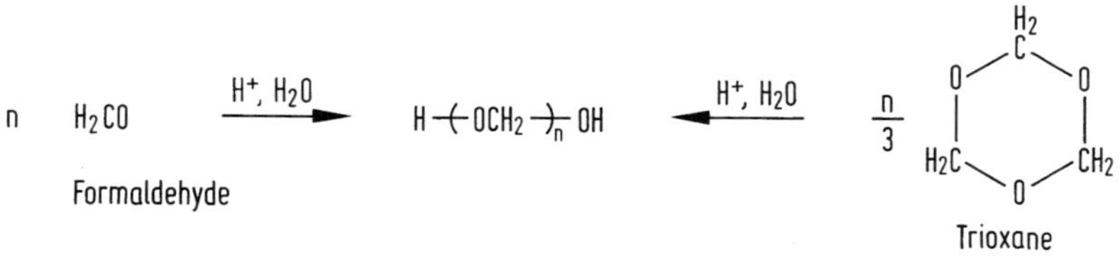

(4.1)

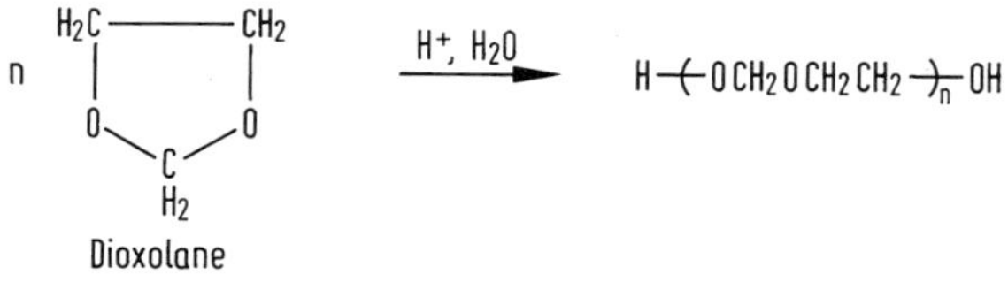

(4.2)

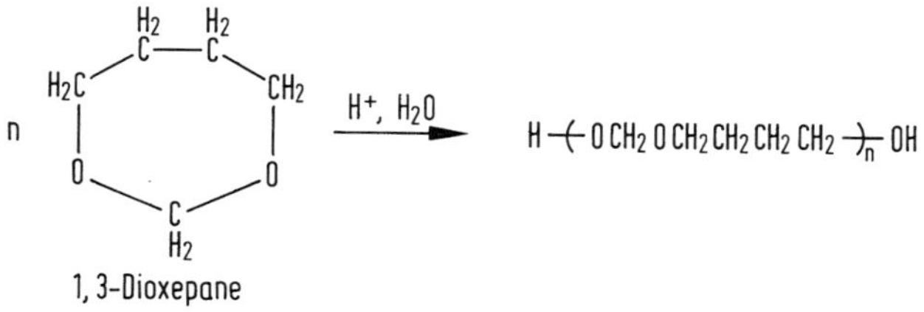

(4.3)

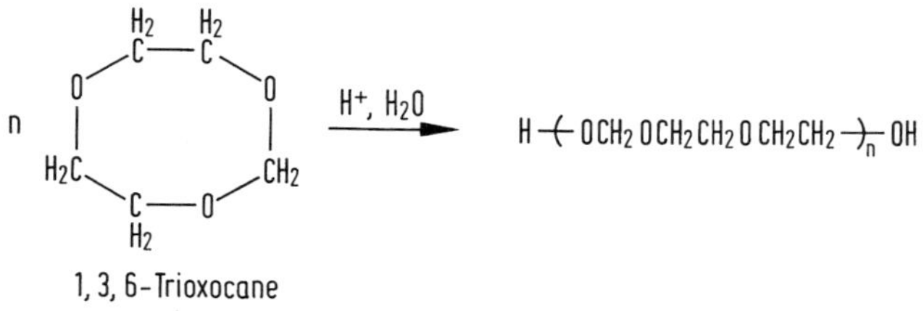

(4.4)

Hemiacetalic or hydroxyl end groups originate through transfer reactions of the polymerizing chains with water. In addition to hemiacetal groups, methoxy end groups can be introduced in a similar way with other transfer reagents such as methanol.

4.2.2 Copolymers

If another polymerizable monomer or prepolymer having an acetal structure is present during the polymerization of formaldehyde or trioxane then it will be built into the polymer chain according to its copolymerization parameter. Block or statistical incorporation is possible. In the special case of polyacetals, comonomers initially in a block can be statistically redistributed over the polymer chain through a transacetalization reaction which occurs alongside the growth reaction [1]. Depending on their kind and amount, comonomers alter the chemical and physical properties of the polymers to a greater or lesser extent.

The incorporation of suitable comonomers for increasing the stability of chain under thermal load has decisive technical significance [1 to 3]. A copolymer of trioxane and dioxolane has, for example, the following structure:

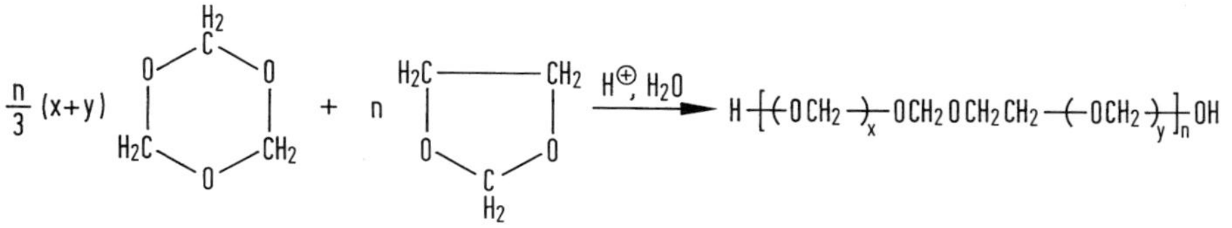

(4.5)

4.2.3 Terpolymers

The possible variations of the polymerization of trioxane allow for the incorporation of a third component by polymerization. In particular, the use of termonomers with two polymerizable groups (bifunctional monomers) leads to technically interesting products having modified rheology. As a rule only small amounts of termonomers are used, leading to partially crosslinked or branched products.

Use of butanedioldiglycidyl ether can lead to, among others, the following structure.

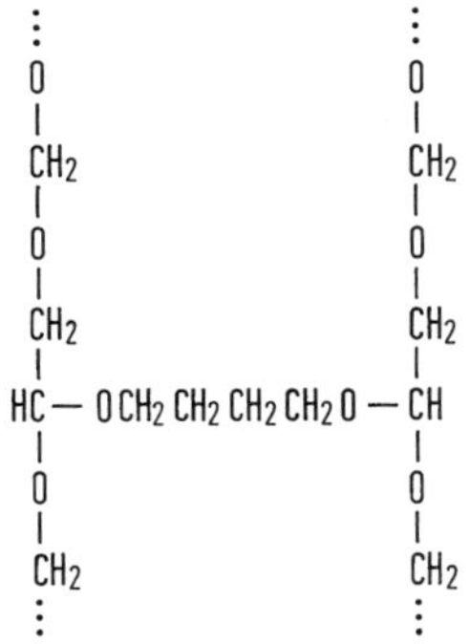

References for Section 4.2

1. *Weissermel, K.*, et al.: Kunststoffe 54 (1964) p. 410
2. *Kern, W.*, et al.: Angew. Chem. 73 (1961) p. 183
3. *Schweiker, G.C.*, et al.: Gordon Research Conf. Boston (1962)

4.3 Preparation of Polyacetals

H. Schlaf

4.3.1 Monomers

4.3.1.1 Formaldehyde

4.3.1.1.1 Properties

Formaldehyde forms the basis for the preparation of polyacetals. At room temperature, monomeric formaldehyde is a colorless gas with a pungent odor and is severly irritating to mucous membranes. The odor detection limit lies between 0.5 and 1.0 ppm. A maximum work-place concentration (MAK-value) of 0.5 ppm has been established [1]. The listing of formaldehyde on the MAK-list under Group III-B by the Senate Commission of the Deutsche Forschungsgemeinschaft (German Research Association; see Section 4.7.10) indicates the existence of findings which presume a carcinogenic potential but which require further clarification. In this connection, the carcinogenic effect of bis(chloromethyl ether), which can be formed when formaldehyde and hydrochloric acid come in contact, ought to be pointed out [2].

Formaldehyde condenses at −19 °C to a colorless liquid having a density of 0.815 g/cm^3 and solidifies to a white mass at −118 °C.

Formaldehyde is distinguished form homologous aldehydes by its extradordinary reactivity [3]. Handling of gaseous formaldehyde is difficult. At temperatures about 15 °C formaldehyde heterogeneously decomposes to yield predominantly methanol and CO. At temperatures above 350 °C the principal products are H_2 and CO [4, 5]. Below 100 °C formaldehyde tends to undergo polymerization. This is strongly accelerated by traces of water vapor or other polar compounds. Formaldehyde dissolves in water with the formation of hydrates (glycols) which, according to the concentration of formaldehyde, principally consist of either methylene glycol, $HO–CH_2–OH$, or polyoxymethylene glycol, $HO–[CH_2O]_n–H$.

4.3.1.1.2 Preparation

Essentially all of the processess currently used for the commercial production of formaldehyde employ methanol as the starting material. Other processes, which start from natural gas (methane), dimethyl ether, saturated or unsaturated hydrocarbons, CO or CO_2, have lost significance.

In the course of the 50s, 60s, and 70s, a series of firms developed processes for the preparation of formaldehyde via partial catalytic oxidation of methane in the gas phase. In spite of the high selectivity of the catalysts and in some cases good yields of formaldehyde, these remain without economic significance [6 to 8].

The development and use of the gas phase oxidation of propane and butane to formaldehyde, principally promoted in the USA [9 to 12], has lost considerable significance on technical and economic grounds. It is principally the separation and purification of the product mixture obtained as well as corrosion problems which make the process unprofitable.

Other processes, such as the preparation of formaldehyde from dimethyl ether [13] or by partial reduction of CO or CO_2 with H_2 [14], are only used regionally or have no practical application.

The commercially employed processes for the preparation of formaldehyde from methanol are based on the catalytic dehydration or oxidation of methanol with air outside the explosive limits (pure methanol-air mixutres with a methanol content between 7 and 8 or 35 and 36 vol-% are explosive) [16]. These include silver-contact and Formox processes.

Silver-contact process

Silver-contact processes comprise the partial oxidative dehydration of methanol in the presence of crystalline or a silver mesh with a deficit of air at temperatures between 600 and 720 °C. Depending on the reaction temperature, the conversion of methanol is from 77 to 87% or 97 to 98%. The reactions taking place on contact with silver can be simplistically described with equations 4.6 and 4.7.

$$CH_3OH \rightleftharpoons CH_2O + H_2 \qquad \Delta H = +84\ kJ/mol. \tag{4.6}$$

$$H_2 + 1/2\ O_2 \rightleftharpoons H_2O \qquad \Delta H = -243\ kJ/mol. \tag{4.7}$$

Summing these two equations yields

$$CH_3OH + 1/2\ O_2 \rightleftharpoons CH_2O + H_2O \qquad \Delta H = -159\ kJ/mol. \tag{4.8}$$

Since the dehydration reaction 4.6 is exothermic, the equilibrium is displaced to higher formaldehyde concentration at higher temperatures. At 400 °C 50% of the methanol employed can be dehydrated, at 700 °C up to 90%. The increase in conversion with increasing temperature is limited since side reactions which lead to, among others, CO and CO_2, noticeably decrease the yield of formaldehyde.

In addition to the contact temperature, the catalyst loading and the amount of water introduced to the reaction mixture influence the conversion of methanol and the yield of formaldehyde.

Thus, in the case of the *process with complete methanol conversion* (BASF process) [18], a mixture of 40 parts by weight water and 60 parts by weight methanol is vaporized and, together with compressed air, converted to formaldehyde on silver cyrstals of varying particle size at 700 °C. After absorption of the reaction gases, a 40 to 42% formaldehyde solution having 1 to 1.5% methanol and a little formic acid is obtained. By particular conduct of the reaction (partial replacement of the water with inert gast) a 50 to 55% formaldehyde solution can be obtained; see Figure 4.1.

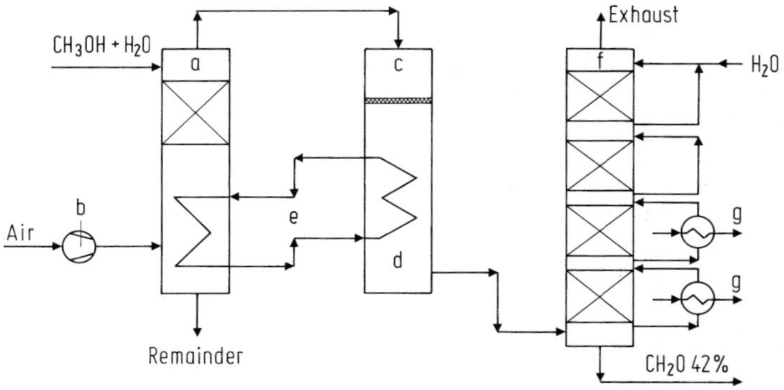

Figure 4.1. Schematic of the BASF formaldehyde process [17]
a, Vaporizer; b, Blower; c, Reactor; d, Heating kettle; e, Heating and cooling system with vapors; f, Absorption tower; g, Condenser

A formaldehyde solution obtained by the *process with incomplete methanol conversion* [19, 20] has a substantially higher methanol content of 5 to 15%. Here, methanol is converted with crystalline silver on a fixed-bed catalyst at temperatures between 590 and 650 °C. The high methanol content in the end product necessitates distillation of the initially 37 to 42% formaldehyde solution. As a rule, a 50% formaldehyde solution containing 0.5 to 1.0% residual methanol is obtained after distillation; see Figure 4.2.

More recently, numerous patent applications from leading manufacturers of formaldehyde attest to efforts to increase the selectivity and yield of the reaction by improving the catalyst and the process [21 to 25].

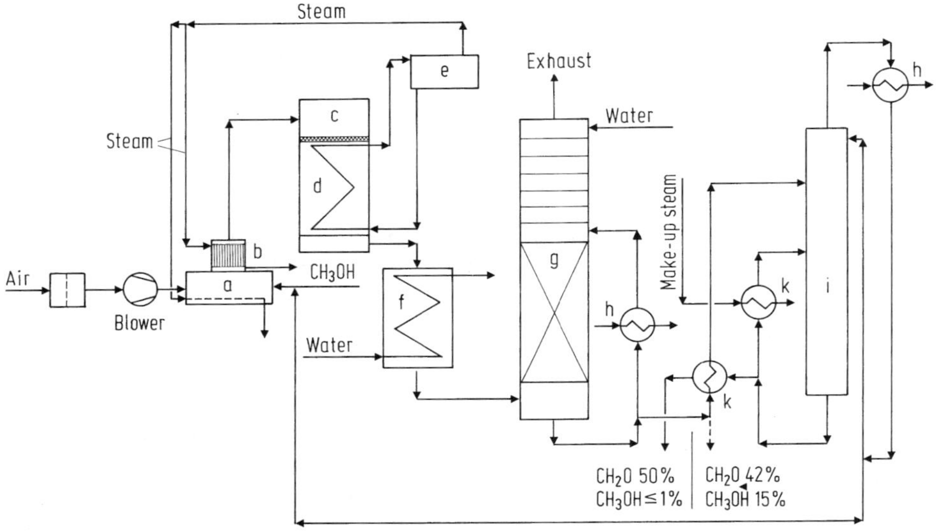

Figure 4.2. Process with incomplete conversion of methanol [17]
a, Vaporizer; b, Superheater; c, Reactor with catalyst; d, Gas condenser; e, Vapor collecting drum; f, Condenser; g, Absorption column; h, Condenser; i, Fractionation column; k, Heat exchanger

Formox process

The Formox process (Figure 4.3), developed by Reichhold Chemical Inc. [26] in the USA and Perstorp in Europe, employs mixtures of iron oxides and molybdenum oxide in place of silver contact for the oxidation of methanol to formaldehyde. The mixtures are, as a rule, doped with small amounts of chromium, cobalt, copper, vanadium, bismuth, or other metals.

Numerous firms in the USA, Europe, and Japan have recently concerned themselves with the improvement of the process, above all in relation to the activity, selectivity, and life of the catalyst. More than 40 patent applications from notable firms attest to the activity in this field [27 to 36].

The thermodynamic equilibrium of the strongly exothermic reaction (see equation 4.8) favors the formation of formaldehyde. Since in this case the high temperatures of the silver-contact process are not necessary, the reaction exhausts nearly all its potential even at 275 to 330 °C. In an important side reaction, the formaldehyde formed is oxidized to CO and H_2O:

$$CH_3OH + 1/2\ O_2 \rightleftharpoons CH_2O + H_2O \qquad \Delta H = -159\ kJ/mol. \tag{4.9}$$

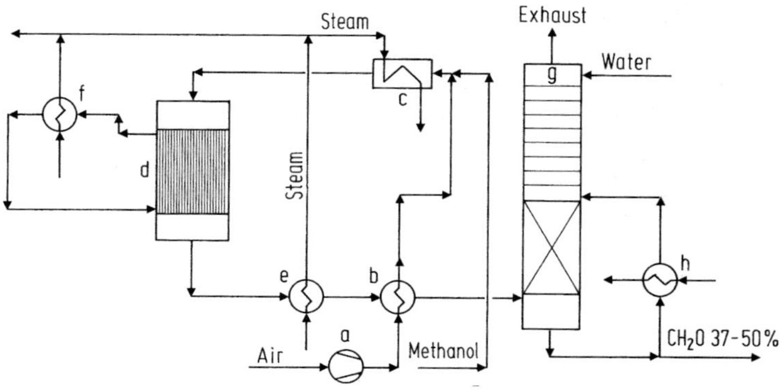

Figure 4.3. Formox process [17]
a, Blower; b, Air pre-heater; c, Vaporizer; d, Reactor; e, Gas condenser; f, Steam generator; g, Absorption column; h, Condenser

An excess of air is used at temperatures from 250 °C to a maximum of 400 °C. The content of methanol in the methanol-air mixture is maintained below the explosive limit of 7 to 8 vol-%. The conversion lies between 95 and 98%.

4.3.1.2 Trioxane

4.3.1.2.1 Properties

In addition to formaldehyde, 1,3,5-trioxane, the cyclic trimer of formaldehyde, serves as a very important starting material for polyacetals. Below its melting point, which lies between 60 and 61 °C, depending on the degree of purity, trioxane forms colorless, needle-like crystals which readily sublime (heat of sublimation 628 kJ/kg). As a rule, trioxane is handled above its melting point as a mobile, water-clear liquid (dynamic viscosity 2.05 mPa · s at 65 °C). It boils at 115 °C at normal pressure. The formation of an azeotropic mixture of 70% trioxane and 30% water that boils at 91.3 °C is a disadvantage for purification by distillation.

Trioxane is soluble in numerous organic solvents, e.g. alcohols, ketones, ethers, aromatic hydrocarbons, and chlorinated hydrocarbons. Its solubility in water depends on temperature. As a rule of thumb, dissolution temperature in °C ≈ solubility in weight-% (at 18 °C up to 17.2 wt-%, at 45 °C up to 45 wt-%, unlimited solubility at 100 °C).

In dealing with trioxane it must be kept in mind that this derivative of formaldehyde, when combined with acids, liberates formaldehyde. Since trioxane attacks skin and mucous membranes, inhalation as well as skin and eye contact must be avoided. A maximum workplace concentration (MAK value) has not been established in Germany. Other important data: flashpoint 35 °C, ignition temperature 424 °C, lower and upper explosion limits respectively 3.6 and 28.7 vol-% at 1013 mbar.

4.3.1.2.2 Preparation

Trioxane is obtained in a suitable reactor by trimerization of formaldehyde from a 60 to 65% aqueous formalin solution with acids, in particular sulfuric acid [37], or ion exchange resins [38, 39], as catalysts at temperatures between 105 and 120 °C. Trioxane, which is present in equilibrium with formaldehyde only at about 3%, is driven out together with water and formaldehyde via distillation and concentrated to 40 to 45% on a downstream column. Trioxane is enriched through extraction with a water-insoluble solvent (methylene chloride [37], benzene [38]). The extract is subjected to an alkaline wash before the extractant is separated by distillation [42]. Trioxane, freed of extractant, contains so-called heavy and light boilers which are separated in fractional distillation. The trioxane obtained in "polymerization grade" can be stored under inert gas at 80 °C for 24 to 36 hours without a loss of quality; see Figure 4.4.

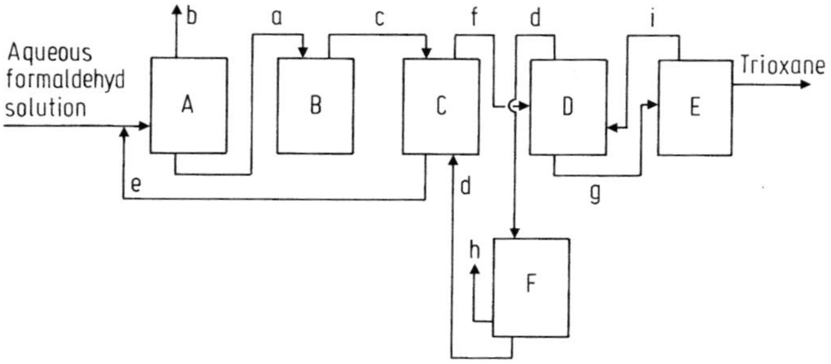

Figure 4.4. Trioxane manufacture [17]
A, Upgrading; B, Trioxane synthesis; C, Trioxane extraction; D, Distillative separation of trioxane; E, Purification distillation; F, Solvent purification.
a: 60–65% formaldehyde solution; b, Diluted formaldehyde solution; c, Mixture of trioxane, formaldehyde, and water; d, Methylene chloride; e, Formaldehyde-water mixture; f, Trioxane and methylene chloride mixture; g, Raw trioxane; h, Impurities; i, Front-ends

There has been no shortage of experiments to improve the process for the preparation of trioxane. Thus, a series of patents describe the use of heteropoly acids to suppress the formation of side products and obtain trioxane in high purity [43 to 45].

Processes for increasing the conversion to trioxane are based on displacement of the chemical equilibrium in the reaction phase or destruction of the physical equilibrium in the trioxane-water-formaldehyde vapor phase [46 to 52].

References for Section 4.3.1

1. Maximale Arbeitsplatzkonzentration und biologische Arbeitsstofftoleranzwerte der Senatskommission zur Prüfung gesundheitsschädlicher Arbeitsstoffe der Deutschen Forschungsgemeinschaft. Publication of BMA, 10. Nov. 1988
2. Ullmanns Encyklopädie der technischem Chemie. 4. Ed. vol. 8. Verlag Chemie, Weinheim, 1974, A. 156
3. *Walker, J. F.* (Ed.): Formaldehyde. 3. Ed. R. E. Krieger, Malabar, Florida 1975, p. 231
4. *Calvert, J. G.; Steacie, E. W. R.:* J. Chem. Phys. 19 (1951) p. 176–182
5. *Bone, W. A.; Smith, H.J.:* J. Chem. Soc. London (1905) p. 910–916
6. Standford Research Institute, California, Rep. No. 23 (1967) p. 113–114
7. Standford Research Institute, California, Rep. No. 23 (1967) p. 113–114
8. DE-PS 2201 249 (1972) Chemische Werke Hüls AG
9. Celanese Corp: Chem. Eng. 55 (1948) p. 136–139
10. US-PS 2570216 and US-PS 2570217 (1949) Celanese Corp.
11. Petroleum Refiner 32 (1953) p. 144
12. *Powers, E. T.:* Oil Gas J. 54 (1955) p. 74
13. *Tademuna, H.*, et al.: Hydrocarbon Process. 45 (1966) p. 195
14. *Ipatieff, V. N.; Monroe, G. S.:* J. Amer. Chem. Soc. 67 (1945) p. 2168–2171
15. Akita Petrochemicals, Japan.: Chem. Week, Dec. 1 and 2 (1966) p. 14
16. Ullmanns Encyklopädie der technischen Chemie. Vol. 7. Munich 1956, p. 656
17. *Diem, H.; Hilt, A.* in: Ullmanns Encyklopädie der technischen Chemie, 4. Ed. Vol. Verlag Chemie, Weinheim 1976, p. 687

18. DE-PS 2442231 (1976) BASF AG
19. *Sleemann, D. G.:* Chem. Eng. 75 (1968) p. 42–44
20. *Weimann, M.:* Chem. Eng. 77 (1970) p. 102–104
21. US-PS 4390730-A (1981) Celanese Corp.
22. DE-OS 3143704 (1981) Bayer AG
23. US-PS 4424397-A (19882) Du Pont de Nemours
24. DE-OS 3213227 (1982) BASF AG
25. DE-OS 3402995 and EP-PS 0150436 (1984) BASF AG
26. US-PS 3978136 (1976) Reichhold Chemicals
27. US-PS 3408309 (1964) Lummus Comp.
28. CEP-PS 1144252 (1966) Montedison
29. JA-AS 44-14926 (1969) Mitusbishi Gas Chemicals Comp.
30. GB-PS 1154058 (1969) Perstorf AB
31. US-PS 3952361 (1974) BASF AG
32. US-PS 3855153 (1974) Reichhold Chemicals
33. US-PS 3978136 (1976) Degussa AG
34. US-PS 4024074 (1977) Società Italiana Resine
35. US-PS 4425260 (1982) Monsanto Co.
36. JA-OS 3022536 (1986) Mitsubishi Chem. Ind.
37. US-PS 2304080 (1940) Du Pont de Nemours
38. DE-PS 1135491 (1960) Hoechst AG
39. DE-PS 1593990 (1966) Società Italiana Resine
40. GB-PS 1012372 (1963) ICI
41. FR-PS 1449000 (1965) Houilleres du Bassin du Nord
42. FR-PS 1419675 (1965) Pechiney Saint Gobain
43. DE-PS 3106476 (1981) Asahi Chem. Comp.
44. JA-OS 59025-387 (1984) Asahi Chem. Comp.
45. JA-OS 57200-392 (1982) Sankyo
46. DE-OS 1919495 (1969) BASF AG
47. US-PS 4110298 (1976) Celanese Corp.
48. DE-OS 3445921 (1986) Hoechst AG
49. DE-PS 1543390 (1969) BASF AG
50. DE-PS 1643256 (1971) BASF AG
51. DE-OS 3421300 (1985) Hoechst AG
52. DE-OS 2853091 (1980) Hoechst AG

4.3.2 Comonomers

Monomers capable of copolymerization with formaldehyde or trioxane comprise in particular small ring-sized oxacyclic compounds having the structural characteristics of an ether [1, 2], acetal [1, 3 to 5], or lactone [6, 16], including spiro ethers [17, 18] or dioxabicycloalkanes [19]. In addition, there are many indications in the literature of cationically copolymerizable monomers with sulfide [1, 7] or olefinic features [1, 6, 8, 9].

Of the cyclic ethers, the epoxides and especially ethylene oxide [1, 2, 7] have achieved the most significance. Copolymers of trioxane and ethylene oxide have been sold since 1961 under the tradenames Celcon® and Hostaform®.

Propylene oxide, styrene oxide, epichlorohydrin, 3,3-bischloromethyl oxacyclobutane, and also compounds such as glycidyl phenyl ether [10] or glycidic esters [11] can be mentioned as further examples of polymerizable ethers. Bifunctional monomers with two polymerizable groups, which are successfully used for the preparation of weakly crosslinked terpolymers, take a certain special place. Diepoxides, diacetals, or compounds with one cyclic acetal group and one epoxy group are included in this group.

In addition to the cyclic ethers, the cyclic acetals (and also oligomeric acetals) are exceptionally well suited for cationic copolymerization with formaldehyde or trioxane. The following compounds from this class ought to be mentioned: dioxolane, 1,3-dioxane, 1,3-dioxepane (butanediol formal), 1,3,5-trioxepane, 1,3,6-trioxocane as well as substituted dioxolanes. The dioxolanes substituted in the 2-position exhibit a peculiarity in copolymerization. These lose the 2- substituent in the form of the corresponding aldehyde and polymerize as unsubstituted dioxolane.

1,3-dioxepane has particular technical significance. Since 1967 its copolymer with trioxane has been on the market under the tradename Ultraform®.

A particular form of the copolymerization with cyclic ethers and acetals is described by Ube Industries KK in Japan. To the end of stabilizing formaldehyde homopolymers comonomers such as ethylene oxide or dioxolane are built into the polymer chain in the presence of Lewis acids in a step subsequent to the prepolymerization step [10, 12].

Styrene [13], vinyl acetate, vinyl ether, vinyl methyl ketone, isobutylene, acrolein, and acrylonitrile belong to the class of cationically polymerizable unsaturated compounds. Up to now corresponding copolymers of the above mentioned compounds with formaldehyde or trioxane have not achieved technical significance. In a similar way to copolymers prepared from formaldehyde and CO [14] remain so far without a useful application.

The possibilities of anionic copolymerization of formaldehyde with the mentioned monomers [15] are less numerous than those of cationic polymerization.

References for Section 4.3.2

1. *Weissermel, K.*, et al.: Kunststoffe 54 (1964) p. 410
2. JA-OS 59015240 (1984) und JA-AS 87031726 (1986) Ube Industries KK
3. JA-AS 88022212 (1988) Ube Industries KK
4. DE-PS 1124703 (1959) Hoechst AG
5. JA-PS 604086 (1959) Celanese Corp.
6. *Iwabuchi, S.*, et al.: Makromol. Chem. 177 (1967) p. 1927
7. BE-PS 603848 (1961) Houillieres du Bassin du Nord
8. *Kern, W.*, et al.: Angew. Chem. 73 (1961) p. 183
9. *Higashimura, T.*, et al.: J. Polym. Sci. 5 (1967) p. 1927
10. *Hermann, H. D.; Burg, K. H.:* Angew. Makromol. Chem. 15 (1971) p. 219
11. *De Mejo, L.*, et al.: Polymer 19 (1978) p. 956; Polym. J. 11 (1979) p. 15
12. JA-AS 83051012 (1983) Ube Industries KK
13. *Höhr, L.*, et al.: Makromol. Chem. 52 (1962) p. 59
14. US-PS 3383364 (1964) Esso Research and Eng. Co.
15. US-PS 3520850 (1964) Diamond Shamrock Corp.
16. DE-PS 1250124 (1968) British Industrial Plastics
17. US-PS 3418278 (1968) Montecatini
18. JA-AS 69008232 (1968) Tokyo Rayon Comp.
19. JA-PS 70006278 (1970) Tokyo Rayon Comp.

4.3.3 Purification of Monomers

The technical acceptance of polyacetals presupposes high molecular weight product. Thus, highly purified monomers are required for production processes. Limited by their synthesis, formaldehyde and trioxane contain impurities which act as chain transfer agents and which thereby hinder the preparation of high molecular weight product. Water, methanol, formic acid, methyl formate, dimethyl formal (methylal), and their homologues are the impurities that most frequently disrupt the course of the polymerization. The required high purity of the monomer is only obtained with expensive, energy-intensive processes.

4.3.3.1 Formaldehyde

Alternative processes for the preparation of high purity and in particular anhydrous formaldehyde are described in a series of patent applications. A two-stage drying is appropriate. The following methods are worthy of mention:

- Washing of monomeric formaldehyde with solvents containing a drying agent, as for example phosphorous pentoxide suspended in organic solvents such as methylene chloride [1] or drying agents like anhydrous $CaCl_2$, Na_2SO_4, or $CaSO_4$ suspended in organic solvents such as benzene, toluene, cyclohexane, pentane, or hexane [2].
- Washing with solvents which take up impurities, above all water, but in which formaldehyde is practically insoluble. To this group belong cyclic carbonates [3], diesters of polyethylene glycols or monoesters of polyethylene glycol monoalkyl ether [4], and hydrocarbons (e.g., toluene) [5]. Also included are mixtures of liquids which take up water such as cyclic ethers, amides, and lactones together with hydrocarbons whereby the mixture of liquids is recirculated and dried through a drying agent [6]. In the counter-current process, 99.99% gaseous formaldehyde is obtianed via solutions of formaldehyde hemiacetals (in particular solutions of the hemiacetals (in particular solutions of the hemiacetal with cyclohexanol) [7].
- Low-pressure, thermal decomposition of paraformaldehyde suspended in paraffin oil [8].
- Pyrolysis of cyclohexane hemiformal at about 140 °C [9].
- Multiple fractional distillation of aqueous formaldehyde solution [10 to 12].
- Decomposition of trioxane vapor on acidic catalysts at 220 °C [13].

The nearly anhydrous formaldehyde obtained in the first step is brought to a high purity state, suitable for polymerization, in the second step, for example through "prepolymerization". This can be carried out for example on vessel walls or packing at about −10 °C [14] or in organic solvents (for example cyclohexane) [14]; or, at low temperatures, in the solvents to be used for polymerization whereby the prepolymer is separated before the actual polymerization. The impurities are bound in the prepolymer.

4.3.3.2 Trioxane

In comparison to formaldehyde, the purification of trioxane to "polymerization grade" quality is easier to control but the process is indeed energy-intensive. The classic method, already described in Section 4.1.2.2, involves extraction of an aqueous formalin-containing trioxane solution. The extraction is carried out with water-immiscible solvents. Alkaline washing of the extract frees the trioxane of formic acid. After separation of the extractant by distillation, the already quite pure trioxane is subjected to a multi-step finishing distillation to remove so-called light and heavy boilers.

An alternative route to the upgrading and purification of trioxane that does not employ a solvent consists of the crystallization of trioxane from its aqueous solution followed by a multi-step dynamic crystallization of the raw trioxane [16, 17].

A number of additional process steps for bringing trioxane into a polymerizable condition are described in the patent literature. An often employed method is treatment of trioxane with zeolites [8], aluminum oxides [9], molecular sieves [20], or ion exchangers [21] for separation of impurities that permanently disrupt the polymerization. Also recommended are: purification of trioxane melt with KOH or $BA(OH)_2$ [22], alkali metals [23], possibly in the presence of phase transfer catalysts [24], or with amines [25] that are soluble in trioxane but which themselves do not vaporize in the subsequent distillation. A very mild and effective method involving the extraction of trioxane from its aqueous solution with supercritical CO_2 is described in [26].

References for Section 4.3.3

1. DE-PS 1104934 (1959) BASF AG
2. BE-PS 612396 (1962) Showa Denko K. K.
3. BE-PS 604954 (1961) BASF AG
4. BE-PS 601892 (1961) Compagnie de Saint Gobain
5. SU-PS RA 137909 (1960) A. N. Levin
6. AT-PS OE 220608 (1960) VEB Leuna-Werke
7. DE-PS 1090191 (1958) Du Pont de Nemours
8. DE-PS 1037127 (1953) Du Pont de Nemours
9. DE-PS 1057086 (1954) Du Pont de Nemours
10. US-PS 2790755 (1953) Du Pont de Nemours
11. US-PS 2824051 (1954) Du Pont de Nemours
12. De-PS 1052971 (1957) Du Pont de Nemours
13. US-PS 2714616 (1955) Celanese Corp.
14. US-PS 2841570 (1953) Du Pont de Nemours
15. DE-PS 1000152 (1953) Du Pont de Nemours
16. EP-PS 162252 (1985) Hoechst AG
17. DE-PS 2855710 (1979) Mitsubishi Gas Chemical Ind.
18. JA-OS 55129-218 (1980) Asahi Chem. Ind.
19. De-OS 1543819 (1969) Pechiney Saint Gobain
20. NL-OS 6415198 (1966) Perstorf AB
21. JA-AS 17915-69 (1969) Tokyo Rayon Co.
22. EP-PS 0036552 (1981) BASF AG
23. DE-OS 1906846 (1969) Celanese Corp.
24. US-PS 4423-238 (1983) Conoco Inc.
25. DE-PS 1543528 (1971) Degussa AG
26. JA-OS 59186-975 (1984) Mitsui Petrochem. Ind.

4.3.4 Polymerization and End Group Stabilization

In the polymerization of formaldehyde and trioxane, two types, homo- and copolymerization, need to be distinguished. Both types of polymerization are initiated with ionic initiator systems. Formaldehyde can polymerize by anionic as well as by cationic mechanisms, while trioxane is capable of polymerizing only under cationic conditions. In any case, the polymerization processes require monomers of high purity. The homo- and copolymerization of both monomers is principally carried out in the gas or liquid phase; the polymerization of the cyclic homologues of formaldehyde (trioxane, tetroxane, etc.) can also be carried out in the solid state. Further distinctions are those among mass, solution, and suspension polymerization processes, which have attained different technical significances for the two monomers. In the realm of technical applications, homopolymers are predominantly prepared from anionically polymerized formaldehyde, while trioxane is processed exclusively to cationically initiated copolymers.

4.3.4.1 Formaldehyde

4.3.4.1.1 Polymerization

Pure, very reactive formaldehyde, by the use of suitable catalysts, is easily polymerized to high molecular, technically useful polyoxymethylenes. As a consequence of the high heat of polymerization of ca. 60 kJ/mol [1], mass polymerization is relatively difficult to control by process technology [2, 3]. Thus, processes for suspension polymerization that work with inert organic solvents have established themselves in practical use. Nonpolar, aliphatic hydrocarbons such as heptane or cyclohexane are preferentially used as the suspension medium. Reaction temperatures vary between −20 °C and 80 °C. Normal pressures are used.

Purified formaldehyde is introduced into the gas space of a reaction vessel which has been charged with the suspension medium and initiator. Effective agitators provide a thorough mixing of the gas and liquid phases. With increasing molecular weight, formaldehyde polymerizing in the liquid phase separates out of the suspension medium as a finely divided solid. The achievable molecular weight lies between 15,000 and 100,000. Dispersing aids are added in an attempt to obtain a very small-grained polymer. This simplifies the subsequent esterification or etherification of the end groups. To insure good removal of the heat of polymerization, the solids content in the polymerization kettle is limited to a maximum of 20%.

A detailed description of the continuous process with subsequent work-up of the polymer is contained in [84].

It would be nearly impossible to list all the initiators mentioned in the patent literature for the polymerization of formaldehyde; the following can be cited as exemplary: amines, amides and ammonium salts [4 to 9], metal chelates [10 to 12], organometallic compounds (for example those of Al [13 to 15] or Sn [17, 17]), metal carbonyls [18], oxides, hydroxides, and carbonates [9 to 21], and Lewis acids [22 to 24].

Data are also available on radiation-induced polymerization of formaldehyde [25], initiators comprising combinations of organometallic compounds with metal chelates [26], and initiators immobilized on a carrier [27].

The range of molecular weight of polyformaldehyde of interest for processing technology lies between M_n 30,000 and 50,000. Thus addition of chain transfer agents is necessary for the control of molecular weight. Water, methanol, and formic acid have proven to be particularly effective [28] but polyfunctional alcohols [29] as well as dicarboxylic acids and their esters [30] have also found use.

Besides the customary process of polymerization in suspension, polymerization processes in the gas phase [31], in a reaction medium of liquid CO_2 [32, 33], or in the homogeneous phase at high pressure [34] are described.

In contrast to the technically significant homopolymer, copolymers of formaldehyde have up to now not played a role, although numerous patent applications point to efforts to develop economical processes for formaldehyde copolymers [35 to 38].

4.3.4.1.2 End Group Stabilization

For many years it was thought that high molecular weight polyoxymethylene could not be employed in engineering use. Methods first worked out in the 50s and 60s (see Section 4.1.2), through which the unstable hemiacetal end groups are protected against thermal degradation by means of esterification or etherification, opened up the possibility of processing polyoxymethylene homopolymer at high temperatures without a depolymerization process, thereby paving the way to practical applications. Of the many methods described since that time, it is the acetylation of the hemiacetal end groups with acetic anhydride that has established itself in practical use [39, 40]. The rate of reaction as well as the yield of encapped polymer can be substantially increased by the use of catalytic amounts of alkali acetates or tertiary amines [39, 41]. The use of an isocyanate or a carbodiimide is suggested to trap the by-product acetic acid that would otherwise lead to degradation of the polymer [42]. The reaction is preferentially carried out in suspension [40, 43, 44] but can also be carried out in the liquid phase or in the gas phase [45]. Depending on the method, the transformation is effected at temperatures between 50 °C and 200 °C with reaction times from 5 minutes up to 3 hours.

While the method of end group stabilization of polyformaldehyde with acetic anhydride is certainly the most important, other methods of esterification have been described. In addition to acetic anhydride, the use of other aliphatic or aromatic anhydrides [46] or carboxylic acid imides [47], also in the presence of acid catalysts such as Lewis acids [46] or sulfonic acids [48], is indicated in the literature. Besides alkali acetates and tertiary amines, polyamides, polypyrrolidone, or quarternary ammonium and phosphonium salts [49] are mentioned as basic catalysts.

An alternative method for obtaining end group stabilized homopolymers proceeds through etherification (more precisely complete acetalization) of the hemiacetal end groups. Alkylating agents such as dimethyl sulfate or orthoformates [50, 51] and epoxyalkanes such as ethylene oxide or propylene oxide [52, 53], as well as acetals in the presence of Lewis acids [54] or cationic exchange resins [55], are suitable for this purpose. In comparison to the esterified polymers, the end group etherified homopolymers have the additional advantage of being stable to alkali in the way that has been observed in copolymers blocked by comonomer building blocks.

A particular method for the stabilization of homopolymer is a subsequent incorporation of a comonomer in the already formed polymer chain. This occurs in the presence of Lewis acids and without degradation of molecular weight [56, 57, 83].

4.3.4.2 Trioxane

4.3.4.2.1 Polymerization

As described in the literature by many authors [58 to 62], the ring-opening polymerization of trioxane follows a cationic mechanism:

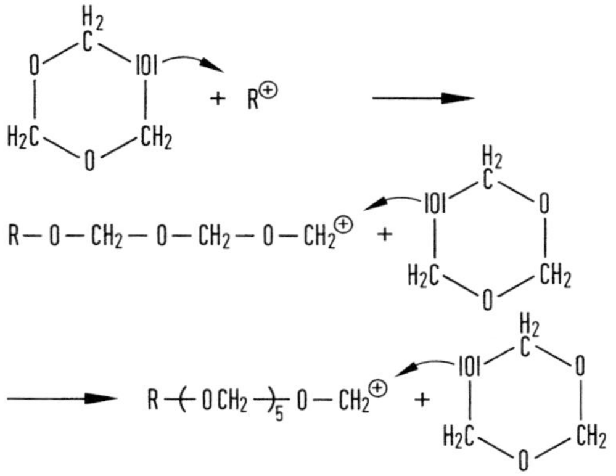

(4.10)

The generation of monomeric formaldehyde up to its temperature-dependent equilibrium concentration (ceiling concentration) [63] is a prerequisite for the actual start of the polymerization.

The polymerization of trioxane can be brought about by a multitude of initator systems, such as:

by protonic acids such as perchloric acid, perfluoroalkane sulfonic acids, or perfluoralkane carboxylic acids, as well as their alkyl esters and anhydrides

by Lewis acids such as BF_3, PF_5, AsF_5, SbF_5, $SnCl_4$, $TiCl_4$ or their complexes

by heteropoly acids such as molybdophosphoric acid or tungsten phosphoric acid

by carbenium salts such as triphenylmethyl hexafluoroarsenate, tropylium tetrafluoroborate or t-butyl perchlorate, oxocarbenium salts such as acylium hexafluoroantimonate and benzylium tetrafluoroborate by onium salts such as trimethyloxonium tetrafluoroborate or *p*-nitrobenzene diazonium tetrafluoroborate

by metal chelates such as molybdenum acetylacetonate, or compounds such as nitryl or nitrosyl salts [64].

In addition to these, the electrochemically initiated [65] and the radiation induced [66] polymerization of trioxane, as well as the photo-initiated polymerization of trioxane in the presence of diaryliodinium or triarylsulfonium salts as photo-initiators [67] are indicated in the literature.

Side reactions that lead to chain termination, chain transfer, transacetalization, or hydride transfer compete with the actual growth reaction [62].

The so-called regulators, which intervene in the polymerization process as chain transfer agents, are important for the adjustment of molecular weight. Essential here is that not only is the growth of one chain ended but a new chain is simultaneously started. Suitable chain transfer agents are: acetals such as formaldehyde dimethylacetal (methylal), orthoesters, carboxylic esters, and also ethers such as di-isopropyl ether. The mechanism of chain transfer can be clarified using methylal as an example:

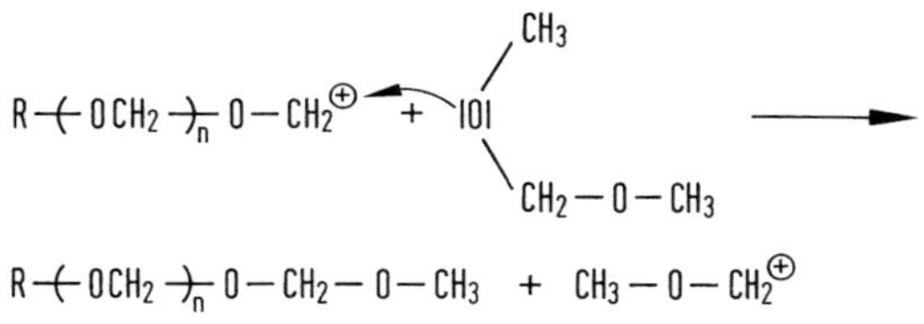

(4.11)

Polymerization of trioxane can be carried out in bulk, solution [68], or suspension [69, 82]. For solution and suspension polymerization, the same solvents as in the case of formaldehyde are used.

Bulk polymerization has been described in the gas phase [70, 71], in the homogeneous liquid phase [72], in the liquid-solid phase (slurry) [73, 74], or in the solid-crystalline phase [75]. The bulk polymerization starting from liquid trioxane with a heterogeneous phase transition to solid, semi-crystalline polymer has established itself in the industrial arena. The hard, crystalline polymer that develops and the liberated heat, which consists of the heat of polymerization and the heat of crystallization in the amount of 28 kJ/mol, allow a continuous polymerization only with devices that conduct away the heat and macerate the polymer, as for example extruders, kneaders [76], or roll mills [77]. Reaction temperatures lie above the melting point but below the boiling point of trioxane, that is between 60 °C and 115 °C. In order to minimize the loss of monomer and thereby the loss of conversion, the polymerization is carried out at the lowest possible temperature.

A process that deviates from the customary technology is described in [77].

Amines or phosphines are preferentially used for deactivating the polymerization initiator [78, 79].

4.3.4.2.2 End Group Stabilization

As cited in Section 4.3.4.1.2, polyoxymethylenes possess a suitable stability for processing only after blocking of the unstable hemiacetal end groups.

Compared to the method of esterification or etherification of homopolymers, copolymerization takes a different path to arrive at thermally stable polymers. Under the conditions of cationic polymerization, small amounts (0.5 to 5 wt-%) of a suitable comonomer are incorporated into the polymer chain (see Section 4.3.2). Compared to end group stabilized homopolymers, copolymers, after removal of their unstable ends, have the additional advantage of being stable against alkaline agents. Furthermore they exhibit the advantage that, after an acidolytic or oxidative cleavage of the polymer chain, the ensuring depolymerization by an unzipping mechanism comes to a stop at a –C–C bond.

$$\begin{aligned} &R\!-\!(OCH_2)_m\!-\!O-CH_2-CH_2\!-\!(OCH_2)_n\!-\!O-CH_2-CH_2\!-\!(OCH_2)_l\!-\!O-CH_2-OH \\ &\xrightarrow{-(CH_2O)_{l+1}} R\!-\!(OCH_2)_m\!-\!O-CH_2-CH_2\!-\!(OCH_2)_n\!-\!O-CH_2-CH_2-OH \end{aligned} \qquad (4.12)$$

Removal of unstable chain ends of a copolymer is achieved in practice through a deliberate thermal or alkali treatment (or a combination of these) in the melt [78] or solution [79]; see Figures 4.5 and 4.6.

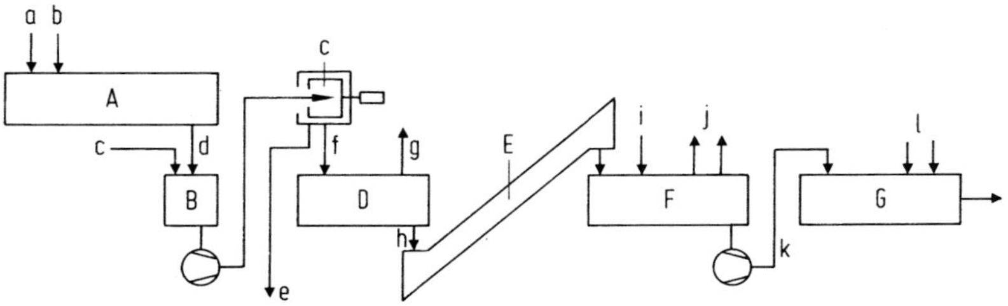

Figure 4.5. Copolymerization of trioxane, work-up in solution
A, Polymerization device; B, Quench tank; C, Solution reactor; D, Precipitator; E, Centrifuge; F, Dryer
a, Monomer; b, Initiator; c, Deactivator/methanol/water; d, Raw polymer; e, Polymer in methanol/water suspension; f, Solution of polymer in methanol/water; g, End group stabilized polymer in methanol/water; h, Methanol/water to work-up; i, Dewatered polymer; j, Methanol/water from drying; k, Polymer for compounding

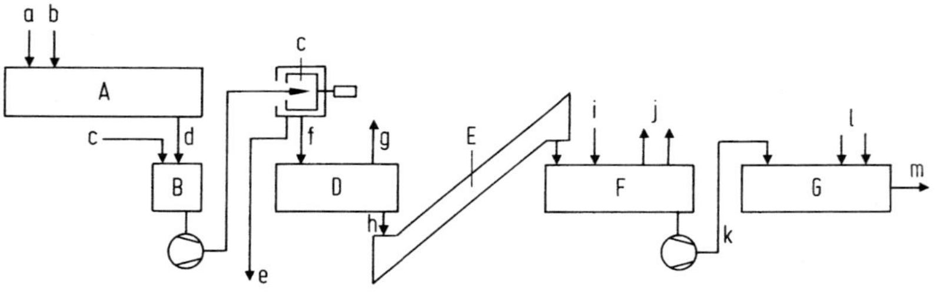

Figure 4.6. Copolymerization of trioxane, work-up in the melt
A, Polymerization unit; B, Quench tank; C, Centrifuge; D, Dryer; E, Conveying screw; F, Melting and degassing unit; G, Finishing extruder
a, Monomer; b, Initiator; c, Deactivator/methanol/water; d, Raw polymer; e, Water/monomer for work-up; f, Dewatered raw polymer; g, Water/monomer for work-up; h, Dried raw polymer; i, Reagent for removal of unstable end groups; j, Degassing; k, Polymer melt; l, Stabilizer, additives; m, Polymer granulate

Fundamental scientific studies on the copolymerization of trioxane were carried out by Kern and his coworkers [58]. It was the Celanese Corporation in the USA and Hoechst AG in western Europe that developed trioxane copolymers to technical maturity in the late 1950s.

References for Section 4.3.4

1. *Walker, J. F.* (Ed.): Formaldehyde. 3. Ed. R. E. Krieger, Florida 1975, p. 180
2. BP-PS 929260 (1963) Kanegafuchi Spinning Co.
3. FR-PS 1386500 (1965) Pechiney Saint Gobain
4. US-PS 2841560 (1958) Du Pont de Nemours
5. FR-PS 1254844 (1961) Bayer AG
6. JA-AS 67019260 (1967) Sumitomo Bakelite Co.
7. BP-PS 793673 (1958) Du Pont de Nemours
8. DE-OS 3044940 (1982) Società Italiana Resine
9. US-PS 4539387 (1985) Du Pont de Nemours
10. JA-AS 67000629 (1967) Ube Industries KK
11. JA-AS 67007629 (1967) Ube Kosan KK
12. EP-OS 029449 (1981) Ube Industries KK
13. US-PS 2848437 (1958) Du Pont de Nemours
14. JA-AS 87013971 (1987) Mitsubishi Chemical Ind.
15. JA-OS 54016954 (1979) Mitsubishi Chemical Ind.
16. DE-PS 1595275 (1970) Asahi Chemical Ind.
17. DE-OS 1946077 (1970) Asahi Chemical Ind.
18. US-PS 2734889 (1956) Du Pont de Nemours
19. JA-AS 4011391 (1965) Teijin Ltd.
20. JA-AS 428935 (1967) Kanegafuchi Spinning Co.
21. US-PS 3007897 (1961) Degussa AG
22. AU-PS 2462967 (1963) Hoechst AG
23. DE-OS 2953911 (1982) Grishin, B. P. Translation to WO 81/1556
24. SU-PS 905226 (1982) B. P. Grishin
25. *Kirynklin, B. P.*, et al.: Vysokomol. Soedin., Ser A 14 (1972) p. 2115
26. JA-AS 07013971 (1987) Mitsubishi Chemical Ind.
27. JA-OS 53035791 (1978) Mitsubishi Chemical Ind.
28. US-PS 3017389 (1962) Du Pont de Nemours
29. DE-PS 3050137 (1982) Asahi Chemical Ind.
30. US-PS 3275603 (1968) Tenneco Chemical Ind.
31. BE-PS 631085 (1963) Kurashiki Rayon Co.
32. *Kagiya, T.*: Bull. Chem. Soc. Jap. 42 (1969) p. 1688
33. US-PS 3491063 (1970) Sumitomo Chemical Co.
34. EP-PS 108324 (1984) BASF AG
35. JA-AS 83051012 (1983) Ube Industries KK
36. JA-AS 69011142 (1968) Sumitomo Bakelite Co.
37. JA-AS 72031435 (1972) Japan Atomic Energy Research
38. JA-AS 72010062 (1972) Kurray Co.
39. *Schweitzer, C. E.*, et al.: J. Appl. Polym. Sci. 1 (1959) p. 158
40. GB 748856 (1956) Du Pont de Nemours
41. GB 770717 (1955) Du Pont de Nemours
42. GB 864403 (1961) Bayer AG
43. DE 1089048 (1960) Du Pont de Nemours
44. US-PS 2964500 (1960) Du Pont de Nemours
45. DE-PS 1133552 (1960) Du Pont de Nemours
46. DE-AS 2027563 (1986) Ube Industries KK, taken back
47. JA-AS 68020308 (1968) Mitsubishi Rayon Co.
48. US-PS 4191839 (1990) Suntech Incorp.
49. NL-OLS 710670 (1971); NL-OS 7106271 (1971) Soecietà Italiana Resine
50. GB 877256 (1960) Du Pont de Nemours
51. US-PS 3393179 (1968) Du Pont de Nemours
52. DE-PS 1301544 (1968) Du Pont de Nemours
53. US-PS 3183211 (1965) Du Pont de Nemours
54. JA-AS 87012804 (1987 Asahi Chemical Ind.
55. FR-PS 2233344 (1975), US-PS 4097453 (1978) Società Italiana Resine
56. JA-AS 73003711 (1973) Ube Industries KK

57. De-PS 29215252 (1972) Asahi Chemical Ind.
58. *Kern, W.*, et al.: Agnew. Chem. 73 (1961) p. 177
59. *Leese, L.*, et al.: Polymer 6 (1965) p. 269
60. *Kern, W.*, et al.: Pure Appl. Chem. 12 (1966) p. 371
61. *Miki, T.*, et al.: Bull. Chem. Soc. Jpn 39 (1966) p. 2480
62. *Weissermel, K.*, et al.: Agnew. Chem. 79 (1967) p. 512
63. *Kern, W.; Jaacks, V.:* J. Polym. Sci. 48 (1960) p. 399
64. US-PS 3428635 (1968) British Industrial Plastics
65. *Fleischer, D.*, et al.: Makromol. Chem. 152 (1972) p. 305
66. *Ishigaki, I.*, et al.: J. Polym. Sci., Polym. Chem. Ed 10 (1972) p. 751
67. *Crivello, J. V.*, et al.: Macromolecules 10 (1977) p. 1307
68. US-PS 2989511 (1958) Celanese Corp.
69. US-PS 2989505 (1957) Celanese Corp.
70. US-PS 3087911 (1957) Celanese Corp.
71. *Mejo, L. De*, et al.: Polymer 19 (1978) p. 956
72. EP-PS 80656 (1982) BASF AG
73. DE-PS 1137214 (1958) BASF AG
74. EP-PS 083698 (1983) Degussa Ag
75. *Jannison, S. E.*, et al.: J. Polym. Sci. 1 (1963) p. 51
76. DE-PS 1161421 (1961) Hoechst AG
77. DE-PS 1720300 (1972) BASF AG
78. DE-PS 3311154 (1987) Asahi Chemical Ind.
79. JA-AS 83051014 (1983) Mitsubishi Gas Chemical Ind.
80. GB 602869 (1961) Celanese Corp.
81. FP 1253533 (1960) Celanese Corp.
82. JA-OS 59159812 (1984) Asahi Chemical Ind.
83. DE-PS 1301544 (1969), DE-OS 1645120 (1970) Du Pont de Nemours
84. Acetal resins; Report Nr. 69 (1971), Stanford Research Institut, Menla Park, California

4.3.5 Polymers of 1,3-Dioxolane and 1,3-Dioxepane

The ready availability as well as the favorable polymerization behavior of both cyclic acetals incited a series of teams to take up scientific investigations on the homo- and copolymerization of both these monomers.

Similar initiator systems as for trioxane are described, such as protonic acids [1, 2], Lewis acids [3], onium salts [4], ion exchange resins [5], or iodine [6]. The cited comonomers comprise cyclic acetals, cyclic ethers, and olefinic product groups [3, 7, 8]. The formation of block copolymers with poly(ethylene oxide) as the block component is described in [2].

A contribution concerning the reaction mechanism is found in [4], which discusses the polymerization in terms of an insertion mechanism.

References for Section 4.3.5

1. *Ringbuer, J.*, et al.: Acta Polym. 32 (1981) p. 238
2. *Reibel, L.*, et al.: Polym. Prepr., Amer. Chem. Soc., Div. Polym. Chem. 26 (1985) p. 55
3. *Okada, M.*, et al.: Polym. J. 1 (1970) p. 1
4. *Plesch, P. H.*, et al.: J. Chem. Soc. 3 (1969) p. 1230
5. *Koyama, K.*, et al.: J. Macromol. Sci. Chem. 8 (1974) p. 711
6. *Ishigaki, J.*, et al.: J. Appl. Polym. Sci. 21 (1977) p. 711.
7. *Yamashita, Y.*, et al.: Makromol. Chem. 122 (1969) p. 284
8. *Okada, M.*, et al.: J. Polym. Chem. Ed. 11 (1973) p. 1237

4.3.6 Polymers of Higher Aldehydes

The technical significance of polyacetals based on formaldehyde has been documented in the preceding sections. There has been no shortage of attempts to investigate possibilities with formaldehyde homologues and the simple ketones with regard to their polymerizability and the properties of their polymers. Details of the polymerization of aldehydes can be found in monographs by Furukawa and Saegusa [1] as well as from Sittig [2].

Furukawa and his coworkers first found that liquid acetaldehyde polymerized in the temperature range of −70 °C using aluminum oxide as the catalyst. The polymerization of acetaldehyde on cationically active fixed beds was described by Weissermel and Schmieder [3]. Processes for the preparation of amorphous elastomeric poly(acetaldehyde) by polymerization on activated fixed bed catalysts are contained in the patent literature [4 to 6].

A multitude of additional initiators are described in the literature. These include: Lewis and protonic acids [7, 8] organozinc complexes with water, amines, or amides [9], as well as phosphoric esters [10], amine alcoholates [11], or complexes of alkali metals with ketones [12].

Amorphous, elastomeric poly(acetaldehyde) is soluble in organic and has little endurance. Thermal or oxidative influences lead to quantitative depolymerization which essentially cannot be retarded by blocking the end groups or with stabilizers. Therefore an amorphous poly(acetaldehyde) is not suitable for technical applications.

Copolymers with olefinic or epoxy compounds yield no improvement of thermal or oxidative stability [13].

Natta [14], Vogl [15] and Furukawa [16], independently from each other, found that acetaldehyde polymerized at low temperatures with the use of organometallic compounds to isotatic, crystalline poly(acetaldehyde). Several Japanese firms worked intensively with this method, as shown by a series of patent applications [17 to 19].

In contrast to amorphous polyacetaldehyde, the crystalline polymer is not soluble in most solvents. The polymer melts (with decomposition) at just over 170 °C.

The polymerization of propionaldehyde and butyraldehyde is described in [15, 16, 20]. Their crystalline products, however, are unstable and have no practical significance. The efforts within industry to prepare usable polymers from higher homologues of formaldehyde are shown in a series of patents. A breakthrough has not yet been achieved [21 to 25].

More detailed information on the cationic and anionic polymerization of chloral in inert solvents at temperatures between −40 °C and 20 °C are found in patent applications [26, 27].

Kargin et al. [28] and Okamura [29] dealt with the polymerization of acetone. Polymerization is achieved by irradiation of solid acetone at temperatures between −136 °C and −196 °C. The polyketal, obtained only in small yield, is extraordinarily unstable and depolymerizes instantaneously at room temperature [28]. The ceiling temperature for acetone polymerization has been estimated at approx. −140 °C [29].

References for Section 4.3.6

1. *Furukawa, J.; Saegusa, T.:* Polymerization of Aldehydes and Oxides. Interscience Publ. New York 1963.
2. *Sittig, M.:* Polyacetal Resins. Gulf Publishing Comp., Houston 1963
3. *Weissermel, K.; Schmieder, W.:* Makromol. Chem. 51 (1962) p. 39
4. JA-AS 67003035 (1968) Kurashiki Rayon Co.
5. IT-AS 714099 (1968) Montecatini Soc.

6. JA-AS 73017874 (1973) Nippon Synthetic Chemical
7. *Vogl, O.:* J. Polym. Sci. 46 (1960) p. 261
8. *Vogl, O.:* Macromol. Synth. 3 (1968) p. 71
9. NL-OS 7005705 (1970), NL-OS 6703805 (1968), NL-OS 6708558 (1968) Nederlandse Centrale Organisatie
10. JA-AS 68018150 (1968) Agency of Ind. Sci. and Techn.
11. GB-PS 1030745 (1968) ICI
12. JA-AS 68025517 (1968) Dikyowa Seikiyu Kagaku
13. JA-AS 69011148 (1968) Kurashiki Rayon Co.
14. *Natta G.*, et al.: J. Polym. Sci. 48 (1960) p. 219
15. *Vogl, D.:* J. Polym. Sci., Polym. Chem. Ed. Bd. 10 (1964) p. 4621
16. *Furukawa, J.*, et al.: Makromol. Chem. 37 (1960) p. 149
17. JA-AS 67002577 (1968), JA-AS 69001224 (1968), JA-AS 6900128 (1968), JA-AS 69001997 (1968), Chisso Co.
18. JA-AS 66008879 (1968) Japan Res. Inst. of Chem. Fibres
19. DE-PS 1568264 (1971) Chiyoda Kako Kensetsu
20. *Natta, G.*, et al.: Makromol. Chem. 37 (1960) p. 156
21. JA-AS 68018149 (1968) Nippon Synth. Chem. Ind.
22. GB-PS 1048610 (1968) ICI
23. JA-AS 483232 (1973), JA-AS 73003232 (1973) Asahi Chem. Ind.
24. *Okamoto, M.:* Nippon Kagaku Kaishi 1 (1974) p. 31
25. GB-PS 1164601 (1968) Brit. Ind. Plastics
26. US-PS 3454527 (1968) Du Pont de Nemours
27. JA-AS 68016157 (1968) Agency of Science and Technology
28. *Waywell, D. R.:* J. Polym. Sci., Polym. Phys. Ed. 8 (1970) p. 327
29. *Okamura, S.*, et al.: Isotops and Radiation Bd. 4 (1961) p. 70

4.4 Finishing of Acetal Resíns

H.-D. Sabel

4.4.1 Stabilization

As many organic substances, acetal resins are degraded by high temperatures, oxygen, and/or UV-radiation. Therefore they require effective protection by stabilizers, not only during processing, but also for use at higher temperatures. Since the degradation reactions, including side and consequent reactions, are multifarious, complex stabilization problems result.

4.4.1.1 Degradation Mechanisms

The fact that thermaloxidative degradation begins at about 160 °C has to be taken into account considering processing conditions. Assuming the degradation of end group-stabilized polymers starts with a cleavage of the molecular chain into two fragments, with an unstable end group each. Then further degradation continues at these end groups and progresses by splitting off formaldehyde (depolymerization, reversal of the synthesis reaction) up to the next comonomer unit, in case of acetal copolymers, or, in case of homopolymer, until the chain is completely decomposed. Depolymerization kinetics show that the resulting side products, or their decay products, intervene in the degradation reaction. With formic acid, e.g., the oxidation product of formaldehyde, an acidolytic cleavage, is superimposed on the thermaloxidative degradation leading to fragments with unstable chain ends as well. This raises the number of damaged chains and increases the rate of depolymerization [5].

In case of photolytic degradation (irradiation of polyacetal in vacuum), hydrogen and carbon monoxide are formed in a mole ratio of (3.5 to 4):1 [1, 2], from which basically two reaction paths for the degradation mechanism can be deduced

$$\begin{array}{l} -\ CH_2O(CH_2O)_n\,CH_2OCH_2OH \xrightarrow{h\cdot\nu} \\ -\ CH_2O(CH_2O)_n\,CH_2O\overset{H}{C}{=}O\ +\ H_2 \xrightarrow{h\cdot\nu} \\ -\ CH_2O(CH_2O)_n\,CH_2O\bullet\ +\ CO\ +\ 1/2\ H_2 \end{array} \qquad (4.13)$$

$$\begin{array}{l} -\ CH_2O(CH_2O)_n\,CH_2O\ - \xrightarrow{h\cdot\nu} \\ -\ CH_2O(CH_2O)_n\,\dot{C}HO\ -\ +\ H\bullet \xrightarrow{h\cdot\nu} \\ \left.\begin{array}{l} -\ CH_2O(CH_2O)_n\,\overset{H}{C}-O\ - \\ \qquad\qquad\qquad\ \ | \\ -\ CH_2O(CH_2O)_n\,\underset{H}{C}-O\ - \end{array}\right.\ +\ H_2 \end{array} \qquad (4.14)$$

According to Equation (4.13) a hemiacetal end group is first dehydrogenated to formate ester, from which a H-atom is split off. The C-radical formed is stabilized by decarbonylation forming an O-radical. According to Equation (4.14) a bond on a CH_2 group in the chain is homolytically cleaved, the C-radical is stabilized by combination, i.e. crosslinking occurs.

In case of photooxidative degradation, the primary step is of the formation of peroxy and hydroperoxy groups on the CH_2 groups [2, 3]. Irradiation of polyacetal in the presence of air leads to defined changes of the IR-spectrum in the region of C=O, OH, and CH_2 bands [1, 2, 4]. All these degradation mechanisms have one thing in common: chain cleavage is the dominating reaction. The chain fragments formed are unstable and fall apart separating formaldehyde.

4.4.1.2 Stabilization Against Thermaloxidative Degradation

Corresponding to the mentioned variety of reactions occurring in case of thermaloxidative degradation, multi-component systems are used without exception for the stabilization of polyacetals. Their individual components can be classified according to their effect as follows:

- antioxidants/radical traps,
- compounds binding the secondary products of oxidation (e.g. formaldehyde, protons).

In principle, as for the stabilization of other polymers, sterically hindered phenols are considered as antioxidants. Because of their lower migration tendency, multi-nuclear hindered phenols are preferred. Examples of these are:

- 4,4′-butylidenebis(6-t-butyl-3-methylphenol)
- 2-2′-methylenebis(4-methyl-6-butylphenol)
- N,N′-hexamethylenebis(3,5-di-t-butyl-4-hydroxyhydrocinnamide)
- 1,6-hexamethylenbis(3,5-di-t-butyl-4-hydroxyhydrocinnamate)
- N,N′-bis(3,5-di-t-butyl-4-hydroxyhydrocinnamoyl)hydrazine
- tetrakismethylene(3,5-di-t-butyl-4-hydroxyhydrocinnamate)methane
- 1,8(3,6-oxacoctamethylene)bis(3-methyl-5-t-butyl-4-hydroxyhydro-cinnamate
- sulfur-containing compounds

In addition, the HALS-stabilizers described below work as radical traps.

Compounds of different types are effective as stabilizers protecting against the secondary products of autoxidation. One of their features is the fact that they bind the released formaldehyde chemically, and thus prevent its oxidation to formic acid. They may also work as proton acceptors.

Some examples of these types of stabilizers are:

- compounds with a multiplicity of hydroxyl groups
- urea and its substitution products
- condensation products of urea derivatives
- amides
- low-melting polyamides
- higher-melting polyamides dispersed in a carrier
- condensation products from melamine and formaldehyde
- salts of carboxylic acids
- alkaline earth oxides

The above mentioned stabilizers allow multiple combinations often having synergistic effects.

4.4.1.3 Stabilization Against Photooxidative Degradation

Photooxidative degradation of polyacetals set off by UV-rays results in two different effects: first chalking is observed. A continuing influence of light leads to a loss of mechanical poperties. The smaller the wall thickness of the part, the more pronounced the effect (in case of a larger wall thickness the formed deposit acts as a protective layer). Both effects can be delayed with the help of stabilizers. UV absorbers and sterically hindered amines (HALS stabilizers) are considered as stabilizers. They can be distinguished by the way they work.

Effective UV absorbers include:

- benzophenone derivatives such as
  2,4-dihydroxybenzophenone
  2-hydroxy-4-methoxybenzophenone
  2-hydroxy-4-alkoxybenzophenone (n-octyl, i-octyl and dodecyl in the alkoxy group)
- benzotriazole derivatives such as
  2-(2′-hydroxy-5′-methylphenyl)benzotriazole
  2-(2′-hydroxy-5′-t-octylphenyl)benzotriazole
  2-(2′-hydroxy-3′,5′-di-t-amylphenyl)benzotriazole
  2-(2′-hydroxy-3′,5′-di{2″-phenyl}-i-propylphenyl)benzotriazole
  2-(2′-hydroxy-3′,5′-di-t-butylphenyl)-5-chlorobenzotriazole
  2-(2′-hydroxy-3′-methyl-5′-t-butylphenyl)-5-chlorobenzotriazole
  2-ethylhexyl-2-cyano-3-3′diphenyl acrylate
- benzoates such as
  p-(t-butyl)phenyl salicylate
  n-hexadecyl-3,5-di-t-butyl-4-hydroxy benzoate
- oxanilides such as
  N-(2-ethylphenyl)-N′-(2-ethoxy-5-t-butylphenyl)oxalamide
  2-ethyl-2′-ethoxyanilide
- UV absorbing pigments, in particular carbon black, but TiO_2 as well has a stabilizing effect against the action of light. When used at a sufficient concentration, they also have a stabilizing effect in the presence of colored pigments.

Combinations of UV stabilizers often have synergistic effectiveness. HALS stabilizers (hindered amine light stabilizers) function as radical traps and, in contrast to UV absorbers, do not follow the Beer-Lambert law. Examples of HALS stabilizers are:

4-acetoxy-2,2,6,6-tetramethylpiperidine
4-stearoyloxy-2,2,6,6-tetramethylpiperidine
4-(phenylcarbamoyloxy)-2,2,6,6-tetramethylpiperidine
bis(2,2,6,6,-tetramethylpiperidyl)succinate
bis(2,2,6,6,-tetramethylpiperidyl)sebacate
bis(1,2,2,6,6-pentamethylpiperidyl)sebacate
bis(2,2,6,6-tetramethyl-4-piperidyl)hexamethylene-1,6-dicarbamate
tris(2,2,6,6-tetramethyl-4-piperidyl)-1,3,5-benzene tricarboxylate
polycondensates from dimethyl succinate and 1-(2-hydroxyethyl)-4-hydroxy-2,2,6,6-tetramethylpiperidine

Because of their different mechanisms of operation, UV absorbers and HALS stabilizers exhibit a pronounced synergy. Therefore, combinations of both stabilizer types are frequently used in formulations to protect polyacetals from light.

4.4.2 Nucleation

The properties of the polyacetals are essentially based on their crystallinity.

Polyacetals exhibit a high rate of crystallization so that nucleants are not required for the solidification to take place. However, their crystal texture is coarse and exhibits a wide distribution of spherulite sizes, so that the mechanical properties are not optimal.

The use of a nucleating agent leads to a uniform, fine texture, increasing the mechanical strength properties [6]. In addition, nucleated polyacetal molding compounds often allow an earlier demolding, that is, a shortening of the cycle time in injection molding [7] (see also Sections 4.7.1 and 4.10.6).

Talc [8], boron nitride [9], 2,3-dioxyquinoxaline [10], branched and crosslinked acetal copolymers [11], acetal block copolymers [12], as well as special melamine-formaldehyde resins [13] are effective nucleants for polyacetals.

4.4.3 Lubricants and Release Agents

In general polyacetals do not require lubricants for processing, neither to increase their flowability nor to improve plastification. In individual cases, however, the melt flow can be increased using a polymeric flow aid. The use of a mold release agent may be appropriate for complex injection molded parts that are difficult to demold, e.g. non-tapered cylindrical shapes or parts with threads, see also Sections 4.10.2, 4.10.3, 4.10.5, and 4.10.6.

Suitable mold release agents are: esters of montanic acid, montanic acid salts, salts of fatty acids, fatty alcohols, polyethylene waxes, N,N′-ethylenediamine bis(stearamide).

References for Section 4.4

1. *Grassie, N.; Roche, R. S.:* Makromol. Chem. 112 (1968) p. 34–39
2. *Schmidt, H.; Wolters, E.:* Kunststoffe 61 (1971) p. 261–265
3. *Kern, W.:* Chem. Z. 91 (1967) p. 255–262
4. *Kellerher, P. G.; Jassie, L. B.:* J. Appl. Polym. Sci. 9 (1965) p. 2501–10
5. *Kern, W.; Cherdron, H.:* Makromol. Chem. 40 (1961) p. 177
6. *Leugering, H. J.,* et al.: Makromol. Chem., Suppl. 1 (1975) p. 621–636

7. *Morlock, G.:* Int. Symp. Macromol. 6 (1978)
8. DE-PS 1247645 (1964) Bayer AG
9. DE-AS 2508126 (1970) Hoechst AG
10. DE-PS 1694078 (1966) Hoechst AG
11. DE-PS 2101817 (1971) Hoechst AG
12. DE-PS 2142091 (1971) Hoechst AG
13. DE-PS 2540207 (1975) Degussa AG

4.5 Standard Products

H.-D. Sabel

All polyacetal manufacturers offer a similar product range: basic grades and various modified grades. The basic grades can be divided into:
- high viscosity, slow-flowing grades of low melt index (MFI 190/2,16 between 2 and about 6 g/10 min)
- middle and low viscosity, easier to easy flowing grades with a melt index (MFI 190/2.16) of over 8 g/10 min

This arrangement of products accounts for the two principal processing methods for polyacetal, injection molding and extrusion. The high molecular weight grades are preferred for extrusion whereas the middle to low molecular weight molding materials serve quite well as injection molding grades.

It ought to be stressed, however, that this division is not absolute and can only serve as a reference point, transitions are fluid. Thus, grades with a melt index (MFI 190/2.16) of around 9 g/10 min on the one hand represent standard injection molding types; on the other hand they can be extruded to semi-finished products, although economically this only makes sense, if the parts are thin-walled. Conversely, the high molecular weight grades often offer advantages in injection molding of thick-walled parts since, due to their high viscosity, they allow an expanding flow desired for filling the mold cavity, and thus have less tendency to form voids.

Branched copolymer of particularly low melt index (MFI 190/2.16 ca. 1 g/10 min) are a specialty in most acetal copolymer product ranges. These grades distinguish themselves by their high melt strength and were developed for extrusion blow molding. Branched, high molecular weight acetal copolymers exhibit better flow under injection molding conditions, i.e. under the influence of high shear rates, than do linear copolymers having a MFI 190/2.16 of about 2 g/10 min. For this reason, they are suitable for injection molding of thick-walled parts which have high strength and are relatively free from voids, for example chain links. The standard grades of all polyacetal product ranges are primarily characterized by their melt index. This is explicitly expressed in the product nomenclature. Differences in mechanical and thermal properties between the various grades are relatively small.

The basis for characterizing various grades according to their melt index relies on the fact that the selection of the grade of polyacetal best suited for a particular case depends on its flowability. Long flow lengths and thin walls require easy or very easy flowing grades. They allow manufacturing of molded parts that are relatively free of orientation and stress, i.e. optimal utilization of the excellent properties of polyacetals.

The standard properties of the individual grades given in the tables of properties supplied by the producers, see also Tables 4.1 and 4.2, are only useful for characterization and cannot be used for selection of a particular grade. Thus, in the tables, the high molecular weight grades indeed exhibit the highest values for toughness. However, thin-walled parts molded from such grades can exhibit a poorer toughness behavior in a practical test than do parts molded from easy-flowing grades. The orientation of the molecular chains caused by the high viscosity destroys the toughness advantage of the high molecular weight polyacetal grades.

The evaluation of microtome sections in polarized light has proved worthwhile for the investigation of molded parts for possible orientation and stress [1]. Information concerning morphology is obtained simultaneously. Commercial polyacetals are, without exception, uniformly finely crystallized.

Most acetal copolymer product ranges comprise a melt index range from 1 to about 50. These correspond to flow lengths in spiral flow molding tests between about 400 mm and 1400 mm (spiral thickness 3 mm, injection pressure 1000 bar, material temperature 205 °C, mold temperature 80 °C) [2]. Grades of acetal homopolymer have melt indices (MFI 190/2.16) between 2.5 and 22 g/10 min [3].

Polyacetals with melt indices between 30 and 50 g/10 min are not only exceptionally well suited for the manufacture of particularly thin-walled parts or for filling of complicated molds (long flow lengths or multiple cavities), they can also be processed at lower material temperatures thus saving energy and allowing shorter cycle times. See also Sections 4.10.3 and 4.10.6.

The polyacetal product ranges also contain a few modified basic grades that should be counted among the basic grades. As for acetal homopolymer there are: high molecular weight grades for extrusion of semi-finished products and for sanitary as well as for injection molding grades with better mold release, slightly better flow, and higher productivity, the latter with acceptance of slightly lower mechanical properties [4].

For acetal copolymers too, there are high molecular weight grades for particular end uses, furthermore injection molding grades with higher stiffness and hardness as well as grades with antistatic formulation [2, 5].

Numerous grades of polyacetal are also offered in a UV-stabilized version.

The range of possible colors comprises standard shades as well as a multitude of special colors, including color concentrates having a corresponding high pigment loading for coloring of natural pellets.

References for Section 4.5

1. *Wolters, E.*, et al.: Plastverarbeiter 24 (1973) p. 549
2. Kunststoffe Hoechst Hostaform. Company brochure Hoechst AG, Frankfurt 1987
3. Acetal-Homopolymer. Company brochure Asahi Chemical Ind. Co., Ltd., Tokio 1978
4. Delrin-Polyacetale. Company brochure Du Pont de Nemours 1985
5. Ultraform-Polyoxymethylen. Company brochure BASF AG, Ludwigshafen 1986

4.6 Modified Products

U. Ziegler

The product range of polyacetals of all suppliers consists of various products with differing molecular weights and thus differing flowability. The property range and thus the breadth of application of polyacetals can be substantially expanded by modification of these basic grades with various additives.

4.6.1 Filled and Reinforced Products

4.6.1.1 Reinforced Products

Of the modified special grades, the polyacetals reinforced with glass fibers have achieved the greatest significance [1 to 15]. Mechanical properties, in particular stiffness, strength, and hardness, are improved by reinforcement with glass fibers. The extent of this improvement depends upon the proportion and length of the glass fibers, the homogeneity of mixture, and bonding between the glass fibers and the polyacetal matrix. The content of glass fibers in commercially available products amounts to up to 40 wt-% [13]. While stiffness, strength, hardness, and resistance to deformation caused by heat of such products are substantially raised relative to those of unmodified products (see Section 4.7), a loss of toughness properties must be set against these advantages [15].

A decreasing of the linear expansion coefficient and of shrinkage is a further positive consequence of glass fiber reinforcement. When glass fibers are used, however, the decrease in the linear coefficient of expansion in injection molded parts is anisotropic (Figure 4.32). For this reason the use of glass beads or mixtures of glass beads and glass fibers is an alternative to the reinforcement with glass fibers [16 to 18]. The anisotropy in properties, for example the directionally dependent shrinkage that is caused by fiber orientation during injection molding of glass fiber reinforced thermoplastics, does not exist in products with glass beads. In this way, complicated molded parts having high stiffness and design-dependent strength can be manufactured with a low warpage [13]; see also Section 4.10.3. The extremely high strengths that are achievable with glass fiber reinforcement are, however, not attained. The loss of toughness properties compared to unmodified polyacetal is similar [13].

4.6.1.2 Products with Improved Friction and Wear Behavior

A further, important group of modified polyacetals are those which contain modifiers for better friction and wear [12 to 15, 19]. The blends with PTFE powder or fibers or with ultrahigh molecular weight polyethylene that are discussed in Section 4.6.3 are included in this category.

For special application cases, such as slowly running, highly loaded planar bearings, products modified with molybdenum disulfide are used. Compared to PTFE-containing products, the products containing molybdenum disulfide offer advantages in toughness [12, 13]. They are useful whenever slow speeds, high surface loading, and small frictional paths occur. At higher frictional speeds, the PTFE-grades have lower coefficients of friction than do the molybdenum disulfide products [19].

Because of their favorable wear behavior, chalk-filled products [20] are used even in cases of dry friction. The addition of silicone oil or oil/graphite combinations as internal lubricants can, even at low concentrations, improve the friction and abrasion behavior [12].

4.6.1.3 Miscellaneous Fillers

In recent times, products deliberately formulated to be antistatic [14, 21], for example by addition of conductive carbon black, have been developed for special applications; for instance, for the cassette market or the furniture industry.

Products reinforced with carbon fibers [22, 23] have indeed been listed in the product offerings of various raw materials producers but up to now large applications are not known [13].

4.6.2 Mixtures with Elastomers

Mixtures (blends) of engineering thermoplastics with elastomers are essential for impact modification [24]. The engineering plastic thereby looses hardness and stiffness but profits exceptionally in stability against impact or tensile loading. Although non-modified polyacetals already have high inherent toughness, they exhibit a certain notch sensitivity. That is, by virtue of the stress concentration that arises, sharp notches effect a substantial drop in impact strength. Mixing with elastomers brings a noticeable improvement not only in the notched impact strength but also in the ability to bear multiaxial impact loading. See also Section 4.10.2.

4.6.2.1 Mixtures with Polyurethane Elastomers

Among the elastomers, polyurethanes have the greatest significance for the impact modification of polyacetals [25 to 42]. The polyurethane elastomers are thermoplastic elastomers, that is fusible, not permanently crosslinked, products. The rubbery elasticity of these materials is achieved via thermo-reversible physical crosslinking.

In the case of impact modification of polyacetals with polyurethane elastomers, multiphase systems without thermodynamic miscibility are obtained. The morphology, that is to say the distribution (size and form) of the elastomer phase in the polyacetal matrix, must be achieved in the compounding step by viscous shearing and extension. The relative amounts of the components, the viscosity relationship of both polymers in the melt, their compatibility with each other, and last but not least the compounding conditions are important quantities bearing on the creation of this morphology [29]. In order to obtain the optimal impact strength, the morphology should be maintained as far as possible in further processing operations.

Interpenetrating blend morphologies have a particular role in the impact modification of polyacetals with polyurethane elastomers (Figure 4.7) [29]. Here we are concerned with special, multiphase structures in which one cannot differentiate between dispersed and continuous phases, since both phases form a continuous network (co-continuous phases). This special structure leads to a steep, stepwise increase in toughness properties in passing from a matrix/particle morphology to the interpenetrating blend morphology (Figure 4.8) [29].

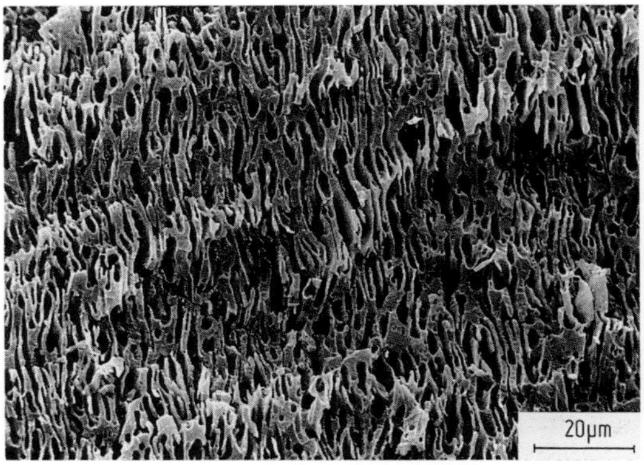

Figure 4.7. Scanning electron microscope image of a mixture of acetal copolymer and a polyurethane elastomer containing 50 wt-% polyurethane (co-continuous morphology) [29]

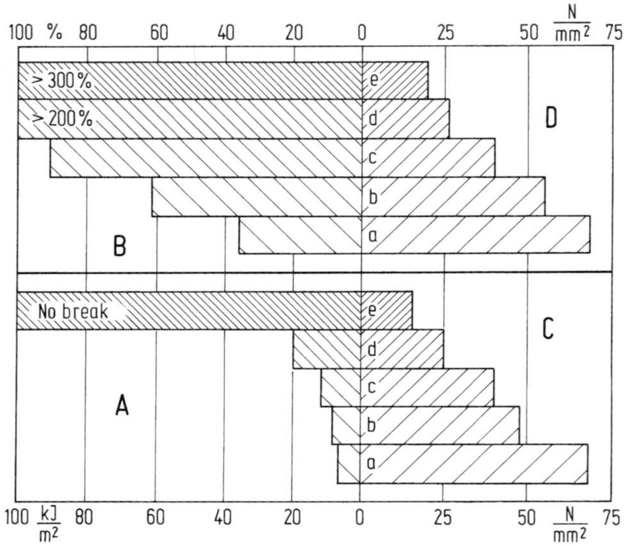

Figure 4.8. Stiffness-toughness relationship of impact modified acetal copolymer [29]
A, Notched impact strength; B, Elongation at break; C, 3.5% Flexural stress; D, Yield stress
a, Non-modified acetal copolymer, Hostaform C 27021; b to d, Acetal copolymers with increasing polyurethane content (Hostaform S 27063, S 27064, and S 27076); e, 1:1 mixture of polyurethane and acetal copolymer

A decrease in mechanical properties such as hardness, stiffness, and strength, which are primarily derived from the semicrystalline structure of polyacetal accompanies the increase in impact strength. The stiffness-toughness relationship of these product types is thus always a compromise that is optimized for the particular application. A large portion of the impact modified polyacetals goes into the manufacture of parts joined with snap-fits and also clips where the combination of the elastic recovery of polyacetals with better impact strength is essential [15].

A further important possibility for modification with polyurethane elastomers is to offset the loss of toughness brought about by reinforcements (e.g. glass fibers) and return it to the level of that of unmodified polyacetals [21, 37, 38]. In so doing one must deal with the loss of stiffness compared to the pure reinforced material. The low coefficient of thermal expansion, also a fundamental property of glass fiber reinforced products, is, however, maintained.

4.6.2.2 Mixtures with Butadiene Copolymers and Acrylate Copolymers

Elastomers based on poly(butadiene) or polyacrylates are also important modifiers [43 to 49].

Here, an optimal grafting or block copolymerization of the elastomers is above all important in order to guarantee an adequate compatibility with the matrix. The essential advantage of modification with polybutadiene is better cold impact strength. In the case of modification with polyacrylates the advantages are better heat-ageing resistance and better weathering resistance.

At this time, however, polyurethane elastomers are the main commercial modifiers for polyacetals.

4.6.2.3 Mixtures with Ethylene-Propylene Elastomers (EPM, EPDM)

To date, knowledge of blends of polyacetals with EPM or EPDM elastomers largely has to be sought in the patent literature [50 to 52]. The principal reason is the poor compatibility of these non-polar elastomers with the polar polyacetal.

In the case of modification of polyamides or polyesters, this incompatibility can be circumvented through the use of olefin elastomers grafted with acid anhydrides. This method founders in the case of polyacetals because of their inadequate reactivity. Neither is the use of the corresponding esterified products or of ionomers [53, 54] entirely successful. Grafting of the olefin elastomer with methyl methacrylate or acrylonitrile and styrene leads to improved compatibility only if the degree of grafting is very high. In the end result this requires a high loading of grafted elastomer in order to secure improvements in the impact strength. It appears from the patent literature that further development work is concentrating on improvement of the compatibility.

4.6.3 Mixtures with Other Polymers

4.6.3.1 Mixtures with Polyolefins

As is the case with mixtures of ethylene-propylene elastomers, the extent of knowledge concerning blends with common, standard polyolefins such as polyethylene or polypropylene is, owing to their incompatibility with polyacetals, to a large extent limited to the patent literature [55 to 60].

Ultrahigh molecular weight polyethylene (UHMW-PE) having a preformed, small-particle morphology (micropowder), however, has a special position here. Since this extremely high molecular weight material, owing to its rheological properties, practically does not melt, the polyethylene phase does not flow together and there is no delamination. Polyacetals modified with UHMW-PE have a fundamental advantage in friction properties [21]; see also Section 4.6.12. Because of the poor phase adhesion that results from the incompatibility between polyacetal and polyethylene, however, the impact strength of the starting polyacetal is decreased.

4.6.3.2 Mixtures with Fluoropolymers

Like ultrahigh molecular weight polyethylene, perfluorinated polymers also lead to non-melting compounds. Fluoropolymers with pre-determined morphology can thus be incorporated into polyacetals without damaging delamination. PTFE micropowders and PTFE microfibers are used [61 to 64].

The fundamental advantages of these blends are their improved friction and abrasion properties (see Section 4.6.1.2) [15]. However, just as in the case of UHMW-PE blends, the toughness deteriorates.

4.6.3.3 Mixtures with Further Polymers

Additional mixtures of polyacetals with various polymers are described in patents. The following can be mentioned here: ethylene-vinyl acetate copolymers [65, 66], polystyrene [67], elastomeric block copolymers having polystyrene blocks [68, 69], polyesters [70], polycarbonate [71, 72], or also polyester-polyether block copolymers [73, 74].

Up to now these blends have not attained commercial significance. Following the trend in the market development for blends, a stronger activity in the area of development of customized combinations of materials for specific application problems is to be expected here also. In particular, ternary and other multiple blends can be expected.

References for Section 4.6

1. GB 793744 (1956) Du Pont de Nemours
2. DE-OS 1720644 (1967) Hoechst AG
3. DE-PS 2121559 (1970) Celanese Corp.
4. DE-AS 2162345 (1970) Celanese Corp.
5. DE-OS 2129229 (1971) Hoechst AG
6. DE-PS 2262859 (1972) BASF AG
7. JA-OS 51052455 (1974) Mitsubishi Monsanto
8. *Roos, G.:* Kunststoffe 57 (1967) p. 421
9. *Roos, G.:* Ind. Anz. 41 (1968) p. 5
10. *Roos, G.; Meisel, H.:*Kunststoff-Berater 16 (1971) p. 954
11. *Wolters, E.; Sabel, H.-D.:* Kunststoffe 62 (1972) p. 632
12. *Sabel, H.-D.:* Kunststoffe 70 (1980) p. 641
13. *Schütte, W.:* Kunststoffe 74 (1984) p. 575
14. Kunststoffe 76 (1986) p. 856
15. Kunststoffe Hoechst, Hostaform. Company brochure Hoechst AG, Frankfurt 1987
16. Hostaform aktuell, Neue Hostaform-Typen. Company brochure Hoechst AG, Frankfurt 1985
17. Mod. Plast. Int. 17 (1987) p. 96
18. EP-OS 273657 (1986) Hoechst Celanese Corp.
19. *Schütte, W.:* Kunststoffe 66 (1976) p. 604
20. DE-PS 154492 (1965) Hoechst AG
21. Eur. Plast. News 16 (1989) p. 35
22. JA-OS 49033947 (1972) Nippon Telecom
23. JA-AS 76014542 (1972) Sumitomo Bakelite KK
24. *Bucknall, C. B.:* Toughened Plastics. Applied Sci. Publ., London 1977
25. *Neidernberg, D.:* Plastverarbeiter 34 (1983) p. 1466
26. *Sternfield, A.:* Mod. Plast. 61 (1984) p. 98
27. *Flexman, E. A.:* Mod. Plast. 62 (1985) p. 72
28. *Flexman, E. A.:* in: Antec '84, New Orleans 1984, p. 558
29. *Kloos, F; Wolters, E.:* Kunststoffe 75 (1985) p. 735
30. *Kloos, F.:* Angew. Makromol. Chem. 133 (1985) p. 1
31. *Wolters, E.; Sabel, H.-D.:* Kunststoffe 77 (1987) p. 997
32. *Chiang, W. Y.; Lo, M. S.:* J. Appl. Polym. Sci 36 (1988) p. 1685
33. *Flexman, E. A.*, et al.: Polym. Prepr. 29 (1988) p. 189
34. *Pecorini, T. J.*, et al.: Polym. Prepr. 29 (1988) p. 136
35. DE-PS 1193240 (1962) Bayer AG
36. DE-OS 2051028 (1969) Celanese Corp.
37. DE-PS 2209985 (1973) BASF AG
38. EP-OS 115846 (1983) Hoechst AG
39. EP-OS 115847 (1983) Hoechst AG
40. EP-PS 116456 (1983) Du Pont de Nemours
41. EP-PS 117664 (1983) Du Pont de Nemours
42. EP-OS 167369 (1984) Celanese Corp.
43. DE-OS 2408487 (1984) Celanese Corp.
44. DE-OS 2420380 (1974) BASF AG
45. US-PS 4424307 (1982) Celanese Corp.
46. EP-OS 115373 (1983) Celanese Corp.
47. DE-OS 3403576 (1984) BASF AG
48. DE-OS 3505524 (1984) Hoechst AG
49. DE-OS 3441547 (1984) BASF AG
50. JA-OS 58/45252 (1981) Sumitomo Naugatuck Co.
51. EP-OS 124879 (1983) Mitsui
52. US-PS 4652613 (1985) Celanese Corp.
53. DE-PS 1694187 (1967) Hoechst AG
54. EP-PS 135263 (1983) Celanese Corp.
55. DE-PS 1569057 (1964) Igoplat/Faigle
56. JA-AS 7141456 (1968) Toray Ind.

57. JA-OS 50052166 (1973) Asahi
58. JA-OS 59064654 (1982) Toray
59. EP-OS 148743 (1984) Mitsui
60. EP-OS 292746 (1987) Mitsubishi Petrochem
61. US-PS 3005795 (1957) Du Pont de Nemours
62. FR-OS 2144386 (1961) Union Carbide
63. *Kar, M. K.; Bahadur, S.:* Wear 30 (1974) p. 337
64. *Shen, C.; Dumbleton, J. H.:* Wear 40 (1976) p. 371
65. JA-OS 50052165 (1973) Asahi Chemical Ind.
66. DE-OS 2818240 (1978) Hoechst AG
67. JA-AS 947/69 (1964) Sakai Seni
68. US-PS 4079100 (1976) Shell
69. DE-OS 2725663 (1977) Shell
70. DE-OS 2248242 (1972) General Electric
71. JA-AS 12463/64 (1961) Shinsetsu Kagaku
72. DE-OS 3407811 (1982) Mitsubishi Gas
73. DE-PS 2449343 (1974) Hoechst AG
74. US-PS 4243580 (1979) Du Pont de Nemours

4.7 Properties of Polyacetals

H.-D. Sabel, U. Ziegler

The property profile of the polyacetals is primarily determined by their crystallinity. The high stiffness, strength, hardness in combination with toughness, the excellent resistance to deformation under the influence of heat and mechanical load and good creep behavior, the favorable friction and wear properties, and the excellent (compared to many other plastics) resistance to solvents and fuels all result from this crystallinity [1 to 7].

A comparison of homopolymers and copolymers shows that the incorporation of small amounts of comonomer results on the one hand in small changes in the crystallinity compared to the homopolymer, leading in turn to insignificant alteration of the mechanical properties, but on the other to a notable improvement in thermal stability for the copolymer. The characteristic properties are very similar, with a few exceptions that will be discussed as necessary.

Their favorable property profile has allowed the polyacetals to make headway in many application areas in practically all industrial sectors. Special alterations in the property profile are possible through further modification of the polyacetals [8 to 14]; for example, higher stiffness, strength, and hardness through reinforcement with glass, improved sliding behavior by addition of special fillers, or improved toughness by modification with elastomers. The possibilities widen the property spectrum and thereby the field of possible uses. These possibilities for variation will be dealt with further in connection with the corresponding properties.

4.7.1 Morphology and Crystallinity

Because of the linear, non-branched structure of the polyacetal homo- and copolymers, these semicrystalline polymers have a high degree of crystallinity. Polyoxymethylene crystallizes in a hexagonal unit cell with the lattice constants $a = 0.477$ nm and $c = 1.739$ nm [15]. Figure 4.9 shows a solution-grown single crystal of polyacetal [16]. The

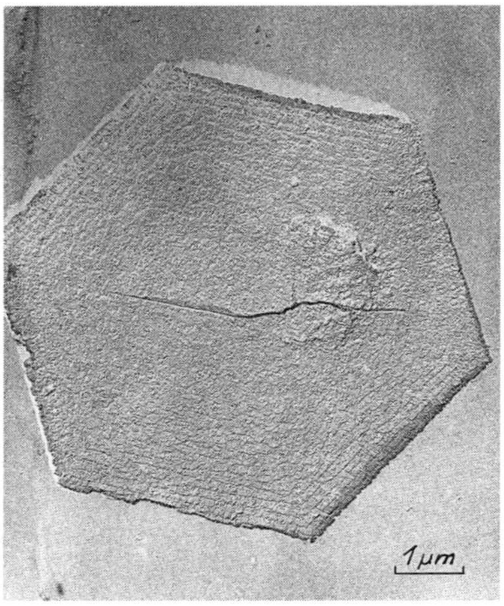

Figure 4.9. Photomicrograph of a hexagonal single crystal of polyacetal grown in solution [16]

chains occur in a helical conformation with an identity period of 9/5 [15]. As a consequence, the density of the crystalline region is 1.49 g/cm^3. According to Wilski [17], the density of the amorphous region of a pure homopolymer is 1.324 g/cm^3 and that of a copolymer with 2 wt-% ethylene oxide is 1.320 g/cm^3. The density of the partially crystalline commercial polyacetals lies between 1.41 and 1.43 g/cm^3.

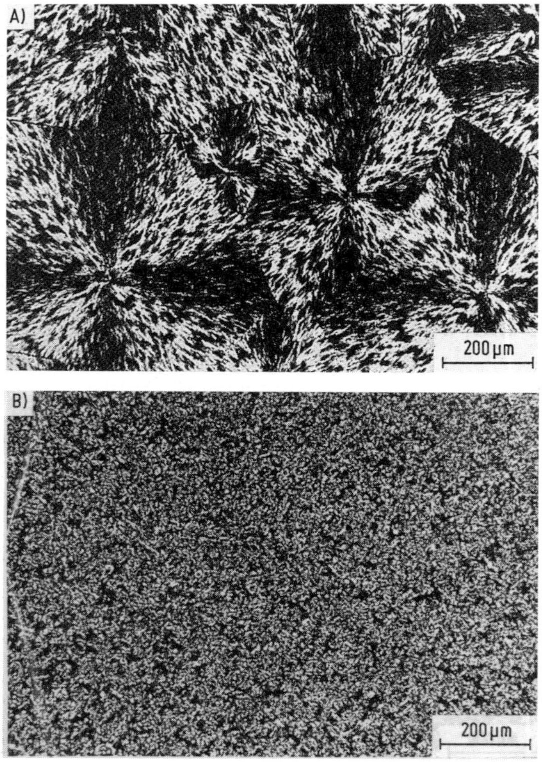

Figure 4.10. Polarized light microscope photographs of the spherulitic structure of (A) acetal copolymer and (B) acetal copolymer nucleated with branched terpolymer [26]

Depending on comonomer content, molecular weight, and sample preparation (slow or fast crystallization), the crystallinity of the products can lie between 50 and 80% [17]. Polyacetals having a lower molecular weight have a higher crystallinity than polyacetals having a higher molecular weight. The crystallinity of copolymers or terpolymers is lower than that of homopolymers. The dependence of density, melting point, and melting enthalpy upon the copolymer composition was intensively investigated by Dröscher et al. with "extended chain" crystals [18 to 20]. The question is still open whether various comonomers are incorporated into the crystalline regions during crystallization or remain preferentially excluded in the amorphous phase.

A spherulitic morphology, observable in the polarized light microscope as the well-known Maltese cross, is produced in the crystallization of polyacetals. The kinetics of spherulitic growth have been investigated in detail by Inoue [25]. The size of the developing spherulite is dependent on the crystallization temperature (larger spherulites develop at higher crystallization temperatures). The rate of crystallization can be increased by addition of nucleants. In this case the number of spherulites formed is increased but their size is decreased; see Figure 4.10. Branched polyacetal terpolymers can, for example, act as such crystallization-accelerating nucleants [26, 27]. The cycle time in processing can be decreased by nucleation. In addition, the homogeneous spherulitic structure in the nucleated materials improves mechanical properties such as strength, stiffness, hardness, and toughness [28]; see also Sections 4.4.2 and 4.10.6.

4.7.2 Mechanical and Technical Properties

4.7.2.1 Short-Term Properties

The short-term mechanical properties of various polyacetals, including non-modified, reinforced, improved friction, and impact modified types, are given in Table 4.1 for homopolymer (using Delrin as the example [5]) and in Table 4.2 for copolymer (using Hostaform as the example [6]). Thermal properties for polyacetal homopolymer can be found in Table 4.3, those for polyacetal copolymer in Table 4.4.

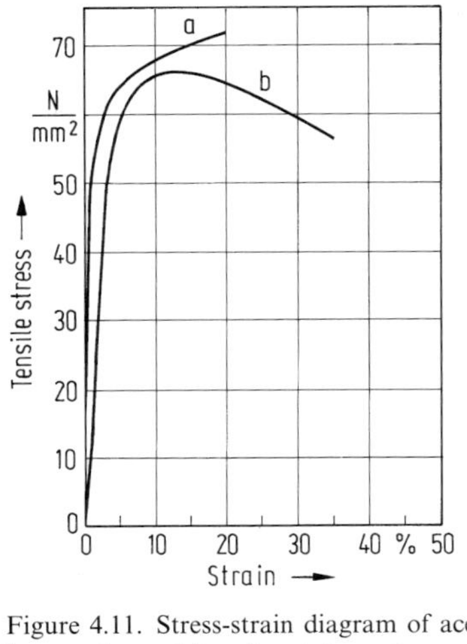

Figure 4.11. Stress-strain diagram of acetal homopolymer (a) and acetal copolymer (b) at room temperature with a drawing rate of 5 mm/min [1]

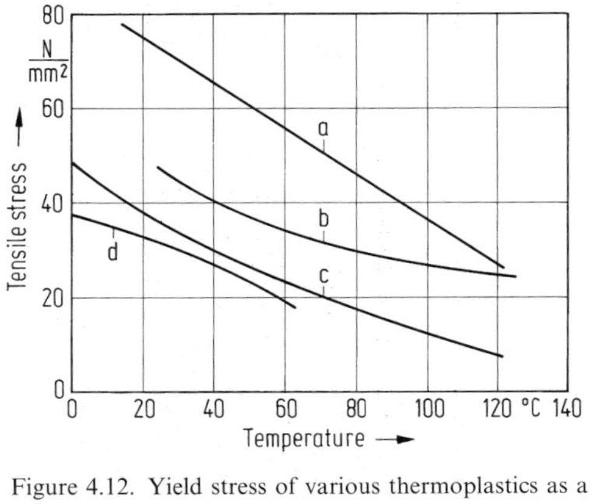

Figure 4.12. Yield stress of various thermoplastics as a function of temperature; deformation speed: 12.5 mm/min [4, 6]
a, Acetal copolymer; b, PA 6; c, PP; d, HDPE

The behavior of materials in continuous, short-term loading is investigated in tensile experiments. The stress-strain diagram, Figure 4.11, gives an overview of the solid-state behavior of polyacetals [1]. In comparison to acetal homopolymers, acetal copolymers exhibit a yield point beyond which the material exhibits plastic flow (necking of the specimen). The ultimate tensile strength of the homopolymers is about 10% higher than the yield strength of the copolymer and conversely the ultimate tensile strain of the copolymer is approximately twice that of the homopolymer.

Figure 4.12 shows a comparison of the yield strength of various thermoplastics as a function of temperature [4, 6]. At 100 °C, acetal copolymer still has a strength like that of polypropylene at room temperature.

Figure 4.13 conveys an indication of stiffness after a 5-second loading as a function of temperature [6].

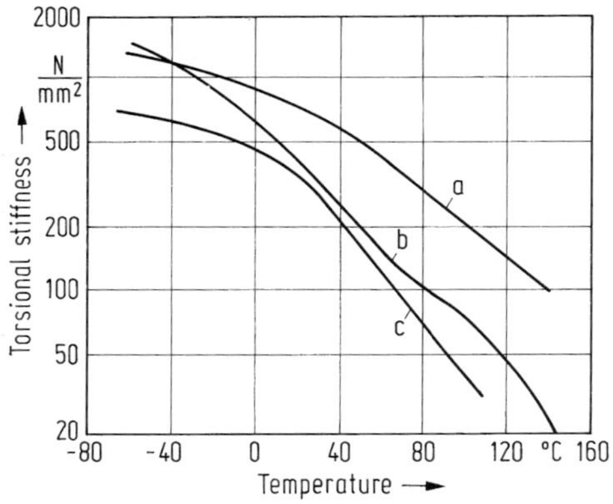

Figure 4.13. Torsional stiffness of various thermoplastics as a function of temperature [6]
a, Acetal copolymer; b, PP; c, HDPE

Polyacetals also exhibit a high surface hardness as a function of temperature as depicted in Figure 4.14 [1].

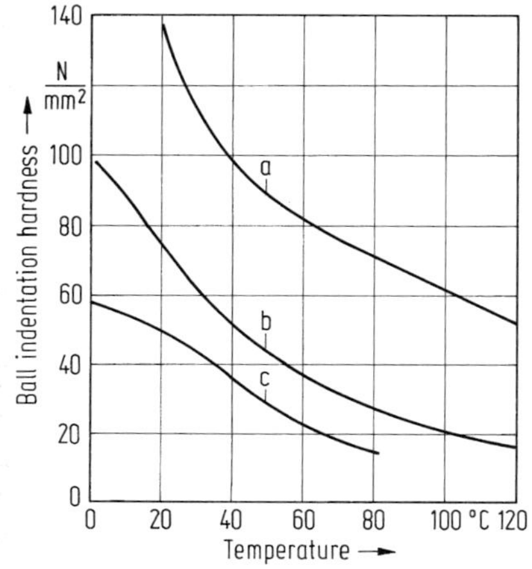

Figure 4.14. Ball indentation hardness of various thermoplastics as a function of temperature [1]
a, Acetal copolymer; b, PP; c, HDPE

Table 4.2. Mechanical Properties of Acetal Copolymers and Modified Grades using Hostaform Grades as the Example [6]

Property	Test Method	Unit	Basic Grades				Basic Easy Flowing Grades					Reinforced Grades					Grades with Improved				High Impact Grades					
			C 9021	C 2521	C 2541	T 1020	C 52021	C 32021 C 32021 AST	C 27021	C 13021	C 13031	C 9021 GV 1/30	C 9021 GV 1/40	C 9021 GV 3/10	C 9021 GV 3/20	C 9021 GV 3/30	C 13026	C 9021 M	C 9021 K	C 9021 TF	S 27063	S 27064	S 27073	S 27076	S 9063	S 9064
Yield Stress	DIN 53455 ASTM D638 Test Speed 50 mm/min	N/mm^2	64	62	62	64	65	63	64	65	71	125	125	54	47	41	65	64	62	49	50	42	50	20	53	42
Strain at Break		%	30	35	35	30	15	20	20	25	25	3	2	14	11	10	25	25	20	15	60	80	60	>150	60	90
Bending Stress	DIN 53452	N/mm^2	67	69	69	72	75	72	72	69	75	120	145	72	72	68	69	60	68	51	50	44	60	21	48	40
Flexural Strength	ASTMD D790	N/mm^2	97	94	94	95	98	98	98	99	102	150	141	87	78	70	99	93	90	75	75	62	75	25	73	58
Tensile-E-Modulus	DIN 53457	N/mm^2	2950	2750	2750	2700	2950	3000	3000	3000	3200	9300	14000	3100	3400	3700	3000	2700	2950	2400	2200	1800	2200	900	2100	1700
Bending Creep Modulus 1-min Value	acc. to DIN 54852 E	N/mm^2	2800	2700	2700	2700	2900	2800	2900	2900	3200	9000	13000	3000	3400	4000	2900	2600	3000	2400	2200	1800	2400	900	2200	1700
6-Day Value	$\sigma_b = 10\ N/mm^2$	N/mm^2	1500	1400	1400	1400	1500	1400	1500	1500	1700	7000	10000	1500	1800	2000	1500	1400	1600	1300	1100	900	1200	n.m.	1050	800
Ball Indentation Hardness	DIN 53456 Force 358 N	N/mm^2	144	144	144	150	148	148	147	143	156	200	215	160	164	167	143	140	145	120	115	90	115	35	110	86
Impact Strength	DIN 53453	kJ/m^2	no break	no break	no break	no break	100	no break	no break	no break	no break	30	15	50	38	26	no break	100-n.br	100-n.br	30	no break	no break	no break	no break	no break	no break
Notched Impact Strength at 23 °C	DIN 53453	kJ/m^2	6	6.5	6.5	6	4	4.5	5	5.5	5.5	4	4	3.5	3	2.5	5	5	5	3.5	6	7	6.5	20	6.5	10
at 40 °C		kJ/m^2	5	5.5	5.5	4.5	3	3.5	4	4.5	4.5	4.5	4.5	–	–	–	4.5	4	4	3	4	4	4	–	5	5
Damaging Energy in Penetration Testing	DIN 53443	J	2	–	–	–	–	–	2	–	–	–	–	–	–	–	–	–	–	–	8	50	–	–	12	60

Table 4.3. Thermal Properties of Acetal Homopolymers and Modified Grades using Delrin Products as the Example [5]

Property	Test Method	Unit	High Viscous Tough	Multipurpose Grades	Fast Flowing		High Productivity		Hydrolysis Resistant	High Stiffness	Low Friction		High Impact	
			100, 107, 150 SA II 100	500, 507, II 500, 517	500 F	900 F, II 900F	500 HP	900 HP	200 PL	570/577	500 AF	500 CL	100 ST	500 T
Melting Point	ASTM D2117	°C	175	175	175	175	175	175	175	175	175	175	175	175
Heat distortion temperature under load at 0.5 N/mm^2	DIN 53461	°C	172	172	170	170	170	170	172	174	168	170	145	169
at 1.8 N/mm^2			136	136	130	130	130	130	130	158	118	124	64	85
Linear Coefficient of Thermal Expansion −40 °C to 30 °C	DIN 53752 (VDE 0304/1)	$1 \cdot 10^{-5}$ K^{-1}	10.4	10.4	10.4	10.4	–	–	–	3.7 bis 8.1	10.4	10.4	12.2	11.0
30 °C to 60 °C			12.2	12.2	12.2	12.2	11.7	11.7	12.2	–	12.2	12.2	12.5	13.0
60 °C to 105 °C			13.7	13.7	13.7	13.7	13.6	13.6	13.7	–	–	–	–	–
105 °C to 150 °C			–	–	–	–	–	–	–	–	–	–	–	–
Thermal conductivity	ASTM D257	W/(m · K)	0.4	0.4	0.4	0.4	0.3	0.3	0.3	–	–	–	–	–
Vicat Softening Point	DIN 53460 B120	°C	162	162	162	162	162	162	162	162	–	162	116	139
	A 50		174	174	174	174	174	174	174	174	174	174	168	171

Table 4.4. Thermal Properties of Acetal Copolymers and Modified Grades using Hostaform Products as the Example

Property	Test Method	Unit	Basic Grades				Basic Easy Flowing Grades					Reinforced Grades					Grades with Improved Frictional Behavior				High Impact Grades					
			C 9021	C 2521	C 2541	T 1020	C 52021	C 32021 C 32021 AST	C 27021	C 13021	C 13031	C 9021 GV 1/30	C 9021 GV 1/40	C 9021 GV 3/10	C 9021 GV 3/20	C 9021 GV 3/30	C 13026	C 9021 M	C 9021 K	C 9021 TF	S 27063	S 27064	S 27073	S 27076	S 9063	S 9064
Heat Distortion Temperature	DIN53461 ASTM D648	°C	104	101	101	97	106	108	107	106	113	153	155	108	110	112	106	100	99	98	91	87	95	57	89	83
Vicat Softening Point	DIN 53460 ASTM D1525	°C	150	151	151	151	149	150	151	151	158	156	159	151	151	151	151	149	150	143	138	126	140	67	137	124
Crystal Melting Range	Polarized Light Microscope ASTM D2133	°C	164–167	164–167	164–167	164–167	164–167	164–167	164–167	169–172	169–172	164–167	164–167	164–167	164–167	164–167	169–172	164–167	164–167	164–167	–	–	–	–	–	–
Linear Coefficient of Thermal Expansion between 20 and 100 °C	DIN 53752 ASTM D696	$1 \cdot 10^{-5}$ K^{-1}	11	11	11	11	11	11	11	11	11	3 (longitud.)	1 (longitud.)	11	[illegible]	8	11	11	11	11	15	16	15	–	15	16

Table 4.1. Mechanical Properties of Acetal Homopolymers and Modified Grades using Delrin Products as the Example [5]

Property[1]		Test Method	Unit	High Viscous Tough	Multi purpose Grades	Fast Flowing		High Productivity		Hydrolysis Resistant	High Stiffness	Low Friction		High Impact	
				100, 107, 150 SA II 100	500, 507 II 500, 517	500 F	900 F II 900 F	500 HP	900 HP	200 PL	570/577	500 AF	500 CL	100 ST	500 T
Tensile strength at break		DIN 53455[2]	N/mm^2	68	70	68	68	68	68	67	62	50	65	42	54
Tensile Strength at Yield	−55 °C 23 °C 100 °C	DIN 53455[2]	N/mm^2	110 72 36	110 72 36	107 70 33	107 70 33	110 70 33	110 70 33	110 70 33	100 62 25	79 50 23	108 70 34	– 43 –	– 57 –
Strain at Break	−55 °C 23 °C 100 °C	DIN 53455[2]	%	38 80 >250	15 50 >250	10 30 >250	10 25 >250	10 20 >250	8 15 >250	25 65 >250	5 12 180	10 15 90	25 45 >250	– 200 –	– 60 –
*Strain at Yield		DIN 53455[2]	%	25	15	12	8	7	5	20	4	5	14	75	25
Shear Modulus		DIN 53455	N/mm^2	790	975	890	890	975	975	730	1340	–	890	370	740
Shear Strength		ASTM D732	N/mm^2	66	66	66	66	66	66	66	66	55	66	34	48
Flexural E-Modulus	−55 °C 23 °C 100 °C	ISO 178	N/mm^2	3600 2600 900	3900 2600 900	3900 2800 900	3900 2800 900	3500 2800 900	3900 2800 900	– 2400 800	6500 5000 2500	3700 2400 800	3800 2800 900	– 1400 –	– 2400 –
Flexural Strength at Yield		DIN 53452	N/mm^2	80	80	84	84	84	76	87	64	61	78	44	78
Tensile E-Modulus		DIN 53457	N/mm^2	3100	3100	3100	3100	3100	3100	3100	6200	2900	3100	–	–
Izod Impact Strength notched unnotched	−40 °C 23 °C 23 °C	ISO 180	J/m	96 133 no break	64 80 no break	– 74 no break	53 69 –	64 75 –	53 70 –	96 123 no break	27 43 –	– 29 450	– 78 2520	– 910 no break	– 135 no break
Charpy Impact Strength notched unnotched	−40 °C 23 °C 23 °C	DIN 53453	kJ/m^2	9.5 13 no break	6.0 8.5 no break	5.5 8.0 no break	4.0 7.0 no break	5.5 8.0 no break	5.5 8.0 no break	9.5 12.0 no break	3.0 3.5 30	3.0 3.5 30	6.0 8.0 no break	11.0 60.0 no break	6.0 14.0 no break
Tensile Impact Strength		ASTM D1822	kJ/m^2	530	420	240	30	170	150	350	65	65	210	1600	760
Compressive Strength at 1% Deformation at 10% Deformation		ASTM D695	N/mm^2	36 124	36 124	31 108	34 21	22 110	22 110	33 117	36 124	31 90	31 107	8 52	16 82
Deformation Under Load		ASTM D621	%	0.5	0.5	0.7	0.7	0.8	0.8	0.7	0.4	0.6	0.7	3.0	0.9
Repeated Flexural Strength 50% r.L., 10^6 cycles		ASTM D671	N/mm^2	32	31	28	28	31	31	29	31	24	28	16	25

[1] All values determined at 23 °C unless otherwise stated

Good frictional properties complement the surface hardness. Measurements on the combination polyacetal copolymer–polyacetal copolymer in the temperature range of 20 °C to 90 °C yield an average value of 0.35 for the static coefficient of friction and an average value of 0.25 for the dynamic coefficient of friction [6]. As shown in Figure 4.15 for the dynamic coefficient of friction against steel, the frictional properties can be further improved in comparison to those of unmodified polyacetal by modification with PTFE or molybdenum disulfide [6].

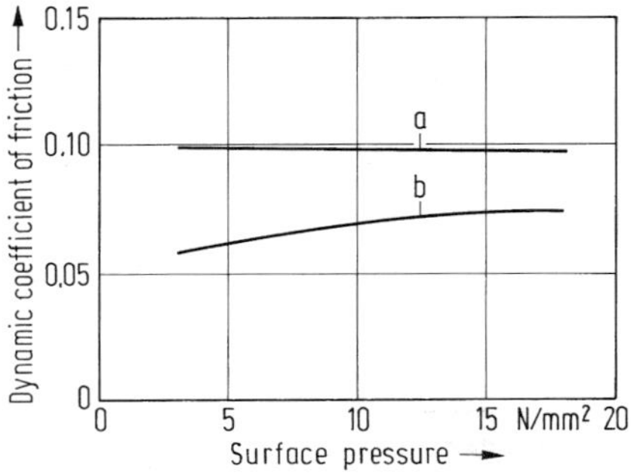

Figure 4.15. Coefficient of dynamic friction for unmodified polyacetal (a) and molybdenum disulfide or PTFE modified grades (b) using Hostaform as the example. Running speed $v = 5$ mm/min, against steel having surface roughness of 2 μm

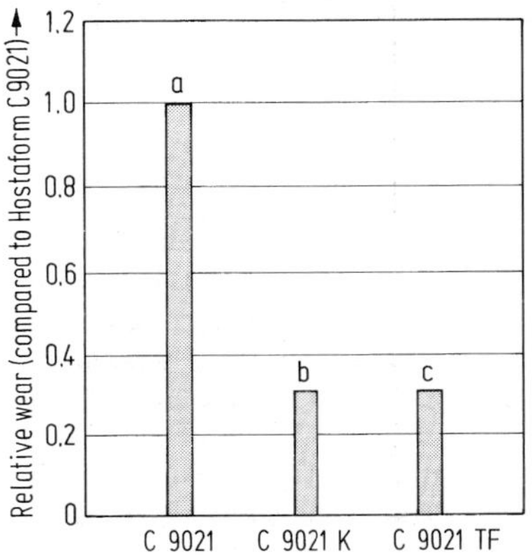

Figure 4.16. Comparison of the wear proclivity of unmodified acetal copolymer (a) with that of chalk (b) or PTFE (c) modified grades using Hostaform as an example. Dry friction on a rotating, polished spindle, surface roughness = 0.8 μm, circumferential velocity = 136 m/min, load = 14 N [6]

The high hardness and low coefficient of friction contribute to the favorable abrasion properties of the polyacetals. Modification with chalk or PTFE can also here lead to improvements compared to non-modified polyacetal; see Figure 4.16 [6].

Polyacetals exhibit good toughness properties under impact loading even at temperatures as low as −40 °C, as Figure 4.17A shows for the example of notched impact

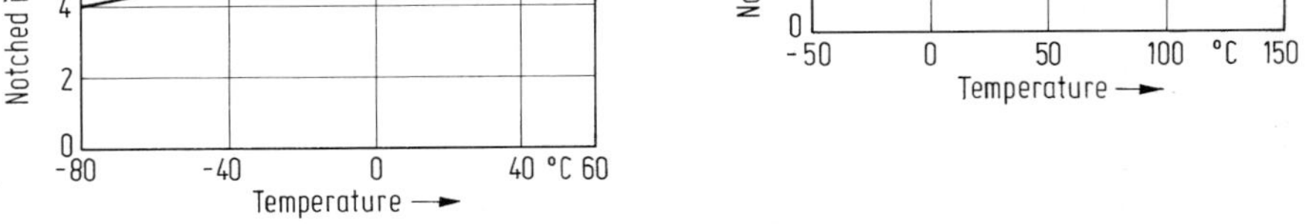

Figure 4.17. A. Notched impact strength according to DIN 53453 of high molecular weight acetal copolymer as a function of temperature [1, 4]

Figure 4.17. B. Notched impact strength according to DIN 53453 as a function of temperature for unmodified acetal copolymer (a) and for elastomer modified copolymer (b) using Hostaform as the example [6, 12, 13]

strength [1, 4]. This can be traced back to the low dynamic glass transition temperature at about −60 °C (γ-relaxation, see Figure 4.33), which is extremely low in comparison to that of other engineering thermoplastics. The level of toughness is generally dependent upon the molecular weight of the polyacetal. The greater the chain length, the greater the energy absorption up to fracture of the material. Through further modification with elastomers, the impact strength of low molecular weight, easy-flowing polyacetals can be improved; see Figure 4.17B [6, 12, 13], albeit with loss of strength, stiffness and hardness. See also Section 4.6.2.

Although unnotched impact strength testing in the case of unmodified polyacetals indeed does not show fracture of the test specimen, reinforcement with glass or filling with PTFE results in substantial losses in toughness [6]. Loss in toughness through reinforcement can again be compensated for by modification with elastomers.

4.7.2.2 Long-Term Properties

Polyacetals are among the thermoplastic synthetic materials and are thus viscoelastic. This is shown by the behavior known as creep (cold flow), a deformation that is dependent upon time, temperature, and load. Besides the short-term properties of the material to be used, it is absolutely necessary that the designer knows the mechanical behavior of the material under continuous loading. The behavior of a material under long-term tensile loading can be determined in two basic types of experiments: in creep tests, that is, measurement of the increase in elongation at constant stress, or in stress relaxation tests, that is, measurement of the decrease in stress at constant strain.

As the creep test on a polyacetal given in Figure 4.18 [4, 6] shows, polyacetals exhibit a relatively small cold flow. From such experiments on a material one can determine the failure envelope (creep rupture strength), that is the stress at which the test bar will fail after a certain time, and also the time-elongation limit, that is the stress at which a particular strain will be surpassed within a certain time, as in Figure 4.19 [4, 6].

The results of creep experiments under uniaxial loading can be extrapolated, within limits, to the multiaxial stress condition. Figure 4.20 shows the creep rupture strength of

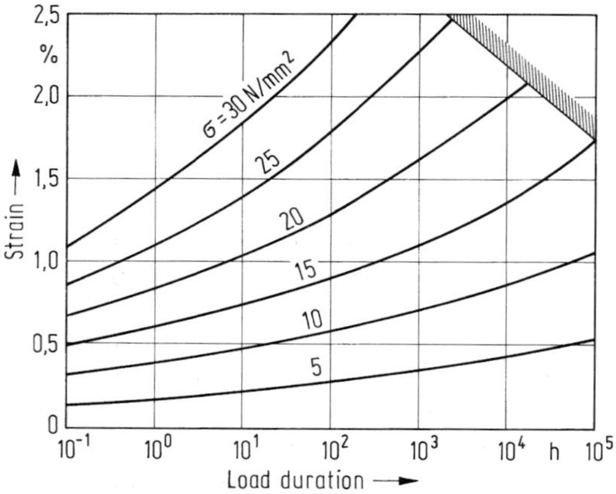

Figure 4.18. Time-elongation lines (creep curves) according to DIN 53444 for acetal copolymer, measured at 20 °C in air [4, 6]

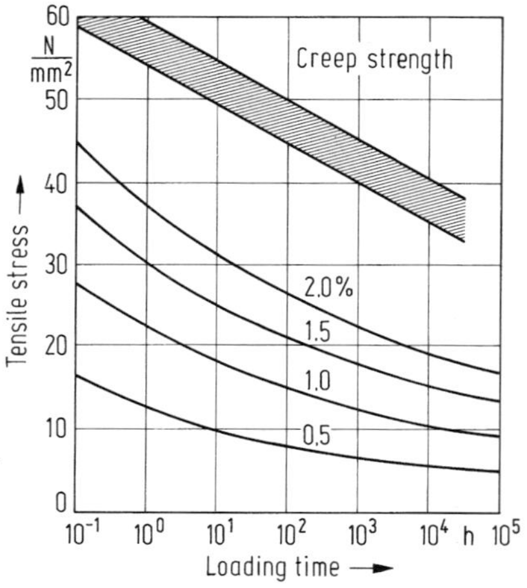

Figure 4.19. Creep rupture strength and time-elongation limits for acetal copolymer at room temperature [4, 6]

tubes of acetal copolymer under internal pressure at various temperatures. These experiments, still in progress, have run over a time period of more than 20 years [6].

In addition to the described creep experiments under tensile loading or internal pressure, a knowledge of the long-term behavior under flexural loading is necessary for dimensioning of many components.

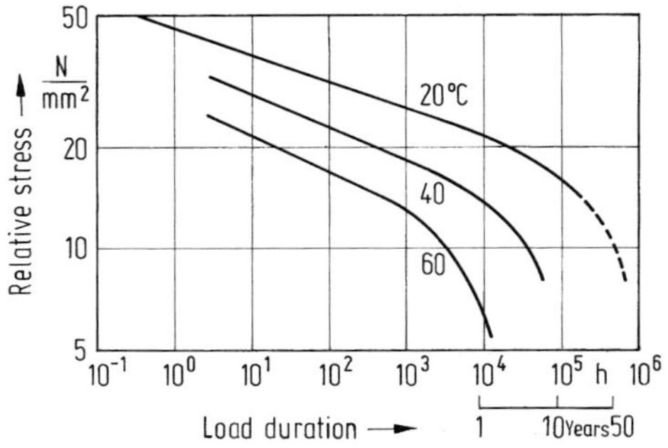

Figure 4.20. Creep strength of tubes made from acetal copolymer at various temperatures immersed in water [6]

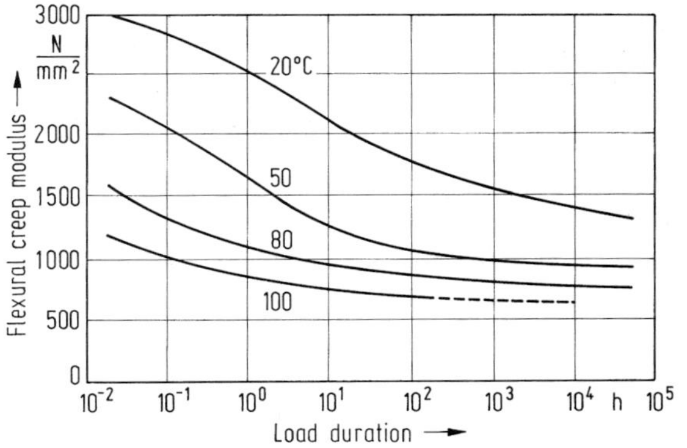

Figure 4.21. Flexural creep modulus of acetal copolymer at various temperatures, measured at an outer-fiber stress of 10 N/mm^2 [6]

Figure 4.21 contains the flexural creep modulus (secant modulus) as a function of loading time at several temperatures for a polyacetal copolymer [6]. Creep can be substantially minimized through the addition of glass fibers. Figure 4.22 shows the flexural creep modulus for an unreinforced and for a 30% glass fiber reinforced polyacetal copolymer [4, 6]. Even at 80 °C, the flexural creep modulus of the glass reinforced material after one year is higher than that of the non-reinforced material at room temperature and short loading times. The influence of glass reinforcement is also reflected in stress relaxation experiments; see Figure 4.23. The relaxation modulus of glass fiber reinforced material lies noticeably above that of the non-reinforced materials and exhibits a less precipitous drop-off, that is, slower relaxation behavior [6].

In addition to a low creep tendency, polyacetals have a good recovery ability. This is depicted in Figure 4.24 for creep and recovery behavior in the case of tensile loading

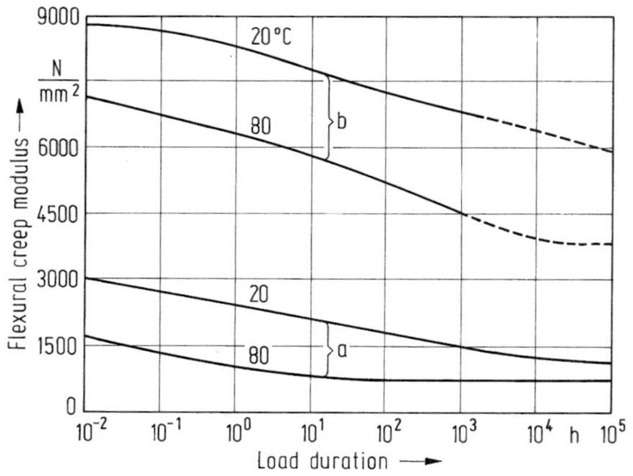

Figure 4.22. Flexural creep modulus of (a) unmodified acetal copolymer and (b) with 30% glass fiber modified copolymer, using Hostaform as the example, measured at an outer-fiber stress of 10 N/mm² [4, 6]

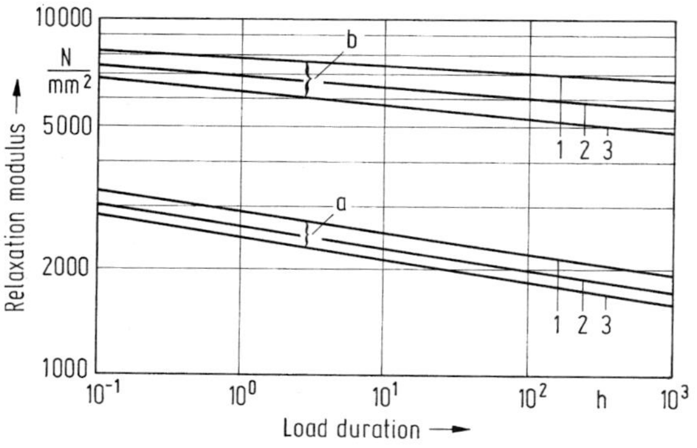

Figure 4.23. Relaxation modulus of unmodified (a) acetal copolymer and (b) with 30% glass fiber modified copolymer, using Hostaform as the example [6] 1, 0.2% strain; 2, 0.5% strain; 3, 1% strain

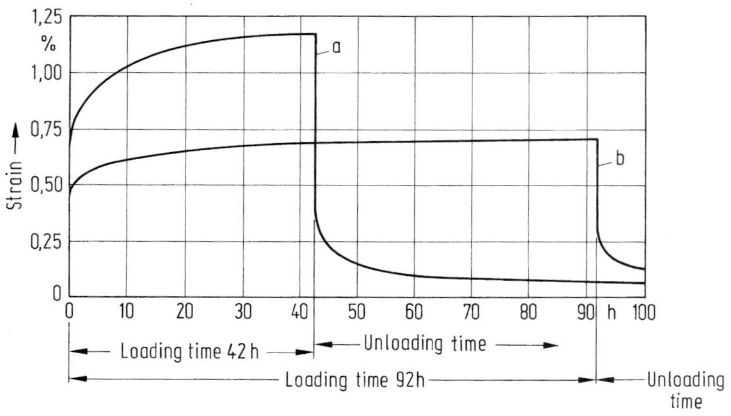

Figure 4.24. Creep and recovery behavior of acetal copolymer under tensile stress at room temperature [1] a, 20 N/mm² tensile stress; b, 13 N/mm² tensile stress

and in Figure 4.25 with a deformation-pressure experiment [1]. At both room temperature and 80 °C, the original state of a sample loaded with a pressure of 20 N/mm² for 24 hours is re-established within a minute of relieving the load. At 100 °C the material

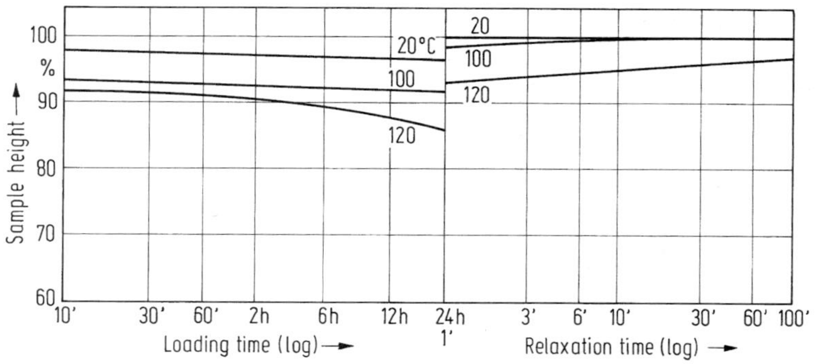

Figure 4.25. Compression creep behavior of acetal copolymer at 20 N/mm^2 stress and various temperatures [1]

completely recovers after 10 minutes. This excellent elastic recovery is an important criterios in many application areas for polyacetals.

In the case of all components that are subjected to a continuous dynamic load, in comparison to a purely static load, one must cope with a decrease in strength; the material fatigues because of the changing load. Components that are periodically loaded must be rated with the help of a measure for fatigue endurance [29], that is, the stress oscillation, measured in a dynamic experiment, that a sample can withstand for a specific number of cycles without fracturing. The fatigue limit can then be read off the Wöhler curve. The fatigue limit in the case of polyacetals lies higher than that in that of many other thermoplastics. In the case of polyacetal copolymers in dynamic tension-compression experiments, it still lies at ca. 30% of yield strength determined in tensile experiments, even after 10^7 cycles; see Figure 4.26 [6]. The fatigue strength decreases with increasing temperature and higher frequency of loading.

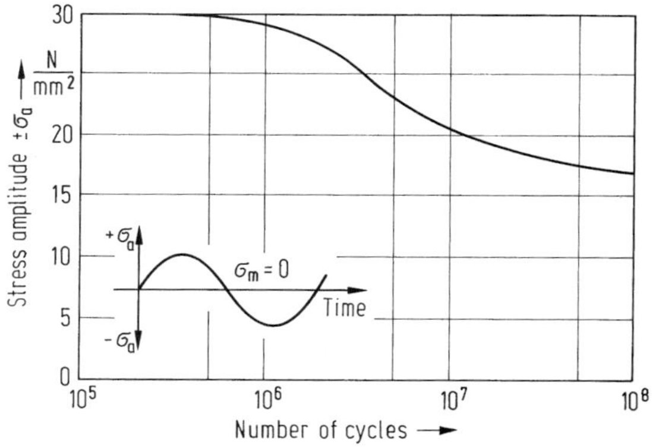

Figure 4.26. Wöhler curve (Stress-N(cycles) graph) of acetal copolymer, measured in alternating tension-compression-range at 23 °C and 10 Hz [6]

If there is no oscillating tensile-compression loading but rather only tensile or compressive loading with intermediate unloading, then the fatigue strength is higher, as depicted in Figure 4.27 for a tensile step experiment [6]. The fatigue strength compared to unmodified polyacetals can be greatly increased through glass fiber reinforcement; see Figure 4.28.

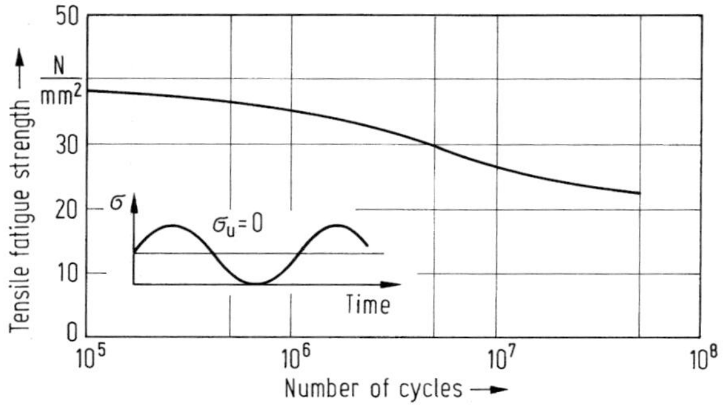

Figure 4.27. Wöhler curve (Stress-N(cycles) graph) of acetal copolymer, measured in a tension fatigue test at 23 °C and 10 Hz [6]

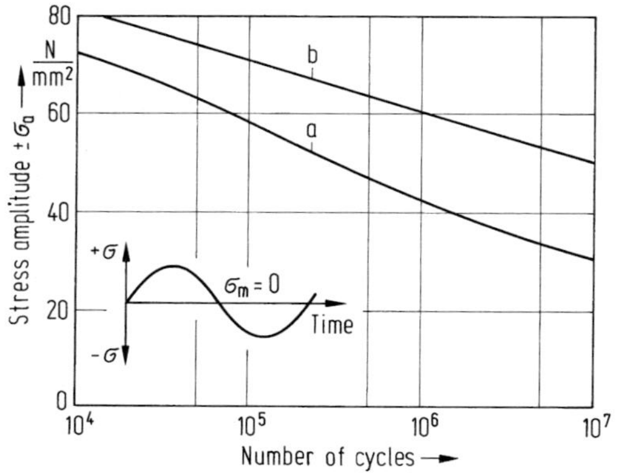

Figure 4.28. Wöhler curves (Stress-N(cycles) graph) of (a) unmodified acetal copolymer and (b) 30% glass fiber modified copolymer, using Hostaform as the example, measured flexural fatigue at 23 °C and 10 Hz [6]

4.7.3 Thermal Properties

4.7.3.1 Short-Term Properties

The thermal properties of the polyacetals are principally determined by the crystalline structure of the material. Melting of the crystalline regions occurs around 175 °C in the case of the homopolymers, about 10 °C higher than in the case of the copolymers where it occurs around 164 °C to 167 °C. The slightly higher values for resistance to deformation under the influence of heat for the homopolymer are based on the differences in the melting temperature and degree of crystallinity; see Tables 4.3 and 4.4 [5, 6].

Glass fiber reinforcement raises the resistance to deformation under the influence of heat. This property is diminished by impact modification.

Figure 4.29 shows the measurement of the temperature dependence of the specific heat, determined with the aid of a differential scanning calorimeter, for acetal homopolymer and copolymer [1]. The differences in enthalpy can be calculated from the temperature dependence of the specific heat capacity; see Figure 4.30 [1, 4, 6]. From this enthalpy difference, one can calculate the amount of heat in thermoplastic processing that is required in order to melt the material or that must be removed by cooling when the material crystallizes.

The thermal conductivity of the polyacetals is temperature dependent and in the case of unmodified types lies around 0.3 to 0.35 W/(m · K). It is increased by glass fiber reinforcement to, depending on glass fiber content, as much as about 0.45 W/(m · K) [6].

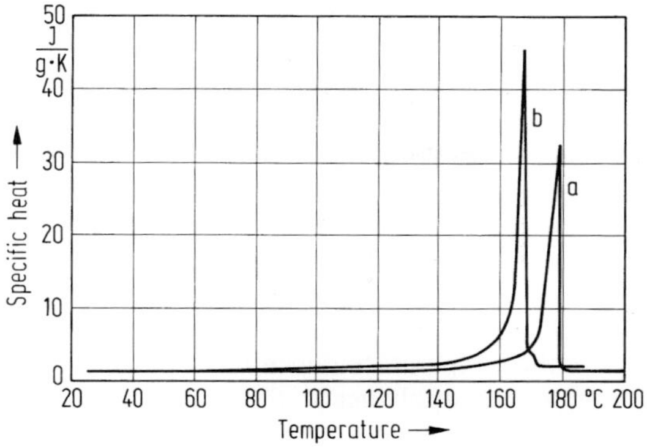

Figure 4.29. Specific heat of (a) acetal homopolymer and (b) acetal copolymer as a function of temperature [1]

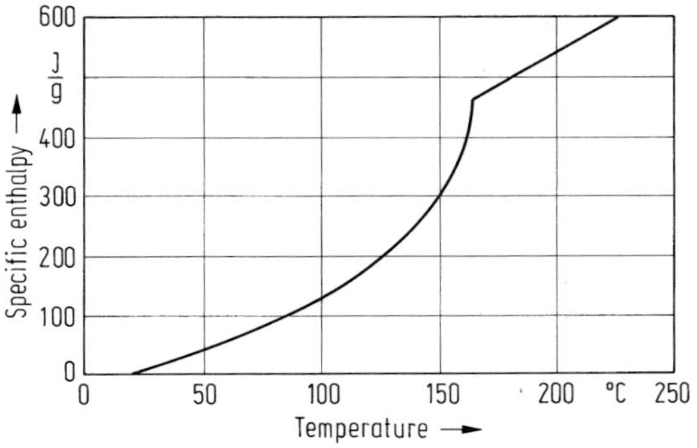

Figure 4.30. Specific enthalpy of acetal copolymer [1, 4, 6]

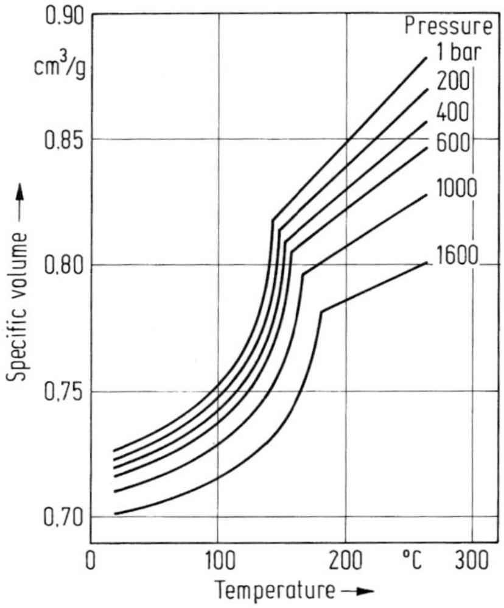

Figure 4.31. Specific volume of acetal copolymer as a function of temperature and pressure [6]

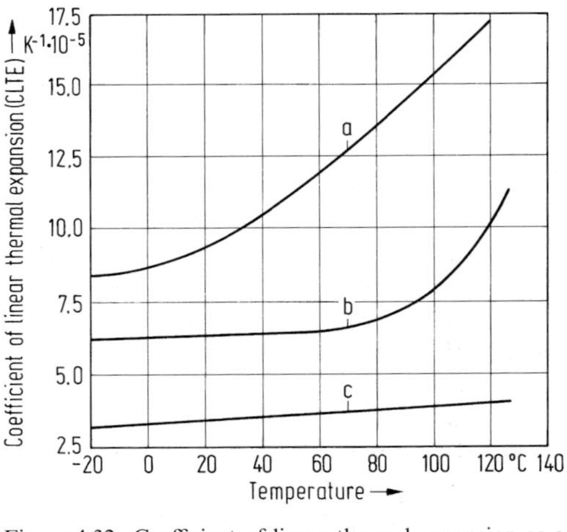

Figure 4.32. Coefficient of linear thermal expansion as a function of temperature
a: unmodified acetal copolymer, b: acetal copolymer reinforced with 30% glass fibers, perpendicular to flow direction, c: acetal copolymer reinforced with 30% glass fibers, in flow direction [4, 6]

Figure 4.31 shows the dependence of specific volume on temperature for various pressure (p-V-T diagram) [6]. The linear expansion coefficient of acetal copolymers is plotted as a function of temperature of Figure 4.32. The value of the linear coefficient of expansion as well as its temperature dependence can be sharply reduced by glass fiber reinforcement. This effect occurs most strongly in the flow direction in injection molded parts [4, 6].

Whereas the good resistance to deformation under the influence of heat is determined essentially by the high crystalline melting point and the high crystallinity, the high toughness of the materials down to temperatures as low as −40 °C (see also Figure 4.17) is based on the low glass transition temperature of the amorphous regions. Figure 4.33 shows the temperature dependence of the shear modulus and the mechanical loss factor for an acetal copolymer [4, 6]. In addition to the melting at 165 °C, a relaxation is found at about −5 °C, and a much more pronounced γ-relaxation at −60 °C to −65 °C. The good low-temperature toughness is based on this pronounced γ-relaxation.

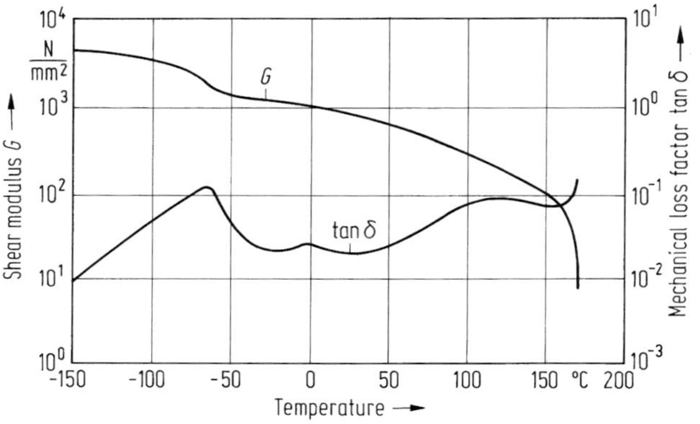

Figure 4.33. Shear modulus G and mechanical loss factor tan δ for acetal copolymer as a function of temperature [4, 6]

4.7.3.2 Long-Term Properties

Polyacetals can be briefly heated to just below the crystalline melting point without damage. For continuous use at elevated temperatures, however, one must also consider the thermal ageing behavior in addition to the creep behavior (see Section 4.7.2.2) [29]. Thermal ageing is not a pure thermal process but rather involves thermaloxidative degradation. The oxidation behavior is firstly dependent upon the chain structure of the polyacetal. Because of the incorporation of the degradation-retarding comonomer, copolymers are more thermally stable than homopolymers. Thermaloxidative degradation can be further suppressed with antioxidants. Numerous influences of the climate in technological applications have different effects on the temporal course of the heat ageing. Terms such as "heat ageing resistance" or "continuous use temperature" do not characterize a material-specific property, but rather must be viewed in connection with the respective application.

For acetal homopolymers and copolymers, the "continuous use temperature" lies in a region from about 90 to 100 °C, according to the type of loading [4]. Figure 4.34 shows the change in the tensile strength for an acetal copolymer that occurs upon ageing at elevated temperatures in air [6]. The good ageing resistance at temperatures up to 100 °C can be seen in the complete retention of strength for over one year. In contrast, the strain at break is more strongly influenced by the effects of temperature than is the tensile strength, see Figure 4.35 [6]. The nascent embrittlement of the material is most quickly observed in the strain at break.

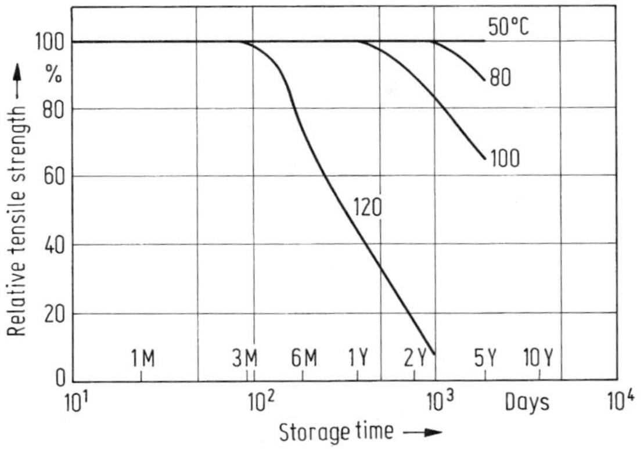

Figure 4.34. Relative tensile strength of acetal copolymer as a function of storing time at higher temperatures in air [6]

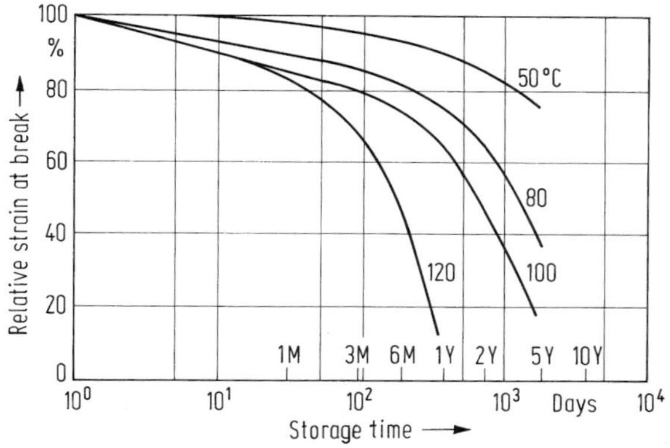

Figure 4.35. Relative strain at break for acetal copolymer as a function of storing time at higher temperatures in air [6]

4.7.4 Electrical Properties

Polyacetals distinguish themselves by good insulation properties and dielectric behavior. These properties, as well as excellent mechanical properties, have opened up numerous electrical and electronic applications. Table 4.5 gives the most important electrical properties using the acetal copolymer Hostaform as an example. Published numbers for other acetal homo- or copolymers do not deviate substantially from those in the table in so far as products of the same composition are compared [7, 31]. It is noteworthy that the grades containing glass or elastomer have slightly higher dielectric constants and that in the case of elastomer-containing grades the electrical resistivity lies minimally lower.

Antistatic formulation effects a reduction of the surface resistivity and the rate of charge dissipation. Addition of electroconductive carbon black of course effects dramatic changes in electrical characteristics. Because of the low water uptake, the dependence of the electrical properties of polyacetals upon environmental moisture is negligible. Except for the surface resistivity, they do not suffer any change even if they are stored in water [32].

Table 4.5. Electrical Properties of Acetal Copolymers, Measured in Standard Climate, 50% RH [30]

Property	Unit	Test Method	Basic Grades								
			C 52021	C 32021, C 32021 AST	C 27021	C 13021	C 13031	C 9021	C 9021 ELS	C 2521	C 2550
Break-through Strength	kV/cm	DIN 53481 VDE 0303 Teil 2	700	700	700	700	700	700	–	650	650
Dielectric Number ε_r		DIN 53483 VDE 0303 Teil 4 ASTM D 150 (Immersion method)	3.7	3.7	3.7	3.7	3.7	3.8	–	3.9	3.9
Dielectric Loss Factor tan δ at 50 Hz		DIN 53483 VDE 0303 Teil 4 ASTM D 150	$5 \cdot 10^{-3}$	$5 \cdot 10^{-3}$	$5 \cdot 10^{-3}$	$5 \cdot 10^{-3}$	$5 \cdot 10^{-3}$	$5 \cdot 10^{-3}$	–	$5 \cdot 10^{-3}$	$5 \cdot 10^{-3}$
Specific Bulk Resistivity	Ω · cm	DIN 53482	$>10^{15}$	$>10^{15}$	$>10^{15}$	$>10^{15}$	$>10^{15}$	$>10^{15}$	10^3	$>10^{15}$	$>10^{15}$
Surface Resistivity	Ω	DIN 53482	$>10^{13}$	$\frac{>10^{13}}{10^9 \text{ to } 010^{11}}$	$>10^{13}$	$>10^{13}$	$>10^{13}$	$>10^{13}$	$5 \cdot 10^2$	$>10^{13}$	$>10^{13}$
Tracking Resistance	Rating	DIN IEC 112	CTI 600	CTI 600	CTI 600	CTI 600	CTI 600	CTI 600	–	CTI 600	CTI 600
Arc Resistance	Rating	DIN 53484	L 4	L 4	L 4	L 4	L 4	L 4	–	L 4	L 4

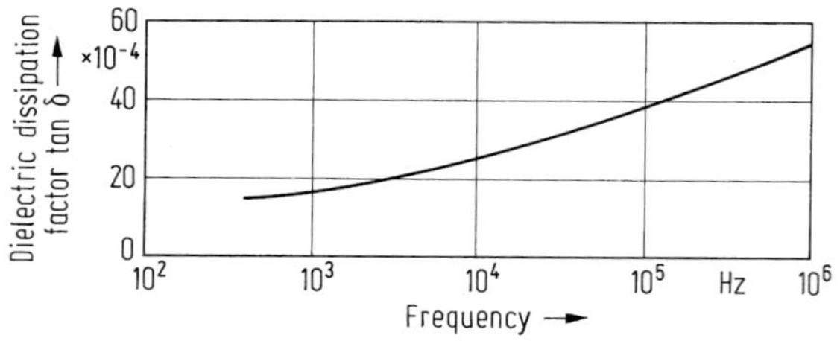

Figure 4.36. Dielectric dissipation factor tan δ as a function of frequency at 25 °C for acetal copolymer (Hostaform C 9021) [6]

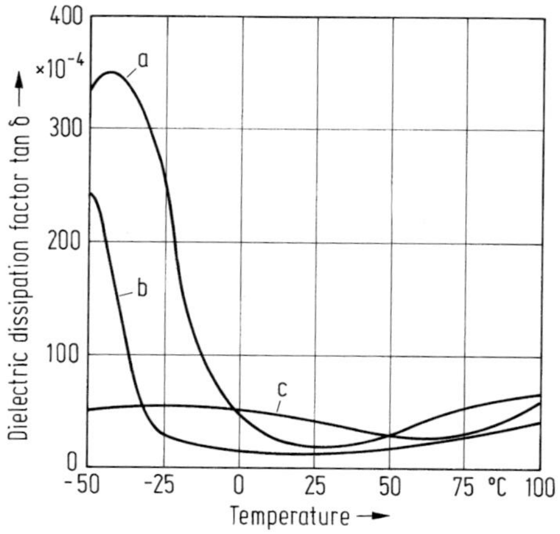

Figure 4.37. Dielectric dissipation factor tan δ as a function of temperature [33]
a: acetal copolymer at 10^5 Hz, b: acetal homopolymer at 10^3 Hz, c: acetal homopolymer at 10^5 Hz

		Reinforced Grades						High Impact Grades						
T 1020	C 13026	C 9021 GV 1/30	C 9021 GV 1/40	C 9021 GV 3/10	C 9021 GV 3/20	C 9021 GV 3/30	C 27021 GV 3/30	S 27063	S 27064	S 27073	S 27073 GV 1/10	S 9063	S 9064	S 27076
610	700	–	–	–	–	–	–	–	–	–	–	–	–	–
3.9	3.7	4.8	–	3.8	4.0	4.5	–	4.6	4.7	4.7	–	4.5	4.6	5.6
$5 \cdot 10^{-3}$	$5 \cdot 10^{-3}$	$5 \cdot 10^{-3}$	$5 \cdot 10^{-3}$	$5 \cdot 10^{-3}$	$5 \cdot 10^{-3}$	$5 \cdot 10^{-3}$	–	$2.5 \cdot 10^{-2}$	$2.5 \cdot 10^{-2}$	$1.5 \cdot 10^{-2}$	–	$2 \cdot 10^{-2}$	$2 \cdot 10^{-2}$	$3 \cdot 10^{-2}$
$>10^{15}$	$>10^{15}$	$>10^{15}$	$>10^{15}$	$>10^{15}$	$>10^{15}$	$>10^{15}$	$>10^{15}$	10^{13}	10^{13}	10^{13}	$>10^{15}$	10^{13}	10^{13}	10^{13}
$>10^{13}$	$>10^{13}$	$>10^{13}$	$>10^{13}$	$>10^{13}$	$>10^{13}$	$>10^{13}$	$>10^{13}$	10^{12}	10^{12}	10^{12}	$>10^{13}$	10^{12}	10^{12}	10^{11}
CTI 600	CTI 600	CTI 600	CTI 600	CTI 600	CTI 600	CTI 600	CTI 600	CTI 600	CTI 600	CTI 600	CTI 600	CTI 600	CTI 600	CTI 600
L 4	L 4	L 4	L 4	L 4	L 4	L 4	L 4	–	–	–	–	–	–	–

While the dielectric constant of polyacetals is practically independent of temperature (−50 to 100 °C) and frequency (10^2 to 10^6 Hz), the dielectric loss factor shows a minuscule increase with frequency (Figure 4.36) [6]. The temperature dependence of the dielectric loss factor is given in Figure 4.37 [33].

The favorable dielectric properties exclude high frequency heating and high frequency welding of polyacetals.

4.7.5 Fire Behavior

As a consequence of their specific degradation mechanism, the cleavage of gaseous formaldehyde under sufficient effect of temperature, polyacetals are combustible. They burn slowly, with dripping, and with a blue flame. Their ignition temperature lies at about 350 °C and their temperature of autoignition lies at about 360 °C.

Underwriters Laboratory has assigned the polyacetals to class HB (horizontal burning) in UL Safety Standard 94.

The maximum allowable burning rate for plastics in automobiles according to MVSS 302 is not exceeded by polyacetals at wall thicknesses ≥ 1 mm.

According to the new, international standards DIN IEC 707/VDE 0304, part 3, the acetal copolymer Ultraform has the following rating: BH 3-25 mm/min (glow plug, horizontal sample position) and FH 3-25 mm/min (Bunsen burner, horizontal sample position).

4.7.6 Chemical Resistance

Based on their high crystallinity, polyacetals exhibit good resistance to numerous chemicals. Extensive experimental results from raw material producers are available [34, 6].

Only hexafluoroacetone sesquihydrate and hexafluoroiso-propanol are known as room temperature solvents. A few strongly polar substances such as benzyl alcohol, γ-butyrolactone, or dimethyl formamide dissolve polyacetals in the vicinity of their melting point.

Polyacetals are not chemically attacked by the customary organic solvents such as alcohols, esters, ketones, aliphatic and aromatic hydrocarbons, and perhalogenated hydrocarbons, even in long-time contact. Their mechanical strength is, if at all, only minimally reduced, and their swelling is slight.

The high resistance against fuels (including M15) as well as against fats, oils, brake fluids, and coolants should be highlighted. Figure 4.38 shows the fuel absorption of the acetal copolymer Hostaform [6]. The low swelling and, as a result, the high dimensional stability are shown clearly [35]. See also Section 4.10.2.

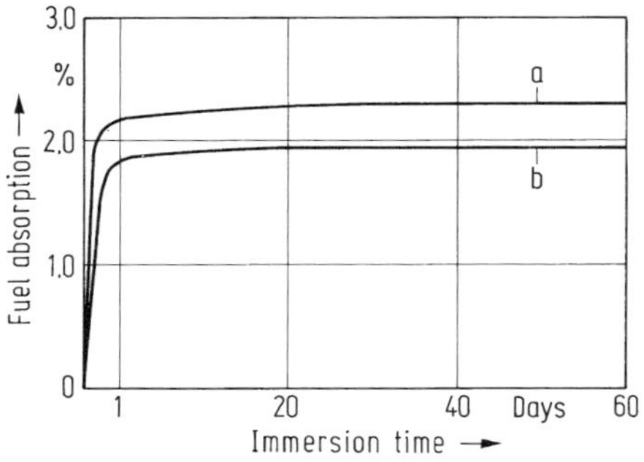

Figure 4.38. Fuel absorption as a function of storing time at room temperature for acetal copolymer [6] a: 85% gasoline (premium) plus 15% methanol; b: gasoline (premium)

Depending on the chemical nature and concentration of the elastomer component, the impact modified polyacetals have a somewhat higher fuel absorption. Addition of methanol to the fuel can bring about a further, minor swelling. A slightly stronger decrease of the tensile strength at break in comparison to the standard grades is connected with this, since the absorbed fuel acts as a plasticizer. By choosing a suitable elastomer component, high impact resistant grades were made available with an adequate fuel resistance which have found their way into applications in contact with fuels [35].

Polyacetals are likewise resistant against neutral, inorganic compounds (salt solutions).

Polyacetals are chemically attacked by strong acids (organic and inorganic) and oxidizing agents; the homopolymer in contrast to the copolymer is additionally attacked by alkali, because of the sensitivity of its endgroups. Because of the possible formation of ZnO, zinc must be considered an alkali. Hence, only acetal copolymers are suitable for application in contact with alkaline cleansers or with zinc.

In the case of pure, weak acids and mixtures that contain aggressive chemicals, the actual damage that occurs is dependent upon concentration, the length of exposure, and temperature. Experiments under actual end-use conditions are imperative in order to evaluate the possibility of using polyacetals.

4.7.7 Hydrolysis Resistance

Figure 4.39 shows that the water absorption of the standard grades of polyacetals is very small [6]. Figure 4.40 shows the longitudinal changes, measured at room temperature, which result from water saturation at various temperatures. A few special basic grades of homopolymer exhibit an approx. 50% higher water absorption; for example Delrin 500 F and Delrin 500 HP compared to Delrin 500 [31]. Compared to the basic grades, the impact modified polyacetals have a slightly higher water absorption, those containing glass, however, a slightly lower.

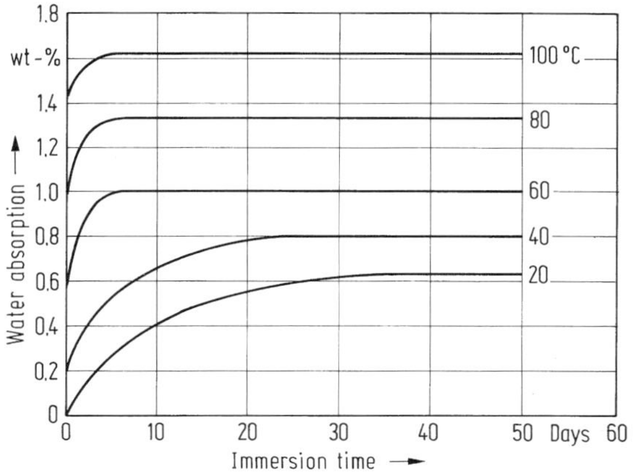

Figure 4.39. Water absorption as a function of storing time at different temperatures for acetal copolymer (Hostaform C 9021) according to DIN 53495 [6]

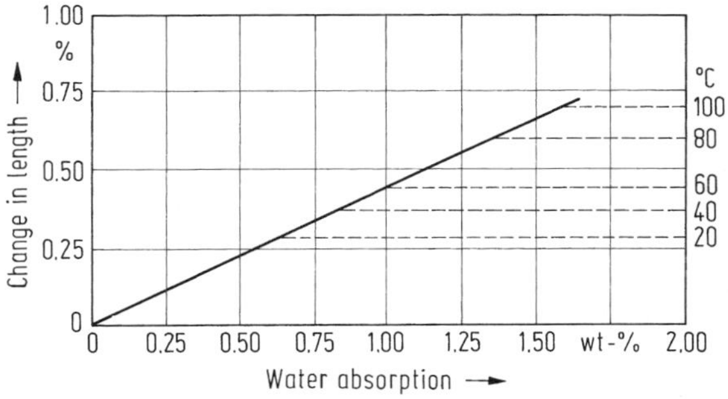

Figure 4.40. Change in length as a function of water absorption for acetal copolymer (Hostaform C 9021) [6]

The hydrolysis resistance of polyacetals depends on their chemical structure and, in a secondary way, also on their stabilization.

As a consequence of their deliberately arranged structure, acetal copolymers exhibit excellent hydrolysis resistance (Figures 4.41 and 4.42). The tensile strength, e.g., decreases only by 25% after two years storage in boiling water. Similarly, the strain at break decreases by only 25% after two years at 80 °C; as expected, it decreases noticeably faster at 100 °C. This behavior results in a long-term application temperature of about 100 °C and is the basis for successful application in laundry and dishwashing machines and in water kettles (Figure 4.76). In case of plumbing applications (see Section 10.5), other demands such as mechanical loads, internal pressure, chlorine concentration, and copper contact are to be taken into consideration.

Acetal homopolymers have hydrolysis-sensitive endgroups. This is not only noticeable in contact with alkali, but also in long-term experiments in hot water. A reduction of the long-term application temperature to about 85 °C must be accepted [10].

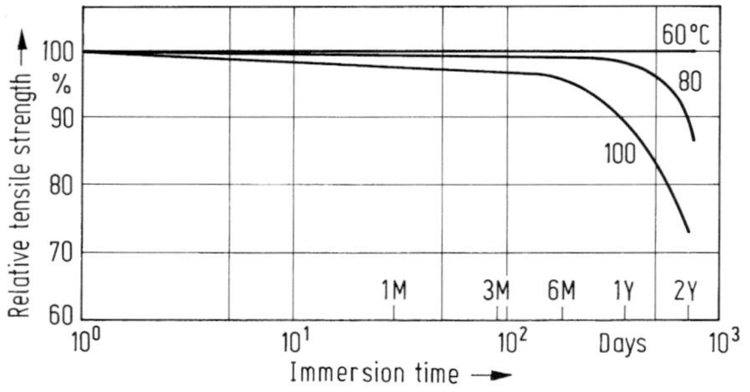

Figure 4.41. Relative tensile strength as a function of storing time in hot water for acetal copolymer (Hostaform C 9021) [6]

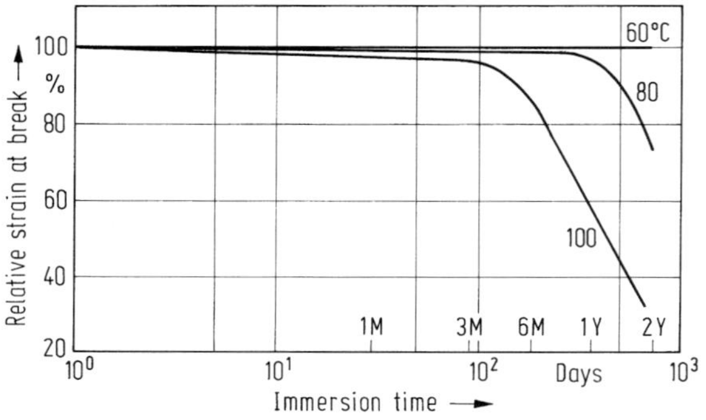

Figure 4.42. Relative strain at break as a function of storing time in hot water for acetal copolymer (Hostaform C 9021) [6]

The atypically stabilized special grade of homopolymer, Delrin 200 PL, does show a somewhat delayed loss of strength compared to Delrin 100. However, in spite of this, it behaves like a homopolymer and does not attain the hydrolysis resistance of the copolymers.

Polyacetals can repeatedly be steam sterilized since the total time of exposure is short.

Assessing the hydrolysis resistance of glass fiber reinforced polyacetals it must be noted that after only a short exposure to hot water, the ultimate tensile strength drops to the level of that of non-reinforced material and remains at this level, provided that the long-term application temperature specific to the matrix is kept in consideration. This is caused by a rapid destruction of the boundary phase between the glass fibers and the matrix. The flexural elasticity modulus shows an analogous but by far less distinct initial drop.

Impact modified polyacetals have a lower hydrolysis resistance than the standard grades.

4.7.8 Weathering and Radiation Resistance

Polyacetals are sensitive to UV radiation and therefore, they are not suitable for outdoor use if not light stabilized. Here copolymers show a somewhat more favorable behavior than homopolymers [36, 37]. The effects of weathering are first manifested in a loss of surface sheen, later then in so-called chalking. This means, a layer of degraded, low molecular weight polyacetal is built up on the surface and is noticed in an increase in melt index with increasing time of weathering [38]. A drop in strength, and in particular a drop in toughness (strain at break), accompanies this course of events. As seen in curve "a" of both Figures 4.43 and 4.44, after five years the toughness amounts to only 40% and the strain at break to only 10% of their respective initial values. The loss of both properties is most severe in the initial stages.

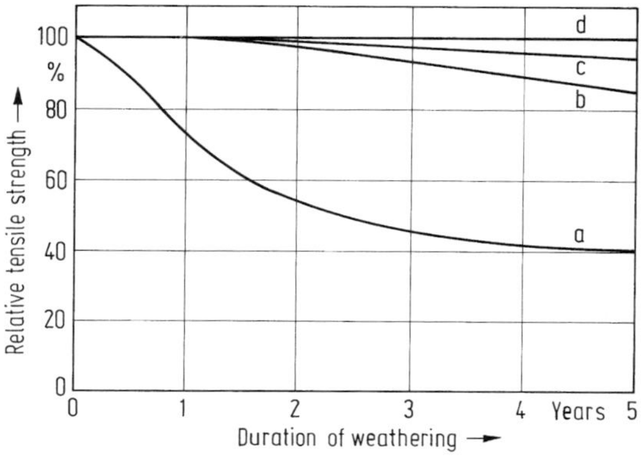

Figure 4.43. Decrease in tensile strength as a function of weathering for unstabilized and stabilized acetal copolymer Hostaform [6]
a: without light stabilizer, b: with UV-stabilizer, c: with TiO_2 pigment, d: carbon black

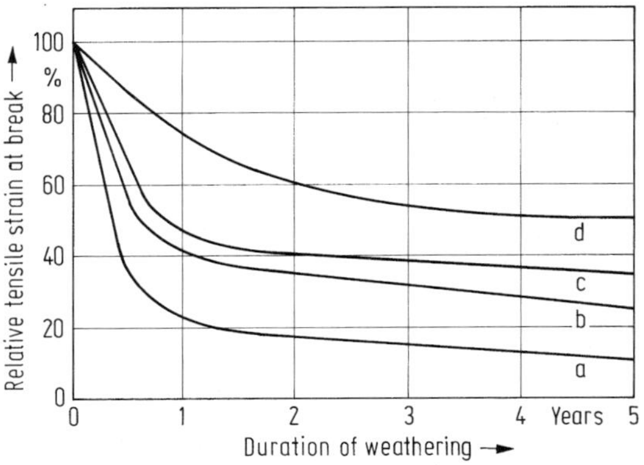

Figure 4.44. Decrease in tensile strain at break as a function of weathering for unstabilized and stabilized acetal copolymer Hostaform [6]
a: without light stabilizer, b: with UV-stabilizer, c: with TiO_2 pigment, d: carbon black

Since the degraded material protects the internal region which has been barely or not at all attacked, the degradation comes more or less to a halt. That is, the tensile strength barely drops further. The level at which degradation stops is strongly dependent upon wall thickness: in the case of a thin-walled part total disintegration can occur, but thick-walled parts can have a respectable retention of shape and strength.

As curves "b" of both Figures 4.43 and 4.44 show, the addition of organic UV-absorbers effects a noticeable improvement in the retention of mechanical properties. According to reference [7] with this improvement a doubling of the utility time can be estimated. Figures 4.45 and 4.46 confirm that this accompanies a decreased surface cracking.

Figure 4.45. Surface cracking of an acetal copolymer Hostaform without UV-stabilizer after 1 year weathering, magnified 120 times [38]

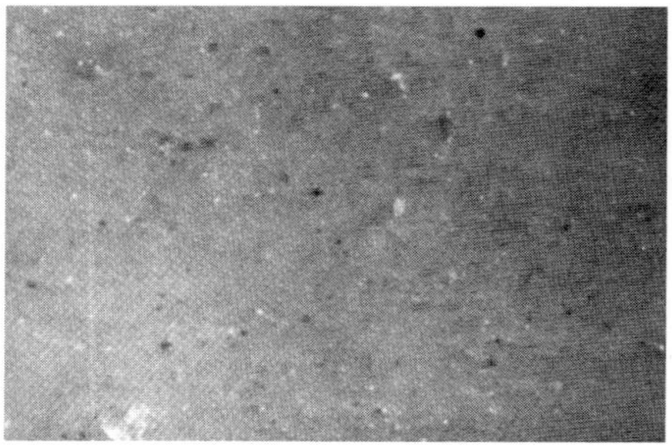

Figure 4.46. Surface cracking of an acetal copolymer Hostaform with UV-stabilizer (benzophenone) after 1 year weathering, magnified 120 times [38]

The effectiveness of various UV-stabilizers increases as follows: Benzophenone derivatives > Benzotriazole derivatives > Combinations with HALS-stabilizers

In most cases, pigmented polyacetals possess a better weathering stability than natural colored polyacetals. During weathering mechanical properties remain unchanged for a longer time, but changes in color caused by chalking or fading of the pigment [38] can occur.

As seen in curves "c" and "d" of Figures 4.43 and 4.44, TiO_2 and, above all, carbon black have a particularly favorable effect. The beginning of chalking is delayed with increasing concentration of carbon black without a significant change in the mechanical properties. The same applies for the addition of UV-absorbers to pigmented polyacetals [38]. Accordingly, a polyacetal with a relatively high loading of carbon black and light stabilizer, e.g. Hostaform C 9025 black 10-1569 [6], shows favorable behavior (see also Sections 4.10.2 and 4.10.5).

The artificial irradiation of acetal copolymers in the Xenon test apparatus and also in a Fadeometer yields the same sequential ranking for all formulations and all mechanical properties studied as for weathering tests [38]. There is, therefore, a basic correlation between the effects of weathering and artificial irradiation, but it is impossible to set up a general correlation factor. In addition, there is no general statement concerning the utility-time, because the requirements on the part vary substantially from case to case.

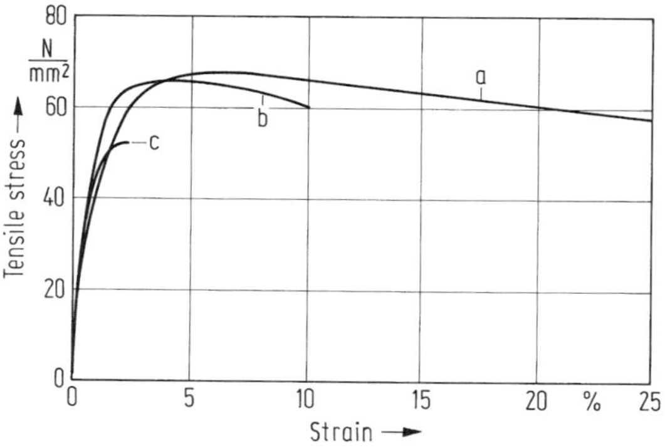

Figure 4.47. Stress-strain curves of irradiated acetal copolymers Hostaform [39]
a: 0 · J/kg, b: 2.5 · 10^4 J/kg, c: 4.0 · 10^4 J/kg

The irradiation resistance of polyacetal is limited. The stress-strain diagrams of the acetal copolymer Hostaform at several dosages is represented in Figure 4.47 [39]. With increasing dosage, first the strain at break and then the tensile strength decreases. In addition, discoloration occurs.

Thus, polyacetals can be sterilized with gamma-rays at the usual dosage of 2.5×10^4 J/kg as long as no high demands are placed on the sterilized devices as far as mechanical properties, in particular toughness, are concerned.

4.7.9 Permeability

Polyacetals exhibit a low permeability for aliphatic, aromatic, as well as halogenated hydrocarbons, alcohols, and esters. The permeability for water and methanol is relatively high, see Table 4.6 [40]. Gas permeability is also low, only for hydrogen and

Table 4.6. Permeability Values for Acetal Homopolymer [40]

Liquids	Permeability $\frac{g \cdot mm}{d \cdot m^2}$	
	23 °C 50% R.H.	38 °C
Ethyl Alcohol/Water (90/10 wt-%)	0.099	–
Ethyl Alcohol/Water (70/30 wt-%)	0.59	3.08
FREON 12/11 propellant (30/70 wt-%)	0.079	0.213
FREON 12/114 propellant (20/80 wt-%)	0.079	0.165
Fuel	0.039	–
Oils (Motor, Mineral, Vegetable oils)	0	0
Perchlorethylene	0.079	–
Tar Remover	0.01	0.0075
Trichlorethylene	9.9	22.1
Toluene	0.236	–

Table 4.7. Gas Permeability and Permeation Coefficient at 20 °C, Measured on 80 μm Films Pressed from Polyacetal [33]

Medium	Gas Permeability *D* (acc. to DIN 53 380 E)	Permeation Coefficient *Q*
	$\frac{cm^3(NTP)}{m^2 \cdot d \cdot bar}$	$\frac{cm^3(NTP) \cdot 100\ \mu m}{m^2 \cdot d \cdot bar}$
Air	8	6
Nitrogen	5	4
Oxygen	24	19
Hydrogen	207	165
Coal Gas	136	109
Carbon Dioxide	471	377
Ethylene	19	15
Methane	12	10
Water Vapor Permeability (acc. to DIN 53122)	$12 \frac{g}{m^2 d}$	

Table 4.8. Permeability of Acetal Copolymer in Comparison to HDPE [35]. Blown Bottles, 1 mm Wall Thickness, Welded, Total Volume 500 ml, Contents 250 ml, Test Temperature 40 °C

Test Fuel	Super Fuel			FAM-DIN 51 604-B (M 15)		
Duration of Experiment	Days					
	120	240	360	120	240	360
Fuel Diffusion HDPE %	– 100	–	–	– 100	–	–
Acetal Copolymer %	< – 0.5	< – 0.5	< – 0.5	< – 10	– 12	– 12.5

carbon dioxide the permeability is higher, Table 4.7 [33]. The extremely low permeability for propane and butane is of significance for the application of polyacetals in the manufacture of lighters and aerosol containers.

The resistance to permeation by fuels is worth mentioning. In storage tests on premium fuels in welded bottles at 40 °C, after 80 days 100% of the fuel diffuses from the HDPE-bottles, but less than 0.5% after 360 days from the acetal copolymer bottles. days. Similarly, M 15 diffuses from HDPE to the extent of 100% in 120 days, but from acetal copolymer it diffuses only to the extent of 12.5% in 360 days (Table 4.8 [35]).

The permeation of the individual components of M 15 into a closed container of acetal copolymer attains a saturation limit at a different time for each component (Figure 4.48 [35]).

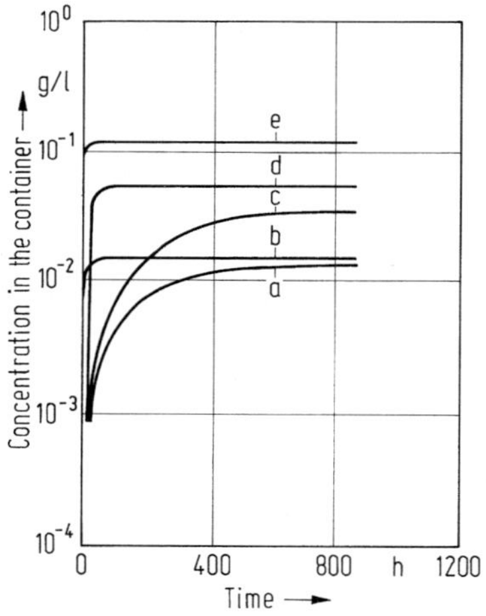

Figure 4.48. Permeation of the individual components of M15 into a closed container from acetal copolymer [35]. Test conditions: permeation surface 180 cm^2, film thickness 80 μm, test temperature 23 °C, test fuel FAM-DIN 51604-B (M15)
a: diisobutylene, b: ethanol, c: isooctane, d: toluene, e: methanol

4.7.10 Physiological Behavior

Polyacetals are odorless, free of taste, and inert. Feeding experiments yielded no clinical findings. Skin compatibility tests, together with numerous applications in which skin contact is given, show that no dermatitis is caused [40].

Processing polyacetals requires strict compliance with the regulations concerning formaldehyde concentratoin at the workplace. At the present time, in the Federal Republic of Germany a value of 0.5 ppm is established as the maximum allowed concentration of formaldehyde at the workplace. Formaldehyde can act as a sensitizer and is classified in Group III B by the Senate Commission of the "Deutsche Forschungsgemeinschaft" (German Research Assocation), see also Section 4.8.2.

The use of polyacetals for the production of essential commodities in the sense of Section 5, Paragraph 1, Number 1 of the Food and Essential Commodities law is regulated in recommendation XXXIII of the Federal Health Office (Germany) which sets the required criteria for purity [41]. In the United States, the use of polyacetals in contact with food is governed by FDA regulations [42, 43].

For the use of polyacetals in the drinking water sector, the Federal Health Office (Germany) has published the KTW-recommendation 1.3.12 [44].

As long as the purity requirements and the suitability for the purpose of the projected use are observed, there are no objections against the use of polyacetals in contact with food and drinking water. For this reason polyacetals have found a multitude of applications in both sectors.

4.7.11 Analytical

H. Schlaf

4.7.11.1 Monomers

4.7.11.1.1 Formaldehyde

Because of their simple procedures and sufficient accuracy, volumetric, wet chemical methods are preferably applied in the analysis of aqueous formaldehyde solutions. The following can be mentioned as examples:

- sulfite method
- peroxide method
- iodometric method
- mercury method

Of the cited methods, the sulfite method is simple to perform and insensitive to impurities. The method is based on the formation of a formaldehyde-bisulfite addition product from formaldehyde and sodium sulfite in aqueous solution whereby an equivalent amount of sodium hydroxide is liberated. One then titrates with sulfuric acid and thymolphthalein.

Detailed descriptions of the above-cited methods, their selectivity, and corresponding examples of use are found in the section on quantitative analysis of formaldehyde in [45].

The most frequent impurities in commercial formaldehyde are methanol and formic acid. Occasionally small levels of iron, aluminum, and copper impurities are also present.

Whereas the presence of formic acid can be determined through titration with sodium hydroxide or selectively through ion chromatography, gas chromatography is the current preferred method of detecting methanol. The earlier applied determination of methanol through oxidation with chromic acid ($CrO_3 + H_2SO_4$) is only seldom used now [46].

Formaldehyde concentrations between 5 and 40 wt-% in aqueous solutions can be simply and conveniently analysed through measurement of density [47].

4.7.11.1.2 Trioxane

The analytical investigation of trioxane has as its first priority the detection and quantification of impurities that permanently disrupt the polymerization. Depending upon the method of manufacture, trioxane may be accompanied by such impurities as water, formaldehyde and its derivatives such as methylal, and its homologues such as dioxymethylene dimethyl ether, formic acid, methyl formate, or homologues of formaldehyde like tetroxane, as well as possible traces of the trioxane extraction agent.

Today, a reliable identification and quantitative determination of most impurities by gas chromatography is possible.

Formic acid can be detected by titration with NaOH and traces of water can be reliably determined by the Karl Fischer method.

4.7.11.2 Polymer

4.7.11.2.1 Residual Monomer Content

The fraction of residual monomer in polyacetal granulate represents an important quality characteristic. As a rule very small amounts of formaldehyde are involved. These are entrapped in the product kernels and are released through diffusion. Thus extractive processes are used in combination with colorimetric methods of analysis for the determination of the residual formaldehyde content. Of the numerous possibilities, the method of Eegriwée using chromotropic acid as the reagent [45] and the acetylacetone method (or lutidine method) have established themselves. This method is used in slightly modified form for the determination of formaldehyde in containers when these must meet the recommendations of the German Federal Health Office [49].

Small amounts of formaldehyde in the ambient air can be determined with the aid of gas detection tubes (sniffer tubes) from the firm Dräger, Lübeck, as well as with test apparatus from the firm Lion Laboratories, Pearlstreet, Cardiff, and the firm Bayer Diagnostic GmbH, Munich (Compur 4100 SD).

4.7.11.2.2 Molecular Weight

The rheological behavior and mechanical properties of a polymer are determined to a large extent by its molecular weight. It is therefore important to have direct or indirect methods that give reliable information about the polymer molecular weight and its distribution.

Determination of the melt index according to DIN 53735 as a measure of the melt viscosity has practical significance. If one measures the melt index at constant temperature but with two different loads, the ratio of the melt index measured at the higher load to that measured at the lower load represents a measure of the molecular weight distribution. If a molded part exhibits a significantly higher ratio than does the starting granulate, this is an indication that damage to the material has occurred during processing; namely an increase in the low molecular weight fraction as a consequence of degradation.

Another practical method is the determination of the solution viscosity in a solvent suitable for poly(oxymethylene) such as dimethyl acetamide, butyrolactone, N-methyl pyrolidone, or octafluoropentanol-1 at temperatures between 110 °C and 150 °C. A direct relationship between the measured "intrinsic viscosity" and the weight average or number average of the molecular weight of the examined sample can be established [50 to 52].

Gel permeation chromatography offers a timely if, from the point of view of the equipment needed, complicated method for the determination of the molecular weight and molecular weight distribution of poly(oxymethylene).

Because of poor room-temperature solubility of poly(oxymethylene) in solvents suitable for GPC, it is necessary to make measurements in the temperature range between 110 °C and 150 °C. Dimethyl acetamide, dimethyl formamide, and butyrolactone are preferred solvents [53].

The use of other methods for determination of polymer molecular weight such as light scattering, osmotic methods, or fractionation methods is difficult, and limited because of the poor solubility of poly(oxymethylene).

4.7.11.2.3 Chemical Composition

As already discussed in Section 4.2, poly(oxymethylene) can be differentiated according to their chemical structure into homopolymers, copolymers, and terpolymers. The qualitative and quantitative questions of the individual structural features are interesting to the specialist.

Besides the so-called building-block analysis as described in [54], ^{1}H- or ^{13}C-NMR spectroscopy are generally applied.

The building-block analysis combines pyrolytic decomposition of the poly(oxymethylene) into fragments in the presence of a catalyst and the subsequent identification of these by gas chromatography.

Numerous authors have reported on the NMR-spectroscopic method of investigation which often includes sequence analysis of the comonomer in addition to the analysis of monomeric building blocks [55 to 59].

Whereas chlorinated hydrocarbons can be used as solvents for copolymers of formaldehylde or trioxane having a high comonomer content, copolyacetals having a low comonomer content demand special solvents and temperatures between 110 °C and 150 °C (Sections 4.7.11.2.1, 4.7.11.2.2, and [60]). For ^{1}H-NMR spectroscopy, one preferentially uses perdeuterated solvents in order to exclude foreign signals. Highly purified 1,1,1,2,2,2-hexafluoro-iso-propanol-d_2 is employed as a solvent at room temperature.

Instead of detailed information on the exact chemical structure of a poly (oxymethylene), one is often simply interested in knowing if one is dealing with a homopolymer or copolymer of formaldehyde. It is possible to differentiate between the two classes of polyacetals on the basis of the alkali sensitivity of the homopolymer by warming a sample with base, for example triethanol amine.

4.7.11.2.4 End Groups

Because of their low concentration, the quantitative determination of end groups in high molecular poly(oxymethylene)s is difficult. Information concerning the determination of acetate and methoxy end groups can be found in [61]. Qualitative identification of an OH or acetate end group can be accomplished with IR-spectroscopy. The quantitative analysis of methoxy end groups can be accomplished by the method of Zeisel; that of the acetate end groups by cleavage with methanolic HCL and determination of the methyl acetate formed. Safety measures must be observed here since the carcinogenic bis (chloromethyl) ether can be formed by the action of HCL on formaldehyde.

4.7.11.3 Additives

Additives can be determined in the polyacetals themselves, in their extracts, in their residues after acidic degradation, or in their combustion residues. Because of the multitude of possible additive systems (see Sections 4.4.1.2 and 4.4.1.3) many different methods or combinations of methods are required.

Infrared-spectroscopy has proven to be reliable for the direct determination of additives. Extracts and residues of degradation can likewise be analysed by IR-spectroscopy. Additionally, colorimetric methods as well as gas and thin-layer chromatography are also suitable and can be used in combination with IR, UV/VIS, or mass spectroscopy. Combustion residues give non-specific information on the content of inorganic additives but is well suited for deliberate trace analysis of metals via atomic absorption, atomic emission, or atomic fluorescence methods.

References for Section 4.7

1. *Schmidt, H.; Spankeren, U. van*, in: Kunststoff-Handbuch. Vol. XI. Hanser, Munich 1971
2. *Schmidt, H.; Geyer, H.:* Kunststoffe 54 (1964) p. 282
3. *Erhard, G.; Schütte, W.:* Kunststoffe 61 (1971) p. 235
4. *Domininghaus, H.:* Die Kunststoffe und ihre Eigenschaften 2. Ed. VDI-Verlag, Düsseldorf 1986, p. 307–334
5. Delrin-Acetalhomopolymer. Company brochure Du Pont de Nemours 1988
6. Kunststoffe Hoechst, Hostaform. Company brochure Hoechst AG, Frankfurt 1987
7. Ultraform-Polyoxymethylen. Company brochure BASF AG, Ludwigshafen 1986
8. *Wolters, E.; Sabel, H.-D.:* Kunststoffe 62 (1972) p. 632
9. *Schütte, W.:*Kunststoffe 66 (1976) p. 604
10. *Sabel, H.-D.:* Kunststoffe 70 (1980) p. 641
11. *Schütte, W.:* Kunststoffe 74 (1984) p. 575
12. *Kloos, F.; Wolters, E.:* Kunststoffe 75 (1985) p. 735
13. *Wolters, E.; Sabel, H.-D.:* Kunststoffe 77 (1987) p. 997
14. Kunststoffe 76 (1986) p. 856
15. *Uchida, T.; Tadokoro, H.:* J. Polym. Sci., Polym. Phys. Ed. 5 (1967) p. 63
16. *Wegner, G.*, et al.: Makromol. Chem. 181 (1980) p. 1763
17. *Wilski, H.:* Makromol. Chem. 150 (1971) p. 209
18. *Dröscher, M.*, et al.: Makromol. Chem. 177 (1976) p. 1695
19. *Dröscher, M.*, et al.: Makromol. Chem. 177 (1976) p. 2793
20. *Dröscher, M.:* Makromol. Chem. 178 (1977) p. 1195
21. *Wolters, E.*, et al.: Plastverarbeiter 24 (1973) p. 549
22. *Kohlhepp, K. G.; Mohnberg, J.:* Kunststoff-Berater 19 (1975)
23. *Leugering, H. J.*, et al.: Makromol. Chem., Suppl. 1 (1975) p. 621
24. *Schmidt, H.; Meinhard, I.:* Plastverarbeiter 29 (1978) p. 483
25. *Inoue, M.:* J. Appl. Polym. Sci. 8 (1964) p. 2225
26. *Sextro, G.*, et al.: Makromol. Chem. 177 (1976) p. 1815
27. DE-PS 2101817 (1971) Hoechst AG
28. *Wolters, E.; Gleisberg, P.:* Plastverarbeiter 27 (1976) p. 201
29. *Wolters, E.; Racke, H. H.:* Kunststoffe 63 (1973) p. 608
30. Technische Kunststoffe Hoechst—Hostaform. Company brochure Hoechst AG, Frankfurt 1988
31. Delrin-Polyacetale. Company brochure Du Pont de Nemours 1984
32. Delrin-Acetalharze. Company brochure A-20731-2, Du Pont de Nemours
33. Werkstoffe der Feinwerktechnik—Polyacetal-Formstoffe. VDI/VDE 2477. VDI-Verlag, Düsseldorf 1974
34. Verhalten von Ultramid, Ultraform und Ultradur gegen Chemikalien. Company brochure BASF AG, Ludwigshafen 1980
35. *Kohlepp, K.; Orth, R.:* Kunststoffe 77 (1987) p. 686
36. *Vesely, R.; Kalenda, M.:* Kunststoffe 59 (1986) p. 107
37. *Kelleher, P. G.; Gesner, B. D.:* Poly. Eng. Sci. 10 (1970) p. 38
38. *Schmidt, H.; Wolters, E.:* Kunststoffe 61 (1971) p. 261
39. *Wolters, E.; Rösinger, S.:* Kunststoffe 63 (1973) p. 605
40. Delrin-Polyacetal. Company brochure Du Pont de Nemours
41. Mitt. Bundesgesundheitsblatt 27 (1984) p. 291
42. Code of Federal Regulations, Title 21, Section § 177, 2470
43. Code of Federal Regulations, Title 21, Section § 177, 2480
44. Mitt. Bundesgesundheitsblatt 22 (1979) p. 264
45. *Walker, J. F.* (Ed.): Formaldehyde. 3. Ed. R. E. Krieger, Malabar, Florida 1975, p. 469–485

46. *Blank, O.; Finkenbeiner, H.:* Chem. Ber. 39 (1906) p. 1326
47. *Natta, G.; Baccaredda, M.:* Giorn. Chim. Ind. Appl. 15 (1933) p. 273
48. DE-PS 2840679 (1980) Degussa AG
49. *Franck, H.:* Kunststoffe 74 (1984) p. 559–564
50. *Wagner, H.*, et al.: Makromol. Chem. 81 (1965) p. 14
51. *Baker, S. J.; Price, M. B.:* Polyacetals. Iliffe Books, London 1970
52. *Major, J.:* Makromol. Chem. 86 (1965) p. 253
53. *Ishigaki, I.*, et al.: Appl. Polym. Sci. 18 (1974) p. 1927
54. *Burg, K. H.*, et al.: Makromol. Chem. 103 (1967) p. 268
55. *Yamashita, Y.*, et al.: Am. Chem. Soc., Polym. Prepr. 11 (1970) p. 185
56. *Pfab, W.; Bernhardt, K.:* Z. Anal. Chem. 251 (1970) p. 25
57. *Boltze, R.*, et al.: J. prakt. Chem. 312 (1970) p. 596
58. *Yamashita, Y.*, et al.: Makromol. Chem. 129 (1969) p. 1
59. *Fleischer, D.; Schulz, R. C.:* Makromol. Chem. 152 (1972) p. 311; Makromol. Chem. 162 (1972) p. 103; Makromol. Chem., Suppl. 1 (1975) p. 235; Makromol. Chem. 176 (1975) p. 677
60. *Alsup, R. G.*, et al.: Appl. Polym. Sci. 1 (1959) p. 185
61. *Koch, T. A.; Lindvig, P. E.:* Appl. Polym. Sci. 1 (1959) p. 164

4.8 Processing of Polyacetals

P. Unger

4.8.1 General

Polyacetals can be processed on all machines commonly used for thermoplastics, for example injection molding machines, extruders, injection and extrusion blow molding equipment, and presses. No particular pretreatment is required as a rule. However, should the granulate become moist through inexpert handling or come in contact with water, then drying in a circulating air oven at 100 °C to 120 °C for 3 to 6 hours (according to the thickness of the layer) is necessary. The layer thickness should not be more than 40 mm. In order to avoid a buildup of water condensation, the polymer must be at room temperature before processing. This is most easily accomplished by storing bags or pallets in the production area for a sufficient period of time.

Likewise, one must pay attention to drying and the avoidance of dust, which promotes moisture absorption, when adding regrind to virgin resin since small amounts of moisture detract from the thermostability in processing.

Thermally damaged material, which can be recognized by, for instance, a brownish discoloration in the case of natural-colored resin, cannot be reprocessed since the physical properties of a molded part can be substantially impaired. Cross-contaminated or otherwise dirty regrind can produce a similar effect.

Furthermore, inept processing of regrind can lead to a buildup of mold deposit and in extreme cases even to corrosion of the molding machine and/or the mold.

From the point of view of process technology, the direct regranulation of sprues and runners at the molding machine and direct addition to original material at about the 20% level has proved to be most favorable. In cases where such a process is followed, one can in general forego drying.

In the case of injection molding machines having screws of small diameter, a certain minimum granulate size of the regrind cannot be exceeded in order to avoid variations in feeding.

Table 4.9. DIN Safety Sheet (Sheet 1)

Company				
Tradename	Hostaform C xxxx, C xxxx			
1.1	Chemical Characterization Polyoxymethylene Copolymer			
1.2	Form: Granulate	1.3 Color: div.		1.4 Odor: –
2	Physical and Technical Safety Details			
2.1	Transformations			
	Melting Temperature	163 to 168	°C	
			°C	
2.2	Density	(20 °C)	ca. 1.4 g/cm^3	
	Bulk Density	ca. 850	kg/m^3	
2.3	Vapor Pressure	(°C) n.a.	mbar	
		(°C)	mbar	
2.4	Viscosity	(°C) n.a.		
2.5	Solubility in Water	(20 °C) insoluble	g/l	
		(100 °C) insoluble	g/l	
2.6	pH-valve (at g/l H_2O)	(°C) n.a.		
2.7	Flash Point	n.a.	°C	
2.8	Ignition Temperature	FIT 320 bis 340	°C	ASTMD D 1929
2.9	Explosive Limits	lower: upper:		
2.10	Thermal Decomposition		over 240 °C	
2.11	Dangerous Decomposition Products	Overheating during processing can cause formaldehyde to be produced		
2.12	Dangerous Reaction	n.a.		
2.13	Further Details			
3	Shipping	GGVSee/IMDG-Code: GGVE/GGVS:	UN-Nr: RID/ADR:	ICAO/IATA-DGR: ADNR:
	Miscellaneous Details: No danger classification.			
4	Regulations The material is not a dangerous working material in the sense of the dangerous material degree in the resolution of 26. 8. 1986. Observe the MAK-Value for Formaldehyde (0.5 ppm; as for 1987) Formaldehyde is included in Group III B of the MAK-List. MAK = maximum allowable workplace concentrations			

Statements are based on the latest status of our knowledge and describe our product in regards to safety requirements and should not be constructed as specifying particular properties.

Table 4.9. DIN Safety Sheet (Sheet 2)

Tradename Hostaform C xxxx, C xxxx

5 Safety Measures! Storage and Handling
5.1 Technical Safety Measures
None

5.2 Personal Safety Equipment | Respiratory Protection: | Eye Protection: | Hand Protection: | Other:

5.3 Work Hygiene Provide adequate ventilation in the work area: adherence to the MAK-Value requires local exhaust in the area of the machine during thermal processing and machining.

5.4 Fire and Explosion Protection
None, it is recommended to follow the general rules of preventive company. Internal Fire Protection

5.5 Dispose of in accordance with the legal regulations for inert waste.
Dispose of as inert waste in accordance with legal regulations.
For example suitable landfill or incineration.

6 Steps in case of accidents and fires

6.1 Increase of spills, discharge, escape
Mechanically remove (danger of slipping).

6.2 Extinguishing Agent Suitable:
All customary fire extiguishing agents.
Not to be used:

6.3 First Aid
After inhalation of formaldehyde provide fresh air and keep warm, seek medical help.
Burns to the skin resulting from contact with molten material must be medically treated.

6.4 Further Details

7 Toxicological Details
The material is not damaging to health.

8 Environmental Details
The material has no effects which injure the environment
Waste Code: 571 221
WGK* = 0

*WGK = Wassergefährdungsklasse (Water endangering classification)

9 Further Advice

4.8.2 General Safety Precautions for Processing

In the processing of polyacetals, the temperature of the melt should not exceed 240 °C. The allowable residence time in the cylinder must also be taken into consideration. In case of exceptionally high thermal stressing, polyacetals will decompose with the generation of formaldehyde.

At the present time in the Federal Republic of Germany, a value of 0.5 ppm is fixed as the maximum allowable workplace concentration (MAK-value; see Section 4.7.10) for formaldehyde. For most other European countries a value of 2 ppm is in effect. The MAK-value is the average value from multiple measurements taken over the course of the work day or shift. This can be carried out with, for example, a Dräger gas detector and the accompanying measuring tubes. Samples should be taken at breathing height in the direct vicinity of the workers.

Exhaust hoods should be installed directly over the machines.

Furthermore, in the case of excessively high thermal stressing, the pressure of the developing gaseous decomposition products in a closed or frozen cylinder can become so great that it equilibrates via the feed opening. If this is not possible then damage to the machine and injuries to personnel may result. It is therefore important to take care that injection nozzles or extruder throats not become clogged with, for example, slugs of cold material.

If thermal decomposition in the cylinder is suspected or detected, the material must be pumped out at reduced settings for the cylinder heating. In order to avoid unnecessary odor generation, it is advisable to immerse the thermally degraded material in water.

Plastics, like all organic products, are combustible. It is in the best interests of the processor to take first prevention measures in the storage, processing, and finishing of plastics. Regulations valid in individual countries are the final authority.

Instruction for the safe handling of polyacetals is given by the raw material manufacturers in their safety data sheets. As an example, a DIN safety data sheet for an acetal copolymer is reproduced in Table 4.9, sheets 1 and 2. From this data sheet it is apparent that polyacetals are not classified as dangerous, do not represent a dangerous material with which to work, do not have environmentally damaging effects, and are not damaging to the health (see also Section 4.7.10).

At the beginning of work with an initially empty injection molding cylinder, the heating zones are set at 190 to 200 °C.

After the plasticizing cylinder has been filled, several shots are taken into the open (so-called pumping out). Here particular attention must be paid to the nozzle temperature. If the temperature is too low the melt will freeze and blocks the nozzle.

In case of short interruptions in the work, the cylinder temperatures should be slightly reduced; the temperature of the nozzle, however, can be maintained.

The following process is followed in case of longer interruptions in the work:

- discontinue addition of granulate
- turn off cylinder and nozzle heating
- pull back the cylinder from mold
- completely expel the melt from the cylinder (purge)

The heating is also switched off if hot-runner molds are used, no matter what particular system is being used.

If various thermoplastics are processed on the same machine, cleaning of the screw and cylinder is always recommended before polyacetal is processed. Cleaning can be accomplished with a purge compound as, for example, an easy-flowing polyolefin, or by mechanically removing the melt after pulling the screw.

4.8.3 Injection Molding

4.8.3.1 Machine Requirements

Polyacetals can be processed on all presently customary standard screw injection molding machines. Devolatilizing machines have, in general, not proved to be worthwhile. In the case of small machines, for example for the manufacture of micro-parts, piston machines instead of screw machines are used for reasons of strength. The use of micro-granulate is recommended for the smallest machines.

Special screws have in general not proved to be necessary; outfitting of the machine with so-called standard screws according to the recommendations of the machine manufacturer is sufficient. So-called three-zone screws with a length-to-diameter ratio of $L/D \geqslant 20$ have proved themselves as standard screws.

The breakdown of the zones often follows the following scheme:

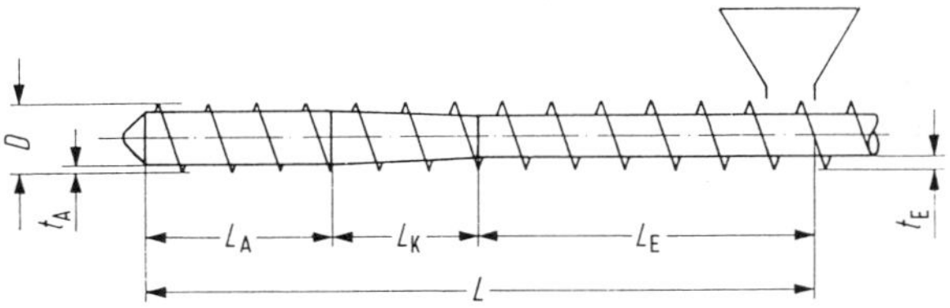

Scew length	$L \geqslant 20D$
Feed zone	$L_E = 0.5L$
Compression zone	$L_K = 0.3L$
Metering zone	$L_A = 0.2L$
Compression ratio	$t_E/t_A = 2$ to 2.5

Cylinders for processing of polyacetals are customarily equipped with open, heated nozzles. Shut-off nozzles should be so hydrodynamically constructed that there are no dead zones with the attendant danger of thermal degradation of the stagnant melt.

Screw decompression is used in order to decompress the melt, for example after the screw is filled, in order to avoid so-called drooling. The screw travel must be adjusted so that the drawing in of air through the (open) nozzle is reliably avoided. Otherwise, burning of the melt as a consequence of the so-called dieseling effect cannot be ruled out.

In processing acetal homopolymer the melt should not be brought into contact with copper, brass, bronze, or other copper alloys in order to preclude catalytic decomposition. In the case of acetal copolymers, such a negative influence is not observed.

The cylinder should not be filled to below about 20% or above 80% of capacity.

For processing glass fiber reinforced polyacetals, wear resistant outfitting of the injection molding equipment, which is part of the normal product offering of machine manufacturers, is recommended.

4.8.3.2 Requirements for Molds

No particular demands are placed on the selection of steel for molds for processing of non-modified polyacetals. Hardened steel, heat treated steel, and penetration hardened steel have equally proved themselves. In the case of processing glass fiber reinforced polyacetals, suitable measures must be taken to prevent wear. Wear resistant tool steels, for example material No. 12601 with hardness HRC 60, or surface coatings, for examples electroless nickel–multilayer chromium or TiN coating, have proved out well. Great importance is to be attached to optimal annealing of the mold.

Air displaced by the melt must be able to escape the cavity. To this end provision is made for air vents whose dimensions are influenced by the melt viscosity. Extremely easy-flow polyacetals, those with MFI of 50 g/10 min, necessitate a vent opening of a few μm, materials with average flow, MFI of 9 g/10 min, necessitate vents of a few hundredths of a millimeter. A non-vented or poorly vented mold promotes a dieseling effect, buildup of mold deposit, and in the extreme case favors corrosion.

Processing of polyacetals in hot-runner molds is standard technology. However, not all systems on the market are equally suitable. It is recommended to consult the experience of the system supplier.

For the mechanical, thermal, and rheological dimensioning of a mold in critical situations, of modern calculation processes should be used to support one's own experience. The same holds true for the design of complex and complicated parts.

4.8.3.3 Flowability

The wide spectrum of applications—from injection molded precision parts to extruded profiles—covered by the polyacetal molding materials requires a product series that lives up to the demands of flowability of the melt and the properties of the part. Thus today's range of polyacetals extends from extremely easy-flowing injection molding grades (MFI = 50 g/10 min) for complicated, thin-walled, precision parts or parts with unfavorable of flow length to wall thickness, to molding grades for thick-walled, sink-free parts (MFI = 1 g/10 min).

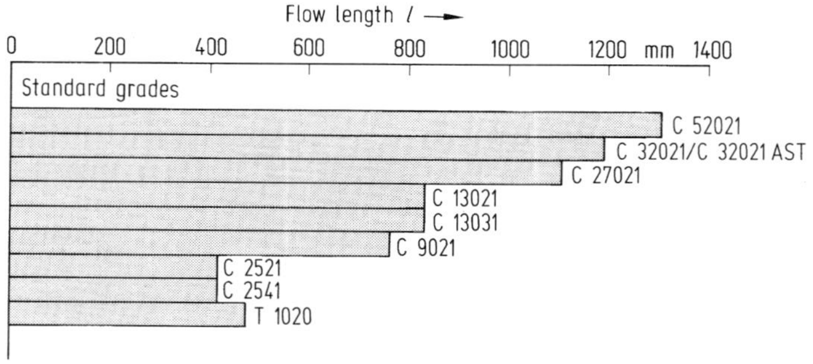

Figure 4.49. Length of the flow path l (spiral test) of different polyacetal grades, thickness of the spiral 3 mm, injection pressure 1000 bar, melt temperature 205 °C, mold temperature 80 °C

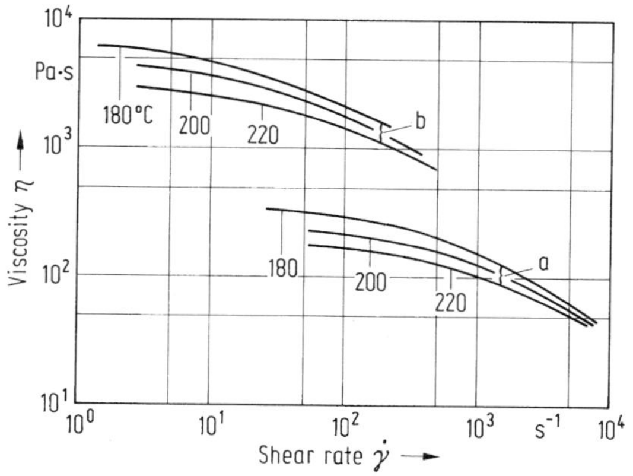

Figure 4.50. Viscosity η as a function of the shear rate
a: Hostaform C 32021, b: Hostaform C 2521

The length of the flow path at a wall thickness of 3 mm for constant processing conditions is given in Figure 4.49 for certain grades of polyacetal.

The dependence of viscosity η upon shear $\dot{\gamma}$ at various melt temperatures for an easy-flowing injection molding grade (a) and an extrusion grade (b) is represented in Figure 4.50.

This clearly shows the possibility of deliberately influencing, within limits, the flow behavior of a melt by variation of processing parameters.

4.8.3.4 Shrinkage, Post-Shrinkage, Tolerances

The term *shrinkage* is used to denote processes that lead to a reduction in sizes of the molded part as compared to the cold dimensions of the mold.

Shrinkage strongly influences the dimensional precision of the molded part and, in the case of non-uniform shrinkage, the warpage. Beyond that the shrinkage, particularly when constrained, affects the type and level of internal stresses and therefore also the strength of the design.

Shrinkage processes begin during processing, when the mass of material transforms from a plastic to a solid state (i.e., upon cooling) and the amorphous melt changes into a semi-crystalline plastic with an attendant reduction in its volume.

The difference between the cold dimensions L_W of the mold and the dimensions of the cooled part (24 hr. after production, DIN 16901) is designated processing shrinkage (mold shrinkage). As a consequence of the mentioned structural transformation in the case of semi-crystalline polymers discussed here, the shrinkage is generally larger than in the case of amorphous plastics.

The overall shrinkage of the injection molded part does not end with processing shrinkage. It continues to shrink more in the course of time, toward a final value. The higher the ambient temperature, the faster this end value is attained. This process is principally determined by post-crystallization and is called post-shrinkage (postmold shrinkage).

In the case of semi-crystalline plastics, the postmold shrinkage is always less than the mold shrinkage. The sum of mold shrinkage and postmold shrinkage is known as total shrinkage.

In addition to the plastic itself, all shrinkage processes depend upon a multitude of processing, use, and part-specific factors. An exact prediction of shrinkage is thus not possible.

ISO-tolerances from series IT 7 (ISO-basic tolerances DIN 7151) cannot be attained using injection molding technology. It is possible but very costly to fulfill the IT-8 rules for precision injection molding.

The following influences are, to a variable degree, most important in determining shrinkage properties:

- *surface temperature of the mold*
  As the temperature of the mold surface T_W increases, the processiing shrinkage increases but the postmold shrinkage decreases (Figure 4.51). This statement, important for precision injection molding, means that the mold temperature, measured at the wall, must lie as close to $T_W = 120$ °C as possible in order to obtain parts of invariable dimensions. A greater mold shrinkage must be accepted.
- *pressure*
  During injection molding, the material being molded experiences various pressures (injection pressure, holding pressure, etc.). It is generally valid that the mold shrinkage and the total shrinkage decrease with increasing pressure (Figure 4.52).

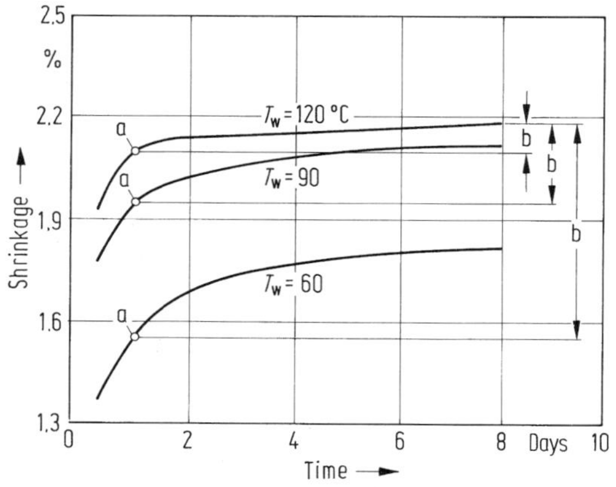

Figure 4.51. Mold shrinkage (a) and postmold shrinkage (b) of Hostaform C 9021 as a function of time and toolwall temperature; T_w = melt temperature 205 °C

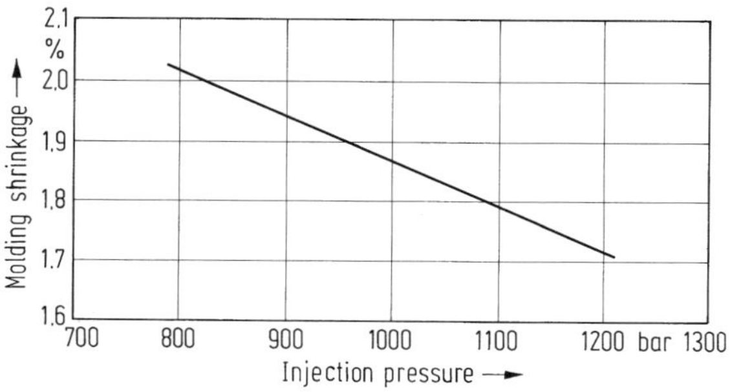

Figure 4.52. Mold shrinkage VS of Hostaform C 9021 as a function of injection pressure; wall thickness 3 mm, tool wall temperature 90 °C, melt temperature 205 °C

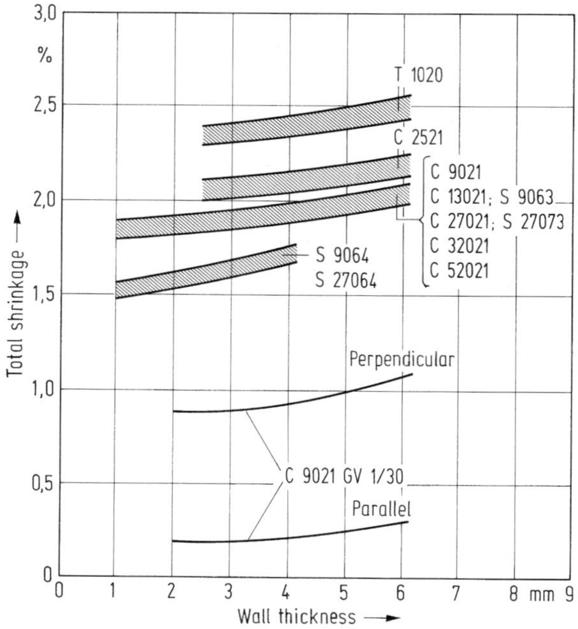

Figure 4.53. Total shrinkage GS of Hostaform as a function of wall thickness s
T_w = 90 °C, $P_{sp} = p_N$ = 1000 bar, T_M = 205 °C

- *part thickness*
  The shrinkage increases with increasing part thickness; see Figure 4.53. The high shrinkage during processing in the case of greater wall thicknesses can lead to warpage in the molded part, especially if the part has significantly variable wall thickness.
- *filler orientation*
  Because of the orientation of the glass fibers in the flow direction, the total shrinkage in this direction for glass fiber reinforced polyacetals is substantially smaller than that in the cross-direciton. This difference is negligible in the case of non-modified polyacetals. The reader is referred to the curves for C 9021 GV 1/30 in Figure 4.53.

In order to maintain warp-free parts, one must strive to minimize differences in shrinkage. The difference in shrinkage is essentially zero in the case of mineral modified, and especially glass sphere modified polyacetals.

In the manufacture of an injection mold, alterations (e.g., re-working) must be allowed for since deviations in the shape and dimensions of the molded part are often unavoidable. These deviations can be compensated for by suitable corrective measures in the cavity. In investigating the causes of, for example, dimensional deviations, one must also consider the influence of the type and placement of the gate as well as the runner system.

The type and positioning of the gate is influenced by various factors such as wall thickness, flow path, flow direction, knitlines, and sink marks.

The size of the gate depends on the wall thickness of the part. If the gate is too generously sized then the cooling time and therefore the cycle time can be uneconomical.

A too-small gate can excessively shorten the required holding time by a freeze-off effect or lead to high shear heating in the melt.

As an approximate value, the diameter of the gate can amount to about two-thirds of the maximum wall thickness. The gate should be planned where the wall thickness is greatest.

4.8.3.5 Processing Conditions

The range of the melt temperature lies between 180 °C and 230 °C, most advantageously between 190 and 220 °C. In the case of impact modified materials, a melt temperature of 210 °C should not be exceeded.

The required melt temperature is attained on the one hand through heating of the cylinder (external heat transfer) and on the other hand through friction (heat introduction by internal and external friction, determined by the rotation of the screw and back pressure).

In the case of polyacetals, the contributions of shear and frictional heating to the total heat supplied should be held as small as possible. This requires that the screw speed and the temperature program settings be controlled.

The cylinder and nozzle temperatures (T_1 to T_n, T_D) are generally set on the controllers in an increasing pattern. A possible temperature program might look as follows:

$$T_1 = 170\ °C,\ T_2 = 180\ °C,\ T_3 = 190\ °C,\ T_4 = 200\ °C,\ T_5 = 200\ °C.$$

Where necessary, the ca. 10 °C higher melting point of acetal homopolymer compared to acetal copolymer is to be taken into consideration.

The back pressure should lie between 0 and 20 bar. By the action of the back pressure the homogenization of the melt during the plasticating stage is, on the one hand, improved; on the other hand, the introduction of heat through friction and shear is increased.

The injection pressure should equal the holding pressure and should lie between 600 and a maximum of 2000 bar. The injection time, t_s, is dependent upon the wall thickness of the finished part. A short time is selected in the case of thin-walled parts but a longer time is selected with increasing wall thickness. The holding time, t_K, and the cooling time t_P, are selected according to empirical values. The clamping force, F, is determined by the injection pressure, the projected surface of the part and runner system, and the injection speed. The clamping force must be high enough so that the mold cannot be forced open. Otherwise there is a danger of forming flash.

The temperature of the mold, measured at the wall (T_w), is adjusted to accomodate the required finished tolerance. The following are generally valid:

$F_T \leqslant 1\%$ (normal injection molding) $T_w = 60$ °C
$F_T \leqslant 0.6\%$ (technical molding) $T_w = 90$ °C
$F_T \leqslant 3\%$ (precision injection molding) $T_w = 120$ °C.

4.8.4 Extrusion

4.8.4.1 Machine Requirements

The customary single-screw extruders are used for the extrusion of polyacetals. Cooling or heating of the screw is not required.

A short compression-zone screw has a suitable geometry. Lengths of 25 D (diameter) are used by preference. Shorter screws frequently lead to surging.

Processing on vented extruders is not recommended.

When extruding polyacetals one must above all keep the narrow melting range as well as the rapid solidification process in mind.

Even in the case of effective cooling, a relatively low output should be chosen to remove the amount of heat, in spite of the low thermal conductivity, from the extrudate. Experience teaches that with simple extrusion dies, a throughput of from 7 to 10 kg/hr should not be exceeded.

In order to utilize the much higher output potential of the extruder, rod stock, as an example, is frequently extruded with the aid of multiple dies.

4.8.4.2 Material Types

As a matter of basic principle the melt viscosity for extrusion is so chosen that a secure handling of the melt is guaranteed. In most cases that means that, with complete homogenization, the melt should be as viscous as possible. Therefore copolymer grades such as Hostaform C 2521 and C 2250 or homopolymer grades such as Delrin 100 and Delrin II 150 SA are offered.

4.8.4.3 Manufacture of Semi-Finished Goods

Extrusion of Rods

The design and functioning of a setup for extruding rods with circular cross section is represented in Figure 4.54. The difficulties in manufacturing void-free round rods increase with increasing diameter since a uniform solidification of the extrudate across the cross section is not possible.

Braking of the emerging rod with brake grippers or a special haul-off places the solidifying melt under as high a pressure as possible so that no sinks can form and so

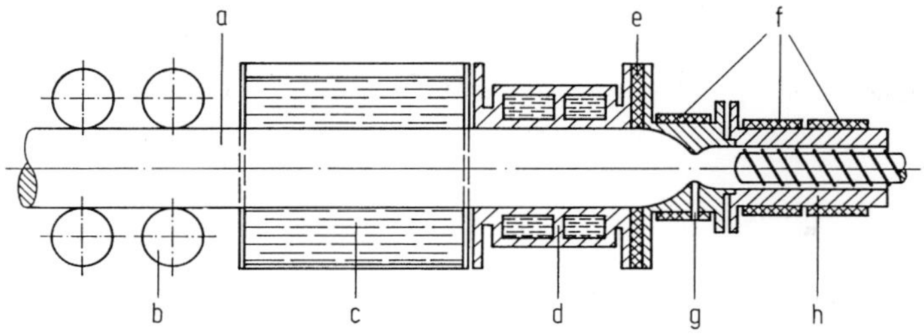

Figure 4.54. Scheme of an extrusion line for rods
a: extruded rod, b: take-off unit, c: water cooling, d: calibrating tube, e: insulation, f: heating bands, g: pressure control, h: extruder

that the stresses set up by volume contraction can be avoided as far as possible. The opposing pressure should be measured and regulated.

After exiting the calibration device, the already solidified exterior of the profile must be thick enough to oppose the induced internal pressure. This is achieved with an external cooling chamber and/or direct water cooling (Figure 4.54). In order to shorten the cooling press leading up to solidification of the melt, it is evident that a low melt temperature should be selected. Thus one works as temperatures normally between 180 °C and 185 °C.

Rods of circular cross section measuring from 3 mm to 200 mm in diameter are standardized according to DIN 16975.

Extrusion of Slabs and Flat Stock

Extruders having screw cross sections of 90 mm or 120 mm are used for slabs having a width of about 1000 mm. The flat rollers are chromed and polished. The temperature of the rolls is controlled with oil or steam. In the case of calendered slabs, throughputs of about 100 kg/hr are known.

In order to produce slabs that are as free from stress as possible and having a shiny surface the nip rolls should be heated to 130 to 135 °C and the run-out rolls to 120 to 125 °C. Auxiliary radiant (IR) heaters installed on the rollstack promote a uniform cooling, including the edge areas of the slab, and thus promote production of a uniform, stress-free profile. As a rule, the edges of the slab are cut rectilinear after it leaves the rollstack.

Slabs can also be prepared by compression molding. Here the extruder is used to pre-plastify the working material, which is then ejected into a preheated compression mold, compressed, and then held a constant pressure and cooled to the demolding temperature.

Slabs having a thickness of 1 mm and 6 mm and a width of 1000 mm are standardized according to DIN 16977. Flatstock having a thickness of 5 mm up to 100 mm and a width of 300 mm to 500 mm is standardized according to DIN 16974.

Extrusion Pipes and Hollow Profiles

Up to now, extrusion has been used chiefly for pipes having outside diameter for about 5 mm up to 8 mm. They are used as protective pipes for Bowden cable.

Pipes having an outside diameter of 10 mm up to 12 mm can be manufactured with the aid of vacuum sizing, while hollow profiles of larger dimensions can only be made with the help of vacuum sizing in combination with internal pressure systems. These points of view are to be considered in light of the various designs for dies and sizing equipment.

An extrusion line for pipes is built up from the following units: extruder and die, sizing equipment, cooling bath, and take-up or saw.

Post-Treatment, Annealing

DIN 16979 (technical terms for delivery for semi-finished goods made of polyacetal) specifies that the semi-finished goods are to be made essentially stress free by the manufacturer by, for example, annealing.

In those cases where exceptional demands are placed on the dimensional stability of the finished part, it is also advisable to post-treat the parts to eliminate residual stresses. Parts can then go to final assembly. This post-treatment (annealing) is carried out in air or nitrogen in the oven or alternatively in liquids (waxes, oils) at 140 °C over a period of time of 10 minutes per millimeter of wall thickness or diameter. At higher temperatures for longer times the material can be thermally damaged which, under some conditions, can be noted as discoloration.

In order to avoid the possible development of new stresses during heating or cooling, both procedures should be carried out slowly and uniformly. This is done, for example, by allowing the semi-finished part to warm up and cool down in and together with the oven. The time necessary for this is added to the so-called secondary operation time.

4.8.4.4 Requirements for the Properties of Semi-Finished Goods

According to DIN 16979, semi-finished goods made from polyacetals should provide certain minimal properties, see Table 4.10.

The standard mentions the possible influences of colorants and fillers on these properties.

Table 4.10. Standards for Semi-Finished Goods Made of Polyacetal according to DIN 16979

Property	Unit	Specification	Testing Norm
Tensile Strength	N/mm^2	$\geqslant 65$	DIN 53455*
Stain at Break	%	$\geqslant 20$	DIN 53455*
Melt Index MFI 190/2	g/10 min	$\leqslant 3{,}5$	DIN 53455

* Testing in tensile experiment, specimen No. 3 having a maximum thickness of (6 ± 0.05) mm, machined from extruded stock and stressed in the extrusion direction.

4.8.5 Blow Molding

4.8.5.1 Extrusion Blow Molding

This process is a two-stage process. The first step comprises the making of a blowable preform, the second comprises the blow molding and cooling of the article in the mold.

Polyacetals having relatively high melt strength are required for extrusion blow molding. Thus, for example, Hostaform C 2521 is suitable for the manufacture of small hollow shapes having a maximum volume of 1 liter. For larger volumes, Hostaform grade T 1020, which exhibits a much higher melt strength, must be used. Shrinkage in the case of blow molded parts lies between 2 and 4%.

Hollow shapes having a volume of 5 litres are made from polyacetal on machines without a melt accumulator; that is, the tube-like preform appears continuously. Larger blow molded parts require machines with accumulator heads in which the plastified material collects until the desired shot volume is attained. The plastic mass is then relatively quickly extruded.

Slow-running, single-screw extruders equipped with screws having a length of 20 to 25 D are suitable for plastifying the material.

Upon exiting the nozzle, the material should have a temperature of about 185 °C. However, one can limit sagging of the melt by careful reduction of the temperature (however not to below 170 °C). In order to achieve a uniform distribution of wall thickness along the axis, a program-controlled regulation of the wall thickness of the part in the axial direction is necessary. This is particularly true in the case of non-cylindrical parts because of their non-uniform radial blow ratios. The die area of the head outlet must be sufficiently heated ($T_D \geqq 170$ °C), otherwise an irregular swelling caused by temperature differences in parison walls set in, precluding continuous processing.

Defect-free containers with glossy surface are only obtained with blow molds heated to about 90 °C.

4.8.5.2 Injection Blow Molding

Injection blow molding is, like extrusion blow molding, a two-step process. The first step is injection molding of a blowable preform in a mold consisting of cavity and core rod. The second step comprises blow molding and cooling in the closed blow mold; see Figure 4.55.

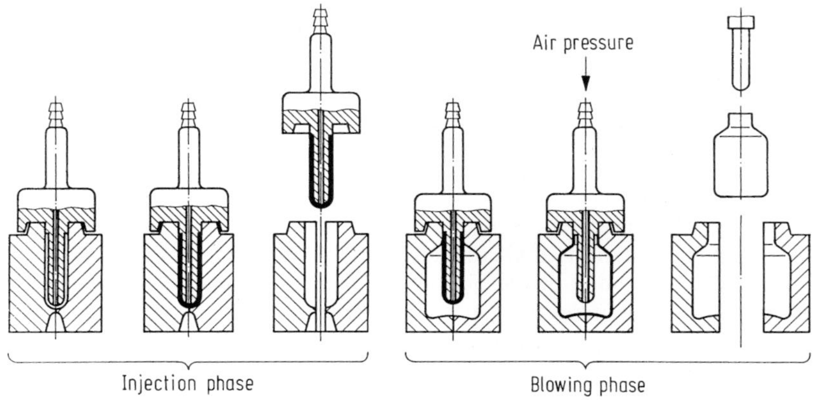

Figure 4.55. Basic principle of injection blow molding

A principal feature of the injection blow molding process is the waste-free manufacture of hollow shapes. Besides that, the process exhibits a series of additional advantages:

- high precision
- uniform wall thickness
- minimal variation in weight and volume
- no weld lines

- optimal surface quality
- improved mechanical properties

This mostly results in a minimal need for material and optimum properties in the finished part.

Against these advantages must be counted the limitation that, in the injection blow molding process, containers having multiple or non-symmetrically juxtaposed openings and blown handles cannot be made. Also, containers having extreme axial or longitudinal cross sections (for example flat, rectangular cross sections with aspect ratio a:b of $>1:2.5$) cannot be realized.

Because of its stiffness and impact strength, polyacetal is suitable for the manufacture of packaging having an internal pressure. Most of the polyacetal consumed in blow molding is processed into aerosol containers.

4.9 Working and Secondary Operations

P. Unger

4.9.1 Machining and Post-Treatment

Most plastic parts for technical applications are manufactured by injection molding. In general, a mechanical treatment or shaping is not carried out. For particular purposes, however, a shaping operation, in many cases machining, is required if:

- parts from solid plastic must be made—this is customary for samples or small manufacturing runs where cutting a mold is not economical or when parts with very thick walls must be made;
- prototypes are to be prepared—these individually manufactured parts, which have the form of the part to be injection molded, are used to carry out tests for practicality or suitability for use;
- injection molded parts need to be further shaped after molding because, for example, certain contours are uneconomical, difficult, or cannot be precisely made using injection molding technology.

When machining plastics, factors which can substantially deviate from those in metal working must be considered.

The extremely small thermal conductivity impedes the removal of the generated heat and can lead to softening of the plastic and thence to smearing of the shaped surface, particularly when the quantity of material to be removed is large. In this case it may be advatageous to cool the piece being worked with compressed air. Liquid coolants are, in general, not required. An immediate removal of the turnings so that these can remove as much heat as possible from the piece being worked is important.

Compared to most other plastics, polyacetals allow a high rate of material removal; at high cutting speeds higher feed rates and deeper cuts are possible. The low heat evolution in the case of polyacetals can be traced back to its good frictional behavior against steel in dry air. On the other hand, the turnings produced are relatively small and are easily removed with the aid of, for example, compressed air. Much heat is thereby led away so that cooling of the piece being worked is only required in particular cases.

As a matter of basic principle, only parts which are largely but not completely free of internal stresses can be manufactured. Residual stresses can be largely eliminated by posttreating (annealing) the part [2].

Machine tools like those customary for wood and metalworking are used for machining.

In general, cutting tools from high-speed steel (SS or HSS) are sufficient. In special cases, above all for machining of glass fiber reinforced polyacetal, hardened or diamond-tipped tools are advantageous.

Under equivalent conditions, the same tolerances can be achieved in the machining of polyacetal as can be achieved in the machining of metals. In the case of high standards, this means that the machining must be carried at constant room temperature. ISO-tolerance of Series IT-7 according to DIN 7175 at surface roughness of a few μm are possible without great effort.

In principle, all of the machining techniques known from metal processing can be used for the machining of polyacetals. This holds true, for example, for:

- filing, rasping, shaving
- planing, turning, drilling, milling
- threading

Comprehensive information on knives and cutting conditions for the machining of polyacetal can be found in the literature [3].

4.9.2 Joining of Molded Parts and Semi-Finished Goods

4.9.2.1 General

In connection with rationalization and the economical manufacture of plastic components, joining and fastening techniques have taken on an increasing measure of importance. Because of the limitations of assembly and manufacturing technology, injection molded parts must be joined to each other after molding. Parts molded of polyacetal can be efficiently and securely joined. Various processes are suitable for this joining. The most important joining methods are discussed below.

4.9.2.2 Snap Connections

Snap connections are positive-locking connections with multiple design possibilities. Peculiar to all is that a protruding bead, thickening, nub, or hook is formed on one mating partner, which protruberance engages a corresponding depression, gap, or undercut on the corresponding mating partner. During the joining process the parts should, in general, be only elastically deformed. The connection can be either permanent or detachable, depending on the design of the mating partners. Permanent snap connections can accept permanent load, even at elevated temperatures.

In the mechanically non-loaded condition, the snap connection is subjected to at most only small stresses and is not, as a rule, leak-tight. Through the use of additonal sealing devices, for example O-rings, or where necessary by the use of adhesives, a leak-tight joint can be achieved.

Because of its simple assembly and the absence of additional fasteners, the snap connection is one of the most economical possibilities for the joining of plastic components.

Snap connections should fix parts in a particular relative position. In some cases a certain amount of play between the connected parts is required, for example in automotive applications as so-called rattle-free connections. The longitudinal stresses to be transferred are comparatively small. In the overwhelming majority of applications, no permanent load, such as, for example, those caused by internal pressure, needs to be absorbed.

Special connecting elements such as snaps-fits, clamps, and clips function on the snap connection principle. They must be easy to assemble, offer the possibility for blind assembly, require small assembly forces, and be able to reconcile the tolerances of the receiving hole [4].

4.9.2.3 Joining with Metal Screws

Screw connection with metal screws is a frequently used possibility for the non-permanent joining of plastic parts. This offers a mechanically strong joint that can bear a permanent load even at elevated temperatures. Through the use of O-rings or other seals, gas and liquid tight joints can be obtained.

Screw connections should permanently maintain the juxtaposition of the parts. This requires a pre-stress that is applied when the screw is tightened and which must be maintained at a sufficient level for a long period of time. The pre-stress must be greater than the forces that arise through the normal functioning of the part and must also be greater than all accidental loadings that occur, for example, during handling and transport of the part.

Screw connections should be easy and economical to manufacture. Screws which cut the female threads while they are being rotated in (so-called self-tapping screws) meet this requirement particularly well [5].

4.9.2.4 Press Connections

A press-fit is a simple and thrifty possibility for achieving a friction-fit connection between various functional elements [6].

This technology finds application in precision mechanics for shaft-hub connection of gears, carriers, couplings, and in addition air wheels or pump impellers.

Securing of plastic bearing bushings is an additional wide area of application. Customarily, one element of the pair is made of metal.

Press connections are distinguished from other rigid connections through simplicity of design, simplicity of assembly, and non-destructive disassembly.

Low assembly and manufacturing costs associated with press connectors in plastics is countered by a lower load-bearing ability compared to similar connections in metals.

Press connections should transfer external stresses or torques between the deformed mating surfaces without slipping up to the pull-out limit. To achieve this, an initial tension (a so-called joint pressing) is a necessary prerequisite. This is produced through the elastic material properties by assembling the parts to be joined with interference, see also DIN 7190.

The allowable transferable forces are directly dependent on the join compression and the coefficient of static friction between the mating surfaces.

4.9.2.5 Ultrasonic Welding

A complete uniting of parts injection molded or blow molded from polyacetal by ultrasonic welding is an economical joining technology which, because of its short welding cycle, particularly in the case of large piece counts, has shown its worth [7].

Ultrasonic welding is employed to advantage in joining, among others, tanks for cigarette lighters, valve bodies, and small containers, as well as in the finishing of components of electronic equipment, in automotive manufacturing, and in the household appliance industry. Connections capable of bearing high loads and that are also gas- and watertight can be made by ultrasonic welding.

According to the present state of knowledge, parts having a welded joint length of about 120 mm can be welded using standard equipment having generator outputs of up to 4 kW. Special apparatus with combined oscillators makes it possible to weld even greater lengths.

It must be noted that the preferred type and execution of a weld depends not only on the material in quesiton but also on the design of the finished part and the demands that will be made on it.

4.9.2.6 Welding with Hot Elements

Welding with hot elements has proved itself for joining parts injection molded from polyacetal, independent of pigments and additives. This process is preferentially applied when the joint will be mechanically loaded, a large joint must be made, or when the shape of the part precludes use of another joining technology.

The surfaces to be joined are heated on a PTFE-coated heating element and subsequently mated to each other under pressure. The temperature of the heating element lies between 220 and 240 °C. The time of contact with the heating element, typically 5 to 30 seconds, depends on the shape of the part and, of course, on the melt viscosity of the grade of polyacetal being used. When joining the two warmed joint surfaces, it is advisable to work with secure guides or jigs to guide the travel of the parts.

4.9.2.7 Adhesive Bonding

Because of its high solvent resistance, polyacetal is difficult to join with conventional adhesives. Only joining with pressure sensitive adhesives is possible. The surfaces must be activated in order to achieve adhesive joints having higher strength. For example, the use of etching solutions, application of primers, and corona-discharge are suitable.

The following adhesive systems can be used after thorough activation of the surface.

Adhesive Type	Chemical Basis
Contact adhesive	Isocyanate crosslinked polychlorobutadiene
Two-component adhesive	Epoxy resin, polyurethane, nitrile rubber/phenolic resin, methacrylate
Hot-melt adhesive	Vinyl copolymers
One-component polymerizable adhesive	Cyanoacrylate

Joints made with these adhesive systems have a sufficiently high strength for many applications.

A room temperature solvent for polyacetal, hexafluoroacetone sesqui-hydrate, has been known for some time [8]. It is suitable for diffusion adhesion of polyacetal, that is the so-called solvent welding. The strengths obtained with this solvent lie above those attainable with the customary adhesive systems. To this end, a single pass over the joint suffices.

It is, however, to be kept in mind that this adhesive system is poisonous. Evolved vapors attack the respiratory tract and one should therefore only work under extraction hood. Furthermore, a dropwise application is recommended for reasons of cost.

4.9.3 Surface Decoration

4.9.3.1 General Requirements for Part Surfaces

A glossy, defect-free surface is absolutely necessary for the final treatment or upgrading of plastic parts. Unevennesses or scratches, flow marks, or an otherwise perturbed surface will not be eliminated in the converting process but rather remain visible on the decorated surface and detract from the surface appearance. This can be corrected not only through appropriate care in polishing the mold but also by adhering to optimal processing conditions. This is particularly true when setting the surface temperature of the mold, the melt temperature, injection pressure, and injection speed.

Visible weld lines cannot be concealed through surface upgrading procedures.

Since parts with a soiled surface cannot be avoided, a cleansing process should precede the surface upgrading procedure.

Special chemical or mechanical treatments are often necessary. Coatings that are applied without a pretreatment require in any case a heat treatment, whether this is immediate as in the case of foil imprints or afterwards as in the case of adhesive layers or stamped inks.

4.9.3.2 Mechanical Pretreatments

Roughening the molding surface by sandblasting, polishing, or similar treatments activates the surface and promotes the adhesion of the subsequently applied layers. Since such procedures require considerable effort they are infrequently applied.

4.9.3.3 Acid Etching

The same effect as in the case of mechanical pretreatment can be achieved by controlled etching of the surface in an acid bath. Here, uniformly distributed anchor points are created by the action of a weakly acidic solution. After such a pretreatment the parts must be thoroughly rinsed with warm water. The paint coatings or adhesives bind mechanically with these anchoring points so that adhesion between the layer and the molded part is achieved.

4.9.3.4 Adhesive Lacquers

Adhesive lacquers are among the coatings that stick without pretreatment to the surface of parts molded of polyacetal. In contrast to block foils or stamping inks, these adhesive lacquers, also known as primers, serve as aids for surface upgrading in that they promote the adhesion of the layer that is to be added afterwards.

Two types of adhesive lacquers are offered: a matte lacquer, which serves as a basecoat for the common commercial topcoats, as well as a lacquer that dries glossy, which allows a subsequent glossy vacuum metallization. Both types are non-pigmented, one-component systems. They are applied to the clean, grease-free surfaces by spraying or dipping and subsequently burned in for 30 minutes at up to 155 °C. Warpage problems can occur at these temperatures.

4.9.3.5 Printing and Screening

Special printing inks are available for non-tacky, smudge-proof printing of polyacetal parts. These are predominantly applied by screen printing or pad printing techniques. Except for cleansing, a pretreatment of the surface is not required.

According to its type, the printing ink is burned in for between 5 and 50 minutes at 140 to 155 °C.

Certain printing inks are suitable for application via a partial etching process. After a drying time of one hour, the excess ink is washed off with spirit and subsequently burned in 3 to 5 minutes at 140 to 150 °C. Here, one must also contend with warpage problems, particularly at longer burn-in times.

4.9.3.6 Thermal Diffusion Printing

Another way to decorate the surfaces of articles molded from polyacetal is dyeing with water emulsifiable colors under the action of heat.

Only natural-colored and light, slightly muted pigmented surfaces can be decorated by this process. The water emulsifiable pigments are preferentially brought into a screen printing formulation and thereafter applied to the clean, untreated surface. Thermally fixable colors are also offered in a corresponding formulation for offset printing. Furthermore, under certain conditions, they can be applied by spraying, brushing, or dipping.

Fixation of the coloration takes place through a heat treatment of from 3 to 5 minutes. Here, the surface of the molded parts attains a temperature of about 140 °C. Excess die is rinsed off with water after the part has cooled in air. During this heat treatment, the die penetrates into the surface by diffusion.

Parts decorated by this process should only be used at room temperature. The higher the use temperature, the faster the die diffuses further into the plastic.

4.9.3.7 Painting

After priming with a bonding primer, molded parts of polyacetals can be well painted. Conventional systems can be used as finish coats. They are above all selected according to the demands on the finish itself, for example resistance to weathering or chemicals or scratch resistance. The bonding strength between the bonding partners polyacetal/primer/finish coat is thus good. Acid pre-treatment results in good bonding with certain paints. In contrast, mechanical pre-treatment does not give satisfactory results. Impact modified polyacetals can also be treated with primers.

4.9.3.8 Metallization

4.9.3.8.1 Metallization in Vacuum

Parts molded of polyacetal whose surfaces have been pre-treated with acid can be vacuum metallized. Selection of the primer and topcoat is based on the experience of those practicing the procedure as well as the properties required. In consideration of the good dimensional heat stability of the material, topcoats that demand a relatively high curing temperature and which have a very good wear resistance can be employed in metallization just as they can in painting.

4.9.3.8.2 Galvanization

After the customary electrochemical treatment steps, parts of polyacetal can be coated with a conductive layer and subsequently glavanized. The surface can be roughened with chemical etching. Molding materials modified with a special chalk, for example, have worked out well for galvanization. A very strong anchoring of the metal layer on the plastic cannot, however, be achieved. The galvanized layer should therefore be thick enough to be self-supporting.

4.9.3.9 Hot (Foil) Stamping

Hot stamping of polyacetal parts is a frequently employed printing process. For hot stamping the surface receives no pre-treatment but, just as with other plastics, it must be clean. The procedure is widely used and the number of suitable foils is correspondingly large. The selection of foils accommodates the individual stamping procdures: positive stamping, negative stamping, flat stamping, relief stamping, and stamping with brass or silicone rubber stamps. The selection of foils also accommodates the requirements that may be placed on the imprint such as scratch and shear strength, resistance to chemicals and weathering, and, of course, the desired color tone including the type of surface (matte, shiny). The required temperature for the stamp depends on the type of foil and lies between 100 °C and 200 °C.

Pressure, temperature, and dwell time of the stamp have the greatest influence on the results of stamping. Depending on the temperature of the stamp, dwell times between 0.2 and 2 seconds can be chosen.

4.10 Applications

P. Unger

4.10.1 General

The demands that are often placed on parts molded from polyacetals can draw upon, among others, the folloiwng properties:

- ability to bear a mechanical load
- dimensional stability
- friction and wear properties
- thermal and electrical properties
- chemical resistance

The time and temperature of use must always be taken into consideration. In all cases, the particular qualities of polyacetal should be kept in mind to ensure that the part is suitable for use under a given set of climatic conditions.

In many cases the object is to replace a common material, thermoset plastic or even metal. Numerous other applications became possible for the first time through the use of polyacetal. Further requirements are:

- little or no secondary operations
- ability to integrate many functions into one part
- simple assembly
- appealing surface quality
- decoration capabilities
- colorability

In many circumstances the suitability for use of a part can be estimated with the help of materials strength theory on the basis of information on the material from which the part is made. However, the suitability for use should always be verified by simulated end-use tests. Finally, the irreversible process of ageing, brought about by external influences and internal changes, must be taken into consideration.

For example, as a consequence of long-time loading under the technically understood climatic conditions, the strength and deformation properties can so change that damage to the component occurs.

Since the life expectancy of a technical component is generally set very high, the demands of the application should be comprehensively known. For verification, accelerated tests can then be carried out in order to be able to judge the suitability for use. Acceleration is achieved for example by determining the creep behavior at temperatures higher than those that occur in practical use. With the help of suitable extrapolation procedures, the behavior of the component under conditions encountered in practice can be forecast to a good approximation.

4.10.2 Uses in the Automotive Industry

Automobile manufacture is one of the most important industrial sectors. In 1987, the automotive industry accounted for 32% of the polyacetal consumption in western Europe [9]. As required, the entire spectrum of polyacetal products such as:

- basic grades
- easy-flowing grades
- glass fiber and glass spheres reinforced grades,
- grades modified for sliding, mating surfaces
- impact modified grades

are used in specific applications.

The so-called thin-wall technology that predominantly involves parts with large flowlength-to-thickness ratios, driven by the automotive industry's need to save weight, was the starting point for the development of easy-flowing polyacetal grades having melt indices up to 50 g/10 min. These molding materials allow filling of the mold cavity with very low orientation; see also Section 4.5. With these grades, parts with very low warpage can be molded even in the case of unfavorable shapes.

The easy-flow grades of polyacetal also allow higher productivity. In many cases it is possible to work with a lower melt temperature or lower mold temperature. Associated therewith is a reduction in the cooling time and thus in the cycle time. Because of the low melt viscosity, the injection pressure can also be reduced. This is advantageous in injection molds with long, thin cores since worries over shifting of the core can be avoided; see also Section 4.5.

The disadvantage of several polyamides whose toughness behavior depends upon ambient conditioning was an influence in the development of impact modified grades of polyacetal. In this case we are dealing with multiphase polyacetal-elastomer systems whose mechanical properties and dimensional behavior, as with other polyacetal types hardly change, owing to the practically negligible moisture absorption. The type of elastomer modification allows substantial latitude, so that the property profile can be adapted to the particular function of the part in a way that is specific to the application. See also Section 4.6.2.

The favorable behavior of polyacetal in the presence of methanol-containing fuels is a further important consideration [10]. See also Section 4.7.6. These and other properties result in an exceptionally large number of applications.

Figure 4.56 shows a gasoline pump whose upper part is made of polyacetal. Compared to the metal version, a more economical manufacture was possible.

Fuel reservoirs, Figure 4.57, are made of impact modified polyacetal. The container is installed outside the fuel tank and can be exposed to impact by stones. Besides high strength, the material must have good reserve elasticity and recovery.

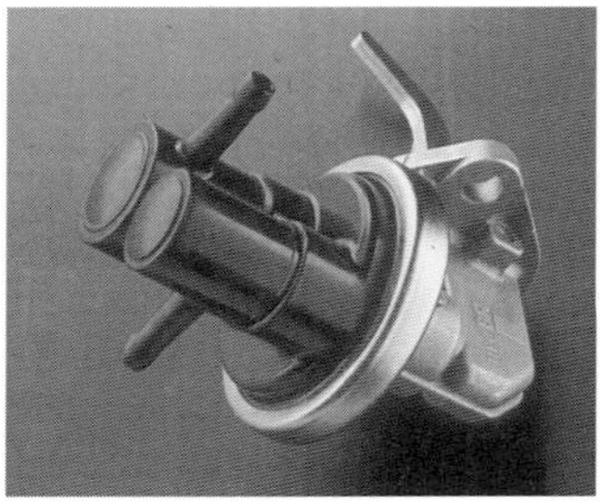

Figure 4.56. Upper part of a gasoline pump

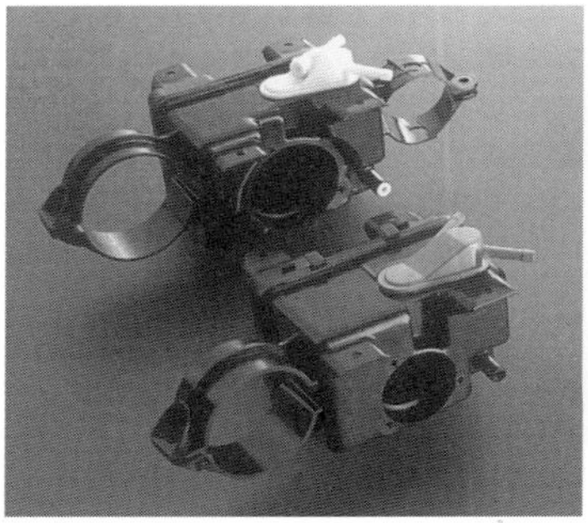

Figure 4.57. Fuel reservoir

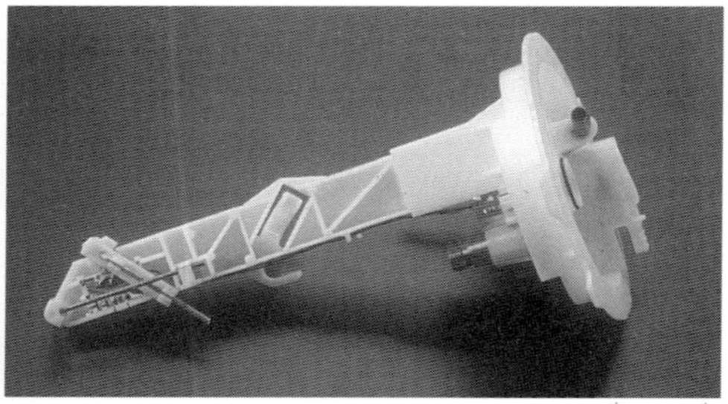

Figure 4.58. Gasoline level indicator

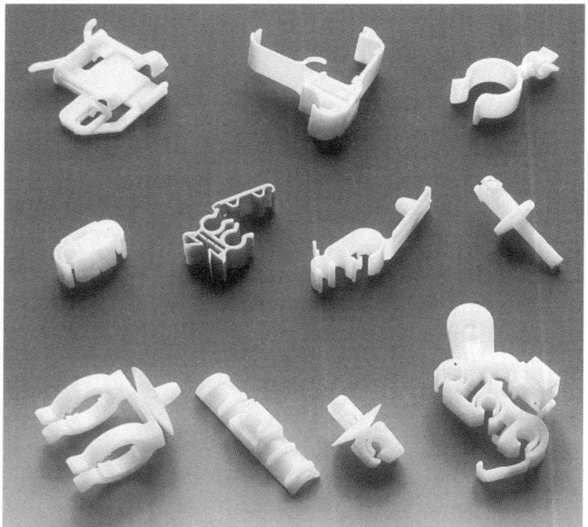

Figure 4.59. Guide clips

Gasoline level indicators, Figure 4.58, are mounted in the fuel tank and for years were predominantly manufactured of metal. Thanks to the integration of many functional parts, senders are today made almost exclusively from polyacetal.

Clips accomplish a variety of tasks, although a number of requirements remain the same:

- ability to bear loads, under some conditions also at high rates of deformation and low temperature
- high resiliency, that is good spring properties
- high holding power

Figure 4.59 shows clips with different spring and snap elements, for example for the secure and play-free retaining of brake guides.

Carburetor floats are hollow bodies, having wall thicknesses less than 1 mm, with which the introduction of fuel is regulated in the float chamber, Figure 4.60. The floats are molded in two parts and subsequently welded. Carburetor floats made from polyacetal have been able to extensively displace cost-intensive metallic floats.

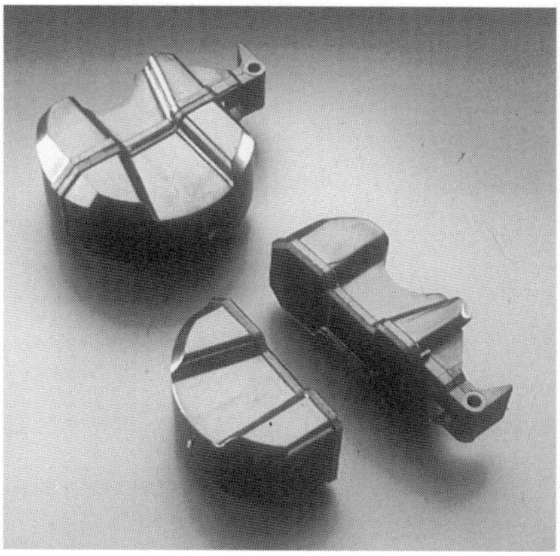

Figure 4.60. Carburetor floats

Figure 4.61. Brake power accumulator housing

In housings for power brake accumulators, thermoplastics are increasingly displacing the thermoset materials which dominated this application for years. Figure 4.61 shows a housing of polyacetal that was designed for a use temperature from −40 °C to +100 °C. Chain-gears for the drive train of two-wheelers, for example mopeds racing motorcycles, are manufactured in polyacetals in order to reduce noise and especially to increase service life, Figure 4.62.

Because of installation considerations, caps for windshield wiper mounts must be extremely flexible. During installation the cap is snapped on and momentarily severely deformed. Thus, for this application a highly impact modified polyacetal is used, Figure 4.63. High UV-resistance and also the ability to form a matte, non-reflective surface, independent of the surface finish of the mold, are special characteristics of this formulation.

The coolant drain screw, Figure 4.64, testifies to the high hydrolysis and chemical resistance of the copolymers. In the case of high performance motors, cooling water temperatures of up to 125 °C can be encountered; see Section 4.7.7.

Further examples of applications are:

- cranks for windows and sunroofs, crank guides
- gasoline immersion pumps
- bubble separators
- luggage carriers and fasteners
- air turbines
- crossheads for gear-type auto jacks

- torsion bar linkages
- fuel expansion tanks
- tank closures
- headrest grommets
- lock housings
- functional parts for automatic safety belts
- water separators

Figure 4.62. Moped chain gear

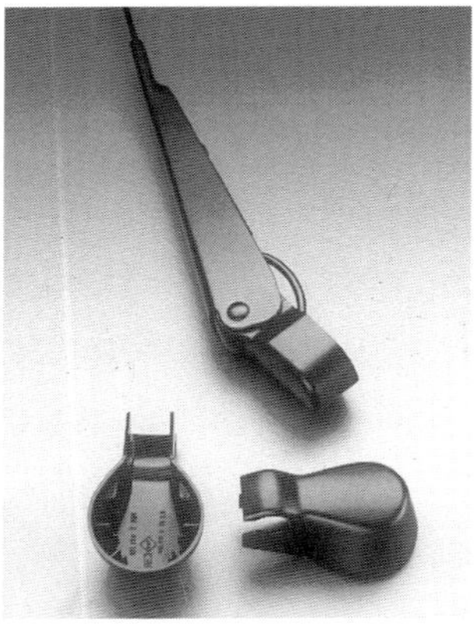

Figure 4.63. Windshield wiper mount cap

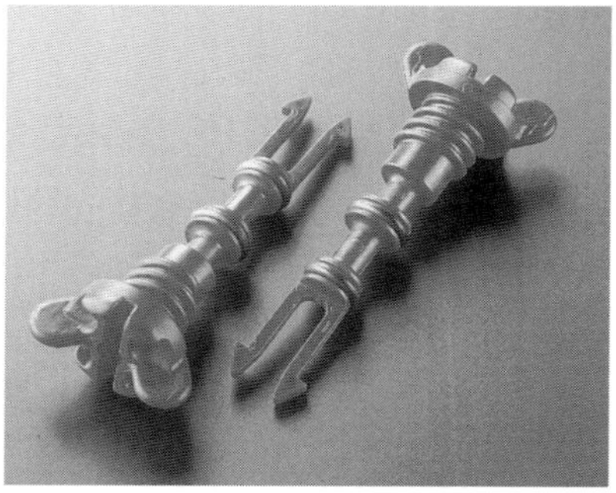

Figure 4.64. Coolant drain screw

4.10.3 Electrical and Electronic Industry

Next to the automotive industry, the electrical industry is an important polyacetal user. The electrical, electronic, and electrical device industries all belong to this family. In western Europe, about 17% of the demand for polyacetal comes from the electrical industry [9].

The so-called outsert-technology plays an important role in this market sector. Besides its electrical insulation properties, the following properties of polyacetal are particularly valued (see also Section 4.7.4):

- high load carrying capacity
- high hardness and stiffness
- good lubricity
- high dimensional and shape stability

The applications in this sector are also correspondingly large.

Polyacetal has largely displaced metals for transferring torque and motion and thus is used in:

- gears such as spur, conical, and screw (with or without lubrication)
- bearing bushings
- cams
- followers

Drives in electrically driven kitchen appliances (Figure 4.65) and pump drives for oral hygene devices (Figure 4.66) that are completely made of plastic are today state of the art.

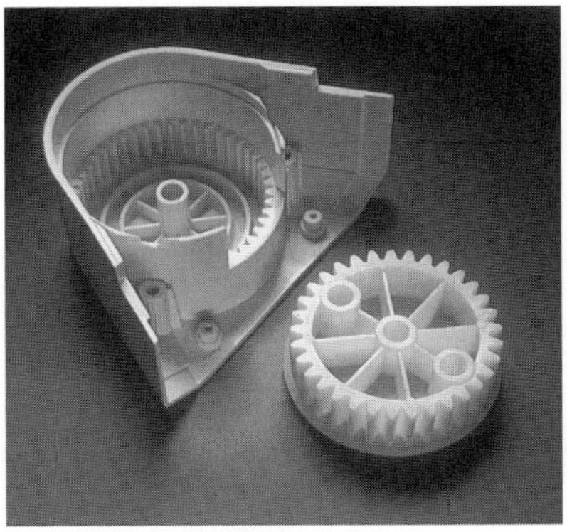

Figure 4.65. Drive for a kitchen appliance

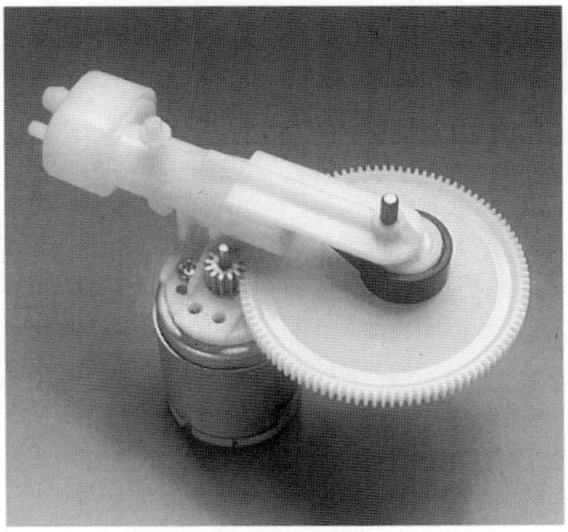

Figure 4.66. Pump drive for oral hygien device

Figure 4.67. Seconds pinion of a clock movement

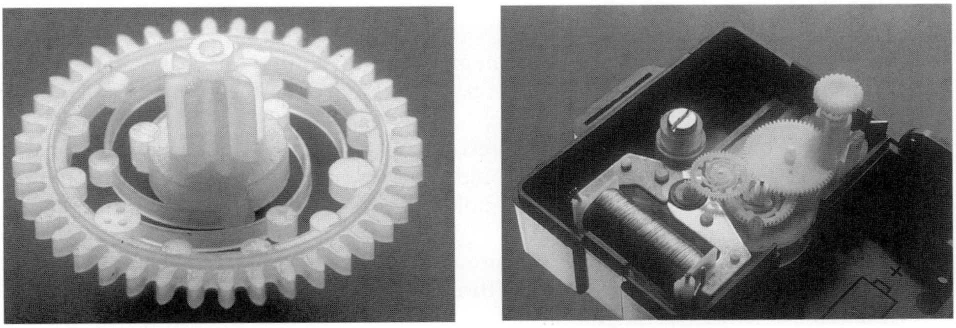

Figure 4.68. Gear with spring elements: left, individual part; right, assembly

A micro-gear having a diameter of 1.32 mm, a bore of 0.27 mm, and weighing 0.0006 g is used for the pinion of the seconds gear in a clock movement, Figure 4.67.

The gear shown in Figure 4.68, in which the hub and crown are joined to each other through spring elements, serves for the smooth, almost noise-free transfer of rotation in a quartz clock. By fully exploiting the elasticity, the gear-play in the drive can be practically completely compensated.

In the outsert technique, plates, customarily metal, are fitted with one or more functional elements in one step. The elements are immediately ready for service. The outsert technology process allows the exacting manufacture of complicated, precision technical subassemblies. Elements produced in this fashion include:

- bearings
- spring hooks
- sliding bearings
- pegs
- axles
- housings

A tachograph, which measures motor speed and RPM on a display chart, is built on a plate on which more than 40 functional elements are molded in one step using an easy-flow polyacetal.

The chassis of a VCR-drive (Figure 4.70) consists of a metal plate on which 55 different functional elements such as retainers for winding plates and threading motors, cable guides, and bearing bushings are injection molded. The binding between the metal plat and the molded parts is accomplished by consciously exploiting the shrinkage of the polyacetal. However, rotatable parts can be realized with the outsert technique.

Figure 4.69. Outsert plate of a tachograph

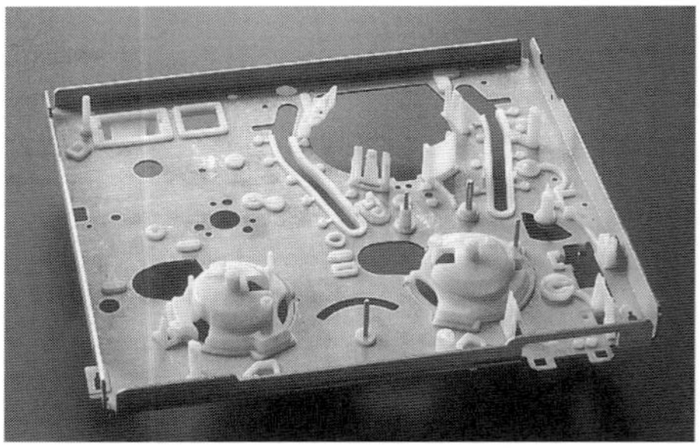

Figure 4.70. Chassis of a video drive

The planetary gearbox of the drive component of a food processor for mixing, kneading, and slicing is composed of 4 ball-gears and a bias-toothed pinion gear, Figure 4.71.

A particular requirement here was resistance to caustic automatic dishwashing detergent since the entire subassembly should be placed in the dishwasher for cleaning. This requirement was fulfilled by acetal copolymer without interfering with the suitability for use.

In a video recorder, a ring gear (Figure 4.72) made from glass spheres reinforced polyacetal handles the tape transport and threading functions for precise running of the tape. Curved elements, ridges, guides, stops (dogs), and cut-outs are integrated into the molding.

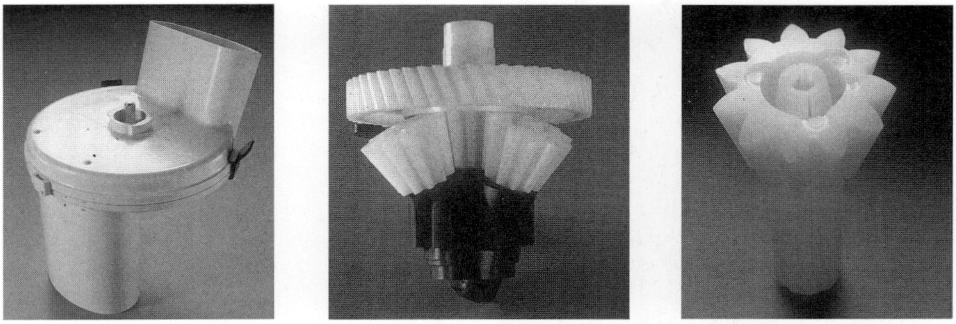

Figure 4.71. Planetary gearbox of a food processor: left, assembly; center, drive; right, individual part

The complicated geometry of this part, required by its functions, places great demands on mold technology. Molding materials having a very low shrinkage and the lowest possible shrinkage anisotropy must be resorted to in order to hold to the narrow tolerances. Thus, a polyacetal reinforced with 30% glass spheres was used.

Screw drives are characterized by quiet running, high transfer in one step, and comparatively small and simple construction. These qualities have made the screw drive the preferred type of drive in small electrical devices such as multi-slicers, grain mills, electric knives, and devices for cleaning carpet or upholstery.

Screw gears are economically made from polyacetals by injection molding. At the present time, maximum transferable loads are up to 400 W. Polyacetal grades modified with chalk or PTFE have proven themselves in power transmission. Figure 4.73 shows the screw drive of a multi-slicer having a drive output of 140 W. The screw gear with integral synchronous belt wheel is injection molded from polyacetal having a melt index (MFI) of 9 g/10 min. The gear is so toothed that the thickness of the teeth on the screw is favorably limited to the thickness of the teeth of the driven gear.

In ground-fault switches, Figure 4.74, the housing, terminal holder, coil bobbin, and reset button are manufactured from polyacetal. This safety-system application requires a part in absolute working order.

The keyboard of a personnel computer is equipped with key guides that are molded in a mold with 58 cavities (Figure 4.75).

Because of the very long flow length, an extremely easy-flowing polyacetal having a melt index (MFI) of 30 g/10 min is used. The complete injection molded piece consisting

Figure 4.72. Ring gear in a video recorder

Figure 4.73. Screw drive of a multi-slicer

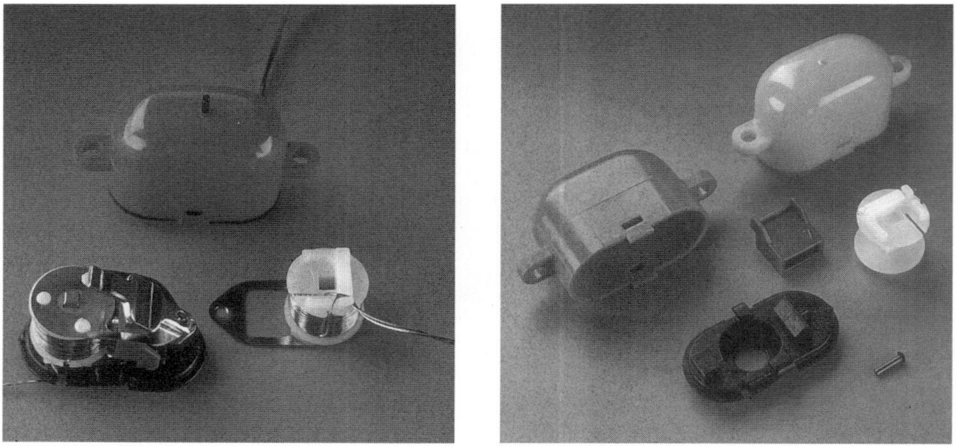

Figure 4.74. Ground-fault switches

Figure 4.75. Keyboard of a personal computer with key guides (left); key guides and runners removed (right)

of the key guides and runner system is inserted into the keyboard panel and all keys are added in one manufacturing step. After removal of the runner system, the part is ready to function.

The pot, top, handle, buttons, and closure piece for a water kettle are manufactured from an easy-flowing acetal copolymer. The parts with tongue and groove fittings require a construction having tight tolerances, that is a very low secondary shrinkage (Figure 4.76). The molded parts are distinguished by resistance to hot water and hydrolysis, a practically negligible tendency to build mineral deposits, and harmlessness according to food regulations. The copolymer used for this application conforms to the recommendations of the German Federal Health Office and the US Food and Drug Administration (see also Section 4.7.10).

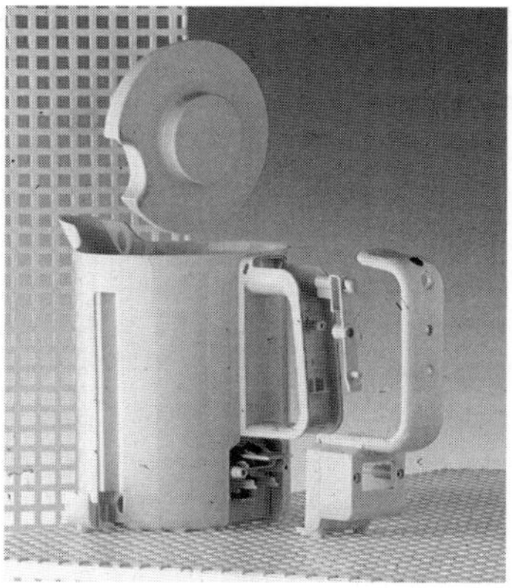

Figure 4.76. Water kettle

4.10.4 Manufacture of Machinery

Manufacture of machines and equipment accounted for ca. 8% of the polyacetal consumption in western Europe in 1987 [9]. In the manufacture of machinery, parts are made as prototypes or individual parts in long and short production runs.

If, in the case of low piece counts, a decision is made to build an injection mold, then molds made of aluminum or cast resin might be used for the purpose of cost reduction.

A piston is used as one component of a shut-off valve in a drink bottling facility. The piston is toothed to fit on a threaded steel shaft. On the basis of good lubricity against steel, dimensional stability, and resistance to chemicals and alkali, the cylinder is machined from acetal copolymer (Figure 4.77).

A ball that executes lifting and turning motions in a transmission is machined out of polyacetal stock (Figure 4.78). An exactly spherical ball having maximum wall thickness as in this example cannot be made by injection molding.

Figure 4.77. Piston for a shut-off valve

Figure 4.78. Ball with external diameter of 100 mm

Oscillating spray nozzles are used for cleaning sieves and filters in the paper and water purification industries and in other facilities as well. As a result of legal regulations and environmental and cost consciousness, facilities are today built that achieve an optimum cleaning effect with minimum water consumption. The oscillating spray nozzles are supplied in two parts (Figure 4.79). The internally cylindrical but externally spherical bearing shells, which are supplied in two parts for assembly, insure that motions follow their course with little friction.

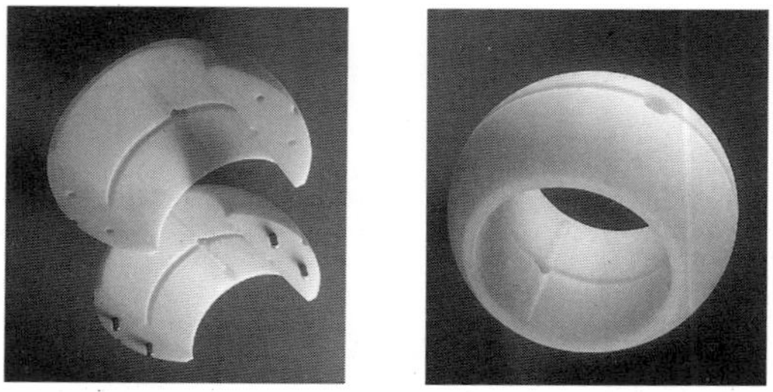

Figure 4.79. Bearing shell for an oscillating spray nozzle: left, in two parts; right, assembled

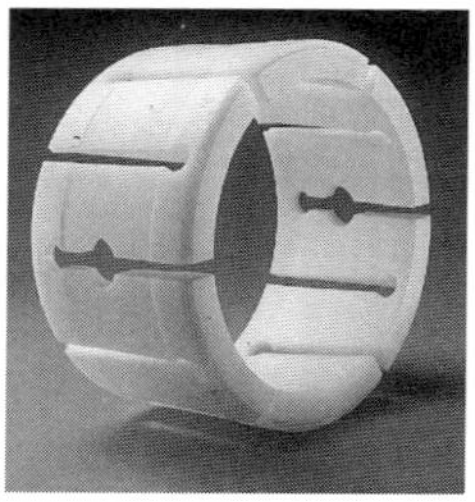

Figure 4.80. Spring collet in winding machine

Slots and holes are provided in the shells for special lubrication since the bearings continuously run dirty and wet, which can lead to scaling and wear.

Spring collets (Figure 4.80) are designed for use in a winding machine for polyester threads. The spools are tensioned in the center of the collet. This obtains a uniform, straight winding. Line speeds of up to 3000 m/min can occur. The winding process runs smoothly and is easy on the equipment. Metallic versions of these collets have not worked well in practice. The spring collets are machined out of sections of tubing (semi-finished stock).

Figure 4.81 shows a partial crown gear having internal and external teeth that is used in building equipment. The gear is machined from a section of large tube which has an outer diameter of 600 mm.

Dimensional sensors for transmitting measurements to a central controller are installed on rotational printing machines. Actuation is accomplished with a screw gear drive that consists of a single-flighted screw and a gear with 14 teeth (Figure 4.82). The drive, whose individual parts are machined, runs without lubrication.

Figure 4.81. Partial crown gear for building equipment

Figure 4.82. Screw gear drive for sensors in printing machines

Pneumatically actuated valves made from polyacetal represent the latest technology in production control technology. The extremely complicated valve bodies (Figure 4.83) require intricate injection molds since rods and cores must be pulled on different levels. A chalk-modified polyacetal has worked well in this application. The maximum working pressure is 12 bar.

Ball bearing cages are used for guiding rolling bodies. The requirements for wear resistance in the area where the balls rotate are met by polyacetal (Figure 4.84). The cages are injection molded in various sizes and are used directly without further working.

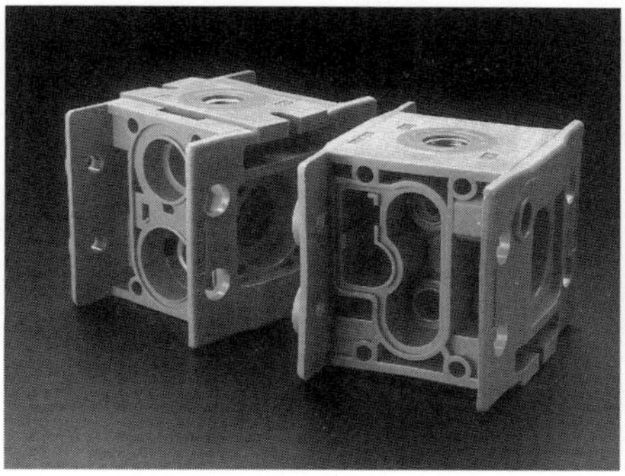

Figure 4.83. Valve bodies

Figure 4.84. Ball bearing cages

4.10.5 Sanitary and Water Installations

The requirements for the plumbing sector are:

- stress crack resistance
- hydrolysis resistance
- very low water absorption
- good dimensional stability
- good resistance to chemicals
- good behavior at long loading times
- low tendency to creep
- low tendency to build up mineral deposits

Parts made from polyacetal have proven themselves in many years of service under special end-use conditions such as high water temperature or when disinfectants are used in the water. However, in the case of functional elements that are subjected to continuous internal stress it must be noted that plastics, in general, tend to creep and this fact must be correspondingly taken into consideration in design and use in individual applications. Values for strength determined in tests of time-to-rupture under stress can give a measure of security for this.

Direct contact with copper and brass in the presence of hot water should be avoided in order to avoid a catalytic degradation of the polyacetal. Chlorinated water can restrict the suitability for use of parts molded from polyacetal.

Only a few examples out of the great number of applications should be mentioned.

The hand-held shower shown in Figure 4.85, which has large, designed-in differences in wall thicknesses from 2 to 9 mm, is manufactured from an extremely easy-flowing polyacetal having MFI = 30 g/10 min. Such a material is attractive when a brilliant surface that shows up dirt is necessary. Large differences in wall thickness without excessively large molded-in stresses can be manufactured without problems through injection molding technology when using these easy-flowing material formulations.

The flushing action in a toilet flush can be initiated with a button. It can also, according to the amount of water needed, be interrupted by a similar motion. The regulating flap with integrated float chamber and diverse spring elements shown in Figure 4.86 plays an important role in one such device.

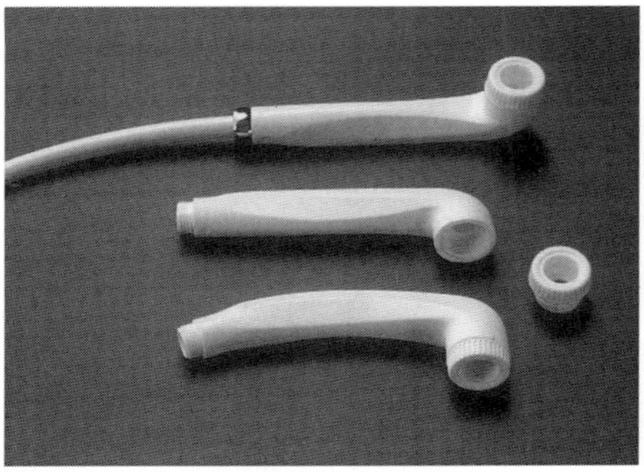

Figure 4.85. Hand-held shower

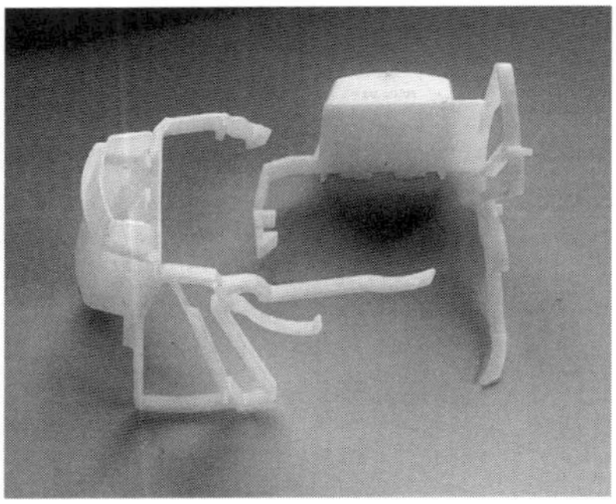

Figure 4.86. Regulating flap for a toilet flush

Figure 4.87. Fittings for water supply networks

Figure 4.88. Sprinker for irrigation

Figure 4.89. Water meter with functional parts of polyacetal: left, overview; right, individual parts

Fittings made from high molecular weight polyacetal (Figure 4.87) are made for water supply via potable water networks or for sprinkling installations. Parts are tested for 60 min at 30 bar at a water temperature of 20 °C. Guaranteed harmlessness according to food regulations is a requirement for contact wit potable water. The fittings are colored black to protect them against UV radiation.

Sprinkling devices are used in great numbers in argiculture. Figure 4.88 shows a sprinkler that rotates stepwise. This part is colored black for UV protection. The loading in long-term testing was 4 bar.

All of the parts shown in Figure 4.89 for a domestic water meter are made out of polyacetal. The seal plate is made of glass fiber reinforced polyacetal and is designed for a pressure of 50 bar for short times. The gears transfer the revolutions of the impeller wheel to the counting mechanism.

Fill valves for flushing tanks in toilet flushes (Figure 4.90) are continually under line pressure. Extensive short-time bursting tests and tests of time-to-fail under stress support the good end-use suitability of these parts made of polyacetal.

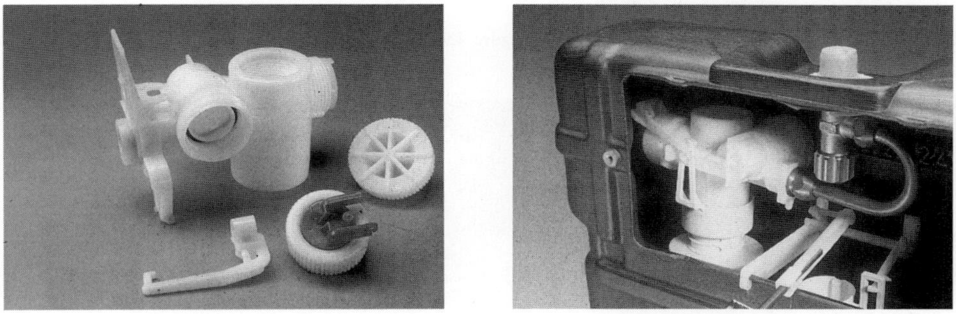

Figure 4.90. Fill valve for a toilet flush: left, individual parts; right, overview

4.10.6 General Applications

Applications for polyacetals are distinguished by their great multiplicity. The high level of properties are economically utilized in almost all imaginable domains, as illustrated by the following examples.

Medical Technology

Figure 4.91 shows a lancet, typical in medicine and cosmetology, that is made out of an extremely easy-flowing polyacetal having a melt index (MFI) of 30 g/10 min. A snap-off cap for protection of the point is molded on.

The pyramid-shaped point places extreme demands on the design and construction of the injection mold. The microstructure must be uniformly fine-crystalline to the outermost regions of the point and, conversely, the quasi-amorphous edge zone must be reduced to a minimum. This can only be achieved with a nucleated, extremely easy-flowing polyacetal; see Section 4.42.

Sports Applications

Ski bindings made from high molecular weight acetal homopolymer have almost completely displaced those made of metal by the zinc diecasting technique (Figure 4.92).

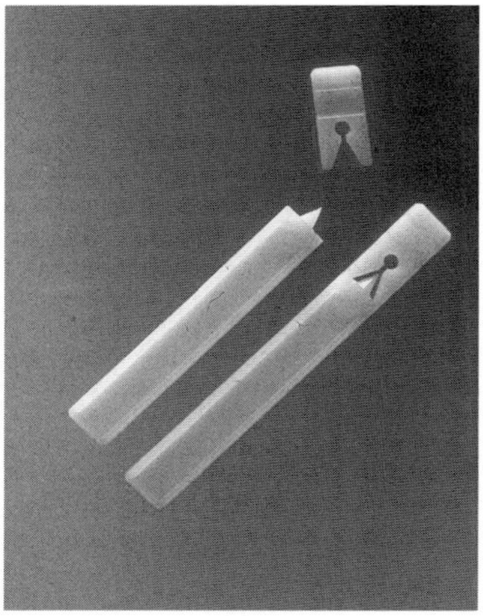

Figure 4.91. Lancet

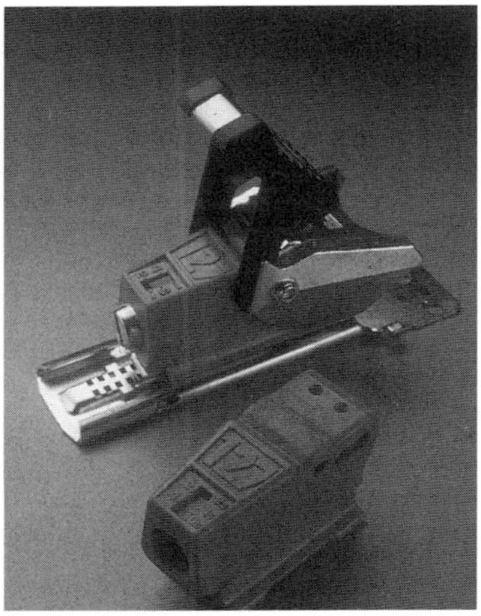

Figure 4.92. Ski binding

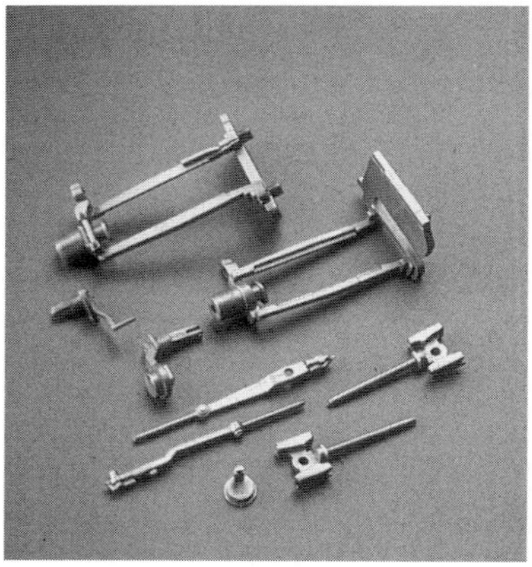

Figure 4.93. Parts for a model train locomotive: left individual parts; right, overview

Toys

The filigree molded parts for the tender locomotive of a model train are injection molded out of a polyacetal formulated with metallic color, Figure 4.93.

Flow-Through Measuring Devices

Polyacetal has long since proved its capabilities for functional parts of gas meters, the so-called Balgen–gas meter movements (Figure 4.94). Already in 1969, investigations at the Technische Universität in Karlsruhe confirmed the resistance of acetal copolymer to fuel gases. For example, the permeability is extremely small in the case of gases and aliphatic hydrocarbons. The requirements for gas meters in DIN 3374/1981 are met by this material.

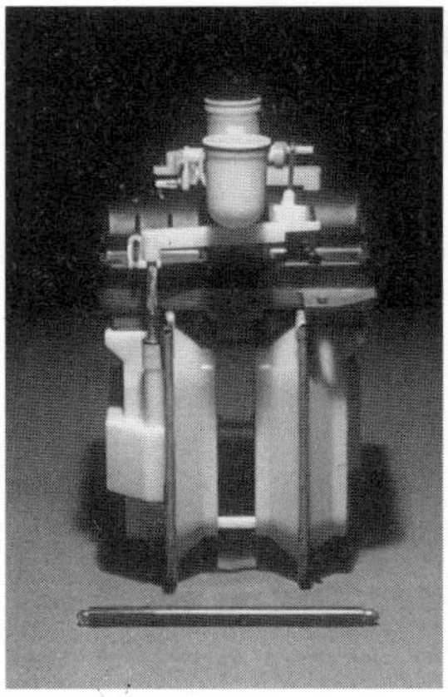

Figure 4.94. Gas meter movement

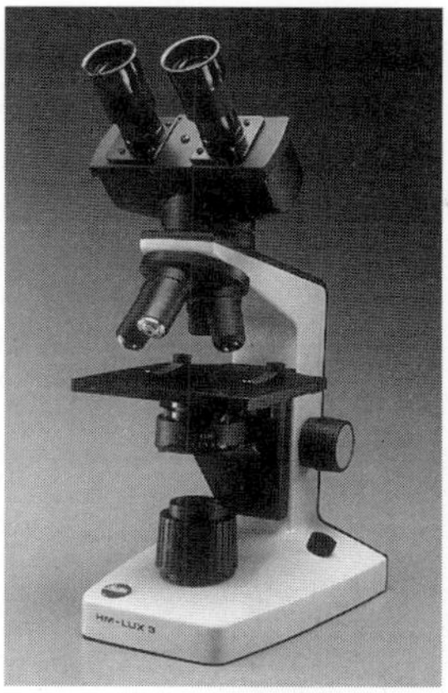

Figure 4.95. Teaching microscope

Optics

A total of 23 components made out of black-colored polyacetal are used in a teaching microscope that has high optical and mechanical capabilities. Thus, for example, the condenser mount and ring illumination aperture are made from this material with the required high precision. The low thermal conductivity, good corrosion resistance, and electrical insulating properties of polyacetal are simultaneously exploited (Figure 4.95).

Figure 4.96. Cutlery handles

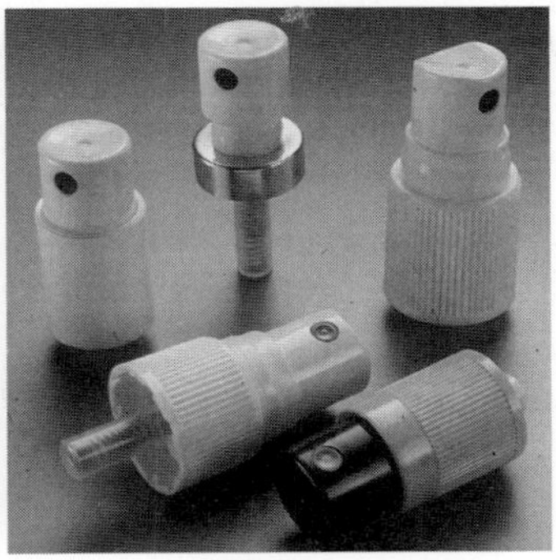

Figure 4.97. Atomizer pumps

Cutlery

On grounds of its dishwasher resistance, insensitivity to scratches, and its good colorability, polyacetal is used for handles of cutlery. The handles are bound to the utensile to make a contignuous form (Figure 4.96).

Cosmetics

Polyacetal has been used for years in the cosmetics sector for the manufacture of atomizer pumps (Figure 4.97). Besides dimensional stability, other properties of this material, like stiffness and resistance to the materials being dispensed such as perfumes, are significant. The atomizers, which work on the principle of a piston pump, produce pressures up to 8 bar.

Writing Implements

By fulfilling the corresponding demands on spring-like properties, hardness, and sliding properties, polyacetal is also suitable for writing implements such as ballpoint pens and fountain pens. For design reasons, the two-color injection is also used (Figure 4.98).

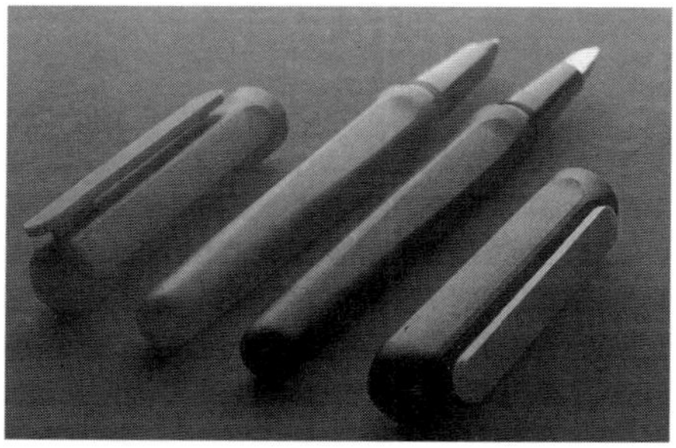

Figure 4.98. Fountain pen and marker

References for Section 4.8 to 4.10

1. Company brochure Drägerwerk AG, Lübeck 1985
2. *Klein, N.:* Plastverarbeiter 3 (1971) p. 163
3. Technische Kunststoffe, C.3.1. Company brochure Hoechst AG, Frankfurt 1984
4. Technische Kunststoffe, B.3.1. Company brochure Hoechst AG, Frankfurt 1986
5. Technische Kunststoffe, B.3.2. Company brochure Hoechst AG, Frankfurt 1988
6. Technische Kunststoffe, B.3.4. Company brochure Hoechst AG, Frankfurt 1988
7. Technische Kunststoffe, B.3.7. Company brochure Hoechst AG, Frankfurt 1987
8. Company brochure Dr. Schuchardt & Co., Hohenbrunn 1973
9. *Wolters, E.; Sabel, H.-D.:* Kunststoffe 77 (1987) p. 997
10. *Orth, R.:* Kunststoffe 73 (1983) p. 2

4.11 Commercial Products

H. Schlaf

In comparison to other technical polymeric materials such as polyester or polyamide, polyacetals are manufactured and sold only by a small group of companies which is easily surveyed; see Table 4.11.

In the application sector the basic differentiation of homo-, co-, and terpolymers is according to their individual property profiles. The terpolymers offered by several manufacturers are especially suited for processing into hollow shapes by extrusion blow molding.

Both homo- and copolymers are equally suitable for processing by injection molding and extrusion. A comprehensive selection of products covering a wide range of melt

Table 4.11. Producers and Trade-Names of Polyacetals

Producer	Trade Name	Polymer	Components
Europe			
Ticona Polymerwerke GmbH	Hostaform	Copolymer	Trioxane/Ethylenoxide
(Hoechst AG/Hoechst Celanese Corp.)		Terpolymer	/Diepoxide
Ultraform GmbH	Ultraform	Copolymer	Trioxane/Butanediolformal
(BASF AG/Degussa AG)			
Tarnow, Poland	Tarnoform	Copolymer	Trioxane/Dioxolane
USA			
Du Pont de Nemours and Co.	Delrin	Homopolymer	Formaldehyde
Hoechst Celanese Corp.	Celcon	Copolymer	Trioxane/Dioxolane/Diepoxide
		Terpolymer	
	Kematal*	Copolymer	
Japan			
Asahi Chemical Industry	Tenac	Homopolymer	Formaldehyde
		Copolymer	Trioxane/Ethylenoxide
Mitsubishi Gas Chemical Co.	Jupital	Copolymer	Trioxane/Ethylenoxide
Polyplastics Co.	Duracon	Copolymer	Trioxane/Ethylenoxide
(Daicel/Hoechst Celanese Corp.)		Terpolymer	/Diepoxide

* Tradename for Celcon in Europe

viscosities allows the manufacture of thick-walled and stable as well as complicated and finely-structured parts by a number of manufacturing techniques.

To a certain degree, the manufacturers kept their product assortments in line with one another, even though the product offerings of the individual producers are variably broad. For example the product offering of Ticona Polymerwerke GmbH comprises 29 basic grades. If one considers colored products, the number of product variants is in the hundreds.

4.12 Market Development

H. Schlaf

Polyoxymethylenes have been on the market since the end of the 1950s. On the basis of their property profile they have penetrated many application sectors as "engineering construction materials". The automotive, home entertainment, machine construction, and general electrical and electronics sectors can be mentioned. The spectrum of applications is reported on in detail in Section 4.10. Understandably, the consumption is concentrated in the highly-industrialized regions of western Europe, the USA, and Japan.

In 1986, the world-wide nameplate capacity for homopolymer and copolymer combined amounted to about 340,000 tons per year [1]. Before 1986, the capacity inceased by 30% within three years, thanks to debottlenecking and new plants. In contrast, a worldwide demand of a maximum of 260,000 t/yr existed in the 1985/86 period. Although, since 1975, the annual increase in demand has been greater than the annual increase in capacity, it can therefore be assumed that by the jump in capacity (30% in three years) those producers who are active in the market have guaranteed a problem-free satisfaction of the requirements for the coming years as well.

Of the three regions mentioned above, the largest consumer in 1985 was western Europe (80,000 t/yr), followed by Japan (75,000 t/yr) and the United States (53,000 t/yr).

In recent years, interesting product developments have opened up further fields of application in the market. Easy-flowing grades, some formulated to be anti-static, were developed for the much noted cassette market and the electronics industry. Here, large production volumes in multiple-cavity molds can be realized with exceptional manufacturing efficiency. Similarly, filled grades were developed for nearly warp-free parts.

The development of grades with low mold deposit has been successfully promoted in recent years. Since 1982, elastomer modified polyoxymethylenes have been offered on the market. The toughness properties, especially the low temperature impact strength, have been improved several times over those of the standard grades. Although manufacturing and processing require a high degree of technical knowledge, the polyacetals can be described as a multifarious class of contrustion materials with a high dynamic of development.

Therefore, middle-term market projections assign to the polyacetals a growth rate of 5 to 6% for the next five years. However, exclusively regional differences in growth may occur.

Reference for Section 4.12

1. Der Kunststoffmarkt in Europa-POM. Kunststoffe 76 (1986) p. 856–857

5 Organic Cellulose Esters Thermoplastic Molding Compounds

F. Müller, Ch. Leuschke

5.1 Introduction

Organic cellulose esters, as modified natural products, have gained special technical importance because of their wide sprectrum of properties. The carboxylic acid esters of cellulose, prepared from C_2–C_4 fatty acids, are mainly used (Figure 5.1). Cellulose acetates (CA) are used as triesters (primary acetates) for photographic films, electric isolation film, and fibers, while thermoplastic cellulose acetate molding compounds can be prepared from partially saponified cellulose acetates (secondary acetates) with the addition of plasticizer.

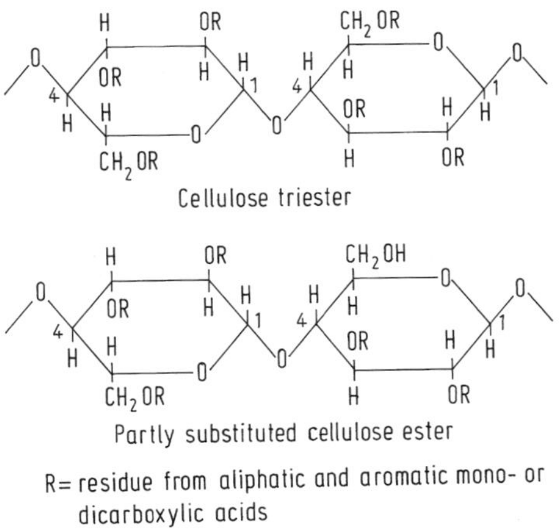

Figure 5.1. Conformation structure for cellulose ester (Haworth)

Whether as thermoplastic molding compounds with plasticizers or as polymer blends, mixed cellulose esters made from acetic and propionic or butyric acids improve such properties as saponification, weatherability, plasticizers acceptance and thermostability, and reduce water absorption. Because of the improved compatibility with organic solvents, mixed cellulose esters are also used as lacquer additives, preferably as viscosity enhancers.

5.1.1 Historical Development

The role of chemically modified natural products in the development of today's plastics dates as far back as the middle of the last century. The efforts to produce materials with new properties from existing natural materials focused very early on cellulose and its products such as parchment paper, vulcanized fiber and celluloid. Attempts were made to overcome such disadvantages as poor wet bonding strength and high flammability by synthesizing new derivatives.

In 1865, Schützenberger [1] succeeded in preparing cellulose acetate from cellulose and acetic acid anhydride; in 1879 Franchimont [2] was able to obtain cellulose acetate with high molecular weight by sulfuric acid catalysis. At first there was no practical interest in cellulose triacetate as its insolubility rendered it difficult to work into a state in which it could be modeled.

An important advance for the use of cellulose acetate was made in 1904 when Eichengrün [3] and Miles [4] produced a soluble cellulose acetate (so-called secondary acetate) exhibiting plastic properties by partial saponification of the triacetate.

This facilitated the production of films and foils that was described in a patent of Farbenfabriken Bayer [3] in 1905. Plastic molding compounds made from secondary acetates with plasticizers and camphor, developed by Eichengrün in 1909 [5], introduced a replacement for easily flammable cellulose. In 1919, Lonarit GmbH [6] patented injection molding compounds of platicized secondary acetate. This initiated a dynamic further development technically based on the use of cellulose esters for molding compounds. The number of patents and publications reached a high point between 1920 and 1935 [7].

Cellulose acetobutyrates (CAB) and cellulose acetopropionates (CP or CAP) [8, 9] were developed between 1946 and 1956, and broadened the property profile of the cellulose esters regarding their use in thermoplastic molding compounds and lacquers. After the introduction of mixed cellulose esters, the development of new products became less significant than product modification. From that period to the present, the development has concentrated on product optimization and treatment, together with improvements from the product-quality and economic standpoints. In 1974, Bayer AG [10] brought the first thermoplastic cellulose ester molding compounds with polymer modification to the market for use in automobiles.

5.2 Chemical Structure of Organic Cellulose Esters

5.2.1 Organic Cellulose Esters in General

The chemical structure of cellulose, with three hydroxyl groups per anhydroglucose unit, limit the extent of ester formation with organic monocarbonic acids to triesters. The synthesis can be accomplished by well-known methods of preparative chemistry. It should be noted that the reaction of cellulose starts in a heterogeneous reaction medium, and is therefore influenced by diffusion processes during dissolution of the native cellulose structures.

Cellulose is mainly a linear polymer made of D-anhydroglucopyranose units with 1,4-β-glycosidic bonds (Figure 5.2). Native cellulose, from cotton plants, for example, has a degree of polymerization (DP) as high as 18,000. This can be decreased to approximately 2500 for cellulose derivatives. The hydroxyl groups of cellulose form intra- and intermolecular hydrogen bonds that hinder the dissolution of the material, hence restrict the chemical reactivity.

Additionally, supramolecular structures of the cellulose influence the process. The reactivity is also reduced because of the semi-crystalline nature of cellulose. The degree of crystallinity can amount up to 60 vol-%.

In general, cleavage of the less acid-resistant glycosidic bond of the cellulose takes place as a consequence of the acid catalyzed esterification reaction. Therefore, the average degree of polymerization of 3000–7000 decreases to 100–400 during the cellulose ester formation.

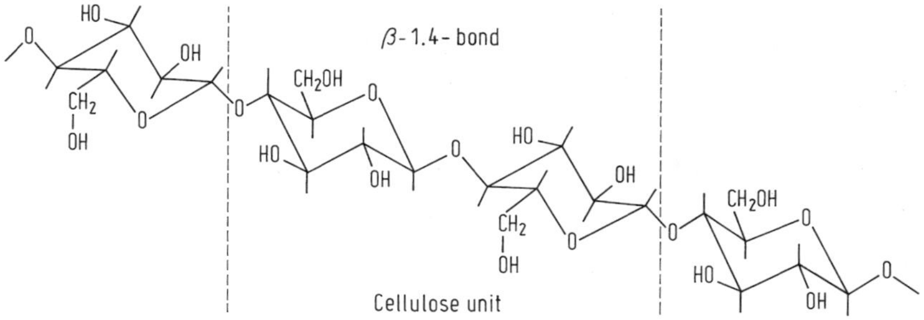

Figure 5.2. Conformation structure of cellulose

5.2.2 Organic Cellulose Esters of Technical Importance

The technically produced cellulose esters are the combined result of polymer analogue and polymer homologue reactions. The organic cellulose esters are characterized according to the acid(s) used in ester formation, as well as their content in cellulose ester. The extent of acylation is described either in wt-% calculated from the acid, or in wt-% calculated from the acyl content. The average degree of substitution (DS) gives an indication of the number of esterified OH-groups per anhydroglucose unit. It should be noted that partially esterified cellulose cannot be directly obtained by a heterogeneous reaction, but only by partial saponification of the triester.

A statistical distribution of the acyl groups can be assumed for the esters obtained by partial hydrolysis. The average degree of polymerization (DP) is investigated by relative viscosity measurements.

Table 5.1. Characteristics and Technical Application Areas of Organic Cellulose Esters

Cellulose Esters	Degree of Polymerization (DP)	Content of Acetic Acid (AA) %	Propionic (PiA) or Butyric (BA) %	Degree of Substitution (DS) (AA)	(PiA) or (BA)	Main Application
Triacetate	ca. 150 to 360	60.0 to 61.5	–	2.8 to 2.9	–	Photofilm, foils Isolation foils
Secondary acetates	ca. 100 to 200	56.0 to 56.5	–	approx. 2.5	–	Thermoplastic compounds
		54.0 to 54.6	–	approx. 2.4	–	Artificial silk, foils
		53.5 to 54.0	–	approx. 2.4	–	Thermoplastic compounds
		53.0 to 53.5	–	approx. 2.3	–	Lacquer raw materials Plates
Acetopropionate	ca. 150 to 200	5	57	0.3	2.3	Thermoplastic compounds
Acetobutyrate	ca. 100 to 150	43	18	2.1	0.6	Isolated lacquer raw materials, foils
		40	22	2.0	0.7	Foils, films
		20	44	1.0	1.6	Thermoplastic compounds
		9	59	0.5	2.3	Melt dip resins

For an adequately differentiated quality orientel treatment of cellulose esters, the purity, reactivity, and kind of cellulose used, as well as the production conditions, must be included alongside the measurements discussed above.

The technically important cellulose esters of C_2–C_4 carboxylic acids can exhibit a broad spectrum of properties, making them usable for many different applications. The degree of substitution and polymerization, as well as the kind and composition of the acyl groups, can be modified. For cast films, amongst others, the physical properties of the product also depend on the conditions under which the polymer was dissolved.

Cellulose esters can be characterized by their light resistance, good mechanical properties, stress crack resistance and resistance to aliphatic petrochemicals. Characteristic data of technically important cellulose ester types and their typical applications can be found in Table 5.1 [11].

References for Sections 5.1 and 5.2

1. *Schützenberger, P.:* C. R. Séances Acad. Sci. 61 (1865) p. 485
2. *Franchimont, A.:* C. R. Séances Acad. Sci. 89 (1879) p. 711
3. DE-PS 252 706 (1905) Farbenfabriken Bayer
4. FR-PS 358 079 (1905)
5. DE-PS 393 873 (1919)
6. DE-PS 441 023 (1919) Lonarit GmbH
7. *Hofmann, R.*, in: Ullmanns Enzyklopädie der technischen Chemie. 3. Ed. Vol. 5, Urban u. Schwarzenberg, Munich, 1954, p. 192–198
8. Mod. Plast. 1 (1957) p. 93
9. Company brochure Bayer AG: Cellidor, Leverkusen, 1963
10. DE-OS 2 46 178 (1974) Bayer AG
11. *Müller, F.*, in: Cellulose. Publication by Fonds der chemischen Industrie. 24 (1985) p. 45

5.3 Production of Cellulose Esters

5.3.1 General Chemical Basics of Cellulose Ester Production

Well-known methods of preparative chemistry can be used to prepare organic cellulose esters. There are, however, some salient points because of the chemical and physical structure as well as the morphology of natural cellulose. For example, since all known reactions of cellulose start in a heterogeneous reaction medium, diffusion processes of the reagents play an important role.

Esterification must be catalyzed by mineral or Lewis acids, so that a statistical decrease of the molecular weight of the cellulose takes place. In Table 5.2 [1], technically important and general preparative production processes are described.

Cellulose formiate can be directly obtained from formic acid and cellulose. CA can only be produced by reaction with acetic anhydride under acid catalysis. Under relatively mild conditions, the esters of linear acids can be synthesized with up to 4 C atoms in an analogous manner to anhydrides. To accelerate the reaction, trifluoroacetic anhydride is used as an impelling reagent. A visible decrease in reactivity can be seen in the case of C_2 to C_4 carboxylic acid anhydride [2]:

$$(H_3C-CO)_2O > (H_5C_2-CO)_2O > (H_7C_3-CO)_2O > [(H_3C)_2CH-CO]_2O$$

Table 5.2. General Preparative and Technical Production Processes for Organic Cellulose Esters [1]

Cellulose Ester	Starting Cellulose	Reagents	Example
Formic acid	Cellulose	Formic acid	Formiate
Aliphatic C_2–C_4 Carboxylic acids	Cellulose	Acid anhydrides, e.g. H_2SO_4, $HClO_4$, $ZnCl_2$	Acetate propionate
Higher aliphatic carboxyl Acids, unsaturated carboxyl acids	Cellulose	Trifluoroacetanhydride as impelling reagent	Butyrate, Mixed esters
		Acid chlorides in pyridine	Valeriate Palmitate Stearate Crotonate
Aromatic carboxyl acids	Regenerated cellulose	Acid chlorides in pyridine/nitro benzene or benzene, acids with chloroacetanhydride as impelling reagent and Mg-perchlorate	Benzoate, nitro-, chlor-, methoxy substituted benzoate
Aromatic dicarboxyl acids	Partly hydrolyzed cellulose acetate	Acid anhydride in pyridine or acetic acid with Na-acette as catalyst	Acetophthalate
Aromatic sulfonic acids Chlorovalerianic acid	Alkali cellulose Cellulose xanthogenate	Sulfonic acid chlorides chloroanhydrides and following hydrolysis of the xanthogenate with mineral acids	Toluolsulfonate Cellulose ester (low degradation)
Carboxyl acids, dicarboxyl acid	Cellulose	Homogeneous reaction medium N-ethyl pyridinium chloride/ pyridine or DMF	Cellulose esters (very low degradation), partial cellulose esters

Isobutyric acid anhydride, on the other hand, only reacts under very intensive reaction conditions leading to strong degradation of the cellulose chains.

During the synthesis of mixed esters of acetic acid and propionic acid, or acetic acid and butyric acid, the acetic acid accelerates the reaction and technically interesting mixed esters are, therefore, easily available. Higher carboxylic acid esters require the use of strongly activated acid derivatives like acid chlorides, or the additional use of chloroacetic anhydride as active co-reagent.

Chemically modified cellulose, the so-called regenerated cellulose, can be used if a sufficient reactivity cannot be obtained by acid derivatives. In newer processes, cellulose esters are obtained by esterification at a phase boundary [3]. For this purpose a reagent, such as chrloroanhydride of an organic acid is reacted with a water-soluble cellulose derivative such as partially substituted cellulose xanthogenate in an organic solvent that is not miscible with water. This reaction takes place quicker than the heterogeneous reaction and without necessarily causing molecular degradation. Disadvantages are the high consumption of chemicals because of side reactions, the incomplete esterification, and the produciton of partially substituted xanthogenates.

Cellulose can be reacted homogeneously in a solvent mixture of N-ethyl pyridinium chloride/pyridine or dimethyl formamide [4]. In this process, the polymer analogous cellulose esters of mono- and dicarbonic acid are extensively obtained with acid anhydrides.

5.3.2 Chemical Basics of the Production of Cellulose Esters

The technically important cellulose esters of the C_2–C_4 monocarboxylic acids or their mixed esters are exclusively obtained by the reaction of cellulose with acid anhydrides in the presence of catalytic acids [5]. Technically, sulfuric acid is especially suitable as a disaggregation reagent for cellulose because of its high effectiveness. The most effective proton catalyst is perchloric acid. It can, however, only be used in small doses synergistically with sulfuric acid because of its corrosive and explosive properties. Other possible acids like methane sulfonic acid, sulfoacetic acid, ammonium sulfate, aniline sulfate, sulfamine acid, and Lewis acids like zinc chloride or orthotitanium acid esters [6] have not been successfully implemented either because they did not offer any advantages compared to sulfuric acid or because they required uneconomically high dosages as in the case of zinc chloride.

Investigations into sulfuric acid catalysis [7, 8] suggest the following mechanism: sulfuric acid semiesters of the cellulose are partially formed, and contribute to solvation. Acetyl cations are formed from acetic acid anhydride and sulfuric acid via an acetyl sulfuric acid intermediate; the acetyl cations are very efficient esterification reagents:

$$H_3C-\underset{\underset{O}{\|}}{C}-O-SO_3H \longrightarrow H_3C-\underset{\underset{O}{\|}}{C}^{\oplus} + {}^{\ominus}O-SO_3H \qquad (5.1)$$

Partially substituted cellulose esters cannot be obtained directly because the reaction proceeds topochemically, and only entirely esterified celluloses (triesters) can be solvated.

The sulfuric acid semiester groups are slowly and incompletely substituted by acyl groups [9]. The remaining amount contribute to the solubility of the triesters, especially during the acetic acid process.

Technical esterification processes generally start as a heterogeneous reaction medium and proceed in a heterogeneous manner to a large extent. The reagents must reach crystalline domains by diffusion. To obtain optimum quality, complete esterification of the cellulose, the relation between the diffusion rate and reaction rate must be empirically coordinated. The exchange of reagents into the inner parts of the fiber and escape of reaction products and side products take place via a diffusion conducted transport. The morphological structures of the fiber—the tertiary and secondary structures as well as the molecular superstructures—are all involved with the pore systems [10]. Because of this complex interplay, the only methods currently available for judging the reactivity of cellulose are empirical, process specific laboratory test methods.

5.3.2.1 Activation of Cellulose

To obtain a sufficient reactivity of the cellulose, the water content should be 4–7wt-% after drying: water causes swelling of the fibers, loosens the hydrogen bonds between the chains, and therefore enlarges the accessible surface. The reactivity increases with increasing amounts of water. The amount of water, however, is limited by the additional consumption of anhydride. Therefore, other additives have been suggested, for example amines, urea, alcohol, and acetone, as well as physical processes like ultrasound treatment [11]. The reactivity can be increased by the introduction of spacer groups [12]. Commercially, only carboxylic acids, preferably acetic acid, have been used as activators to date [13].

5.3.3 Raw Materials for Technical Production

The following raw materials are used for the production of CA, cellulose acetopropionates (or cellulose propionates, as they are generally called outside the English-language literature), and cellulose acetobutyrates:

- ultra-pure cellulose
- acetic acid anhydride
- propionic acid anhydride
- butyric acid anhydride
- proprionic acid
- butyric acid

Acetic acid and methylene chloride are used as solvents.

5.3.3.1 Cellulose

Only specially prepared, ultra-purified cellulose can be used for the production of cellulose esters. Contamination in the form of dirt particles and non-reactive cellulose, which may cause gel particles when applied in solutions, films and molding compounds, must not remain in the starting material. For applications involving thermal strain, chemical structures in the cellulose must be largely removed by chemical treatment during work-up.

Purification of cellulose esters by filtration of the solutions requires high technical standards, and is not effective enough to remove all gel particles.

Suitable celluloses for this kind of work-up are linters and cellulose from wood pulp. These are gaining in importance as raw material.

Linter Cellulose

Linters provide a homogeneous and almost pure natural form of cellulose. Linters are the short staples separated from cotton seed capsules in a second step after the filaments are removed for spin cotton. The linters can then be worked up by mechanical separation (sifting) and chemical purification (pressure-boiling with 3–5% NaOH and bleaching in several steps) to produce ultra-pure cellulose [14] (Table 5.3).

Table 5.3. Analytical Characteristics of Bleached Linters [14]

Composition	Characteristic
α-cellulose	99.7%
β-cellulose	0.2%
γ-cellulose	0.1%
Carboxyl groups	<0.02%
Total ashes	0.02%
Degree of polymerization	1000 to 7000

Cellulose from Wood Pulp

In order to control for fluctuations in the price and availability of linters, which are affected by crop yield and cultivation restrictions, the development of wood pulp cellulose has been accelerated in the last 20 years.

Today, through special processes, it is possible to obtain cellulose esters from pulp that have nearly the same quality as cellulose esters made from linters in terms of tensile strength, color, clarity, light and temperature stability. The isolation and purification processes of wood cellulose demand a higher chemical expenditure than purification of linters because the cellulose must be separated from lignin by distingeration, and purified in an intensive chemical follow-up treatment. Degradation and oxidation harm the product as a consequence of the process. The damage should be kept as small as possible to minimize negative effects on processability and quality (Table 5.4) [14].

Table 5.4. Comparison of Some Chemical and Physical Properties of Cotton Linters and Spruce Sulfite Cellulose [14]

Property	Linters	Spruce Cellulose
Chemical composition	>99% α-cellulose, Rest: β-cellulose γ-cellulose	86 to 97% cellulose, Rest: xylan, mannan, lingnine, ash
Cellulose distribution in the cell walls	evenly over the sectional area	inside > outside
Packing density, ordinary state	high	lower; inside > outside
Maximum degree of polymerization in the plant	ca. 12,000	ca. 10,000
DP of the bleached technical product	ca. 7000	ca. 3500
Crystallinity, IR-spectroscopic (O'Connor-Index)	2.15	1.48
Inner surface against H_2O at 20 °C after BET	150 m^2/g	200 m^2/g
Water retention ability	50%	80%

5.3.3.2 Aliphatic Monocarboxylic Acids

Monocarboxylic acids are generally produced by the following processes [15, 16]:
- carbonylation of methanol: acetic acid
- direct oxidation of naphtha: acetic acid, propionic acid
- oxidation of aldehydes: acetic acid, butyric acid
- Reppe carbonylation of olefins: propionic acid

5.3.3.3 Carboxylic Acid Anhydrides

Acetic Acid Anhydride

The acetic acid that is equivalently produced during the esterification of cellulose must be converted to acetic acid anhydride for further use. In the Wacker process [17], ketene is produced by pyrolytic splitting from acetic acid with triethyl phosphate as catalyst, and reaction with acetic acid.

Propionic and Butyric Acid Anhydride

Propionic and butyric acid can be transformed into their anhydrides by reacting them with acetic hydride. The technically optimal process can be carried out in a distillation column with simultaneous fractional separation of the propionic acid anhydride or butyric acid anhydride and acetic acid components.

5.3.4 Technology of Cellulose Ester Production

5.3.4.1 Process Steps

The technical production of organic cellulose esters follows these process steps:
- cellulose, extraction/production, selection, and work-up
- pretreatment of the cellulose
- esterification of the cellulose
- partial saponification
- precipitation of the cellulose ester
- aftertreatment
- recovery of solvents and reactants

5.3.4.2 Cellulose Pretreatment

Pretreatment is with 30–100 parts of acetic acid to 100 parts of cellulose and, if needed, a small amount of sulfuric acid. The acid is sprayed on the cellulose in a sprinkling vat with intensive mixing at temperatures of up to 50 °C for 1 to 3 hours.

5.3.4.3 Esterification Processes in General

Three different processes are available for the produciton of cellulose esters. They can be modified for many different purposes. In two of these proceses, the acetic acid process and the methylene chloride process, the cellulose esters are obtained in a solvent. In the third process, the fiber acetate process, the reaction takes place heterogeneously in a non-solvent, from which the cellulose triacetate can be isolated.

The acetic acid and methylene chloride processes can be used to prepare CA, cellulose proprionate and cellulose acetobutyrate. The fiber acetate process is limited to producing cellulose triacetate. Furthermore, the triacetates obtained in this process show a much different solvation behavior compared to products obtained in the solvent process.

Continuous processes have been investigated thoroughly [18 to 20] but have generally not yet proven suitable for industrial use. This is partly due to the difficulty of standardizing the reactivity of cellulose on a sufficiently high level by pretreatment, and partly because of the costliness of removing the large quantity of water initially needed.

To date the direct esterification of the cellulose with ketene has not been successful. The reactivity of the activated cellulose is clearly insufficient to obtain an acceptable degree of esterification and to exclude discoloring by-products. Only products by one of the two solvents processes can be used for thermoplastic compounds.

5.3.4.3.1 Acetic Acid Process

As a semi-solid mass the activated cellulose reacts in acetic acid containing an excess of acetic acid anhydride (10–40 wt-%) and 2–15 wt-% sulfuric acid as catalyst, with intensive mechanical mixing. The strongly exothermal reaction (1.03 kJ/g cellulose) causes the temperature to increase. According to the desired viscosity range of the cellulose esters, the temperature must be held to fixed limits by external cooling and by the addition of cooled reaction mixture, to suppress the degradation by glycoside cleavage that occurs at the same time as esterification.

In the acetic acid process, the dissolution of cellulose esters will succeed only if high amounts of sulfuric acid are present because partially formed sulfuric acid semiesters are effective as solubilizers.

When a fiber-free, high viscosity solution has been formed from the initially fibrous mass, the esterification process can be stopped by adding water.

5.3.4.3.2 Methylene Chloride Process

In the methylene chloride process the reaction enthalpy is removed by distilling the low-boiling methylene chloride, which permits a quick and very effective process control even in high viscosity solutions. Furthermore, methylene chloride is an excellent solvent for cellulose esters, especially cellulose triacetates. A small amount of sulfuric acid (1 wt-%) is sufficient as catalyst. Sulfuric acid ester groups are formed in a small amount through which a more favorable solution is created for the eventual thermal stabilization.

Figure 5.3 shows a flow chart for CA fabrication via the methylene chloride process.

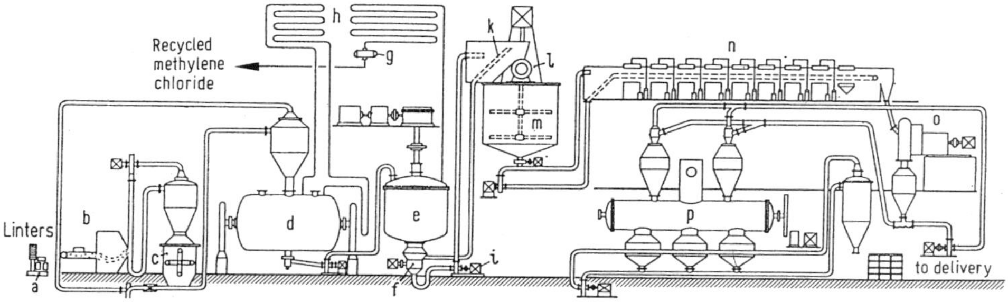

Figure 5.3. Manufacturing of cellulose acetate by methylene chloride process [21]
a, scale; b, unbaling; c, washbox; d, acetylation; e, precipitation; f, crusher g, vacuum receiver; h, tube condenser; i, slurry pump; k, screen belt; l, mill; m, flotation; n, sludge belt; o, pusher centrifuge; p, vacuum paddle-dryer

5.3.4.3.3 Fiber Acetate Process

Only cellulose triacetate is produced via the fiber acetate process. The cellulose is suspended in a non-solvent like benzene, toluene or carbon tetrachoride, and is esterified heterogeneously with acetic acid anhydride, using a perchloric acid as catalyst. The triacetate can be obtained undissolved as a fibrous material.

5.3.4.4 Partial Hydrolysis

The partial hydrolysis of the dissolved tiester takes place in the presence of an acid catalyst, usually sulfuric acid in commerical processes, with a water concentration of 5–10 wt-% resulting from the reaction of the unused anhydride. The sulfuric acid concentration, water amount, and temperature of 40–80 °C allows the hydrolysis process to be controlled in such a way that the molecular degradation can be kept low or to a desired level.

Concurrent with the partial hydrolysis of the organic acid esters, the maximal separation of the sulfuric acid ester groups bound to the cellulose esters is accomplished by using a higher hydrolysis rate of the sulfuric acid esters under optimal process conditions [9]. The amount of cellulose–sulfuric acid ester groups should be minimized because their thermal instability yields sulfuric acid. Under thermal strain the sulfuric acid causes a dramatic molecular degradation, mainly by glycoside cleavage.

A special salt stabilization is necessary for the sulfuric acid ester groups that remain after hydrolysis. The rate of hydrolysis can be observed by monitoring the solubility in empirically determined solvent systems. After obtaining the desired composition, neutralization of the catalyst with salts like sodium or magnesium acetate stops the hydrolysis.

5.3.4.5 Precipitation and Processing

The cellulose esters are isolated from the reaction solution by precipitation into diluted acetic acid. Before this, methylene chloride must be totally removed through distillation.

To obtain easily washable particles, the solution of the cellulose esters should be brought to close to the precipitation point by adding small amounts of diluted acid in a first step. In an empirically optimized dosing and mixing process, the precipitation is completed by adding the bulk of diluted acid, which allows the esters to precipitate out in the form of flakes or powder [22].

The products are ground to the desired particle size and washed intensively. Continuous processes working via the principle of countercurrent achieve the best results. The diluted acids can be partially recovered during the work-up process. After the countercurrent washing the cellulose esters are intensively washed with water to remove residual acid, stabilized with salt, and dried.

5.3.4.6 Bleach and Stabilization

To obtain high quality, heat resistant, color stabilized and lightly colored cellulose esters, the products must be bleached and specially stabilized. They can be bleached with well-known bleaching agents like calcium permanganate or oxalic acid, or technically more advantageously with chlorine oxides in an aqueous suspension of cellulose ester flakes. To obtain a high quality products for thermoplastic molding compounds, sulfuric acid bound in the cellulose esters must be largely removed.

In addition to the hydrolytic separation optimized by the hydrolysis process, the elimination can be continued homogeneously in an acidic aqueous suspension of the flakes by boiling or treating with overheated steam. The bound sulfuric acid remaining in the cellulose esters after this treatment amounts to 0.05–0.05 wt-% (calculated on the basis of cellulose ester). To avoid thermal degradation this residual acid must be neutralized by alkaline, alkaline earth, or earth metal salts of dicarboxylic acids like oxalic acid, or hydroxycarboxylic acids like citric acid or tartaric acid. Epoxidized soybean oil or bisphenol glycidyl ether are used as synergists for salt stabilizing during compounding and processing of thermoplastic molding compounds.

To keep thermal degradation and dicoloration as low as possible, other additives with phosphoric and sulfuric groups like triarylphosphite, especially neopentylgycol phosphite [23], chlorothiophenol [24], 3,3′-thiodipropionic acid dialkylester [25], as well as salts like calcium hydrogen oxalate or citrate and magnesium sterate in combination with phenols [26] are suitable. Stabilization against damage from thermal oxidation is described in Section 5.5.3.

5.3.4.7 Recovery of Acids and Solvents

Recovery of Acetic Acid

Depending on the process, 2 to 6 parts of diluted acetic acid (15–20%) are formed per part cellulose ester. Their recovery is an important factor in the profitability of the process. Continuous technical processes are based on extraction and azeotropic distillation. The diluted acid is extracted with ethylacetate, benzene or ketones employing countercurrent technology. The extract is fractionated and acetic acid is obtained at the bottom of the column with a purity of 99.8%. Ethylacetate and water are drawn off the top azeotropically.

Recovery of Methylene Chloride

Methylene chloride collects during the distillation of the raw solution, and can be separated from the codistilling water. The obtained purity is high enough for reuse as solvent.

References for Section 5.3

1. *Müller, F.*, in: Cellulose. Publication by Fonds der chemischen Industrie 24 (1985) p. 45
2. *Malm, C. J.; Hiatt, G. D.;* in: *Ott, E.; Spurlin, H. M.; Grafflin, M. W.* (Eds.): High Polymers. Vol. 5: Cellulose and Cellulose Derivatives, Part II. Interscience Publ., New York 1954, p. 788
3. *Rogowin, Z. A.; Galbraich, L. S.:* Die chemische Behandlung und Modifizierung der Cellulose. (Ed.) *W. Albreacht*. Thieme, Stuttgart 1983
4. *Husemann, E.; Siefert, E.:* Makromol. Chem. 128 (1969) p. 288
5. *Eischer, Th.; Fischer, W.;* in: Ullmanns Enzyklopädie der technischen Chemie. 4. Ed. Vol. 9. Verlag Chemie, Weinheim 1975, p. 228
6. JA-PS 101 083 (1978) Asahi Chem. Industry
7. *Sihtola, H; Laamanen, L.:* Das Papier 18 (1964) p. 600
8. *Tanghe, L. J.; Brewer, R. J.:* Anal. Chem. 40 (1968) p. 350
9. *Malm, C. J.*, et al.: Ind. Eng. Chem. 38 (1946) p. 77
10. *Jeffries, R.*, et al.: Text. Res. J. 38 (1968) p. 234
11. US-PS 4024335 (1977) Union Carbide Corp.
12. *Koura, A.; Krause, Th.:* Cellul. Chem. Technol. 19 (1985) p. 497
13. *Eicher, Th.*, in: Ullmanns Encyclopedia of Industrial Chemistry, 5. Ed., Verlag Chemie, Weinheim 1986, p. 438
14. *Temming, H.; Grunert, H.:* Temming-Linters. 2. Ed. Temming, Glückstadt 1972 p. 37
15. *Reimenschneider, W.*, in: Ullmanns Encyclopedia of Industrial Chemistry. Vol. A5. Verlag Chemie, Weinheim 1985, p. 235
16. Fatty Acids Synthesis and Applications. Chem. Technology Rw. No. 9. Noyes Data Corp., Park Ridge, New York 1973
17. DE-PS 687 065 (1933) u. US-PS 21 08 829 (1934) Cons. f. Elektrochem. Ind.
18. US-PS 26 03 634 (1952) Celanese Corp.
19. US-PS 27 31 247 (1956) Société Rhodiaceta
20. US-PS 37 55 297 (1973) Celanese Corp.
21. Bios Final Report No. 1850, Item No. 21 H. M. Stationary Office, London 1948
22. US-PS 42 22 276 (1980) Eastman Kodak Co.
23. US-PS 33 05 378 (1967) Eastman Kodak Co.
24. US-PS 29 17 398 (1959) Eastman Kodak Co.
25. US-PS 33 14 808 (1967) E. I. DuPont Nemours Co.
26. US-PS 28 99 315 (1959) Eastman Kodak Co.

5.4 Properties of Organic Cellulose Esters

5.4.1 General Physical and Chemical Properties

5.4.1.1 Melt Temperature

The melting points of cellulose triesters fall drastically in the series linear carboxylic acids with 2 to 6 carbon atoms from over 300 °C for the triacetate to a range of 80–100 °C for ester derivatives with 6 or more carbon atoms in the acyl group (Figure 5.4) [1].

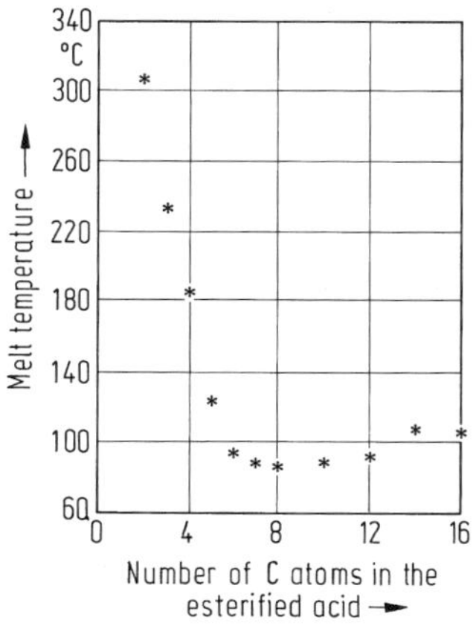

Figure 5.4. Melt temperatures of cellulose triesters [1]

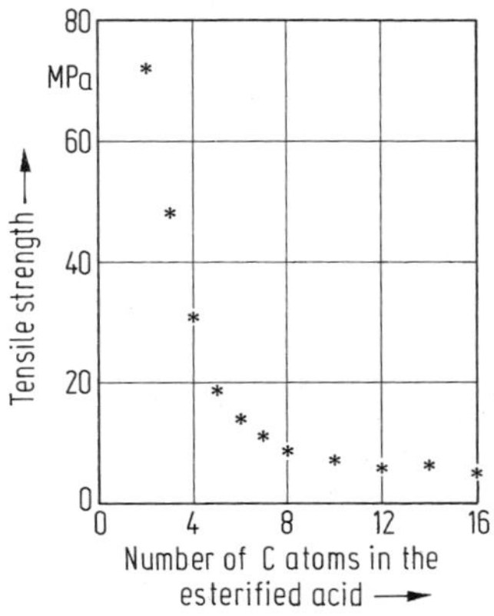

Figure 5.5. Tensile strength of cellulose triesters [1]

5.4.1.2 Tensile Strength

Similarly the values for the tensile strength decrease with increasing chain length of the acid residue from that of acetate to that of capronic acid. Tensile strengths for higher carboxylic acids are very low (Figure 5.5) [1].

5.4.1.3 Water Absorption

As the paraffin character of the cellulose esters increases with the length of the alkyl chains, the solubility of cellulose esters in organic solvents also increases. The moisture absorption, in contrast, decreases. Compared to cellulose acetate, cellulose propionate and butyrate demonstrate a decreased ability to absorb water (Figure 5.6). While the water absorption for cellulose acetate is high and sensitive to the relative humidity, the water absorption for cellulose butyrate is small and almost independent of the relative humidity [1].

The dramatically reduced water absorption of cellulose propionate and especially butyrate is important for preparing thermoplastic compounds with decreased warpage in damp climates. Furthermore, the reduced water absorption and the low saponification rate of cellulose propionates and especially butyrates—compared to cellulose acetates—forms the basis for the production of weather resistant thermoplastics.

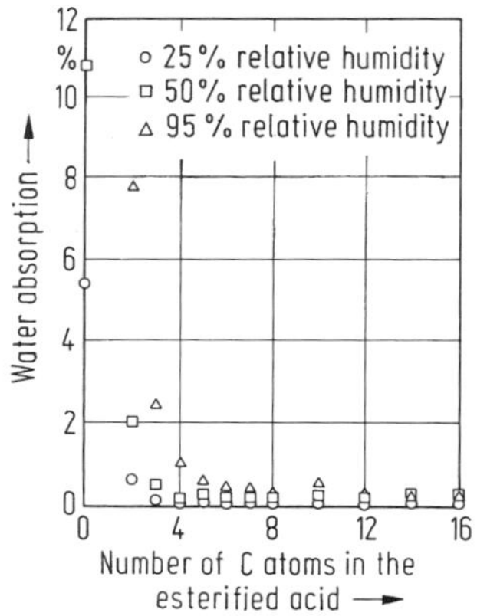

Figure 5.6. Water absorption of cellulose triesters at 25 °C [1]

5.4.2 Chemical and Physical Properties of Organic Cellulose Esters

5.4.2.1 Solubility

The solubility of cellulose esters depends on the kind of substitutents, the average level of substitution, the distribution of substituents along the chain, and their relative distribution among primary and secondary alcohol groups.

5.4.2.1.1 Cellulose Acetates

The choice for solvents for commercially prepared cellulose acetates (prepared by partial hydrolysis) is limited by the level of substitution and the statistical distribution of the substituents. The solubility in hydrogen bonding solvents increases with hydroxyl content. At a substitution level of 0.8 water solubility is achieved (Table 5.5) [2].

Table 5.5. Solutility of Cellulose Acetate with Different Esterification Degrees [2]

Esterification Degree	Bound Acetic Acid %	Chloroform	Acetone	2-Methoxyethanol	Water
2.8 to 3.0	60 to 62,5	soluble	–	–	–
2.2 to 2.7	51 to 59	–	soluble	–	–
1.2 to 1.8	31 to 45	–	–	soluble	–
0.6 to 0.9	18 to 26	–	–	–	soluble
0.6	18	–	–	–	–

In addition to the hydrogen bond formation, the homogeneous statistical distribution of substituents in the C_2, C_3, and C_6 positions of the anhydrocellulose unit plays an important role in solubility. Cellulose esters with a substitution level of 0.8–1.9 that are directly synthesized in a homogeneous solution and in which, as ^{13}C-NMR studies have shown, the acetylation occurs mostly at the C_6 position, are not water soluble [3].

5.4.2.1.2 Cellulose Acetopropionates and Cellulose Acetobutyrates

Compared to the solubility of cellulose acetates, the solubility of cellulose propionates is much higher. Cellulose propionates are soluble in a number of organic solvents starting with the triesters, and the solubility is not as dependent on the degree of hydrolysis. The solubility of cellulose acetoproprionates and acetobutyrates depends strongly on the composition of the product [4].

5.4.2.2 Thermal Characteristics of Organic Cellulose Esters

The cellulose esters have been characterized by Mandelkern [5] as polymers with high melt temperatures and low melt enthalpies. For example, cellulose tributyrates:

Melt Temperature T_m:	207 °C
Melt Enthalpy ΔH:	12600 J mol^{-1}
Melt Entropy ΔS:	33.9 J mol^{-1} K^{-1}.

The low entropy suggests that the cellulose esters form extended chains.

Figure 5.7 shows representative results of TGA, DTA, and TBA (torsional braid analysis) for cellulose triacetate [6]. The values of the glass transition temperature $T_g = 172$ °C and the melting point $T_m = 307$ °C are identical for TBA analysis and dilatometry [7]. In addition to a T_m at 290 °C, DTA shows three second order transitions between 186 and 213 °C that are not yet completely understood [8].

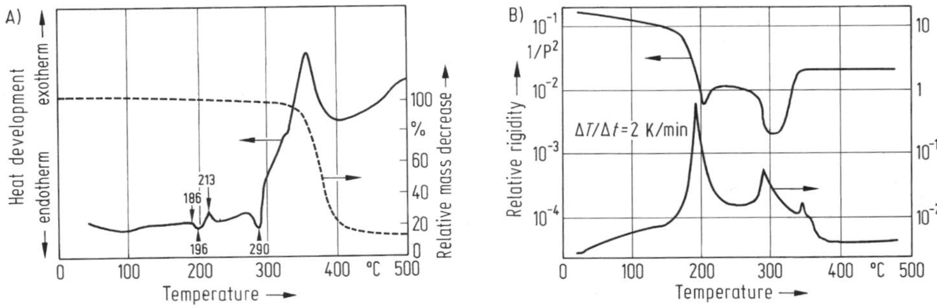

Figure 5.7. Analysis of cellulose triacetate [6], measurement in N_2 atmosphere. A) TGA; B) DTA, TBA

For practical uses with high thermal demands, the TGA analysis shows that a strong degradation of cellulose triacetate begins at 356 °C [6]. Due to the narrow window between the melt temperature and the degradation temperature, processing cellulose esters as workable melt is difficult; impossible in the case of cellulose triacetate and even in the case of the lower melting cellulose acetopropionates inevitably associated with some degradation.

5.4.2.3 Cellulose Ester–Plasticizer Systems

Because of the limited processing window, it is necessary to add plasticizers to achieve thermoplastic molding grades. Monomeric plasticizers reduce the amount of crystallinity and drastically lower the glass transition and melt temperatures. A number of models have been advanced to explain plasticization in polymers: solvation/desolvation theory, gel theory, and the free volume concept [9 to 11].

To obtain technically interesting property profiles in plasticized thermoplastic materials at least a partial compatibility between the base polymer and the plasticizer must be achieved. Compatibility of plasticizers with cellulose ester follows the same general trends as their solvation behavior.

5.4.2.3.1 Cellulose Acetates

Only partially saponified cellulose acetates with an average substitution degree between 2.3 and 2.5 are sufficiently compatible with plasticizers such as phthalic acid esters and arylphosphates.

5.4.2.3.2 Cellulose Acetoproprionate and Cellulose Acetobutyrate

These mixed esters are compoundable with many plasticizers based on aromatic and aliphatic dicarboxylic acids. The amount of external plasticizer needed to achieve a thermoplastic melt is lower because the long side chains of cellulose esters function as internal plasticizers. Slight saponification of the triester leads to a material with an optimum physical property profile. Hydrogen bond formation also plainly plays a role in the interaction between the mixed cellulose ester and the plasticizer.

Figures 5.8 and 5.9 show the miscibility of phthalate esters in cellulose acetate, cellulose acetobutyrate and cellulose butyrate at temperatures at 25 °C and 180 °C [12]. Phthalates from long chain alcohols are not completely soluble with cellulose acetate.

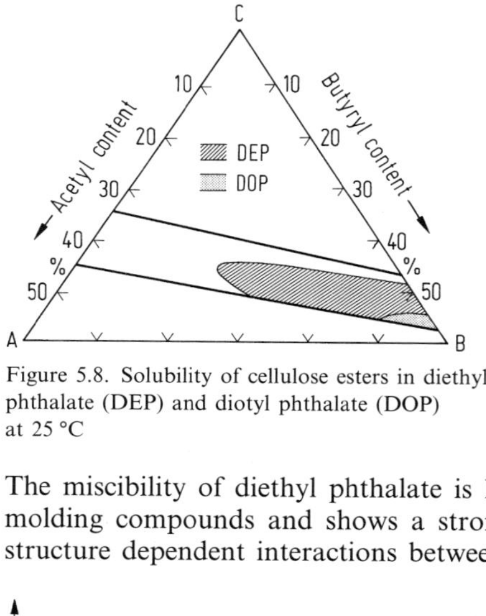

Figure 5.8. Solubility of cellulose esters in diethyl phthalate (DEP) and diotyl phthalate (DOP) at 25 °C

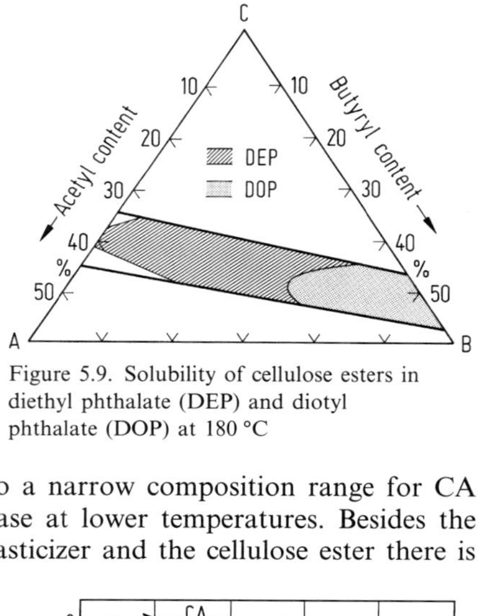

Figure 5.9. Solubility of cellulose esters in diethyl phthalate (DEP) and diotyl phthalate (DOP) at 180 °C

The miscibility of diethyl phthalate is limited to a narrow composition range for CA molding compounds and shows a strong decrease at lower temperatures. Besides the structure dependent interactions between the plasticizer and the cellulose ester there is

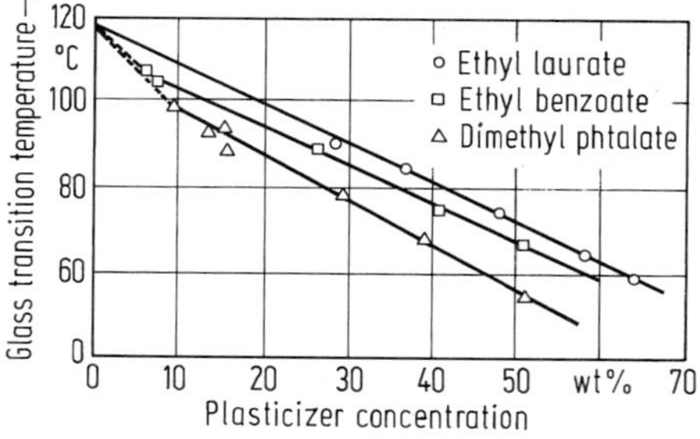

Figure 5.10. Glass transition temperature T_g of cellulose tribuyrate as a function of plasticizer concentration [13]

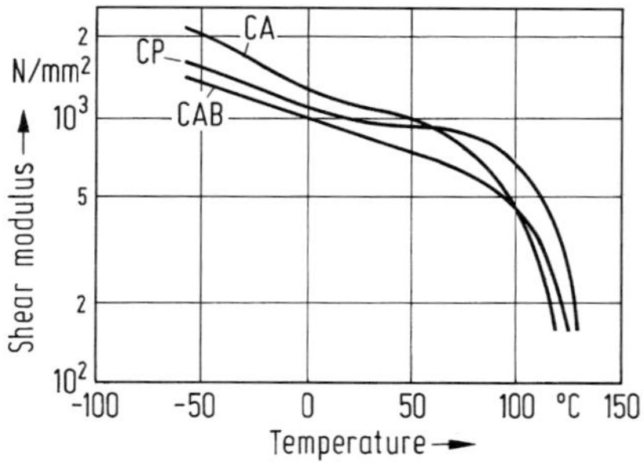

Figure 5.11. Shear modulus temperature curves of cellulose acetate (CA) with 25% plasticizer, cellulose propionate (CP) with 5% plasticizer and cellulose acetobutyrate (CB) with 5% plasticizer

also an influence of the semi-crystallinity of the cellulose ester. Mandelkern and Flory arrived at this conclusion from studies of cellulose tributyrate [13] with various plasticizers (Figure 5.10). The favorable mixing at the plasticizer in the amorphous region of the cellulose ester led to a disproportionate decrease in the glass transition temperature.

Plasticized cellulose esters can be seen as "hard gels" because of the rigidity of the cellulose molecule. As shown by the shear modulus curves in Figure 5.11, relatively high mechanical properties can be maintained over a wide plasticizer composition range [14].

A summary of typical properties of mixed cellulose esters is given in the diagram for cellulose acetobutyrate in Figure 5.12 [4].

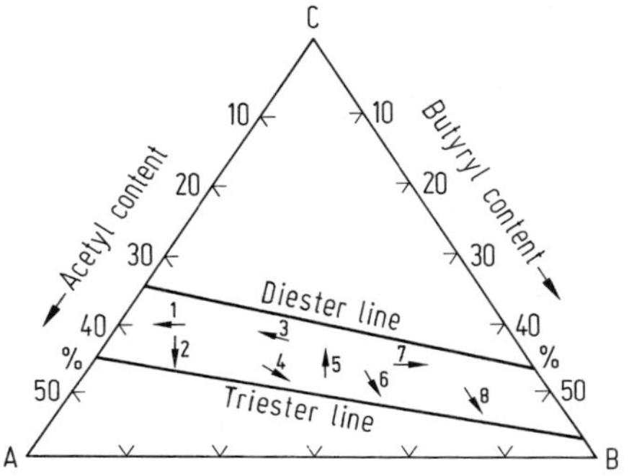

Figure 5.12. Influence of the contents of cellulose acetobutyrates on their physical properties [4]
1: increasing tensile strength and rigidity
2: decreasing water absorption
3: increasing melt temperature
4: increasing plasticizer compatibility
5: increasing solubility in polar solvents
6: increasing solubility in non-polar solvents
7: increasing flexibility
8: decreasing density

5.4.2.4 Basic Investigations on Elasticization with Polymers

Addition of monomeric plasticizers to thermoplastic cellulose esters has some disadvantages beyond the predominantly positive properties of cellulose ester thermoplastic molding compounds; for example, decreased heat distortion temperature, decreased E-modulus, plasticizer evaporation and migration. An improvement can be made by elasticizing the cellulose esters with soft polymers [15]. As with plasticizers, production of polymer blends requires polymers that are at least partially compatible.

Solubility parameters have been used to guide the characterization of such polymer blends. Investigations of cellulose esters indicates that there is strong influence of hydrogen bonds on the miscibility behavior. The results of these investigations are tabulated in Tables 5.6 and 5.7 [16]. Based on solubility parameters cellulose esters should be miscible with butadiene-acrylonitrile (NBR) copolymers but not with ethylene vinyl acetate (EVAC) copolymers. Torsional pendulum studies show, however, that only EVAC 80 forms a homogeneous phase with CAB. Probably the cohesive energy density is further influenced by interaction with hydrogen bonds in this case. In blends of

Table 5.6. Properties of Mixtures of Cellulose Esters and Butadiene Acryl Nitrile (NBR) and Ethylene Vinyl Acetate (EVAC) Copolymers [16]

Component 2 \ Component 1	Cellulose Acetopropionate	Cellulose Acetobutyrate
NBR 28	two phases, opaque	two phases, opaque
NBR 34	two phases, opaque	two phases, opaque
NBR 39	two phases, opaque	two phases, opaque
EVAC 70	two phases, transparent	two phases, transparent
EVAC 80	two phases, transparent	one phases, transparent

Table 5.7. Chemical Composition, Refraction Index and Solubility Parameter of Cellulose Esters, Butadiene Acryl Nitrile (NBR), and Ethylene Vinyl Acetate (EVAC) Copolymers [16]

Polymers	Acetic Acid %	Composition Propionic Acid %	Butyric Acid %	Solubility Parameter $(J/cm^3)^{1/2}$	Refraction Index n_D^{20}
cellulose acetobutyrate (CAB)	18.5	–	45	40.2	1.4804
cellulose acetobutyrate (CAP)	4.5	58.5	–	39.7	1.4757
cellulose acetate (CA)	55.3	–	–	41.0	1.4730
		Acryl nitrile			
NBR 28		28		39.3	1.5150
NBR 34		34		40.6	1.5140
NBR 39		39		41.4	1.5110
EVAC 70		70		37.7	1.4669
EVAC 80		80		38.5	1.4569

cellulose esters and NBR, on the contrary, it appears that the interactions between cellulose ester bonds outweigh the dispersive or polar interaction with NMBR molecules.

Comparisons of mixtures of cellulose esters clearly demonstrate the difficulties in categorizing the solubility parameters of cellulose. While cellulose acetobutyrate (CAB) and cellulose acetopropionate (CAP) (with solubility parameters 9.5 and 9.6 respectively) are miscible with one another in all compositions, neither CAP nor CAB is homogeneously miscible with cellulose acetate (CA), even though the formation of hydrogen bonds is possible.

The properties of these blends as well as blends of cellulose esters with ethylene vinyl acetate or vinyl acetate graft products are described in Section 5.8.

5.4.3 Liquid Crystalline Properties of Cellulose Ester Solutions

Because they are polymers with stiff chains, many cellulose derivatives can form mesophases in concentrated solutions [17]. Anisotropic solutions of cellulose acetate (substitution degree of 2.5–3.0) can be obtained at high concentrations (40–50 wt-% polymer) in various solvents [18]. Investigations of solutions of cellulose acetates (DS = 2.45) in acetic acid and substituted acetic acids indicate that the liquid crystalline critical concentrations depend mostly on the pH of the acid [19]. Increasing the acidity, which may be seen as the measure of interaction between solvent and polymer, decreases the critical concentration. In the most favorable solvent, trifluoroacetic acid, the critical concentration of cellulose acetate is 28 wt-%.

Liquid crystalline solutions have been used in development research to spin fibers that show improved tensile strength, E-modulus, and reduced water absorption.

References for Section 5.4

1. *Malm, C. J.*, et al.: Ind. Eng. Chem. 43 (1951) p. 684
2. *Eicher, Th.; Fischer, W.*: Celluloseeter. In: Ullmanns Enzyklopädie der technischen Chemi. 4. Ed. Vol. 9. Verlag Chemie, Weinheim 1975, p. 227
3. *Miyamoto, T.*, Ed. Vol.: J. Polym. Sci. Chem. 23 (1985) p. 1373
4. *Ott, E.; Spurlin, H. M.; Grafflin, M. W.*: High Polymers. Vol. 5: Cellulose and cellulose derivatives, Part II. 2. Ed. Interscience Publ., New York 1954, p. 798

5. *Mandelkern, L.*, in: Physical Properties of Polymers. American Chemical Society, Washington 1984
6. *Gillham, J. K.;* AICHE J. 20 (1974) p. 1066
7. *Russell, J.; Van Kerpel, R.G.:* J. Polym. Sci. 25 (1957) p. 77
8. *Klarman, A. F.*, et al.: J. Polym. Sci. Polym. Chem. Ed. 7 (1969) p. 1513
9. *Doolittle, A. K.:* Mechanism of plasticization. In: *Bruns, P. F.:* Plasticizer technology. Vol. 1. Reinhold, New York 1965
10. *Thinius, K.:* Chemie, Physik und Technologie der Weichmacher. Leipzig: Dt. Verl. für Grundstoffindustrie 1963
11. *Stepek, J.; Daoust, H.:* Additives for Plastics, Springer, New York 1983, p. 7
12. *Malm, C. J.:* Ind. Eng. Chem. 34 (1942) p. 430
13. *Mandelkern, L.; Flory, P.:* Am. Chem. Soc. 73 (1951) p. 3206
14. Bayer Kunststoffe. 3. Ed. Bayer AG, Leverkusen 1963, p. 186 and 187
15. *Leuschke, Ch.; Wandel, M.:* Plastverarbeiter 33 (1982) p. 1096
16. *Casper, R.; Morbitzer, L.:* Angew. Makromol. Chem. 58/59 (1977) p. 1
17. *Meeten, G. H.; Navard, P.:* Polymer 23 (1982) p. 483
18. *Bheda, J.,:* Coll. Polym. Sci. 258 (1980) p. 1335
19. *Aharoni, S. M.:* Mol. Cryst. Liq. Cryst. 56 (1980) p. 237

5.5 Modification of Organic Cellulose Esters

5.5.1 Plasticizer

The only commercially significant cellulosics for thermoplastic molding applications are cellulose diacetate, cellulose acetopropionate, and cellulose acetobutyrate with the following acid content:

- Cellulose diacetate (CA)
  acetic acid content 51.5–53.5% (for acetate block molding)
- acetic acid content 53.5–56.5% (for injection molding)
- Cellulose acetopropionate (CAP)
  acetic acid content 3–8%, propionic acid content 55–62%
- Cellulose acetobutyrate (CAB)
  acetic acid content 19–29%, butyric acid content 43–47%

These unmodified milled basic materials for thermoplastic cellulose ester molding compounds are also described as CA, CAP, and CAB flakes.

These materials are not thermoplastically worked in practical applications without plasticizers or modifiers. Cellulose diacetate flakes begin to degrade thermally even before the melting point; CAB and CAP form workable melts but can be molded into only very simple forms because of their high viscosity. In addition yellowing often occurs and the properties are compromised.

The use of plasticizers allows for stable melts of reasonable viscosities which can be molded and retain useful physical properties [1]. The plasticizer content varies from 15 to 20 wt-% for CA and from 3 to 5 wt-% for CAB and CAP. The maximum amount of plasticizer depends upon the desired properties. The practical limits are 33–38 wt-% for CA and 20–25 wt-% for CAP and CAB [2 to 4].

Suitable plasticizers [1, 4, 5, 6, 14] include

- *for CA*:
  - diethyl phthalate
  - dimethylglycol phthalate
  - triphenyl phosphate or trichlorethyl phosphate in combination with dimethyl or diethyl phthalate.

(Dibutyl, diisopropyl, or di-2-ethylhexyl phthalate may be used only in combination with the plasticizers described above.)

- *for CAP and CAB:*
  - dibutyl phthalate
  - di-2-ethylhexyl phthalate
  - dibutyl, di-2-ethylhexyl adipate
  - dibutyl, dioctyl sebacate
  - dibutyl, diotyl acelate

  (Citric acid and palmitic acid are used in special appplications, e.g. in the food sector.)

5.5.2 Polymer Modifications

Until 1974, only the above described monomeric plasticizers were used to modify organic cellulose esters and development seemed to have stagnated. After Bayer's successful modification of cellulose acetobutyrate with ethylene vinyl acetate [7], however, research activity intensified [8]. In 1981 ethylene vinyl acetate graft copolymer was introduced as a new polymeric modifier for CAP [9 to 11]. This development is certainly not yet over. Opportunities continue for the development of organic cellulose esters with improved heat distortion temperatures, stiffness, hardness, and barrier properties while retaining the transparency, impact resistance, and stress crack resistance typical of cellulosics. Further discussions of the properties of cellulose ester blends can be found in Section 5.8.3.

No compatible polymer modifiers have yet been found for cellulose diacetate. The best combination of properties in CAP/CAB blends with ethylene vinyl acetate copolymers or graft copolymers is obtained by adjusting the compositions such that there is slight immiscibility without compromising the material's transparency [1].

The electron micrograph in Figure 5.13 shows ethylene vinyl acetate copolymer (EVAC) embedded in a CAB matrix. Between 5 and 30 wt-% EVAC can be used. To obtain the desired high heat distortion temperatures, hardness, and rigidity, the amounts are generally limited to between 5 and 10 wt-%.

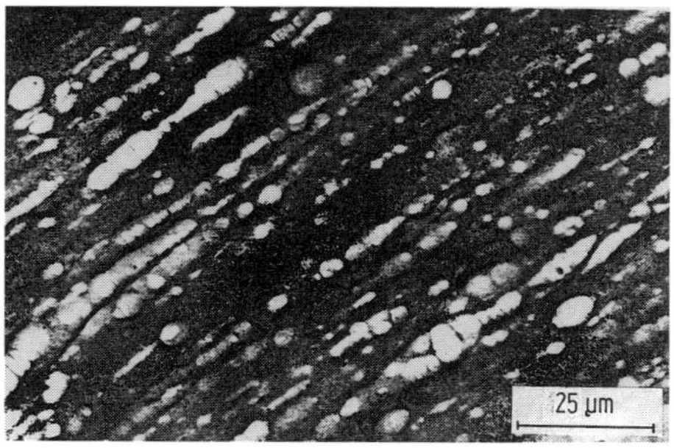

Figure 5.13. Cellulose acetobutyrate with 10 wt-% EVAC as modifier, electron microphotograph [8]

5.5.3 Thermal Stabilizers, Buffers, and Antioxidants

Commonly used stabilizers for organic cellulose esters are epoxides, especially epoxidized soy bean oils, and alkali or earth alkali salts of acetic or carbonic acid, as well as tartaric, oxalic, and citric acids. Effective antioxidants include substituted phenols such as t-butylphenol or preferably sterically hindered phenols such as 2,6-di-t-butylcresol. Diphenylamine and esters of thiodipropionic acid are also effective [12 to 14].

5.5.4 Stabilizers to Improve Weatherability

Only CAB and to some extent CAP molding compounds can be effectively stabilized to achieve long-term weatherability. To date no stabilizers for cellulose diacetate that maintain utility under outdoor weathering over a period of years have been found.

Common stabilizers for mixed esters are salicylic acid esters, resorcinol monobenzoate, UV-absorbers, and various combinations of these compounds. Level of 1–2 wt-% are usually effective [14]. Use of these additives in cellulose acetobutyrate molding materials prevents yellowing, loss of gloss, and loss of transparency; microcracking and loss of mechanical strength are considerably reduced. This allows parts such as automotive styling strips to maintain appearance and utility after outdoor weathering for many years [15].

5.5.5 Processing Aids

5.5.5.1 Mold Release Agents

Mold release agents are not often required for the molding of cellulose esters. Small amounts of paraffin oil (<1 wt-%) are used in cellulose acetobutyrate and propionate extrusion grades to avoid sticking on calender rolls. Butyl sterate is used in injection molding, for example of ball point pens, to aid the removal of material from the screw threads.

5.5.5.2 Resin Feeding Aids

Zinc sterate is used at levels of 0.01 to 0.02 wt-% to improve the feeding of granulate both in extrusion and injection molding. Higher levels lead to discoloration. It also prevents the development of small bubbles in thick-walled parts, working zinc sterate into the molding compound as opposed to the granulate is not effective.

5.5.6 Flame Retardants

Only flame retardants that do not adversely affect transparency and physical properties are commercially interesting. The addition of relatively large quantities of trichloracethyl phosphate (>20 wt-%) against severe flammability is known to attain V0 ratings of UL 94 even at thicknesses of 1.6 mm. Because the resulting deformation temperature (Vicat temperature 70–75 °C) and stiffness are only average, this procedure is of limited use.

5.5.7 Glass Fiber Reinforcement

It is possible in principle to add short glass fibers to cellulose ester resins, and this was widely practiced at one time [16]. The mechanical properties are dependent on a good binding of the fibers with the resin matrix through the use of appropriate glass sizers [17]. Glass fiber reinforced cellulose esters no longer have any significant applications.

5.5.8 UV/IR Absorbers

The relatively low processing temperatures of cellulose esters and the good solubility of UV and IR absorbers in them make cellulose compounds treated this way well adapted as optical filters [18]. These applications must be prepared with special care because the optical sector requires materials of very high purity that do not scatter light. CP resins have proven especially successful in lenses for sun and safety glasses.

5.5.9 Dyes and Pigments

Cellulose ester resins can be dyed in every desirable transparent, translucent, or opaque shade. Because of their low processing temperatures the choice of dyes and pigments is not limited. Testing must be done to prove light fastness especially for soluble dyes and "adjustment additives".

The upper limit of dye or pigment is 0.2 wt-% for transparent preparations and 2–3 wt-% for opaque preparations. The compatibility of commercial color concentrates or master batches with cellulose esters must be ascertained because they often contain incompatible binders.

5.6 Blends with Other Polymers

Except for the ethylene vinyl acetates and their respective graft copolymers described in Section 5.5.2 no other polymers have been found to form homogeneous and transparent blends from the point of view of compatibility and refraction index.

The results of experiments with ethylene vinyl acetate and CAB could not be transferred directly to cellulose acetopropionate since its compatibility with other polymers is limited in the first place. It was not until the development of graft copolymers based on EVAC [7, 9, 10] that materials sufficiently compatible with CAP were obtained. For the development of new blends fundamental questions still remain to be resolved by research on compatibility, which must in turn be informed by preparatory work on the structure of cellulose esters and polymers [1].

5.7 Production of Compounds

5.7.1 Preparation of Raw Materials

Stabilized milled cellulose ester flakes are predried by tumbling and then intensively mixed (slightly damp or after further drying) with plasticizers, stabilizers, and antioxidants. Starting with damp flakes promotes the distribution of plasticizer but the remaining dampness must be removed in roll dryers under slightly reduced pressure.

5.7.2 Addition of Additives

To the dried raw material as described in Section 5.7.1 (maximum water content 5 wt-%) additives are introduced using intensive mixers according to specific requirements: dyes, absorbers, mold release agents, weatherizing stabilizers, and so on; for fine adjustments thermal stabilizers and antioxidants are added. To adjust the hardness, additional plasticizers are introduced.

5.7.3 Compounding

Depending upon the type of cellulose ester and the plasticizer system, compounding occurs at 150–220 °C. For compounding heavy roll mills or single or twin screw kneaders are used.

Depending upon the machinery cubic granulate (pouring density of 400 to 470 g/l) or stand granulate (bulk density 550 to 620 g/l) is used [4]. Plasticizer is unavoidably volatizied during the compounding process and must be removed, condensed, and disposed of by incineration. There are no problems with waste water or other waste gases during compounding.

References for Section 5.5 to 5.7

1. *Müller, F.:* Das Papier 39 (1985) p. 591–600
2. Cellidor Ringbuch Company brochure Bayer AG, Leverkusen 1975
3. Cellidor, organische thermopl. Celluloseeter, Eigenschafte. Company brochures Albis Plastic GmbH, Hamburg 1988
4. *Eicher, Th.; Fischer, W.:* Celluloseeter. In: Ullmanns Encyklopädie der technischen Chemie. 4. Ed. Vol. 9. Verlag Chemie, Weinheim 1975
5. *Gnamm, H.; Sommer, W.:* Lösungsmittel und Weichmachungsmittel. Wiss. Verlagsgem., Stuttgart 1958
6. *Thinius, K.:* Chemie, Physik und Technologie der Weichmacher. Dr. Verlag für Grundstoffind., Leipzig 1963
7. DE-OS 2 426 178 (1974) Bayer AG
8. *Fischer, W.*, et al.: Kunststoffe 67 (1977) 6, p. 348–352
9. *Leuscher, Ch.; Wandel, M.:* Plastverarbeiter 33 (1982) 9, p. 1095–1098
10. DE-OS 2 951 747 (1981) Bayer AG
    DE-OS 2 951 748 (1981) Bayer AG
    DE-OS 2 951 800 (1981) Bayer AG
11. *Alberts, H.*, et al.: Kautsch. Gummi Kunststoffe 38 (1985) 8, p. 689–693
12. Modern Plastics Encyclop. 1970–1971, p. 840 and 876
13. BGBL 19 (1976) 6, p. XXVI
14. *Bogan, R. T.; Brewer, R. J.:* Celluloseesters/Plastics. In: Encyclopedia of Polymer Science and Engineering. 2. Ed. Vol. 3. Wiley, New York 1985, p. 158
15. Cellidor, Zierleisten in der Automobilindustrie. Company brochure Bayer AG, Leverkusen 1979
16. *Fischer, W.:* Kunststoffe 63 (1973) 5, p. 292–295
17. *Fischer, W.; Morbitzer, L.:* Untersuchunger über den Einfluß von Haftvermittlern auf den Verbund zwischen Glasfaser und Thermoplasten. In: Die Angewandte Makromolekulare Chemie. Vol. 9. Hüthig und Wepf, Basel 1969
18. *Wandel, M.; Leuschke, Ch.:* Kunststoffe 74 (1984) p. 589–592

5.8 Properties of Cellulose Ester Molding Compounds

5.8.1 General Properties

Cellulose ester resins show a unique combination of properties which distinguish them from other molding compounds. The broad spectrum of properties [1] is characterized by:
- high transparency
- high impact strength
- low sensitivity to stress cracking
- weather resistance (CAB molding compound)

In addition there are:
- rapid loss of electrostatic charges
- high resistance to surface leakage and good electrical insulating properties
- good sound absorbing properties and texture similar to natural horn
- good resistance to hydrolysis and attack by mineral oils
- unlimited capacity for coloration
- antifogging through surface saponification
- good printing, coating, and bonding properties

Compromises need to be made for the following properties:
- dimensional stability under heat
- stiffness, hardness, and creep behavior (cold flow)
- migration and evaporation of plasticizers
- solvent resistance

Improvement of these properties (with the exception of solvent resistance) may be achieved by polymer modification of cellulose ester molding, compounds [1 to 5].

5.8.2 Special Properties of CA, CAP, and CAB Molding Compounds

In addition to the mechanical and technological properties, the properties summarized in Table 5.8 need to be considered in the selection of cellulose ester molding compounds.

Table 5.8. Special Properties of Cellulose Ester Molding Compounds

Property	Cellulose Acetate	Cellulose Propionate	Cellulose Acetobutyrate
Low temperature impact strength	Brittle at $T < 0$ °C	to −30 °C	to −40 °C
Long term weather resistance	Resistance to stabilizing	Can be stabilized	Easy to stabilize
Plasticizer migration	Possible migration	Low migration	Low migration
Dimensional stability	Medium (high water absorption)	Good (low water absorption)	Good (low water absorption)
Odor	Nearly neutral	Neutral	Butyric acid

5.8.3 Mechanical, Thermal and Rheological Properties

A wide variety of cellulose ester materials with special properties can be produced by varying the methods and contents of modification with plasticizers and polymers and combining them with the right cellulose ester or mixtures of cellulose esters. Hardness, dimensional stability under temperature variation, toughness, bending and tensile strength, and flow characteristics can be adapted according to the specifications.

Hardness is adjusted by manufacturers in two different ways:

(a) adjustment with standard plasticizer contents within permissible deviations [7, 8];

(b) adjustment according to flow grade [9].

Because the melt viscosity depends not only on plasticizer content but also on the degree of polymerization of the cellulose ester batch (flakes) used, in case (a) a constant level of properties and in case (b) a constant flow grade is achieved.

5.8.3.1 Short-Term Properties

Tables 5.9 through 5.15 show the mechanical, thermal and rheological properties of molding compounds made of CA, CAP and CAB as a function of plasticizer content. Deviations of 15–20% for the property levels are possible according to the particular material's tolerance for plasticizers and the type of plasticizer used.

Table 5.16 show properties of CAP and CAB molding compounds containing plasticizers as compared to those with polymer modifiers, such as ethylene vinyl acetate and ethylene vinyl acetate grafted copolymers.

Table 5.9. Mechanical Properties of CA Molding Compounds Depending on Plasticizer Content *W*

Properties	Unit	Molding Compounds with *W*			Specifications
		32% wt-% material	27 wt-% material	22 wt-% material	
Tensile stress	N/mm^2	29	38	50	DIN 53455/ISO R527
Tensile yield	%	3.1	3.3	3.6	
3.5% bending stress	N/mm^2	37	50	65	DIN 53452
Modulus of elasticity	N/mm^2	1500	2000	2700	DIN 53457
Impact strength					
at 23 °C	kJ/m^2	n.b.[1]	n.b.	100 to n.b.	DIN 53453/ISO R179
at −40 °C	kJ/m^2	65	55	45	
Impact strength (notched)	kJ/m^2	18	8	5	
Izod impact strength	J/m^2	260	200	90	ASTM D256/ISO R180
Ball indentation hardness H_{30}	N/mm^2	42	55	78	DIN 53456/ISO 2039

[1] no break

Table 5.10. Mechanical Properties of CP Molding Compounds Depending on Plasticizer Content *W*

Properties	Unit	Molding Compounds with *W*				Specification
		20 wt-% material	15 wt-% material	10 wt-% material	5 wt-% material	
Tensile stress	N/mm^2	20	29	40	48	DIN 53455/ISO R527
Tensile yield	%	3.4	3.5	4.0	4.4	
3.5% bending stress	N/mm^2	30	40	53	60	DIN 53452
Modulus of elasticity	N/mm^2	1000	1400	1900	2110	DIN 53457
Impact strength						
at 23 °C	kJ/m^2	n.b.[1]				DIN 53453/ISO R179
at −40 °C	kJ/m^2	n.b.	n.b.	80 to n.b.	70 to n.b.	
Impact strength (notched)	kJ/m^2	15	10	8	2	
Izod impact strength	J/m^2	560	500	350	60	ASTM D256/ISO R180
Ball indentation hardness H_{30}	N/mm^2	35	45	62	78	DIN 53456/ISO 2039

[1] no break

Table 5.11. Mechanical Properties of CAB Molding Compounds Depending on Plasticizer Content *W*

Properties	Unit	Molding Compounds with *W* 20 wt-% material	15 wt-% material	10 wt-5 material	5 wt-% material	Specifications
Tensile stress	N/mm^2	17	25	34	42	DIN 53455/ISO R527
Tensile yield	%	3.5	3.6	4.0	4.3	
3.5% bending stress	N/mm^2	24	33	45	56	DIN 53452
Modulus of elasticity	N/mm^2	800	1200	1600	1900	DIN 53457
Impact strength						
at 23 °C	kJ/m^2	n.b.[1]	n.b.	n.b.	n.b.	DIN 53453/ISO R179
at −40 °C	kJ/m^2	n.b.	n.b.	90 to n.b.	80 to n.b.	
Impact strength (notched)	kJ/mm^2	25	18	10	4	
Izod impact strength	J/m^2	620	540	370	95	ASTM D256/ISO R180
Ball indentation hardness H_{30}	N/mm^2	25	35	53	70	DIN 53456/ISO 2039

[1] no break

Table 5.12. Thermal Properties of CA Molding Compounds Depending on Plasticizer Content *W*

Properties	Unit	CA Molding Compounds with *W* 32 wt-% material	27 wt-% material	22 wt-% material	Specifications
Coefficient of linear Thermal expansion	$K^{-1}10^{-6}$	120	115	105	VDE 0304/DIN 53752
Thermal conductivity	W/(k · m)	0.22	0.22	0.21	DIN 52612
Specific heat capacity	kJ/kg · K		1.3 to 1.7		
Vicat softening temperature VST/B/50[1]	°C	77	89	100	DIN 53460/ISO 306
Dimensional stability under heat					
Method A: 1.81 N/mm^2	°C	47	59	73	DIN 53461/ISO 75
Method B: 0.45 N/mm^2	°C	59	73	87	

[1] pretreatment of specimens 16 h at 80 °C

Table 5.13. Thermal Properties of CP Molding Compounds Depending on Plasticizer Content *W*

Properties	Unit	CA Molding Compounds with *W* 20 wt-% material	15 wt-% material	10 wt-% material	5 wt-% material	Specifications
Coefficient of linear Thermal expansion	$K^{-1}10^{-6}$	145	131	123	120	VDE 0304 DIN 53752
Thermal conductivity	W/(k · m)	0.22	0.21	0.21	0.21	DIN 52612
Specific heat capacity	kJ/kg · K			1.7		
Vicat softening temperature VST/B/50[1]	°C	69	85	98	108	DIN 53460 ISO 306
Dimensional stability under heat						
Method A: 1.81 N/mm^2	°C	62	79	90	94	DIN 53461
Method B: 0.45 N/mm^2	°C	74	88	97	99	ISO 75

[1] pretreatment of specimens 16 h at 80 °C

Table 5.14. Thermal Properties of CAB Molding Compounds Depending on Plasticizer Content *W*

Properties	Unit	CAB Molding Compounds with *W* 20 wt-% material	15 wt-% material	10 wt-% material	5 wt-% material	Specifications
Coefficient of linear Thermal expansion	$K^{-1}10^{-6}$	148	136	129	119	VDE 0304/ DIN 53752
Thermal conductivity	W/(k · m)	0.22	0.21	0.21	0.20	DIN 52612
Specific heat capacity	kJ/kg · K	1.7				
Vicat softening temperature VST/B/50[1]	°C	65	78	88	102	DIN 53460/ ISO 306
Dimensional stability under heat						
Method A: 1.81 N/mm^2	°C	60	70	78	94	DIN 53461/
Method B: 0.45 N/mm^2	°C	75	84	89	98	ISO 75

[1] pretreatment of specimens 16 h at 80 °C

Table 5.15. Rheological Properties of CA, CP, CAB Molding Compounds Depending on Plasticizer Content *W*

Property	Unit					Specifications
		CA Molding Compounds with *W*	32 material %	27 material %	22 material %	
Melt index[1] MFI (210/2.16)	g/10 min		10	5	2.1	DIN 53753/ ISO R 113
		CP Molding Compounds with *W* 20 wt-% material	15 wt-% material	10 wt-% material	5 wt-% material	
		28.5	11.5	3.7	2.0	
		CAB Molding Compounds with *W* 20 wt-% material	15 wt-% material	10 wt-% material	5 wt-% material	
		30	14.5	6.5	2.0	

[1] pretreatment of granules according to DIN 7742 part 2

5.8.3.2 Long-Term Properties [6]

Figures 5.14 through 5.17 [6] give an overview of the long-term properties of CA, CAP, and CAB platicizer modified molding compounds adjusted in typical hard and soft versions respectively. Figure 5.18 compares creep rupture strength of a hard cellulose ester molding compound containing plasticizers with that of a hard polymer modified cellulose ester compound.

5.8.4 Electrical Properties

Cellulose ester molding compounds are distinguished by high dielectric strength and high tracking resistance. Nevertheless, applications are limited because dimensional stability at elevated temperatures and combusion performance are often inadequate to

Table 5.16. Mechanical, Thermal and Rheological Properties of Polymer Modified and Plasticizer Containing (*W*: Plasticizer Content) CP and CAB Molding Compounds

Properties	Units	CP Molding Compounds with *W*		CAB Molding Compounds with *W*		Specifications
		5 wt-%[1] material	5 wt-%[2] material	10 wt-%[1] material	10 wt-%[2] material	
Tensile stress	N/mm²	48	62	34	50	DIN 53455/
Tensile yield	%	4.4	6.6	4.0	4.7	ISO R 527
3.5% bending stress	N/mm²	60	75	45	68	DIN 53452
Modulus of elasticity	N/mm²	2110	2600	1600	1900	DIN 53457
Impact strength						
at 23 °C	kJ/m²	n.b.[3]	n.b.	n.b.	n.b.	DIN 53453/
at −40 °C	kJ/m²	70 to	n.b.	n.b.	n.b.	ISO R179
Vicat temperature VST/B/50	°C	108	127	88	96	DIN 534560/ ISO R306
Melt index MFI (210/2,16)	g/10 min	2.0	1	6.5	2.2	DIN 53753/ ISO R113

[1] monomeric plasticizer
[2] polymer modified (EVAC or EVAC grafted copolymer)
[3] no break

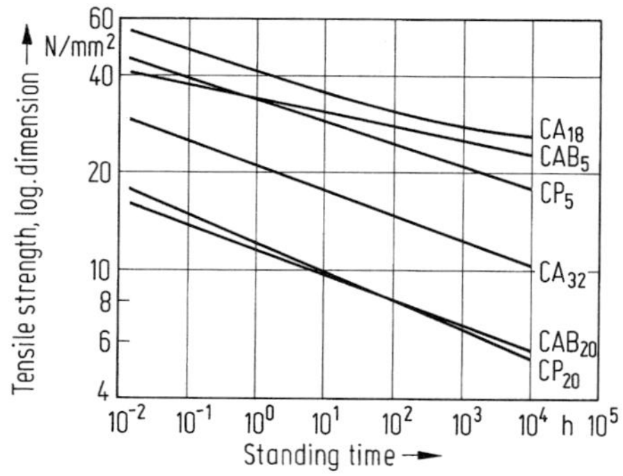

Figure 5.14. Creep rupture strength according to DIN 53444 or ISO/R 899 of cellulose acetate with 18% plasticizer (hard), cellulose acetate with 32% plasticizer (soft), cellulose propionate with 5% plasticizer (hard), cellulose propionate with 20% plasticizer (soft), cellulose acetobutyrate with 5% plasticizer (hard) and cellulose acetobutyrate with 20% plasticizer (soft). Test conditions: 23 °C, 50% relative humidity [6]

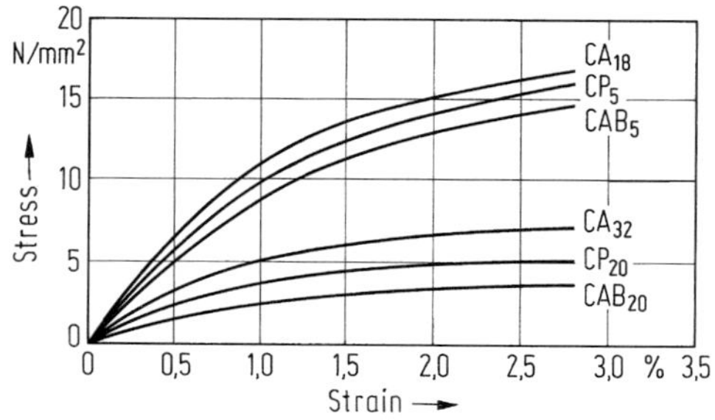

Figure 5.15. Isochronic stress-strain according to DIN 53444 or ISO/R 899 of cellulose acetate with 18% plasticizer (hard), cellulose acetate with 32% plasticizer (soft), cellulose propionate with 5% plasticizer (hard), cellulose propionate with 20% plasticizer (soft), cellulose acetobutyrate with 5% plasticizer (hard), and cellulose acetobutyrate with 20% plasticizer (soft). Test time 1000 h, test conditions 23 °C, 50% relative humidity [6]

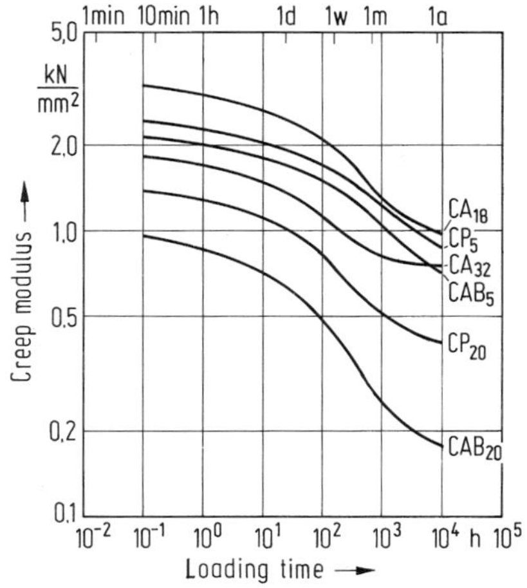

Figure 5.16. Creep modulus according to DIN 53444 or ISO/R 899 of cellulose acetate with 18% plasticizer (hard), cellulose acetate with 32% plasticizer (soft), cellulose propionate with 5% plasticizer (hard), cellulose propionate with 20% plasticizer (soft), cellulose acetobutyrate with 5% plasticizer (hard), and cellulose acetobutyrate with 20% plasticizer (soft). Test conditions: 23 °C, 50% relative humidity [6]

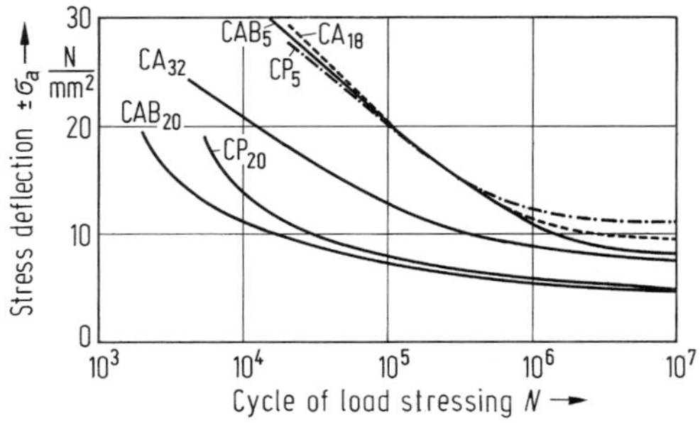

Figure 5.17. Flexural fatigue test by DIN 50100 of cellulose acetate with 18% plasticizer (hard), cellulose acetate with 32% plasticizer (soft), cellulose propionate with 5% plasticizer (hard), cellulose propionate with 20% plasticizer (soft), cellulose acetobutyrate with 5% plasticizer (hard), and cellulose acetobutyrate with 20% plasticizer (soft). Frequency 7 Hz.

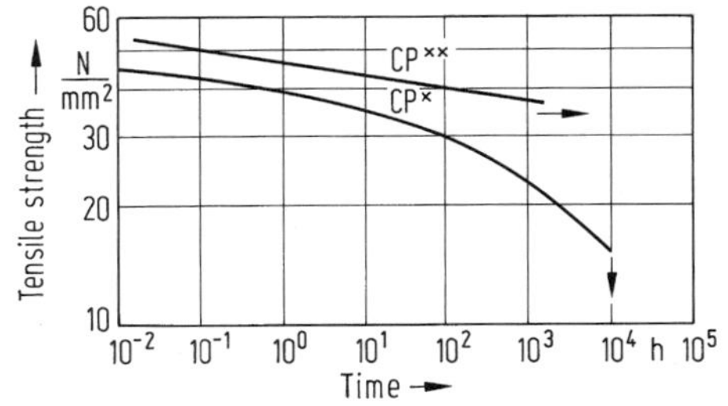

Figure 5.18. Creep rupture strength test according to DIN 53444 or ISO/R 899 cellulose propionate with 5% plasticizer (hard), compared to EVAC-graft-cellulose propionate (5% grafted)

Table 5.17. Electrical Properties of Cellulose Ester Molding Compounds with Medium Hardness (CA, CP, CAB with Plasticizer Content *W*) [6]

Type of Test	Unit	Test Specimen	CA with $W = 22$ wt-% Material	CP with 10 wt-% Material	CAB with 10 wt-% Material	Specifications
Dielectric strength *E*d (50 Hz, 0.5 kV/s)						
dry		Round discs	31.5	35.5	35.0	VDE 0303
4 days at 80% relative humidity	kV/mm	95 mm diamter	29.0	33.0	33.0	part 2,
24 hrs. in water		× 1 mm	28.0	33.0	33.0	DIN 53481
Surface resistance						
dry			$8 \cdot 10^{13}$	$2 \cdot 10^{14}$	$9 \cdot 10^{13}$	VDE 0303
4 days at 80% relative humidity	Ω		$3 \cdot 10^{12}$	$1 \cdot 10^{13}$	$9 \cdot 10^{12}$	part 3,
24 hrs. in water		150 mm × 15 mm × 4 mm	$4 \cdot 10^{11}$	$5 \cdot 10^{12}$	$9 \cdot 10^{12}$	DIN 53482
Resistance between contacts R_{oc}						
dry			$5 \cdot 10^{15}$	$5 \cdot 10^{15}$	$5 \cdot 10^{15}$	VDE 0303
4 days at 80% relative humidity	Ω		$1 \cdot 10^{13}$	$6 \cdot 10^{13}$	$5 \cdot 10^{13}$	part 3,
24 hrs. in water		150 mm × 15 mm × 4 mm	$7 \cdot 10^{11}$	$2 \cdot 10^{13}$	$2 \cdot 10^{13}$	DIN 53482
Volume resistivity						
dry		Round discs	$2 \cdot 10^{15}$	$1 \cdot 10^{16}$	$4 \cdot 10^{15}$	VDE 0303
4 days at 80% relative humidity	Ω/cm	95 mm diameter	$2 \cdot 10^{12}$	$5 \cdot 10^{13}$	$6 \cdot 10^{13}$	part 3,
24 hrs. in water		× 1 mm	$2 \cdot 10^{11}$	$1 \cdot 10^{13}$	$2 \cdot 10^{13}$	DIN 53482
Dielectric constant, dry 1 ε_1 trocken						
at 50 Hz		Round discs	5.1	4.1	4.0	VDE 0303
at 800 Hz		95 mm and 80 mm	4.0	3.9	3.8	part 4,
at 1 MHz		diameter × 1 mm	4.1	3.6	3.4	DIN 53483
Dissipation factor, dry tan δ						
trocken						
50 Hz		Round discs	0.009	0.005	0.006	VDE 0303
at 800 Hz		95 mm and 80 mm	0.019	0.011	0.012	part 4,
at 1 MHz		diameter × 1 mm	0.050	0.026	0.028	DIN 53483
Tracking resistance						
KB test						VDE 030
test solution A		20 mm × 15 mm	>600	>600	>600	DIN 53480/6

meet test requirements. The most important electrical properties of CA, CAP, and CAB at medium hardness are compiled in Table 5.17 [6].

5.8.5 Weathering Performance

No effective weathering stabilizers for CA molding compounds have been found. CAB molding compounds with ethylene vinyl acetate modification keep their gloss longer after weathering than CAB with plasticizers [2]. Cellulose ester molding compounds show almost no yellowing. A surface treatment of celluloe ester molded parts with solutions containing UV absorbers is more effective than working UV absorbers into the bulk.

Figure 5.19 shows the decrease in mechanical strength of sheets of CAB molded parts (medium hardness) after outdoor weathering in Central Europe. In normal practice the demands will lie somewhere in between the values for compression and tensile areas.

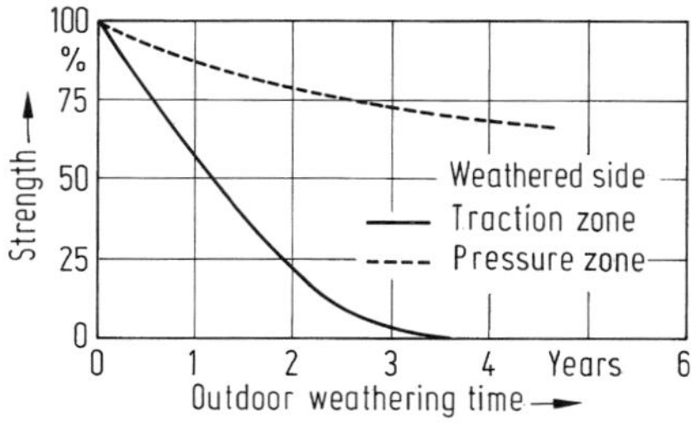

Figure 5.19. Decrease of mechanical strength, after outdoor weathering in Central Europe of extruded sheets of cellulose acetobutyrate (3 mm thickness) with 10% plasticizer plus stabilizer, EKWA procedure after DIN 53373

5.8.6 Resistance to Chemicals

Table 5.18 describes the behavior of CAP or CAB and CA molded parts exposed to various solvents, mineral oils, acids, and bases [7, 8]. It must be emphasized that loss of plasticizers during long-term contact with solvents produces brittleness. Short-term contact, even with swelling solvents, does not result in stress corrosion cracking. After evaporation of the solvents, the initial property level is restored.

- Cellulose ester molding compounds are permeable to water vapor [90 to 120] $\cdot$ 10^{-8} g/cm $\cdot$ h $\cdot$ Torr). Therefore, they are not suitable for packaging of water-containing material.

5.8.7 Combustion Behavior

As compared to nitrocellulose, organic cellulose ester molding compounds burn slowly but are not self-extinguishing. Effective flame retardants that do not influence stiffness, transparency, and dimensional stability under heat are unknown. The combusion behavior according to Limiting Oxygen Index (16.8 to 19.6%) is within the range found in polymers with aliphatic main chains [10]. Table 5.19 gives burning speeds for CA, CAP, and CAB molding parts with medium hardness.

Table 5.18. Chemical Resistance of Cellulose Ester Molding Compounds (Independent of Plasticizer) Rating: 1: Resistant; 2: Short-term Resistant; 3: Unstable; 4: Swelling; 5: Soluble

Solvent	CP and CAB Molding Compounds	CA Molding Compounds
Water	1	1
Alcohols	4	4
Ethyl acetate	5	5
Methylene chloride	5	5
Acetone	5	5
Carbon tetrachloride	5	1
Trichoroethylene	4	4
Perchloroethylene	4	1
Benzene	4	1
Xylene	4	1
Kerosene	1	1
Higher octane gasoline	4	1
Mineral oil (paraffins)	1	1
Linseed oil	1	2
Turpentine	4	1
Lavender oil	4	2
Ether	2	1
Formaldehyde	2	3
Chlorophenol	5	5
Sulfuric acid, conc.	3	3
Sulfuric acid, 10%	2	2
Hydrochloric acid, conc.	3	3
Hydrochloric acid, 10%	3	3
Nitric acid, conc.	3	3
Nitric acid, 10%	3	3
Potassium hydroxide, 50%	3	3
Potassium hydroxide, 10%	2	3

Table 5.19. Burning Rate of Cellulose Ester Molding Compounds (CA, CP, and CAB) after ASTM-D635-74

Molding Compound	Plasticizer Content wt-%	Burning Rate cm/min
Cellulose acetate (CA)	22	2.8
Cellulose propionate (CP)	10	3.2
Cellulose acetobutryate (CAB)	10	3,2

5.8.8 Optical Properties

Cellulose ester molding compounds achieve levels of light transmission similar to window glazing for visible light. The high transmission of UV light is decreased by UV absorbers for applications such as sunglasses. For high quality material the stray light is less than 0.5 $cd/m^2 \cdot lx$ (2 mm thickness). The refraction index n_D is 1.48 for CAP and CAB molding compounds and 1.49 for CA molding compounds, depending on plasticizer content.

Figure 5.20 shows the light transmission of a CAP molding compound of transparency equivalent to that of glass with and without UV absorbers. CA and CAB molding compounds exhibit virtually the same behavior.

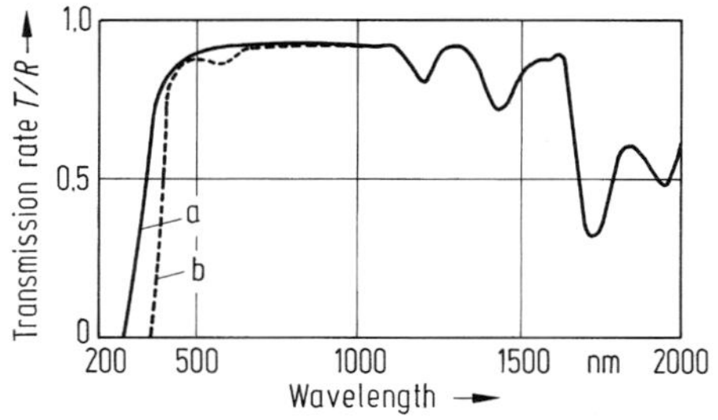

Figure 5.20. Cellulose propionate molding compound (a) with and (b) without UV absorber. Sample thickness: 2 mm. Color: crystal clear

5.8.9 Raw Density

The raw densities given in Table 5.20 refer to non-pigmented molding parts of CA, CAP, and CAB.

Table 5.20. Denisty of Cellulose Ester Molding Compound (CA, CP and CAB) Depending on Plasticizer Conent after DIN 53479 or ISO R1183

Molding Compound	Plasticizer Content wt-%	Density g/cm³
Cellulose acetate (CA)	32	1.26
	27	1.27
	22	1.28
Cellulose propionate (CP)	20	1.19
	15	1.20
	10	1.21
	5	1.22
Cellulose acetobutyrate (CAB)	20	1.17
	15	1.18
	10	1.19
	5	1.21

References for Section 5.8

1. *Leuschke, Ch.; Wandel, M.:* Plastverarbeiter 33 (1982) 9, p. 1095–1098
2. *Fischer, W.*, et al.: Kunststoffe 67 (1977) 6, p. 348–352
3. DE-OS 2 426 178 (1974) Bayer AG
4. DE-OS 2 951 757 (1981) Bayer AG
   DE-OS 2 951 748 (1981) Bayer AG
   DE-OS 2 951 800 (1981) Bayer AG
5. *Alberts, H.*, et al.: Kautsch. Gummi Kunstat. 38 (1985) 8, p. 689–693
6. *Eicher, Th.; Fischer, W.:* Celluloseester. In: Ullmanns Encyklopädie der technischen Chemie. 4. Ed. Vol. 9. Verlag Chemie, Weinheim 1975
7. Cellidor, organische thermopl. Celluloseester, Eigenschaften. Company brochure Albis Plastic GmbH, Hamburg 1988
8. Dexel. Company brochure Courtaulds Speciality Plastics, Spondon-Derby (U.K.) 1988
9. Tenite Cellulosic Plastics, Acetate-Butyrate-Propionate. Company brochure Eastman Chemical Products, Kingsport (USA) 1983/84
10. *Isaacs, J. L.:* J. Fire Flamm. 1 (1970) p. 36

5.9 Analysis and Standards for Organic Cellulose Esters

H. H. Müller, A. Horbach

5.9.1 Analysis of Chemical Structure

The chemical structure of cellulose esters is characterized by the degree of substitution, DS, which is defined as the average number of acyl groups per glucopyranose unit. The degree of substitution can be determined directly by measuring the number of acyl groups or indirectly by measuring the number of OH groups. For mixed cellulose esters such as cellulose acetopropionate or cellulose acetobutyrate, an analysis of the different acyl groups (such as the content of acetyl, propionyl, and butyryl groups) is needed for an exact description.

Because the solubility of cellulose esters depends mainly on their chemical structure for typical molecular weights and molecular weight distributions, the degree of substitution can be estimated from solubility, for example for cellulose acetate.

Generally the content of acidic groups is determined by alkaline saponification (e.g. with aqueous NaOH or ethanolic KOH) and subsequent back-titration of the excess base [1 to 7]. This relatively simple experimental analytical procedure is applied as an ASTM standard method for cellulose acetate [8] and mixed esters such as cellulose acetopropionate and cellulose acetobutyrate [9]. With mixed esters, the C_2, C_3, and C_4 carboxylic acids (as produced by alkaline saponification are distilled from the acidic saponification solution and analyzed either by gas chromatography or by titration after separation of the acids by selective extraction [1 to 7]. In the case of cellulose acetate, the amount of acetic acid can be determined by gas chromatography even without separation from the acidified saponification solution [10]. An elegant procedure to determine the content of acyl groups of cellulose esters is by methanolysis in the presence of BF_3 [11] or sodium or lithium methanolate [12, 13] and subsequent analysis of the methyl esters by gas chromatography. The acetyl or acyl groups can also be split off by phosphoric acid (170 °C, 15 minutes) and detected by gas chromatography [14].

The content of hydroxyl groups in cellulose esters is often determined in the form of an OH number by esterfication with acetic anhydride. The cellulose ester sample is reacted with a surplus of acetic anhydride, then the consumption of the anhydride is determined by hydrolysis of the excess anhydride and acidimetric titration of the acetic acid formed [2, 5, 7]. Another widely used wet chemical procedure to determine hydroxyl groups in cellulose esters, which is also an approved ASTM standard [8, 9], is the reaction with phenyl isocyanate, usually in pyridine solution, to form phenyl urethane groups. The derivatized polymer is precipitated, for example in acetone, isolated, and the content of introduced phenyl urethane groups is analyzed by UV photometry [1, 7, 15]. The content of primary hydroxyl groups in cellulose esters can also be determined selectively by reaction with chlorotriphenyl methane, usually in pyridine solution. After the reaction, the polymer is precipitated with acetone or another suitable medium and the amount of triphenyl methyl ether groups introduced by the derivatization is determined by UV photometry [1, 7 to 9, 16].

As an alternative to wet chemical methods, the contents of acyl or hydroxyl groups in cellulose esters can be determined by proton or ^{13}C-NMR spectroscopy [17 to 20], IR spectroscopy [21], or pyrolysis gas chromatography [7, 22]. In many cases, NMR spectroscopy is capable of determining the distribution of the acyl and hydroxyl groups at the C_2-, C_3-, and C_6-positions of the glucopyranose unit [20]. Infrared spectroscopy and pyrolysis gas chromatography add to the cost of calibration. In practice, these techniques of instrumental analysis are rarely used for routine needs.

5.9.1.1 Analysis of Molecular Weight

Molecular weight determinations of cellulose are generally carried out for routine investigations by solution viscosity measurements in capillary viscosimeters of Ostwald or Ubbelohde type. The selection of solvents depends on the kind of cellulose ester and the degree of substitution [8, 9]. For example, the viscosity of cellulose acetate with acetic acid content $\geq 50\%$ can be measured in a dichloromethane/methanol mixture (9:1 parts by volume) [23]. If the solution viscosity is determined for several diluted polymer solutions with different concentrations, the intrinsic viscosity (η) can be extrapolated by standard procedures [24, 25]. The average molecular weight can then be calculated as an average viscosity $\bar{M}_v$ using the Mark-Houwink-Kuhn-Sakurada function (which is tabulated for cellulose acetate and other cellulose esters in different solvents [20, 26 to 28].

Absolute numbers for average molecular weight of cellulose esters can be obtained by light scattering methods ($\bar{M}_w$) or by membrane or vapor pressure osmometry ($\bar{M}_n$) [20]. Formamide, dimethyl acetamide, acetone, tetrahydrofuran, and chloroform are used as solvents of cellulose acetate for light scattering measurements [19, 20, 29]. Difficulties due to microgel formation in the polymers may occur. For membrane osmosis measurements, dimethyl acetamide, acetone, tetrachloroethane and chloroform [6, 20, 30] are used.

A molecular weight distribution characterization for cellulose esters can be obtained by gel permeation chromatography. Cellulose acetate with DS ≥ 2.5 can be analyzed under standard conditions with gels made from poly(styrene-co-divinylbenzene) with tetrahydrofuran as the eluent with a refraction index detector [32]. In general, the molecular weight determined based on calibration with polysytrene may not result in actual cellulose ester molecular weights, but only in relative numbers. If these relative numbers are not sufficient, actual numbers can be obtained from gel permeation chromatography using a light scattering detector.

5.9.1.2 Analysis of Additives

Additives in cellulose esters such as plasticizers, stabilizers, antioxdants, and UV absorbers are determined by different techniques depending on their kind and amount. They generally include spectroscopic or chromatographic methods (e.g. UV, IR, NMR spectroscopy, gas chromatography, HPLC) [1, 34, 35]. In simple cases involving highly soluble cellulose esters, these additives can be determined directly in solution by using an appropriate solvent. However, sample preparation techniques have to be applied to enrich the additive and/or separate it from the matrix in case of interferences in the subsequent analysis steps: for example, Soxhlet extraction of additives or precipitation of polymers in a solvent that keeps only the additives that are to be specified in solution. The plasticizer content of CA is extracted routinely through Soxhlet extraction in diethyl ether. The extracted, restricted polymer components are then weighed [36].

5.9.2 Standards for Cellulose Ester Molding Compounds

5.9.2.1 International Standards

BS 1524-55:	Cellulose Acetate Molding Material [37]; contents: Specification of hardness and testing
ASTM-D 706-82:	Specification for Cellulose Acetate Molding and Extrusion Compounds [38]

ASTM-D 707-82: Specification for Cellulose Acetate Butyrate Molding and Extrusion Compounds [38]

ASTM-D 1562-82: Standard Specification for Cellulose Propionate Molding and Extrusion Compounds [38]; contents: Specification of hardness and tests

5.9.2.2 German Standards[1]

DIN 7742, Part 1: Plastic Compounds–Cellulose Ester Molding Compounds (CA, CP, CAB) [39]. Classification and Description (Jan. 1988); contents: Classification of cellulose esters according to their chemical compositions, main applications, major additives and most important properties

DIN 7742, Part 2: Plastic Compounds–Cellulose Ester Molding Compounds (CA, CP, CAB) [39]. Manufacturing of test specimens and determination of properties (1989); contents: Conditions for manufacturing of test specimens, description of standard tests and deviations from standard procedures (no numbers for properties provided)

5.10 Health and Environmental Aspects

5.10.1 Health Specifications

Cellulose ester molding compounds are considered to be harmless with respect to human health. In Germany, the Federal Ministry of Health has released upper limits for the amounts of stabilizers and optical brighteners, type and amount of plasticizers, and amounts of other processing additives [40]. CAB was not considered because of its butyric acid odor.

Shipping of cellulose molding compounds is not regulated because those materials are not considered hazardous [5].

5.10.2 Biodegradation of Cellulose Esters

The ever-increasing use of plastics increases waste disposal problems. In 1986, plastics accounted for about 12% of all waste [41]. For many years, various ways to reduce plastic waste have been investigated: recycling, chemical treatment for recovering the monomer units, monomer recycling through pyrolysis, energy recycling through incineration, and development of biodegradable polymers. The expanded use of biodegradable polymers is considered to be a very promising solution [42, 43].

For plastics produced by derivatization of the natural products, the biodegradation solution would seem to be within view already. Natural polymers such as starch, dextran, cellulose, and proteins already have the necessary chemical structure for biological degradation, which generally occurs in two steps:

- hydrolysis of the polymer chain
- oxidation of the fragments

This biodegradability is largely retained in hydrophilic cellulose esters. Cellulose esters can be transformed by partial saponification at DS < 1 into degradable materials [44].

[1] The coding is identical to that of the no longer used system of ISO/TC_{61}

While the degradability itself is determined by the chemical structure, the rate of degradation is determined by the morphology; for example foam-like structures will accelerate the degradation process. The introduction of cellulose units into block copolymers with synthetic polymers produces biodegradability even for such mixed systems. Thin films of block copolymers based on cellulose acetate are promising development in that direction [45].

References for Section 5.9 and 5.10

1. *Schröder, E.; Franz, J.; Hagen, E.:* Ausgewählte Methoden der Plastanalytik. Akademie-Verl., Berlin 1976
2. *Yarsley, V. E.; Flavell, W.; Adamson, P. S.; Parkins, N. G.:* Cellulosic Plastics. Iliffe Books, London 1964
3. *Rouse, B. P.:* Cellulose Ester. In: *Mark, H. F.; Gaylord, N. G.:* Encyclopedia of Polymer Science and Technology. Vol. 3. Interscience Publ., New York 1965
4. *Krause, A.; Lange, A.:* Kunststoff-Bestimmungsmöglichkeiten. 3. Ed. Hanser, Munich 1979
5. *Eicher, T.; Fischer, W.:* Celluloseester. In: Ullmanns Encyklopädie der technischen Chemie. Vol. 9. Verlag Chemie, Weinheim 1975
6. *Tsuzuki, M.*, et al.: J. Appl. Polym. 25 (1980) p. 2567
7. *Müller, H. H.:* Charakterisierung von Polymeren. Chemische Methoden. In: *Bartl, H.; Falbe, J.:* Houben-Weyl, Methoden der organischen Chemie, Makromolekulare Stoffe. Vol. E 20, Thieme, Stuttgart 1987
8. ASTM D 871: Standard Methods of Testing, Philadelphia 1972
9. ANSI/ASTM D 817: Standard Methods of Testing. Philadelphia 1972
10. *Esposito, G. G.; Swann, M. H.:* J. Paint Technol. 43 (1971) p. 60
11. *Wandel, M.; Tengler, H.:* Gummi Asbest Kunstst. 19 (1966) p. 141
12. *Haken, J. K.:* Progr. Org. Coat. 7 (1979) p. 209
13. *Haken, J. K.:* Int. Lab. (1982) p. 26
14. *Williams, R. J.; Siggia, S.:* Anal. Chem. 49 (1977) p. 2337
15. *Voss, W.; Reimschuessel, R.:* Makromol. Chem. 27 (1958) p. 101
16. *Malm, C. J.*, et al.: Anal. Chem. 26 (1954) p. 188
17. *Goodlett, V. W.*, et al.: J. Polym. Sci., 9 (1971) p. 155
18. *Kamide, K.; Okajima, K.:* Polym. J. 13 (1981) p. 127
19. *Kamide, K.; Saito, M.:* Eur. Polym. J. 20 (1984) p. 903
20. *Kamide, K.; Saito, M.:* Adv. Polym. Sci. 83 (1987) p. 5
21. *Dechant, J.:* Ultrarotspektroskopische Untersuchungen an Polymeren. Akademie-Verl., Berlin 1972
22. *Groten, B.:* Anal. Chem. 36 (1964) p. 1206
23. International Organization for Standardization, Norm ISO 1157 (1975)
24. *Hoffmann, M.; Krömer, H.; Kuhn, R.:* Polymeranalytik I. Thieme, Stuttgart 1977
25. *Bohdanecky, M.; Kovar, J.:* Viscosity of Polymer Solutions. Elsevier, Amsterdam 1982
26. *Phillip, H. J.; Bjork, C. F.:* J. Polym. Sci. 6 (1951) p. 549
27. *Rafikov, S. R.; Pavlova, S. A.; Tverdokhlebova, J. J.:* Determination of Molecular Weights and Polydispersity of High Polymers. Israel Program for Scientific Translations, Jerusalem 1964
28. *Bandrup, J.; Immergut, E. H.:* Polymer Handbook. 2. Ed. Wiley, New York 1975
29. *Sharples, A.; Swinton, F. L.:* J. Polym. Sci. 50 (1961) p. 53
30. *Kamide, K.*, et al.: Polym. J. 11 (1979) p. 523
31. *Yau, W. W.; Kirkland, J. J.; Bly, D. D.:* Modern Size—Exclusion Liquid Chromatography. Wiley, New York 1979
32. *Kamide, K.*, et al.: Brit. Polym. J. 15 (1983) p. 91
33. *Crompton, T. R.:* Chemical Analysis of Additives in Plastics. 2. Ed. Pergamon Pr., Oxford 1977
34. *Haslam, J.; Willis, H. A.; Squirrell, D. C. M.:* Identification and Analysis of Plastics. 2. Ed. Heyden, London 1980
35. *Crompton, T. R.:* The Analysis of Plastics. Pergamon Pr., Oxford 1984
36. International Organization for Standardization, Norm ISO 1875 (1982)
37. BS 1542: Cellulose Acetate Molding Material, British Standards House, London 1955
38. ASTM-D 706: Standard Specification for Cellulose Acetate Molding and Extrusion Compounds. Philadelphia 1982

ASTM-D 707: Standard Specification for Cellulose Acetate Butyrate Molding and Extrusion Compounds. Philadelphia 1982
ASTM-D 1562: Standard Specification for Cellulose Propionate Molding and Extrusion Compounds. Philadelphia 1982
39. DIN 7742 Part I: Kunststoff-Formmassen; Celluloseester (CA, CP, CAB)-Formmassen, Einteilung und Bezeichnung: DIN 7742 Part II: Herstellung von Probekörpern und Bestimmung von Eigenschaften. Beuth, Berlin 1988/Entwurf 1989
40. BGBL 19 (1976) 6, p. XXVI
41. *Tallman, J.*: Waste Age. 18 (1978) p. 141
42. *Milgrom, J.*: Polym. Plast. Technol. Eng. 18 (1982) p. 167
43. Mod. Plast. Int. 17 (1987) Juli, p. 64
44. *Reese, E. T.*: Ind. Eng. Chem. 49 (1957) p. 89
45. *penn, B. G.*: J. Macromol. Sci. Chem. A 16 (1981) 2, p. 473

5.11 Processing of Thermoplastic Cellulose Ester Compounds

5.11.1 Preparation

Cellulose esters absorb moisture during storage and therefore should be stored in dry areas. Predrying of the resin with conventional dryers is essential. Without predrying, CA can be processed with a maximum moisture level of 0.15% (injection molding) or 0.05% (extrusion). The drying time depends on percentage of moisture, type of dryer and drying temperature. Drying temperature should be set between 60 and 90 °C, depending on plasticizer content. Softer formulations tend to form lumps at overly high drying temperatures and plasticizers evaporate (especially cellulose acetate material). For medium hard and hard formulations under normal conditions temperatures of 70–80 °C and 3 hours in circulating air drying cabinets, or 1.5 hours in high efficiency dryers, are sufficient.

Predrying is not necessary if a vented screw and the mositure content does not exceed greater than 0.6%. The plasticizer evaporates during predrying and processing, producing fumes that are odorous and should be vented in the interest of workplace hygiene.

5.11.1.1 Cleanliness

Cellulose esters are not only incompatible with most other thermoplastics, but also with each other. Only mixed cellulose esters (CAP and CAB) can be mixed with each other. Therefore, strict cleanliness must be practiced during storage, drying and processing of regrind.

5.11.1.2 Processing of Regrind

Cellulose ester materials can be reused without noticeable loss in properties. Strict cleanliness must be observed and thermally degraded material must be sorted out.

The regrind must be shredded, fines filtered out, and the remaining material dried before reuse. Usually up to 20% regrind is added to virgin material. Separate processing is recommended for sensitive colors. Mixing materials of different hardness formulations should be avoided since this could result in problems such as lack of homogeneity with extruded material.

5.11.2 Injection Molding [1 to 5]

5.11.2.1 Injection Molding Machines

Conventional injection molding machines are suitable for thermoplastic processing of cellulose esters. The following information is, therefore, geared mainly to material-specific details.

Three-zones screws with a L/D ratio of 18:1 to 20:1 and a flight depth ratio of 2:1 to 3:1 are preferred. The cylinder and screw do not need to be chrome plated.

Screw feeding is not only dependent upon the screw geometry but also the feed zone temperature. Heat due to friction generated by high screw speed, that is, high throughput, must be controlled by external cooling or lower cylinder temperature to prevent granulate adhesion to the screw which would interrupt the throughput. Past experiences have shown that this unique cellulose ester resin behavior occurs almost always with screw diameters of greater than 80 mm. The solution is to enlarge the flight depth of the feed zone (compare Table 5.21).

The flight depth ratio should not be adjusted upwards and should be kept within the set limits. In the event the screw geometry is not ideal the material feeding can be improved by adding 0.01–0.02% zinc stearate. The low viscosity of cellulose ester melt absolutely requires a non-return valve. The length of the shear gap should be less than 2 mm if the screw diameter is small, but not longer than 4 mm for larger screws. The use of shut-off nozzles is not recommended due to pressure loss and heat damage.

Table 5.21. Feed Zone Flight Depth as a Function of Screw Diameter

Screw Diameter mm	Conversion Factor A for Intermediate Values at Screw Diameter	Flight Depth in Feed Zone A × D mm
90	0.135	12
120	0.13	15
>120	–	max. 16

5.11.2.2 Injection Molding Die

Demolding can be facilitated by a 1° taper or slant in the mold. All edges on the mold should be rounded because sharp edges reduce mechanical strength with a notching effect. The easy flowability of cellulose ester melt permits the use of pinpoint gates. The danger of free meltstream formation can be avoided by a deflector plate (opposite mold wall) at a maximum 3 mm distance from the gate.

Hot runner molds are also suitable for cellulose esters. Both halves of the mold, especially when molding flat parts, should be temperature controlled separately. The mold temperature, depending on material hardness, should be 40–80 °C. Good venting of the mold during cellulose ester processing is important to prevent plasticizer precipitation (dull spots) and burn marks (bubbles).

Linear moldfilling is very strongly to preferred as centrally located gates lead to warpage. The location of the gate is always a weak spot in the molded part (leading to notching) and has to be placed in an area of no load.

5.11.2.3 Processing Temperatures

The cellulose ester processing temperatures are dependent on the material hardness setting, as well as flowlength and wall thickness of the part. Figure 5.21 shows a

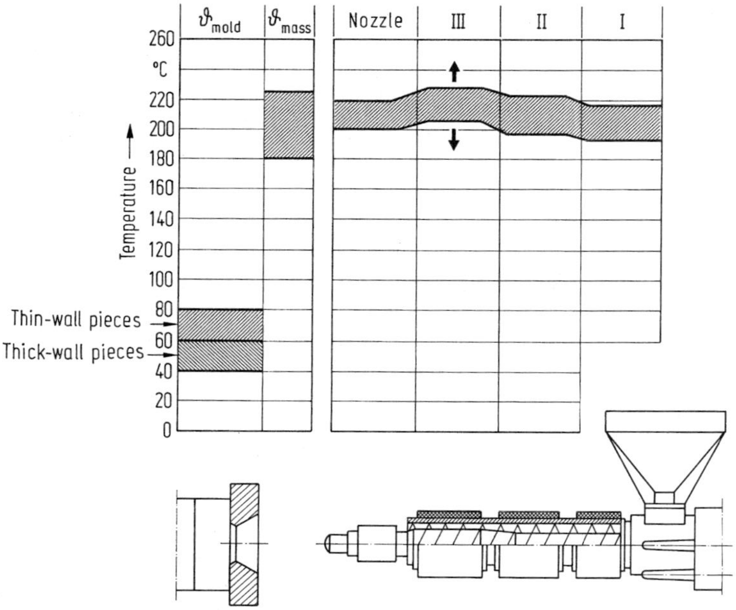

Figure 5.21. Mold temperature and temperature profile from the feeding to the nozzle for cellulose ester with hard and soft settings

temperature profile from the feed zone to the nozzle. The high temperatures are for the harder material settings and the low temperatures for soft settings. The temperature profile can be adjusted (see arrows) to accommodate screw design, heat due to friction, viscosity and required flowlength. Figure 5.22 (same processing temperature as in Figure 5.21 and high injection speed) shows the ratio between flowlength and wall thickness.

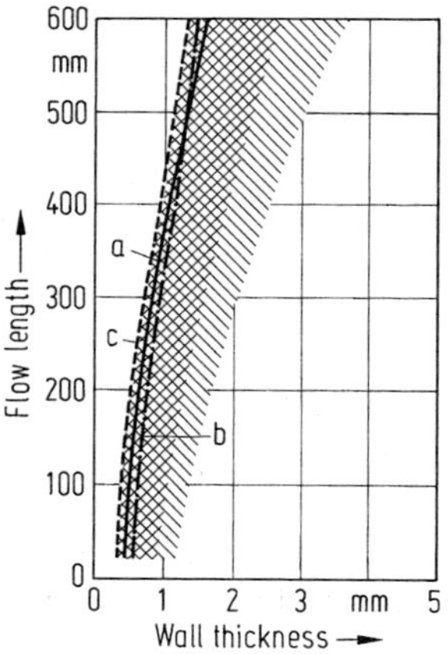

Figure 5.22. Flowlength-wall thickness ratio during cellulose ester (at medium-hard setting) processing a, CA 22%; b, CP 10%; c, CAB 10%

Maximum processing temperatures are required to achieve mechanical strength and surface gloss of the molded part in cellulose ester compounds. Because of the good flowability of the melt, the material is often processed at too low a temperature, which leads to brittleness.

5.11.2.4 Injection Speed

For processing cellulose esters, maximum injection speed is preferred, especially for thin-walled parts and long flowpaths. A lower or graduated injection speed is only practical if the meltstream does not encounter resistance like a deflector plate or a sprung core on entering the hollow space within the mold, or in voluminous thick-walled parts.

5.11.2.5 Injection, Back, and Holding Pressure

Cellulose esters can be processed at a relatively low injection pressure, about 800 bar. To obtain low-stress parts the back pressure is set as low as possible, just enough to avoid bubbles and sink marks. Using a pinpoint gate, a holding pressure time of more than 8 seconds is useless because the gate will freeze up within 3 to 8 seconds. To eliminate entrapped air in the melt, the back pressure should be about 10% of the injection pressure.

5.11.2.6 Shrinkage During Processing and Warpage

Shrinkage and warpage of cellulose ester molded parts varies widely depending on part design and processing conditions. Aside from part design, the most influential factors are wall thickness of the part, holding pressure, cooling time in the mold, melt temperature, and injection speed. Attainable peak values range from 0% for thin parts with high holding pressure and long cooling time in the mold (mesured near the gate) to 0.9–1.1% for thick parts, low holding pressure and short cooling time (measured away from the gate) [6]. Exact shrinkage under these conditions is not predictable. The shrinkage for a part of average design, under normal processing conditions, is between 0.4 and 0.7%. The shrinkage of cellulose ester parts in the flow direction is less than the shrinkage across the flow. This always leads to distortion of flat parts with central gating.

The good flowability of cellulose ester makes linear mold filling possible, even with a diaphragm gate.

5.11.2.7 Metal Embedding

The elasticity and cold flow tendency of cellulose ester makes metal embedding by spraying around the metal part, warm insertion, or ultrasonic insertion possible without danger of stress cracking.

5.11.2.8 Weld Strength

Welding together of cellulose ester flow fronts during injection molding at optimum conditions (high melt, high mold temperatures, and high injection speed) can be performed without difficulty. The tensile strength on these seam areas will reach 80–90% of the original strength.

5.11.3 Injection Blow Molding

Blow molding has not proven a significant technology with respect to cellulose esters to date.

5.11.4 Extrusion

All types of cellulose esters can be extruded into sheets, film, tubing, and profiles according to current norms for processing and machinery. The following descriptions are therefore limited to material-specific directives.

5.11.4.1 Extruders

Cellulose ester is, aside from a few exceptions, processed on air- or water-cooled single-screw extruders with external heating elements. Since there is no corrosion under normal processing conditions, extruders can be made of standard type steels [1, 2, 7]. Screws that have proven to be effective are single-stage, three-zone screws with a length of 2–25 D and constant compression ratio of 1 D.

Zone	Screw Length 20 D	25 D
Feeding zone	9 D	11 D
Compression zone	4 D	6 D
Metering zone	7 D	8 D

The flight depth ratio (compression ratio) of the three-zone screws is between 1:2.4–1:3 (metering zone/feed zone).

It is advisable to furnish the screws with a mixing section at the end of the metering zone. The screw length is thereby extended 2 or 3 D. Generally, processing cellulose esters with deep-cut screws, as shown in Figure 5.23, is advantageous.

The pulsations that frequently occur with single-screw extruders have the following origins:

- inlet zone too short and too shallow, i.e. less than 6 D
- metering zone too short, i.e. less than 5 D
- usage of center progressive screws

Extruders having vented screws can process undried resin with a moisture content of up to 0.6% with the following zone settings:

Zone	Screw Length 25 D	30 D
Feed zone 1	8 D	10 D
Compression zone 1	2 D	2 D
Metering zone 1	4 D	6 D
Feed zone 2	4 D	4 D
Compression zone 2	2 D	2 D
Metering zone 2	5 D	6 D

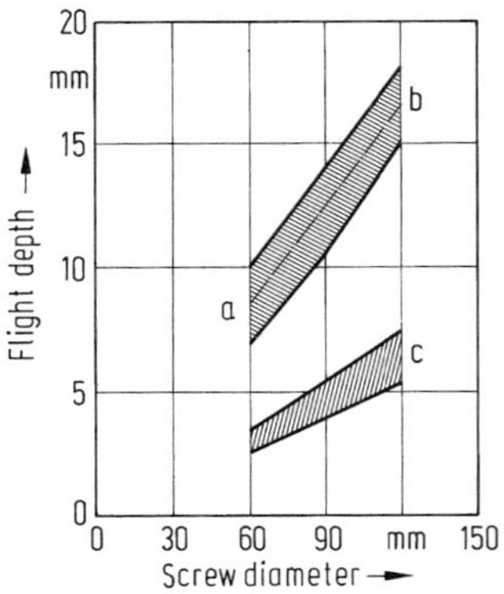

Figure 5.23. Extrusion processing of cellulose ester molding compounds, flight depth of the screw
a, flight depth region for effective feeding;
b, flight depth of feed zone;
c, flight depht of metering zone

It is recommened to include a mixing section when building a new machine with length of 30 D or 33 D. The dimensions for the flight depth of vented screws are similar to those for three-zone screws (Figure 5.23); however the second metering zone should be cut somewhat deeper than the first one.

5.11.4.2 Temperature Control

The temperature profile within the screw section, from feed to metering zone, is usually increased up to the beginning of the metering zone (intake/compression zone) when extruding film, sheets, tubing, or profiles. Temperature should be kept constant within the metering zone (discharge zone and adapter) and variable in the nozzle. To obtain better venting of the melt, the temperature at the first metering zone is often set higher. The grade of cellulose ester and the degree of hardness determine the processing temperature (Figure 5.24). Higher temperature settings are selected for harder resins, films, and small profiles; lower settings for larger profiles and sheets.

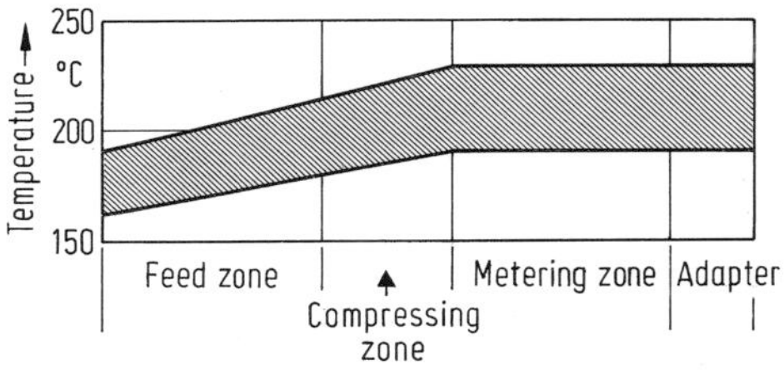

Figure 5.24. Extrusion processing of cellulose ester molding compounds, temperature profile

5.11.4.3 Special Rheological Features

When shear viscosity of the melt is measured as a function of deformaiton speed, it is apparent that compared to other thermoplastics, a sudden viscosity elevation takes place at high shear speeds. No theoretical explanation has been found for this effect; it cannot be explained by a melt fracture. The extruder should not operate in this critical range (Figure 5.25) [7].

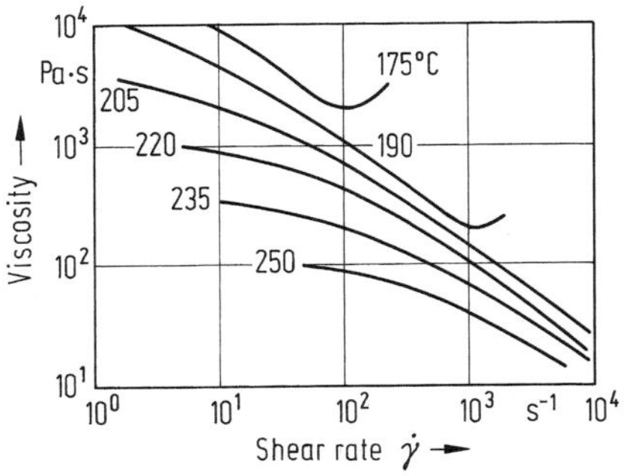

Figure 5.25. Viscosity as a function of shear rate of cellulose ester molding compounds, CAB with 8% plasticizer

All cellulose esters have the same rheological curve characteristics; only the quantitative values change depending on grade and hardness.

5.11.4.4 Film and Sheet Extrusion

Films and sheets are normally extruded with a slot die with adjustable bar and flex lips. The ratio of lip length to sheet thickness must be at least 20:1 for thick sheets. Therefore, the lip pieces must be interchangeable. To obtain higher pressure buildup in front of the screw tip, a perforated plate with a set of screens is inserted between screw tip and die entrance. The set of screens is assembled as follows:

one support screen with 1000 mesh cm^2
two fine screens with 3600 mesh cm^2
one support screen with 1000 mesh cm^2

Film with thickness of less than 0.8 mm is cast onto heated, chromed double chill roll. Film with thickness above 0.8 mm is extruded on a triple polished roll. The vertical pouring down onto the roll is advantageous for cellulose ester melt processing, allowing easy escape of plasticizer vapors to the top. The roll temperature is dependent upon composition, discharge of the extruder, sheet thickness, operating (extruding in the upper or lower roll gap), and roll diameter. The following temperatures are recommended for operating the upper roll gap:

Upper roll 70–90 °C
Middle roll 70–90 °C
Lower roll 100–130 °C

Multicolored sheets, such as those used for eyeglasses, can be produced by lamination or by adding different color cellulose ester batches from between two and five extruders into one common die [8].

5.11.4.5 Pipe Extrusion

Pipe extrusion is not an important application for cellulose ester. The same extruders that are used for film and sheet are usually used for tubing.

Table 5.22 gives values that should be observed for the geometrical design of pipe, nozzle, and calibration for the processing of cellulose esters into tubes. Tubes of less than 6 mm in diameter can be manufactured particularly efficiently using the free take-off method and subsequently quenching the tubular melt in a water tank. The tubular melt usually emerges horizontally from the die, passes through the air over a distance of about 60–100 mm without being guided and then enters the water tank through a plate provided with a suitable aperture.

Table 5.22. Standard Values for the Geometrical Design of Pipe, Nozzle, and Calibration for the Processing of Cellulose Esters into Tubes by Using a Water Tank

Pipe Diameter mm	Nozzle Diameter %	Wall Thickness %	Lip Length	Diameter Calibration %
<30	ca. +15	to 15	15 to 20 × S[1]	ca. +2
>30	ca. +10	10 to 15	20 × S[1]	ca. +2

[1] S: wall thickness

5.11.4.6 Profile Extrusion

This process is primarily used for composite profiles of CAB melt and aluminum foil such as would be used for automobile moldings [9, 10]. Aluminum foil, 50–100 μm thick, is covered with adhesive and coated with CAB melt in a crosshead die (Figure 5.26). The shape of the CAB alulminum composite profile is retained by corresponding nozzle plate shape and post calibration.

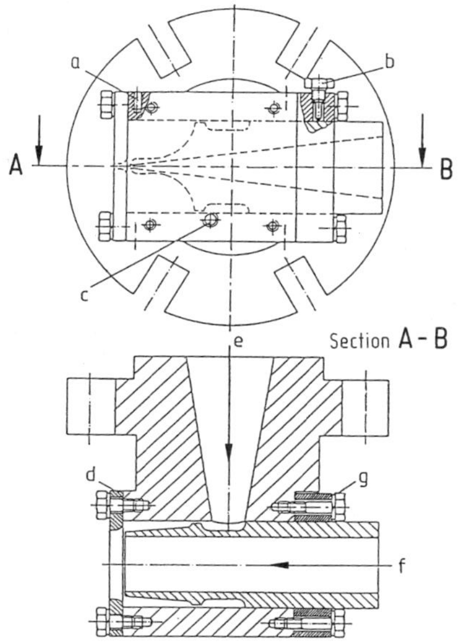

Figure 5.26. Cross-head die for coating aluminum film with CAB-molding compound
a, holes for heater plate; b, adjusting screw; c, hold for temperature sensor; d, die plate; e, inlet; f, foil feed; g, mountable, 3 heating bands, 1 nozzle plate, 1 flange

After the profiles are extruded, further working is usually run through additional processing steps by bending, shaping and sealing the ends. This unproblematic secondary processing of CAB/aluminum composite profiles is primarily used for decorative molding. Stiffer composite profiles can be obtained by coextrusion. for example with ABS. The process is basically the same as with CAB/aluminum profiles. Both thermoplastic resins are coextruded as separate layers onto the aluminum foil. The outer CAB layer can be applied as a full or partial jacketing.

CA and CAP resins are never, or very rarely, used for aluminum composite profiles. CA is not weatherable and has no adhesion to aluminum surfaces. CAP is also not suitable for prolonged exposure to the elements as on automobiles, but due to its neutral odor it is suitable for interior applications such as refrigerator trim.

5.11.5 Rotation Molding

CAB and CAP resins in powder form, grain size less than 1 mm, with soft settings, can be rotation molded into large-volume hollow objects (e.g., lamps) with evenly distributed wall thicknesses. The powder is melted in twin-walled, oil-heated molds which rotate around two vertical and separately controllable axles. Air-heated molds have not proven to be useful. Rinsing with nitrogen is necessary to prevent blow-outs. Rotation molding of cellulose ester material is not a widely used practice.

5.11.6 Coating By Using a Fluidized Bed

Medium to soft CAB and CAP powders, grain size less than 0.8 mm, are suitable for coating metal objects. The objects to be coated are preheated to 160–250 °C, depending on the wall thickness, then immersed for a few seconds in the fluidized bed [11, 12].

5.12 Secondary Operations

5.12.1 Thermoforming of Film and Sheet

Conventional methods for thermoforming cellulose ester film and sheet are vacuum and compressed-air processes. Cellulose ester sheets having thickness greater than 0.3 mm must be predried prior to thermoforming because cellulose ester absorbs moisture after a short storage time. This leads to formation of bubbles. Sheets greater than 0.3 mm thick are hung in a circulating air drying chamber for three to four hours at 60–80 °C.

Sheets, particularly thicker sheets, can be prewarmed by short-wave IR radiation having a maximum radiation of 1.7 μm. Short-wave IR radiation provides a prewarming that is shorter and more even than that obtained from ceramic radiation.

The thermoforming temperature is between 170 and 200 °C, depending on hardness setting. Only in this temperature range do the parts have low stress and therefore the least warpage. They are dimensionally stable, with maximum visual appearance and mechanical strength [13]. Cellulose ester sheet and film have a high ability to stretch, making a positive forming process shaping ratio of $H{:}D = 1.5{:}1$, and a negative forming process shaping ratio of $H{:}D = 0.5{:}1$ obtainable (H, forming height; D, diameter).

5.12.2 Cold Forming

Cold forming of semifinished cellulose ester parts is not practiced, since elongation at greater than 20%, white stretch marks occur near the edges which reduce the mechanical strength.

5.12.3 Imprinting

Cellulose ester parts are hot and cold imprintable via conventional methods without any problems. When using imprinting film it is important that the film should be plasticizer resistant.

5.12.4 Bending and Crimping

The common decorating methods for CAB/aluminum laminated profiles are bending and crimping. The laminated profiles can even be bent into tight configurations. It is remarkable that the elongation of pure aluminum foil is about 10%, but can be as high as 100% for the laminated profile.

Complicated shaping and crimping in excess of 20% elongation at the edges are always executed with heat (see Section 5.12.2).

5.12.5 Machining and Polishing

Conventional tools that are used for machining plastics are also suitable for cellulose ester. Lathing, drilling, milling, stamping, and cutting of inner or outer threads can all be performed on cellulose ester parts. Attention must be paid to the cutting speed and the shape of the tool so that no local overheating of the part occurs. A drill bit with a tip angle of 60–90°, two hollow grooves and freecut angle tools with a rear cutting angle of 12–18°, for example, are suitable for working cellulose ester parts. The drum rotation speed should not exceed 30–60 m/min.

Cellulose ester parts respond well to sanding and polishing. The eyeglass industry employs the drum polishing method, which has been developed to perfection for this application. The type of sanding and polishing mediums, drum rotation speed and polishing cycles, are all dependent upon the type of cellulose ester, the hardness and the drum filling level [14].

For restorative polishing a disc with a wobble according to the particular requirements should be used with the aid of hard polishing wax. To avoid smearing the drum rotation speed should not exceed 5 to 30 m/sec.

Low-stress parts can also be polished while immersed in a solvent such as ethyl acetate.

5.12.6 Welding

Welding of cellulose ester parts with conventional methods is possible in principle, but joining by adhesion is preferred in practice. Ultrasonic welding cannot be performed in the field due to the high attenuation of ultrasonic waves. This process must be performed in-house. Metal parts such as fasteners, however, can be installed ultrasonically without problems.

5.12.7 Bonding

The easiest and most often used method for bonding for cellulose ester compounds is solvent-based bonding. It should be noted that CA parts can adhere only to other CA parts, CAB or CAP parts only to other CAB or CAP parts.

The following solutions and solution combinations have proven to be effective bonding agents:

- CA parts: Methyl glycol acetate with 5–10% solubilized cellulose acetate resin
- CAP and/or CAB parts: 50 parts methyl glycol acetate and 50 parts ethyl acetate with 5–10% solubilized CAP or CAB resin

Strong solvents such as acetone should not be used in concentrated form since it causes brittleness in the bonded area. CA parts that are to be bonded to CAP parts or CAB parts as well as to other thermoplastic parts obtain maximum adhesion with two-component bonding systems based on polyurethane or epoxy resin. It should be noted that bonding CA parts to thermoplastic parts with low stress crack resistance, like polystyrene can result in problems due to migration of the plasticizer.

5.12.8 Secondary Surface Treatment

5.12.8.1 Coating/Printing

Cellulose ester parts are printable and easily coated [15]. When choosing a coating system, especially for cellulose acetate, it should be noted that coatings must be plasticizer resistant and should not contain overly soluble or highly swelling solvents when applied to large surfaces. Injection molded parts should be molded with a minimum of stress in order to avoid distortion and surface imperfections. The strength of coated cellulose ester parts can be reduced by 30–50% if the non-coated surface is stressed (coated side in the tensile range).

5.12.8.2 High-Vacuum Metallization

Only the harder CAP and CAB are suitable for evaporating metallization. Soft settings and CA materials without plasticized precoatings need a very long evacuation time. The steam vacuumed layer gradually becomes dull due to plasticizer migration.

The evacuation time is significantly reduced by predrying the parts to be metallized. Precoating, even for hard CAP and CAB, is recommended to improve adhesion and surface gloss.

5.12.8.3 Surface Tinting

Surface tinting is widely practiced on cellulose ester parts such as eyeglass frames, buttons, buckles, and combs. The dye baths used for this purpose are aqueous alcoholic solutions of dissolved dyes, available on the open market. The piece to be colored are immersed in slightly heated (30–40 °C) baths, for 0.5–2.0 minutes depending on intensity desired, and then rinsed. The light stability of this tinted surface is poorer than that of colored resins.

5.12.8.4 Scratch Resistant Coating

Surface coating with layers of hard-coating improves scratch resistance. Parts made from cellulose ester, due to relatively low heat resistance, cannot match the standards of high heat resistant thermoplastics with scratch resistant polysiloxane-based coating systems baked on at 120–130 °C. Improvement may be obtained through UV cured hardcoatings which are currently being developed.

5.12.8.5 Anti-Fogging (Anti-Condensation) Treatment

The cellulose ester surface can be saponified through alcoholic sodium or potassium solution to create a lasting anti-fogging effect for sheets for safety glasses and visors. The effect has not been obtained by any anti-fogging coating systems.

Solutions for the treatment are as follows:

- *Saponifacation solution*: 20–30% potassium or sodium solution (with distilled water) and 5–10% ethanol (96%)
- *Neutralizing solution*: 10–20% lactic or 5% acetic acid (with distilled water)

The parts to be saponified are immersed in the saponification solution at 40 °C for 15–25 minutes with gentle motion, followed by rinsing in soft water and 30–40 °C neutralizing solution. After neutralization the parts are rinsed at least twice with soft water and then kept in distilled water for about 30 minutes. The clouding that is visible initially disappears upon drying.

This process causes pronounced surface swelling and only parts such as extruded sheet that are largely free of stress and orientation are suitable. Injection moulded parts frequently display surface disturbances, especially near the gate.

5.12.8.6 Electrostatic Surface Charge

Cellulose ester part surfaces are, like all electric non-conductive parts, electrically charged through friction. Those surface charges will reduce within minutes to a level which eliminates the dust particle attraction so common with most thermoplastics. CA is more advantageous in this respect than CAP or CAB because of its higher moisture absorption ability (Figure 5.27). Additional treatments with antistatic additives for cellulose ester material is unnecessary.

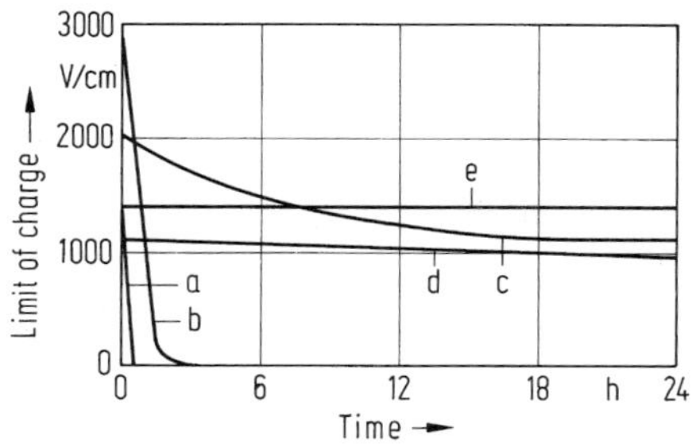

Figure 5.27. Leveling off of electrostatic charges of cellulose acetate and cellulose propionate or acetobutyrate parts compared to some fully synthetic resins
a, cellulose acetate; b, cellulose propionate/cellulose acetobutyrate; c, polyolefin; d, PMMA; e, PS-HI

References for Sections 5.11 and 5.12

1. Cellidor, organische thermopl. Celluloseester-Eigenschaften. Company brochure Albis Plastic GmbH, Hamburg 1988
2. Dexel. Company brochure Courtaulds Speciality Plastics, Spondon-Derby (U.K.) 1988
3. Tenite Cellulosic Plastics. Acetate-Butyrate-Propionate. Company brochures Eastman Chemical Products, INC, Kingsport (USA) 1983/84
4. Celluloses Acetate Plastic. Company brochure Daicel Chemical Ind. LTD (J): Acetyloid—Celluloses Acetate Plastic, Tokyo, 1980
5. Verarbeitung Spritzgießen. Company brochure Bayer AG, Leverkusen 1980
6. Schwindung bei Spritzgußteilen. Company brochure Bayer AG, Leverkusen 1977
7. *Brinkschröder, F. J.; Fischer, W.:* Kunststoffe 64 (1974) p. 635–639
8. Patente Brillenplattenherstellung:
   US-PS 2 803 041 (1957) Celanese
   US-PS 3 274 646 (1966) Celanese
   US-PS 3 422 175 (1969) Celanese
   GB-PS 996 432 (1965) Rowland
   DE-OS 1 904 882 (1965) Rowland
   FR-PS 1 397 602 (1965) Rowland
   DE-AS 1 110 398 (1961) Rehau
   FR-PS 1 404 594 (1965) Bonnet
   DE-AS 1 170 616 (1964) Berstorff
   DE-OS 2 031 668 (1971) Sordelli
   DE-OS 2 122 604 (1971) Sordelli
9. *Fischer, W.*, et al.: Kunststoffe 67 (1977) 6, p. 348–352
10. *Göller, J.*, et al.: Kunststoffe 65 (1975) 6, p. 326–332
11. *Röhm, W.; Glässer, H.:* Kunststoffe 49 (1959) 10, p. 503–505
12. *Leuschke, Ch.:* Product Finishing 21 (1968) 5, p. 48–51
13. *Witt, W.; Bühl, R.:* Plastverarbeiter 19 (1968) 12
14. Acetat Trommelverfahren/Propionatbrillen in Stahlformen gespritzt. Company brochures Menzerna-Werk, Schleif- und Poliermittel GmbH u. Co, Karlsruhe 1985/87
15. Schlußblackierung von Brillengestellen aus CP und PA. Company brochure Ernst Diegel GmbH, Lackfabrik, Alsfeld 1984

5.13 Applications

5.13.1 Traditional Areas of Applications

When looking through older prospectives and literature, it becomes apparent that many cellulose ester applications from 1960 and later have either been replaced by fully synthetic thermoplastic materials and/or their blends or lost their dominance. However, there are some applications which have maintained their position since the time when cellulose materials had almost no competition. The following four representative areas of application illustrate the developments:

- large-area parts
- applications based on technical developments and current fashion trends
- cost-critical applications
- cellulose material applications that have remained typical

Large-Area Parts

This includes switch-box covers for electrical power plants, fluorescent lights/light cover, tubes, window glazings, light bulbs, transparent advertising signs, etc.

Other than transparent advertising signs made, for example, out of thermoformed cellulose acetobutyrate sheets, which are widely used in the U.S.A., none of these applications has gained general acceptance. The reason for failiure of the other projects are tendency to cold flow, inadequate temperature resistance, stiffness, and flammability (not self-extinguishing). Fully synthetic thermoplastics, like polycarbonate, have a distinct advantage here.

Applications Based on Technical Development and Current Fashion Trends

This includes lamp shades, wastebaskets of rolled-up, glued continuous tubing (rotaflex method), heels for women's shoes, implosion safety glasses for television sets, telephone dial discs, automotive steering wheel covers and inside knobs/handles, etc.

Large sales were generated over many years with cellulose material in this application category before technical developments found other solutions which made, for example, the implosion safety glass or telephone dial disc obsolete. Another example is the continuously increasing effort to design safer inside automotive applications; this led to extensive substitutions of steering wheel covers, knobs and handles of CAB and/or CAP material with polyurethane integral-foam material. Changes in fashion trends eventually dried up request for rotaflex lamp shades, waste baskets, etc.

Cost-critical Applications

This includes, among others, stacking boxes, tool boxes, furniture and door knobs, bicycle rear lights, warning light covers, housing and covers of all types, key and control buttons, hair-dryer visual rings, labels, semi-finished products, lighted signage letters and so forth, farm fence insulators, packaging boxes and tubes, and skin and blister packaging, etc.

These applications have been almost completely substituted with the more economical and fully synthetic thermoplastic materials and/or their blends. The cellulose ester materials do not have property deficits, as seen in the examples of housings and covers for kitchen instruments or telephone housings. The soundproof and dust-repellent properties, surface gloss, color brilliance as well as stress-crack resistance, transparency, and toughness are still very interesting properties; however, except for luxury product ranges, they are too expensive. Future ecological discussion may lead to an increase in significance of cellulose materials for packaging, particularly softener-free polymer blends.

Cellulose Material Applications that have remained typical

This includes tool handles, hammer heads, CAB/Al-composite profiles (moldings), chair seats, shield tubes for workshop lights, eyeglass frames (cut and injection molded), letter templates, combs, cosmetic-powder cases, lipstick cartridges, etc.

In these typical applications, cellulose ester material plays a dominant role because of its favorable property combinations such as transparency, toughness, weather stability, stress-cracking resistance, comfortable horn-like grips, good polishing capability, dust resistance, and good resistance against lubricants and oils.

5.13.2 Matching of Applications to Properties

The following shows typical applications of cellulose ester materials with the appropriate combination of property requirements.

5.13.2.1 Automotive Equipment (CAB, CAP)

CAB/aluminum–composite profiles as moldings for special functions, coverings, and decorations; bicycle fenders (Figure 5.28)

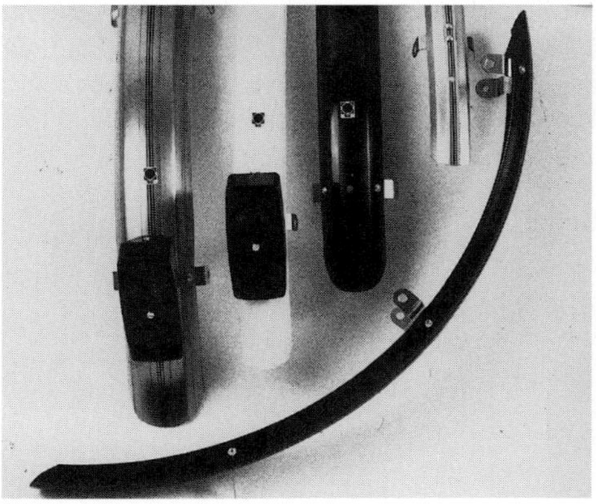

Figure 5.28. Bicycle fenders of CAB/Al-composite profile (Photo: Albis, Hamburg)

Figure 5.29. Insert-molded CAB steering wheels (Photo: Bayer AG)

Stress-cracking resistant metal coating, adhesion to aluminum foil; transparency; toughness; weather stability; moldability by bending, crimping, and stamping
Automotive steering wheels (Figure 5.29), control knobs

Stress-cracking resistant metal-insert molding; horn-like grip; self-polishing effect during use; can be polished or matted with solvents, depending on the melt equipment; toughness; transparency; good flowability; adhesion properties

5.13.2.2 Optical Applications (CA, CAP)

High-grade eyeglass frames:
- cut from CA sheet (Figure 5.30);
- injection molding from CAP melt

Surface colorability; unlimited melt dyeing; sking-perspiration resistance; stress-cracking resistance; self-polishing effect during use; skin compatibility.

Work protection and sunglasses lenses; visors for automotive helmets and work protection; ski goggles (Figure 5.31); swim goggles

Toughness, transparency; long-term anti-fogging effect due to surface saponification; pressure polishable extruded sheet; quick flow-off of electro-static loads

Figure 5.30. Eyeglass frames of cut CA sheet (Photo: Courtaulds Speciality Plastics, Spondon-Derby)

Figure 5.31. Ski goggles of CAB and/or CP (Photo: Bayer)

5.13.2.3 Tool Industry/Machine Construction (CA, CAP, CAB)

Screw driver (Figure 5.32) and plier grips; tension testers; hammerheads; filter cups; machine covers; protective tubing for workshop lights.

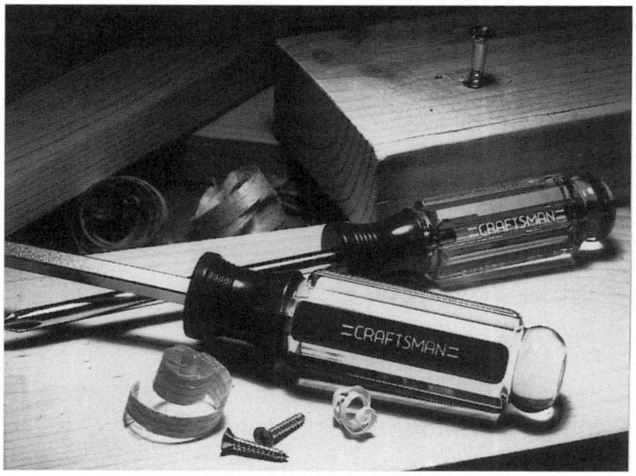

Figure 5.32. Screwdriver handles of CAB (Photo: Eastman Chemical Products, Kingsport/USA)

Toughness; transparency; mineral oil and grease resistance; stress-cracking resistant metal insert molding; good electrical insulation properties; horn-like grip.

5.13.2.4 Household/Furniture/Toiletry Articles (CA, CAP, CAB)

Combs, cosmetic powder compacts (Figure 5.33); lipstick cartridges; silverware grips; haribrush bodies; toothbrush handles; fashion accessories, e.g. buckles, buttons and hair pins; shoe tree sections; chair seats; furniture fittings; CAB/aluminum-composite profiles as decorative trims; waste-paper dispenser covers; small housings

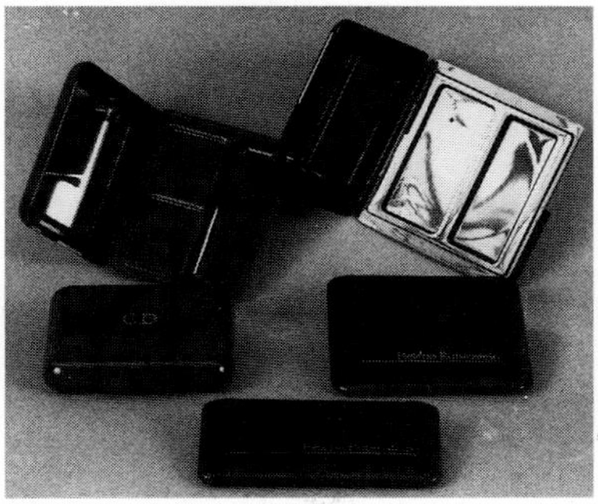

Figure 5.33. Powder compacts of CP (Photo: Albis, Hamburg)

Transparency; toughness; unlimited colorability; stress-cracking resistance; horn-like character; self-polishing effect during use; electrostatic load resistance

5.13.2.5 Office Equipment (CA, CAP, CAB)

Ballpoint pen housings; pressure pins; letter and curve templates; pen holder parts (Figure 5.34)
Toughness; horn-like grip; self-polishing effect during use; unlimited colorability; transparency.
Adhesive tapes
Transparency; color neutrality; writability

5.13.2.6 Advertising/Packaging (CAB, CA, CAP)

Transparent advertising signs (Figure 5.35); signs; packaging profiles; folding boxes; films
Transparency; toughness; weather stability; dust-resistance

5.13.2.7 Sport/Leisure (CAB, CAP)

Compass housings; fishing bait; see also optical sector
Transparency; toughness

5.13.2.8 Medical Articles (CAP)

One-shot molded housings; hearing aid housings
Transparency; stress-cracking resistance; sterilizablility; sound-proof properties

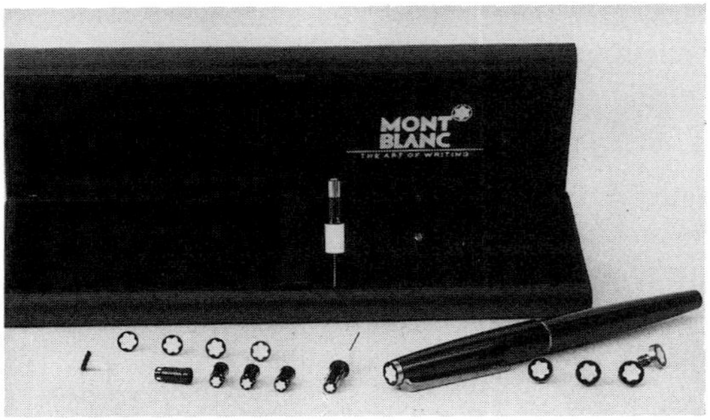

Figure 5.34. Fountain pen parts of CAB (Photo: Albis, Hamburg)

Figure 5.35. Transparent advertisement of CAB sheet, deep drawn (Photo: Bayer AG)

5.13.2.9 Toys

Model train (Figure 5.36) and car housings; model airplanes; doll bodies; model train base plates
Toughness, high imaging precision; unlimited colorability; sound-proof properties

Figure 5.36. Model train constructions of CP (Photo: Bayer AG)

5.14 Other Applications for Cellulose Ester

The property profile of cellulose esters can be varied for a broad palette of application areas by the type of substitutes, substitution degree, and setting of certain viscosities. As well as thermoplastic molding compounds, cellulose esters are used as fibers, films, foils and coating additives.

5.14.1 Fibers

Textile fibers [1 to 3] and cigarette filters [1 to 3] are manufactured from CA. The major ingredient is cellulose-2,5-acetate (54 to 56% acetic acid), which is suitable for textile fibers as well as for cigarette filters. Since 1954 cellulose triacetate is used in minor amounts for textile fibers.

Textile fibers of cellulose secondary acetate and cellulose triacetate show similar textile properties; however, due to its chemical structure, the secondary acetate shows higher water absorption. In comparison to synthetic fibers, the major advantage of cellulose esters is in water absorption capability, whereas other properties such as tear resistance, washability and wearing are on a lower level.

Cellulose secondary acetate is excellently suited for the manufacture of cigarette filters in the form of bundled fibers (cigarette filter tow). A high reduction of cigarette smoke condensate and specific absorption for hazardous materials can be achieved without noticeable reduction or alteration in taste. The filter tow is produced as an unending string of several thousands fibers, which are crinkled and compressed in various cross-section forms, and supplied for further processing to cigarette manufacturers.

The CA fibers are spun of highly cleaned solutions, that is, acetone for cellulose-2,5-acetate and methylene chloride/methanol (90:10) for cellulose triacetate. The purity of the solution and the viscosity and distribution of substituents of the cellulose ester exert a deciding influence on fiber quality. The manufacturing and cleaning processes for the fiber acetates are currently so mature that in practice only highly cleaned wood pulp cellulose is now esterified.

5.14.2 Films and Foils

Since the introduction of suitable technical manufacturing methods for cellulose acetates, the first application of CA films was to replace easily combustible nitrocellulose bases for photographic film. As the term "safety film" suggests, this greatly reduced the danger of fire. Even today CA is still used for film bases, particularly in the amateur area [4]. In the technical sector, where high dimensional stability and minimal water absorption are required, they have largely been substituted by synthetic polymers, particularly polyester.

During the last decade, cellulose triacetate electrical insulation film, manufactured according to the casting method, has almost been completely replaced by synthetic polymers. In view of the combination of properties such as mechanical strength, heat distortion temperature and water resistance at decreasing film thickness, synthetic polymers (e.g., polycarbonate and polyester) have proven successful.

Extruded or cast CA packaging foil has been able to dominate certain applications as decorative packaging material. The reason is transparency, brilliance, quick flow-off of electrical static, printability, simple manufacturing methods, and ability to be combined with other materials.

Membranes of CA, manufactured by special casting methods, can be manufactured for the following applications: cleaning of brackish water by reverse osmosis, ultra filtration, cleaning and processing of fruit juices, and carrier materials for electrophoresis. Since films and foils of CA are biodegradable, one could expect them to gain significance in the future on ecological grounds.

5.14.3 Coatings

Based on their color-fastness, reduced flammability, thermal stability and resistance against aliphatic hydrocarbons, coatings based on cellulose acetate were among the initially favored developments. Except for special applications in wood and paper coating, their use is no longer significant today. On the other hand, cellulose acetopropionate and especially CAB are highly desirable as coating additives in many formulas in combination with synthetic polymers; in special metal coatings they are indispensable [5]. They are used because of their better compatibility with standard coating components, increased water resistance, extensive weather stability, and the possibility of obtaining a rich coating with low-viscosity types. In metallic coatings, addition of CAB achieves and fixes a uniform and controlled distribution of the metallic glitter in the coating base prior to applying the clear coating, which guarantees a uniform appearance.

5.14.4 Special Applications

Special areas of applictions [5] for CA include hollow fibers for blood detoxification; for cellulose acetopropionate and cellulose acetobutyrate, low-profile additives for SMC (sheet molding) and BMC (bulk molding) melts, residual-free optical discs through surface saponification as well as contact lenses with high oxygen permeation and excellent carrier properties.

5.15 Market Development and Economic Aspects

Table 5.23 shows the average cellulose ester volumes consumed worldwide for the period of 1982 to 1985, according to area of application. CA fibers have the largest share. Their total volume during 1970 to 1983 remained practically constant, as shown in Table 5.24. Losses in the textile sector were counteracted by increases in filter tow.

Table 5.23. Average Yearly World Production Volume for Cellulose Esters and Their Areas of Applications (1982 to 1985)

Cellulose Ester	Area of Application	Volume (1000 t/y)
Tri-acetate	textile fibers	280
Secondary acetate	filter tow (cirarette filter)	370
Tri-acetate	photo film	60
Secondary acetate	thermoplastic materials and films	50
Ester mixtures	thermoplastic materials	30
CP + CAB	coating raw materials	5
Total		795

Table 5.24. World Production of Cellulose Acetate Fibers (in 1000 tons/year)

Application	1970	1976	1980	1983
Textile fibers	427	316	337	275
Filter tow	154	260	335	334
Total	581	576	672	609

Up to 1968, cellulose ester materials of cellulose acetate and cellulose ester blends reached a total worldwide volume of over 110,000 tons per year. Since that time, substitution of cellulose ester materials has steadily increased, which has led to a volume reduction of about 70%. Since 1984 a near standstill has been arrived at, which has been characterized by the dominance of cellulose ester materials in certain areas of applications where their special property combination favors them over synthetic polymers.

Limiting cellulose esters to specific high-grade applications increasingly required optimization of quality which, coupled with volume reduction, led to a disproportionate cost increase. As a result, the price doubled in the time period from 1972 to 1984 and still remains on a relatively high level today.

Only in the US have cellulose materials maintained the relatively high consumption of 50,000 tons per year (1985/86) [6] in spite of the general volume reduction, in which, for example, 10,000 tons per year were processed solely for extruded cellulose ester film and sheet.

In the near future the only reason to expect volume increase would be increased demands in applications. An opening up of new applications or return to abandoned ones may result from ecological requirements or a result of petroleum shortages in view of the tendency to "renewable raw materials". However, this would be coupled with the condition that the cellulose raw material should be increasingly available biologically, in the purest possible form, without damage to the environment.

References for Sections 5.14 and 5.15

1. *Steinmann, H. W.*, in: Handbook of Fiber, Science and Technology. Vol. IV: Fiber Chemistry. Dekker, New York 1985, p. 1001
2. *Serad, G. A.*, in: Encyclopedia of Polymer Science and Engineering. 2. Ed. Vol. 3. Wiley, New York 1985, p. 226
3. *Astheimer, H. J.*, in: Ullmann's Encyclopedia of Industrial Chemistry. 5. Ed. Vol. A5. Verlag Chemie, Weinheim 1986, p. 447
4. *Kruck, P.*, in: Ullmann's Encyklopädie der technischen Chemie. 4. Ed. Verlag Chemie, Weinheim 1979, p. 434
5. *Bogan, R. T.; Brewer, R. J.*, in: Encyclopedia of Polymer Science and Engineering. 2. Ed. Vol. 3. Wiley, New York 1985, p. 158
6. Mod. Plast. 12 (1987) p. 22

5.16 Commercial Products

The most important manufacturers of cellulose ester materials and their trade names are summarized in Table 5.25.

Table 5.25. Tradenames and Manufacturers of Cellulose Materials

Tradename	Manufacturer/Country
Cellulose acetate materials	
Acetyloid	Daicel Chemical Industries, Ltd./Japan
Bergacell	Bergmann Kunststoffwerk KG/Germany
Cellolux	La-Es s.p.a./Italy
Dexel	Courtaulds Speciality Plastics/England
Setilithe	Tubize Plastics s.a.-n.v./Belguim
Sicalit	Sic Mazzucchelli/Italy
Tenite Acetate	Eastman Chemical Products, Inc./USA
Cellulose propionate materials	
Cellidor CP	Albis Plastic GmbH/Germany
Tenite propionate	Eastman Chemical Products Inc./USA
Cellulose acetobutyrate materials	
Cellidor B	Albis Plastic GmbH/Germany
Tenite butyrate	Eastman Chemical Products Inc./USA

Index